# Aquatic Ecology and Biodiversity

# Aquatic Ecology and Biodiversity

Editor: Olando Martin

R CALLISTO
REFERENCE

www.callistoreference.com

**Callisto Reference,**
118-35 Queens Blvd., Suite 400,
Forest Hills, NY 11375, USA


Visit us on the World Wide Web at:
www.callistoreference.com

© Callisto Reference, 2017

ISBN: 978-1-63239-821-5 (Hardback)

### Cataloging-in-publication Data

Aquatic ecology and biodiversity / edited by Olando Martin.
  p. cm.
Includes bibliographical references and index.
ISBN 978-1-63239-821-5
1. Aquatic ecology. 2. Aquatic biodiversity. 3. Aquatic biology. I. Martin, Olando.
QH541.5.W3 A68 2017
577.6--dc23

# Table of Contents

**Permissions**

**List of Contributors**

**Index**

# Preface

Aquatic ecology is the study of various interactions of flora, fauna and ecological characteristics in an aquatic environment. It comprises the study of marine as well as freshwater ecosystems. This book presents complex subject of aquatic ecology in the most comprehensible and easy to understand language. It aims to present researches that have transformed this discipline and added its advancement. This book on aquatic ecology and biodiversity consist of topics dealing with preservation of ecologically significant areas, damage caused due to natural calamities and fishing activities, etc. The book would prove fruitful for researchers and students in the area of sustainable agriculture, natural habitat management and wildlife habitats.

The researches compiled throughout the book are authentic and of high quality, combining several disciplines and from very diverse regions from around the world. Drawing on the contributions of many researchers from diverse countries, the book's objective is to provide the readers with the latest achievements in the area of research. This book will surely be a source of knowledge to all interested and researching the field.

In the end, I would like to express my deep sense of gratitude to all the authors for meeting the set deadlines in completing and submitting their research chapters. I would also like to thank the publisher for the support offered to us throughout the course of the book. Finally, I extend my sincere thanks to my family for being a constant source of inspiration and encouragement.

**Editor**

# Piscivore-Prey Fish Interactions: Mechanisms behind Diurnal Patterns in Prey Selectivity in Brown and Clear Water

**Lynn Ranåker[1]\*, Jens Persson[2], Mikael Jönsson[3], P. Anders Nilsson[1,4], Christer Brönmark[1]**

1 Department of Biology, Aquatic Ecology, Ecology Building, Lund University, Lund, Sweden, 2 Swedish Agency for Marin and Water Management, Gothenburg, Sweden, 3 Department of Biology, Functional zoology, Biology Building, Lund University, Lund, Sweden, 4 Department of Environmental and Life Sciences, Biology, Karlstad University, Karlstad, Sweden

## Abstract

Environmental change may affect predator-prey interactions in lakes through deterioration of visual conditions affecting foraging success of visually oriented predators. Environmental change in lakes includes an increase in humic matter causing browner water and reduced visibility, affecting the behavioural performance of both piscivores and prey. We studied diurnal patterns of prey selection in piscivorous pikeperch (*Sander lucioperca*) in both field and laboratory investigations. In the field we estimated prey selectivity and prey availability during day and night in a clear and a brown water lake. Further, prey selectivity during day and night conditions was studied in the laboratory where we manipulated optical conditions (humic matter content) of the water. Here, we also studied the behaviours of piscivores and prey, focusing on foraging-cycle stages such as number of interests and attacks by the pikeperch as well as the escape distance of the prey fish species. Analyses of gut contents from the field study showed that pikeperch selected perch (*Perca fluviatilis*) over roach (*Rutilus rutilus*) prey in both lakes during the day, but changed selectivity towards roach in both lakes at night. These results were corroborated in the selectivity experiments along a brown-water gradient in day and night light conditions. However, a change in selectivity from perch to roach was observed when the optical condition was heavily degraded, from either brown-stained water or light intensity. At longer visual ranges, roach initiated escape at distances greater than pikeperch attack distances, whereas perch stayed inactive making pikeperch approach and attack at the closest range possible. Roach anti-predatory behaviour decreased in deteriorated visual conditions, altering selectivity patterns. Our results highlight the importance of investigating both predator and prey responses to visibility conditions in order to understand the effects of degrading optical conditions on piscivore-prey interaction strength and thereby ecosystem responses to brownification of waters.

Editor: Eric J. Warrant, Lund University, Sweden

**Funding:** This research was funded by a grant from The Swedish Research Council for Environment, Agricultural Sciences and Spatial Planning (grant #223-2005-1393). The funders had no role in study design, data collection and analysis, decision to publish, or preparation of the manuscript.

**Competing Interests:** The authors have declared that no competing interests exist.

\* Email: Lynn.Ranaker@biol.lu.se

## Introduction

Predation by piscivorous fish is an important structuring force in freshwater food webs, where piscivory can cause complex trophic cascades with repercussions at the community and ecosystem levels [1–5]. Piscivore foraging can be divided into different foraging-cycle stages, including encounter, reaction, attack, capture and ingestion of prey [6,7]. Theoretical foraging models suggest that encounter rate is a function of search volume and prey density [8–10], where search volume, in turn, is a function of reaction distance and swimming speed. For visually hunting piscivores, reaction distance is affected by environmental factors such as ambient light levels and turbidity as well as characteristics of both predator and prey species. Several studies of piscivores have shown that reaction distance decreases with increasing turbidity [8,9,11–13]. Further, optical conditions of the water may also affect attack rate [14,15] and prey selection [11,16]. From the prey's perspective, deteriorated optical properties can induce

reduced escape success due to poor timing of escape responses [17], but turbid water may also act as a refuge from visual predators [10]. The optical qualities of water hence hold an important key to our understanding of piscivore-prey fish interactions.

Most studies on how optical conditions affect piscivore foraging success focus on the effects of light intensity or turbidity caused by increasing levels of clay particles or algae [18,19]. However, in recent years it has been recognised that increasing inputs of humic matter makes our lakes browner [20–22]. The loading of humic substances from terrestrial into aquatic systems is expected to increase in the future due to changes in land use, climate change [22] and reduced sulphur deposition [21,23]. This brownification [20] of our inland as well as coastal waters may have far-reaching effects on biotic interactions and ecosystem functions [24,25]. Humic substances have strong effects on the light climate in the water column by attenuating light, mainly in the UV and blue/green region, resulting in a light spectrum that is dramatically different from non-humic waters [26,27]. The reduced light climate will cause negative effect on the contrasts between objects and their background, resulting in a reduced reaction distance of both predator and prey detection [28]. The effect of brownification on piscivore-prey interactions may vary depending on how strictly different species rely on visibility for their performance; brownification could in fact benefit species that are less negatively affected by such changes in visibility conditions.

The pikeperch (*Sander lucioperca*) is a common and naturally occurring piscivore in European freshwaters and is often the dominant piscivore species in turbid or brown lakes [5,29]. Pikeperch is commonly introduced to lakes for biological control of cyprinids and for commercial and game fisheries due to its high economic value. Pikeperch introduction success [30,31], densities [5] and growth [29] correlate positively with high water colour or turbidity. Pikeperch is an actively searching piscivore [32] that forages in open water at low light intensities [33,34], and it could become an increasingly important piscivore with increasing brownification of lakes. In order to evaluate the potential effects of increasing brownification and pikeperch abundance on processes and patterns in freshwater lake ecosystems, it is essential to understand pikeperch predatory behaviour and potential effects on prey fish populations. If the effect of pikeperch on different prey species change with increasing brownification, and the trophic roles of the prey species differ, we should expect altered trophic and ecosystem functions as a result of changes in fish community composition.

Here, we study prey selectivity in pikeperch when foraging on European perch (*Perca fluviatilis*) and roach (*Rutilus rutilus*) in waters with different levels of humic content and during daylight and night conditions. The studies were performed in both the field and laboratory. Further, to approach the mechanisms behind patterns in prey selectivity, we studied the behaviour of pikeperch and prey fish during the different stages of the foraging cycle in laboratory experiments.

## Methods

### Field study

The field study was performed in October 2009. One clear water lake (Lake Västersjön) and one brown water lake (Lake Osbysjön), both located in Skåne, Southern Sweden, were selected for this study. Each lake was sampled with gillnets (standardized multi-mesh bottom gillnets [35]) during day and night on four occasions, resulting in eight fishing occasions for each lake, to obtain catch per unit effort (CPUE) estimates of fish compositions as a result of diel activity patterns in fish. Four multi-mesh nets were used on each sampling occasion, along with two additional nets with a mesh size of 45 mm to select for pikeperch. Nets were placed in the profundal zone at a depth of 3–9 m, the main feeding habitat of pikeperch. Each sampling occasion lasted for eight hours, from 10 to 18 o'clock during day and 22 to 06 during night samplings.

Total length of perch and roach individuals was measured (nearest mm) and the total mass per species and net was used to estimate catch per unit effort. Gut contents from all pikeperch were analysed in the lab and consumed prey were counted and identified to species. Pikeperch selectivity for roach and perch were estimated using Ivlev's selectivity index [36].

Temperature and secchi-depth (Lake Västersjön, $9.6 \pm 0.1°C$, $3.27 \pm 0.02$ m and Lake Osbysjön, $10.2 \pm 0.4°C$, $0.58 \pm 0.01$ m, mean $\pm$ SD) were measured each day of fishing. Absorbance of DOC in lake waters was measured with a spectrometer (Beckman DU 800; Beckman, Fullerton, California, USA) at 420 nm after the water was filtered through a GF/F filter (absorbance Lake Västersjön $= 0.032$, and Lake Osby sjön $= 0.082$).

### Laboratory experiments

**Collection and maintenance of experimental fish.** Six pikeperch (total length 289–341 mm, total weight 198–324 g) were caught in Lake Ringsjön, nearby Lund, Southern Sweden. Pikeperch were acclimatized to experimental conditions ($16.5 \pm 0.5°C$, mean $\pm$ SD, 9:15 h light:dark regime) in 500 l aquaria for five months. Four weeks before the start of the experiment pikeperch were moved and held individually in separate compartments ($50 \times 50 \times 50$ cm) of larger aquaria.

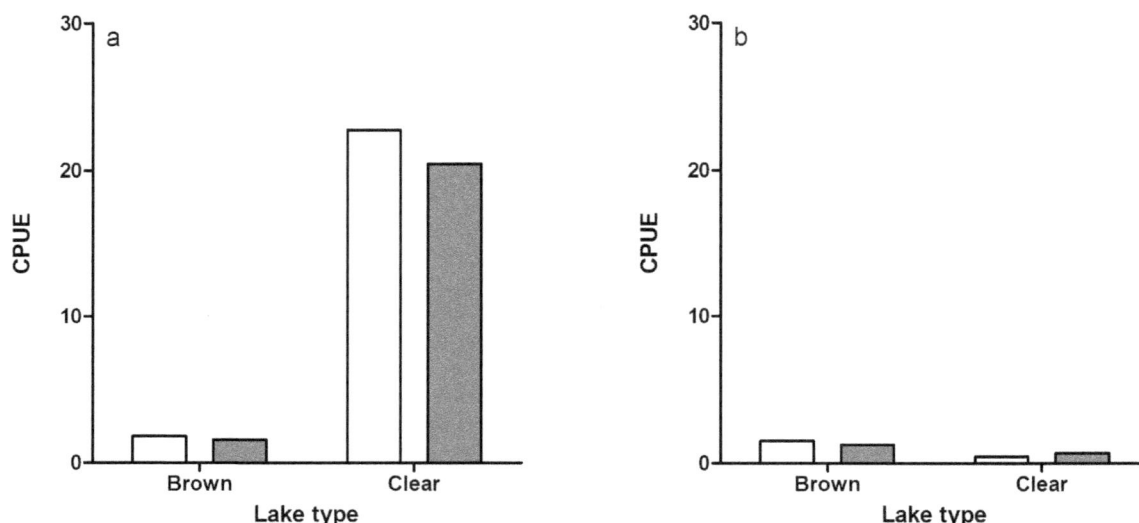

Figure 1. Catch per unit effort (CPUE) of perch (white bars) and roach (grey bars) in one brown and one clear water lake during day (a) and night (b).

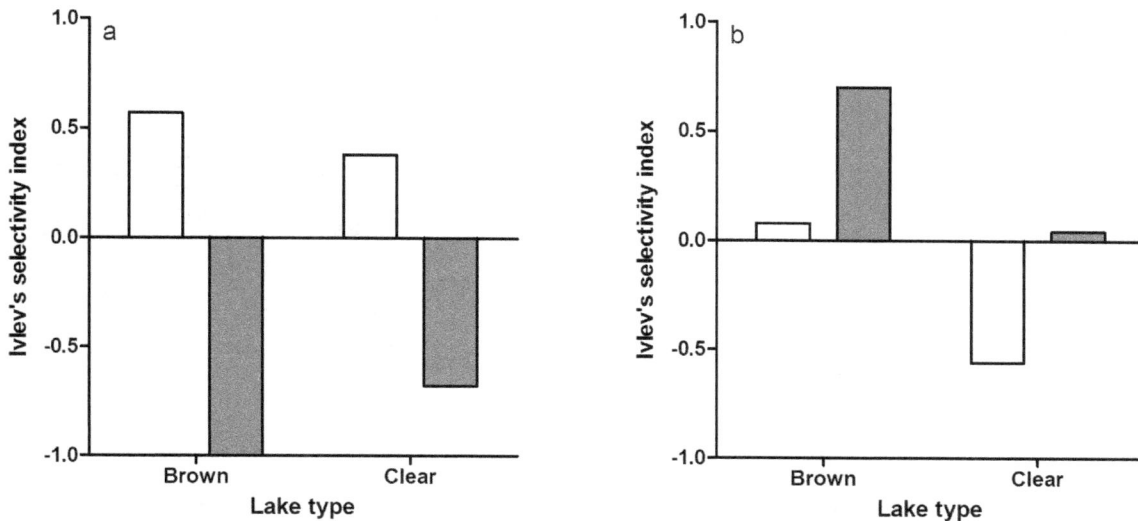

**Figure 2. Selectivity of perch (white bars) and roach (grey bars) in one brown and one clear water lake during day (a) and night (b).**
The horizontal line at 0 correspond to the null hypothesis of equal selection of prey, values close to 1 represent high prey selection and values close to −1 represent low selection.

Perch were caught in Lake Hjärtasjön and Lake Ringsjön and roach were caught in Lake Ringsjön; both species were caught with dip nets. Prey fish were of similar total length (perch: 64±0.3 mm, roach: 64±0.4 mm, mean±SE), body depth (perch: 12.7±0.1 mm, roach: 12.2±0.1 mm) and wet weight (perch: 2.1±0.04 g, roach: 2.1±0.05 g). Perch and roach were acclimatized to indoor conditions for a minimum of one week in two 300 L aquaria before used in experiments. Perch were fed frozen and thawed chironomid larvae and roach were fed dry pellets every second day to maintain their condition. Pikeperch were fed live perch and roach before and between trials. To standardise feeding motivation, all pikeperch were starved for 48 h before trials. The study complies with the current laws in Sweden, no specific permits were needed for the field studies. Ethical concerns on care and use of experimental animals were followed under the permission approved for this study (M165-07) from the Malmö/Lund Ethical Committee.

**Experimental arena and water transparency.** All experiments were carried out in a transparent PVC arena (200×50×50 cm) with a water depth of 35 cm. All corners were rounded with PVC sheets to prevent prey fish from hiding or being cornered. The water temperature in the arena was kept at 16–16.5°C in all experiments. Humic water was collected from the 'Black Pond' (secchi depth 4 cm, pH 6.5–7.0) nearby Lund, and was used to brown colour the experimental water to the desired visual ranges. Visual range, used as a parameter for water clarity, was measured by observing a vertically held black and white secchi disk (Ø: 0.1 m) from one transparent end of the arena, and was set to the horizontal distance at which the human eye could no longer discern contrast between the black and the white disc parts [37,38]. Humic pond water was diluted to the visual ranges 0.25, 0.5 and 2 m to be used in experiments. Absorbance of DOC was measured in the same way as the lake water (absorbance at the different visual ranges were: 0.25 m = 0.319, 0.50 m = 0.081 and 2 m = 0.022). To simulate daylight condition we provided light from two halogen spotlights (500 w), resulting in a light intensity of 500–600 lux at five mm above the water surface. Night treatments were performed in darkness (<0.001 lux).

## Prey selectivity study

Pikeperch selectivity for roach and perch was studied in the laboratory at the three visual ranges at day and night light conditions. Five individuals each of perch and roach were acclimatized in the arena for 15 minutes before a pikeperch was introduced and the experiment was started. Each trial was terminated after 45 minutes and remaining prey were counted. Each treatment combination was replicated five times, where five pikeperch participated in all treatments in a random order using a randomized block design. Individual prey fish were only used once.

Pikeperch prey preference was calculated using the Manly-Chesson selectivity index [39], ranging from 1, indicating positive selection, to 0, indicating negative selection or avoidance. In our case with two prey categories, no selectivity is indicated by an index not significantly different from 0.5.

## Behavioural study

Pikeperch and prey fish behaviours were studied in separate trials. We were neither able to observe behaviours during night conditions, nor in the 0.25 m visual range treatment, why the behavioural study was performed only in daylight conditions and in the treatments with visual ranges of 0.5 and 2 m. Five individuals each of perch and roach were acclimatized for 15 minutes in the experimental arena before one pikeperch was introduced and the trial began. Each trial was terminated after a successful attack by the pikeperch. Number of interests (when pikeperch observed a prey item) and attacks for each species were observed and counted during the trials. Observations were made behind a tarpaulin to minimize disturbances. All trials were also video recorded from the long side of the arena for further analysis of the first attack distance of pikeperch and the first escape distance for each prey species (perch and roach). Each treatment was replicated six times using six individual pikeperch in a randomized block design as above.

## Statistics

The effects of the factors visual range (random) and day/night light conditions (fixed) on pikeperch selectivity between perch and

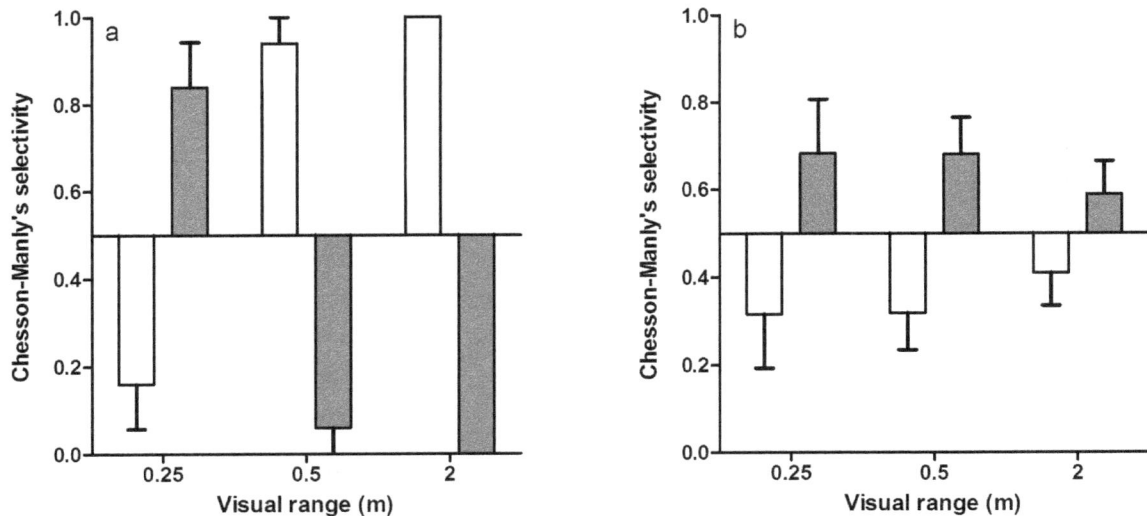

**Figure 3. Pikeperch prey selection of perch (white bars) and roach (grey bars) at different visual ranges at day (a) and night (b) in the laboratory experiments.** The horizontal line at 0.5 represents the null hypothesis of equal prey selection. Error bars denote 1 SE.

roach prey in the laboratory experiment were evaluated in a mixed effect randomized block (rb) ANOVA in SPSS (release 19). As the Manly-Chesson index for one prey species in one trial is always one minus the index for the other species, we only included the selectivity index for perch as a dependent variable. Further, as each pikeperch participated in all treatment combinations, the experimental design was not fully replicated. Including pikeperch individual identities as a blocking factor (random) in the statistical model (but not its interaction with other factors) allows for evaluation of treatment effects compensating for potential differences in levels between pikeperch individuals, as well as adjusting the degrees of freedom according to the recurring use of individuals (Quinn & Keough, 2002). The Manly-Chesson indices were further evaluated for differences from no selectivity in one-sample t-tests with the null hypothesis of an index of 0.5, again using perch indices as the dependent variable. When all prey of one species is consumed the Manly-Chesson index cannot be used. In one trial (visual range 2 m and night) all the roach were consumed, and, hence, this trial was not included in the analysis. Pikeperch and prey behavioural attributes were analysed in mixed effect rb ANOVAs, as above, for the dependent variables number of interests in prey, number of strikes, capture success, attack distance, and prey escape distance. Model residuals were not different from normal distributions (Kolmogorov-Smirnov tests, $z = 0.536–1.281$, $p = 0.075–0.909$).

## Results

### Field study

Net fishing in the clear and the brown water lake showed that there were similar densities (CPUE) of roach and perch in the two lakes (Fig. 1). However, CPUE of both species was higher during the day in the clear lake, suggesting high activity of both species during the day in this lake. The gut content analyses revealed that pikeperch showed a positive selection for perch but not of roach during the day in both lakes, as indicated by Ivlev's selectivity index (Fig. 2a). This pattern changed at night when pikeperch selectively fed on roach in the brown water lake, whereas in the clear water lake there was no selection for roach and a negative selectivity for perch (Fig. 2b).

### Prey selectivity experiment

In the laboratory experiment where we studied pikeperch prey selection for roach or perch we found that there were no main effects of visual range ($F_{2,2} = 1.265$, $p = 0.442$) or light conditions ($F_{1,2.001} = 1.863$, $p = 0.305$) on the selectivity of pikeperch and, further, there was no difference among pikeperch individuals in selectivity ($F_{4,19} = 0.770$, $p = 0.558$). However, there was a significant interaction between visual range and light conditions ($F_{2,19} = 12.394$, $p<0.001$), indicating that pikeperch change their selectivity between day and night conditions and, further, that

**Table 1.** The effects of light condition and visual range on prey selectivity of pikeperch.

| Light conditions | Visual range (m) | df | t | p |
| --- | --- | --- | --- | --- |
| Day | 0.25 | 4 | −3.288 | 0.030 |
| Day | 0.5 | 4 | 7.223 | 0.002 |
| Day | 2 | - | - | - |
| Night | 0.25 | 4 | −1.473 | 0.215 |
| Night | 0.5 | 4 | −2.112 | 0.102 |
| Night | 2 | 3 | 0.816 | 0.474 |

The treatment of daylight conditions at 2 m visibility was excluded from the analysis as all perch and no roach were eaten, resulting in no variance in data.

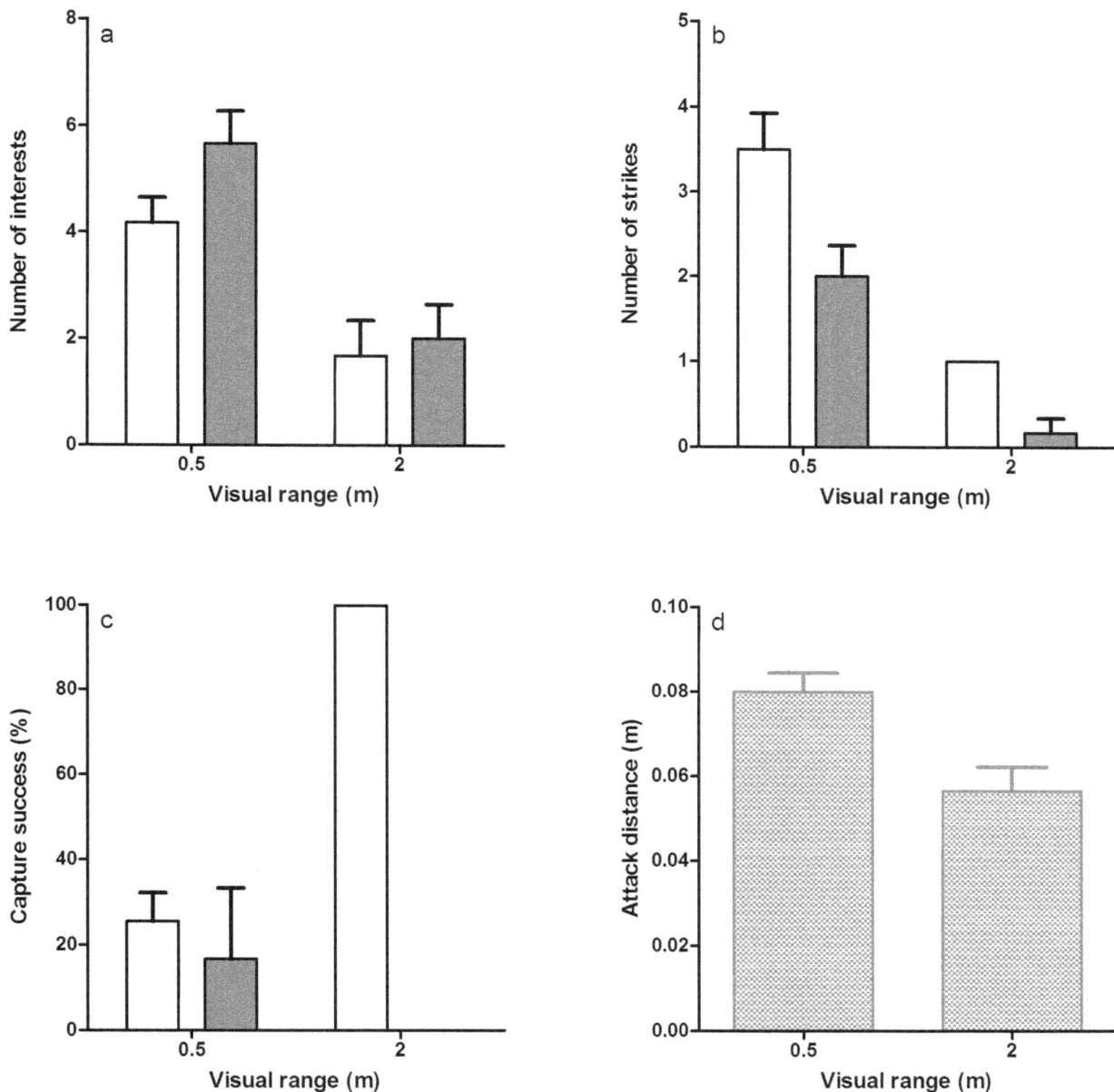

**Figure 4. Behavioural parameters of pikeperch foraging on perch (white bars) and roach (grey bars) at visual ranges of 0.5 and 2 m.** Behavioural parameters include number of interests (a) and strikes (b), as well as capture success (c). In the analyses of attack distances (d), data for the two prey species are pooled. Error bar denote 1 SE.

visual range in daylight conditions affect prey species selectivity (Fig. 3). One-sample t-tests for each combination of visual range and light conditions revealed that pikeperch showed a significant selection for roach at the 0.25 m visual range in daylight conditions, but selected for perch at longer visual ranges (see above; Fig. 3a). During night, pikeperch did not show any significant selection for any prey in any of the visual ranges (Table 1, Fig. 3b).

## Pikeperch and prey behaviours

The rbANOVA revealed a significant interaction between visual range and prey species on both pikeperch capture success and prey escape distance (Table 2). With regards to capture success, the interaction effect originated from a change from a similar capture success on roach and perch prey in water with a visual range of 0.5 m to 100% capture success on perch prey and a 0% capture success on roach at a visual range of 2 m (Fig. 4c). There was no difference between species in prey escape distances in water with 0.5 m visual range, whereas escape distance in roach was substantially longer in water with 2 m visual range (Fig. 5). Pikeperch attack distance on perch prey was significantly affected by visual range (Table 2) with a longer attack distance in water with 0.5 m visual range (Fig. 4d). All other interaction terms as well as number of interests (Fig. 4a) and number of strikes (Fig. 4b), including pikeperch individual as a blocking factor, had no significant effects on the measured behaviours (Table 2).

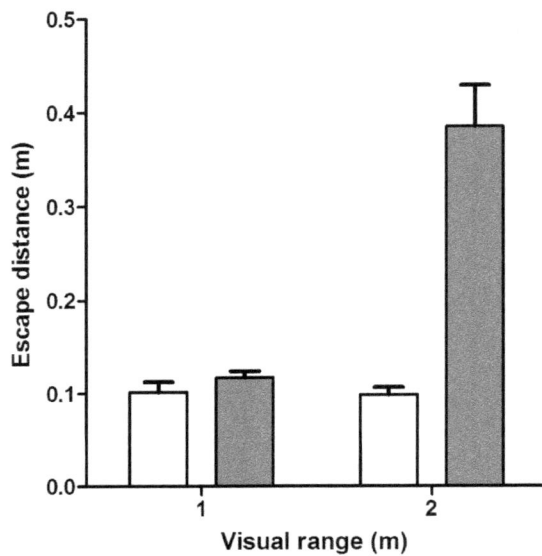

**Figure 5. Escape distance of perch (white bars) and roach (grey bars) at a visual range of 0.5 and 2 m.** Error bar denote 1 SE.

## Discussion

In the field study as well as in the laboratory experiments we found changes in pikeperch prey selectivity between perch and roach as the optical conditions changed, including changes in both light intensity (night/day) and brown coloration. Many studies indicate that degraded optical conditions decrease the possibilities for the predator to choose among prey resulting in absence of selectivity for any prey [11,40,41], whereas our results instead show a change in selectivity from one prey species to another. The field observations show selectivity for perch in daylight conditions in both brown and clear water, but at night this pattern changed and we found that there was selectivity for roach in brown water. The results from our laboratory experiments on pikeperch selectivity provide us with further details on how light and visibility conditions affect prey selectivity in pikeperch. The significant interaction between visual range and light condition in the rbANOVA along with t-tests on prey selectivity highlights two interesting aspects. It corroborates a difference in prey selectivity between day and night conditions, with no significant selectivity among perch and roach prey during night, whereas during day there was a significant prey selectivity. Moreover, the selectivity during day was dependent on the visual range as pikeperch showed a selection for roach in very poor visibility conditions (visual range = 0.25 m), whereas with increasing visual ranges there was a change in selectivity towards a preference for perch. This suggests that there may be a threshold level of degrading visual conditions, where pikeperch selectivity change from a selection for roach to a selection for perch. Such a threshold level of humic contents are to be found in lakes today [42] and may very well be reached in an increasing number of lakes if the documented brownification continues to increase according to predictions. In order to understand and predict the consequences of such potential thresholds for fish community structure it is crucial to increase our understanding of the mechanisms behind prey selectivity in pikeperch. As selectivity can be a function of active preference in the predator, differential encounter rates between prey types or differences in both predator and prey behaviour [43,44], we used controlled behavioural

**Table 2.** The effects of visual range (VR, random factor) and prey species (Prey, roach or perch, fixed) on pikeperch foraging behaviours.

|  | Number of interests | | | Number of strikes | | | Capture success | | | Attack distance | | | Escape distance | | |
|---|---|---|---|---|---|---|---|---|---|---|---|---|---|---|---|
|  | df | F | p | df | F | p | df | F | p | df | F | p | df | F | p |
| VR | 1,1 | 27.939 | 0.119 | 1,1 | 42.250 | 0.097 | 1,1 | 0.402 | 0.640 | 1,5 | 14.412 | 0.013 | 1,1 | 0.952 | 0.508 |
| Prey | 1,1 | 2.469 | 0.361 | 1,1 | 12.250 | 0.177 | 1,1 | 1.428 | 0.444 |  | N/A |  | 1,1 | 1.233 | 0.467 |
| VR*Prey | 1,15 | 1.219 | 0.287 | 1,15 | 0.2153 | 0.281 | 1,15 | 21.884 | <0.001 |  | N/A |  | 1,15 | 32.896 | <0.001 |
| PI | 5,15 | 2.194 | 0.109 | 5,15 | 0.875 | 0.521 | 5,15 | 0.401 | 0.840 | 5,5 | 1.706 | 0.286 | 5,15 | 0.747 | 0.601 |

Pikeperch individual (PI) was included as a random blocking factor.

experiments to evaluate possible mechanisms affecting pikeperch prey selectivity.

An active prey choice in pikeperch should be indicated by different aptitudes for attack between prey species. As neither pikeperch number of interest, nor number of attacks differed between prey species, the behavioural results suggest that active choice is not a major contributor to pikeperch selectivity. Instead, prey selectivity in pikeperch seems to be a result of processes at later stages in the foraging cycle, i.e. at the capture stage, where success can be affected by characteristics of both the predator and the prey in combination with environmental factors. We found that capture success was affected both by prey species attacked and visual range in the water. In the laboratory experiment we found a 100% capture success for pikeperch foraging on perch in clear water and a 0% capture success when foraging on roach, which thus explains why pikeperch show selectivity for perch in daylight conditions and at long visual ranges. Pikeperch attack distances were always shorter than the measured escape distances of roach in clear water, i.e. roach avoid predation by initiating an early escape response at distances that are outside the distance where pikeperch initiate their attacks, a so called safe distance also used by other fish species [45].

At the shorter visual range of 0.5 m, both capture success and escape distances are comparable between prey species. Still, results from both the lakes and the experiments show selectivity for perch at this visual range during day. Roach are known to school tightly as a response to predation threat and rely on a high swimming capacity for predator evasion, whereas perch reduce predation by fine-tuned manoeuvrability, cryptic coloration and spiny rayed fins [46–48]. Perch also commonly adopt an inactivity strategy when facing predation risk, which we also observed in the behavioural experiment. Schooling acts as to reduce encounter rate [49,50] and to dilute individual risks and confuse predators [51]. These behaviours could lie behind the maintained selectivity for perch in spite of comparable capture success and escape distance among species; we observed pikeperch to generally approach their prey slowly and attack from a short distance and inactivity should be an inappropriate measure to avoid pikeperch predation. Further, the results also suggest that schooling and high swimming capacity, as in roach, is more efficient to reduce pikeperch predation rates compared to fine-tuned manoeuvrability and spiny rayed fins [48], as in perch.

During night and in highly brown water (visual range of 0.25 m) there is a relative shift towards selectivity for roach in both the lakes and experiments, although this selectivity was not significant in experiments. However, due to logistic reasons we were not able to quantify the behavioural attributes of pikeperch and prey during night and at 0.25 m visual range why mechanistic explanations of selectivity patterns under these circumstances are more speculative. The results may, however, be interpreted as a reduction of the efficiency of anti-predatory behaviours in roach with decreasing light intensity or increasing water colour, as schooling and timing of fast-start escapes could be impaired with

poor visual information. Vision is a key component in school formation [52], and reduced optical condition due to light limitations or turbid/brown water results in a split up of schools [53,54]. Further, the visual system of pikeperch is adapted for low light intensities with a specific sensory adaptation, *tapetum lucidum* [55,56], which should enhance the ability of pikeperch to detect prey in brown water environments. Differences in visual capacity in pikeperch and roach during poor light conditions may affect the relative disadvantage due to differences in escape and detection distances present in daylight conditions. Further, feeding efficiency of pikeperch has been shown to be unaffected by light condition and turbidity [57], which indicates that pikeperch may use sensory input from the lateral line system when foraging in poor optical conditions [58]. Thus, a combination of reduced efficiency of behavioural responses in prey and a relative advantage for the predator due to better vision/sensory input in poor optical conditions may facilitate the change in selectivity found in pikeperch at night and in very brown waters.

Pikeperch are well adapted for foraging under visually degraded conditions compared with other sympatric piscivores [57,59]. This suggests that pikeperch should be an increasingly important piscivore in a future, browner lake scenario. As piscivore predation rates and prey selectivity can impose far-reaching structuring forces on fish communities and trophic processes, environmentally driven prey selectivity in pikeperch holds important cues to our understanding of lake system processes in a changing environment. This may be especially true for lake systems undergoing major brownification processes, as they are commonly signified by relatively low productivity [24], where selective predation should assert a strong structuring force on lower trophic levels. Our work highlights the importance of considering piscivore-prey interactions and visibility conditions in evaluations and predictions of trophic effects from the increasing brownification of lake ecosystems.

## Supporting Information

**Data S1  Data behind fig. 1–5.**
(XLSX)

## Acknowledgments

Thanks to students and teachers from Osby naturbruksgymnasium for invaluable collaboration in the field, as well as Osbysjöns FVO and Västersjön FVO for their generous permission to include their lakes in our study. Mattias Ekvall helped collecting prey fish, and Pontus Persson lent his boat for field work. Ethical permit for experiments on animals was provided by the Malmö/Lund Ethical Committee (M165-07).

## Author Contributions

Conceived and designed the experiments: LR JP MJ PAN CB. Performed the experiments: LR JP. Analyzed the data: LR JP PAN. Contributed reagents/materials/analysis tools: LR JP MJ PAN CB. Contributed to the writing of the manuscript: LR JP MJ PAN CB.

## References

1. Brönmark C, Miner JG (1992) Predator-Induced Phenotypical Change in Body Morphology in Crucian Carp. Science 258: 1348–1350.
2. Brönmark C, Paszkowski CA, Tonn WM, Hargeby A (1995) Predation as a determinant of size structure in populations of crucian carp (*Carassius carassius*) and tench (*Tinca tinca*). Ecology of Freshwater Fish 4: 85–92.
3. Carpenter SR, Kitchell JF (1988) Consumer control of lake productivity. Bioscience 38: 764–769.
4. Schulze T, Baade U, Dorner H, Eckmann R, Haertel-Borer SS, et al. (2006) Response of the residential piscivorous fish community to introduction of a new predator type in a mesotrophic lake. Canadian Journal of Fisheries and Aquatic Sciences 63: 2202–2212.
5. Kangur K, Park YS, Kangur A, Kangur P, Lek S (2007) Patterning long-term changes of fish community in large shallow Lake Peipsi. Ecological Modelling 203: 34–44.
6. Holling CS (1965) The functional response of predators to prey density and its role in mimicry and population regulation. Memoirs of the Entomological Society of Canada: 1–60.
7. Endler JA (1991) Interactions between predator and prey. In: Krebs J R., Davies NB, editors. Behavioural ecology: an evolutionary approach. Oxford: Blackwell Scientific Publications. 169–196.
8. Beauchamp DA, Baldwin CM, Vogel JL, Gubala CP (1999) Estimating diel, depth-specific foraging opportunities with a visual encounter rate model for

pelagic piscivores. Canadian Journal of Fisheries and Aquatic Sciences 56: 128–139.

9.  Mazur MM, Beauchamp DA (2003) A comparison of visual prey detection among species of piscivorous salmonids: effects of light and low turbidities. Environmental Biology of Fishes 67: 397–405.

10. Utne-Palm AC (2002) Visual Feeding of Fish in a Turbid Environment: Physical and Behavioural Aspects. Mar Fresh Behav Physiol 35: 111–128.

11. Reid SM, Fox MG, Whillans TH (1999) Influence of turbidity on piscivory in largemouth bass (*Micropterus salmoides*). Canadian Journal of Fisheries and Aquatic Sciences 56: 1362–1369.

12. Miner JG, Stein RA (1996) Detection of predators and habitat choice by small bleugills: effects of turbidity and alternative prey. Transactions of the American Fisheries Society 125: 97–103.

13. De Robertis A, Ryer CH, Veloza A, Brodeur RD (2003) Differential effects of turbidity on prey consumption of piscivorous and planktivorous fish. Canadian Journal of Fisheries and Aquatic Sciences 60: 1517–1526.

14. Engström-Öst J, Mattila J (2008) Foraging, growth and habitat choice in turbid water: an experimental study with fish larvae in the Baltic Sea. Marine Ecology-Progress Series 359: 275–281.

15. Jonsson M, Ranaker L, Nilsson PA, Bronmark C (2013) Foraging efficiency and prey selectivity in a visual predator: differential effects of turbid and humic water. Canadian Journal of Fisheries and Aquatic Sciences 70: 1685–1690.

16. Abrahams M, Kattenfeld M (1997) The role of turbidity as a constraint on predator-prey interactions in aquatic environments. Behavioral Ecology and Sociobiology 40: 169–174.

17. Meager JJ, Domenici P, Shingles A, Utne-Palm AC (2006) Escape response in juvenile atlantic cod *Gadus morhua* L.: the effects of turbidity and predator speed. The journal of experimental Biology 209: 4174–4184.

18. Nurminen L, Horppila J (2006) Efficiency of fish feeding on plant-attached prey: Effects of inorganic turbidity and plant-mediated changes in the light environment. Limnology and Oceanography 51: 1550–1555.

19. Radke RJ, Gaupisch A (2005) Effects of phytoplankton-induced turbidity on predation success of piscivorous Eurasian perch (*Perca Fluviatilis*): possible implications for fish community structure in lakes. Naturwissenschaften 92: 91–94.

20. Graneli W (2012) Brownification of Lakes; Bengtsson L, Herschy, R. And Fairbridge, R., editor. New York: Springer Science.

21. Erlandsson M, Buffam I, Fölster J, Laudon H, Temnerud J, et al. (2008) Thirty-five years of synchrony in the organic matter concentrations of Swedish rivers explained by variation in flow and sulphate. Global Change Biology 14: 1191–1198.

22. Roulet N, Moore TR (2006) Browning the water. Nature 444: 283–284.

23. Monteith DT, Stoddard JL, Evans CD, de Wit HA, Forsius M, et al. (2007) Dissolved organic carbon trends resulting from changes in atmospheric deposition chemistry. Nature (London) 450: 537.

24. Karlsson J, Byström P, Ask J, Ask P, Persson L, et al. (2009) Light limitation of nutrient-poor lake ecosystems. Nature 460: 506–510.

25. Wissel B, Boeing WJ, Ramcharan CW (2003) Effects of water color on predation regimes and zooplankton assemblages in freshwater lakes. Limnol Oceanogr 48: 1965–1976.

26. Davies-Colley RJ, Vant WN, Smith DG (1993) Colour and clarity of natural waters: science and management of optical water quality. New Jersey: The Blackburn Press. 310 p.

27. Ranåker L, Nilsson PA, Brönmark C (2012) Effects of degraded optical conditions on behavioural responses to alarm cues in a freshwater fish. Plos One 7.

28. Aksnes DL, Utne ACW (1997) A revised model of visual range in fish. Sarsia 82: 137–147.

29. Keskinen T, Marjomaki TJ (2003) Growth of pikeperch in relation to lake characteristics: total phosphorus, water colour, lake area and depth. Journal of Fish Biology 63: 1274–1282.

30. Svärdson G, Molin G (1973) The impact of climate on scandinavian populations of the zander *Stizostedion lucioperca*. Stockholm: Institute of Freshwater Research Drottningholm. 112–139 p.

31. Lehtonen H, Miina T, Frisk T (1984) Natural occurence of pike-perch *Stizostedion Lucioperca* and success of introduction in relation to water quality and lake area in Finland. Aqua Fennica 14: 189–196.

32. Turesson H, Brönmark C (2004) Foraging behaviour and capture success in perch, pikeperch and pike and the effects of prey density. Journal of Fish Biology 65: 363–375.

33. Greenberg LA, Paszkowski CA, Tonn WM (1995) Effects of prey species composition and habitat structure on foraging by two functionally distinct piscivores. Oikos 74: 522–532.

34. Horky P, Slavik O, Bartos L (2008) A telemetry study on the diurnal distribution and activity of adult pikeperch, *Sander lucioperca* (L.), in a riverine environment. Hydrobiologia 614: 151–157.

35. Appelberg M (2000) Swedish standard methods for sampling freshwater fish with multi-mesh gillnets. Fiskeriverkets Information 1: 1–32.

36. Ivlev VS (1961) Experimental ecology of the feeding of fishes. New Haven, Connecticut: Yale University Press. 302 p.

37. Davies-Colley RJ (1988) Measuring water clarity with a black disk. Limnology and Oceanography 33: 616–623.

38. Jönsson M, Hylander S, Ranåker L, Nilsson PA, Brönmark C (2011) Foraging success of juvenile pike *Esox lucius* depends on visual conditions and prey pigmentation. Journal of Fish Biology 79: 290–297.

39. Chesson J (1978) Measuring preference in selective predation. Ecology 59: 211–215.

40. Shoup DE, Wahl DH (2009) The effect of turbidity on prey selection by piscivorous largemouth bass. Transactions of the American Fisheries Society 138: 1018–1027.

41. Abrahams M, Kattenfeld M (1997) The role of turbidity as a constraint on predator-prey interactions in aquatic environments. Behavioral Ecology and Sociobiology 40: 169–174.

42. Ranåker L (2012) Piscivore-prey fish interactions, consequences of changing optical environments: Lund University. 100 p.

43. Lima SL, Dill LM (1990) Behavioral decisions made under the risk of predation: a review and prospectus. Can J Zool-Rev Can Zool 68: 619–640.

44. Turesson H (2003) Foraging behaviour in piscivorous fish: mechanisms and pattern [Doktoral thesis]. Lund: Lund University. 134 p.

45. Einfalt LM, Grace EJ, Wahl DH (2012) Effects of simulated light intensity, habitat complexity and forage type on predator-prey interactions in walleye *Sander vitreus*. Ecology of Freshwater Fish 21: 560–569.

46. Svanbäck R, Eklöv P (2011) Catch me if you can - predation affects divergence in a polyphenic species. Evolution 65: 3515–3526.

47. Eklöv P, Hamrin SF (1989) Predatory efficiency and prey selection - interactions between pike *Esox Lucius, Lucius*, perch *Perca Fluviatilis* and rudd *Scardinus erythrophthalmus*. Oikos 56: 149–156.

48. Eklöv P, Persson L (1995) Pecies-specific antipredator capacities and prey refuges - interactions between piscivorous perch (*Perca fluviatilis*) and juvenile perch and roach (*Rutilus rutilus*). Behavioral Ecology and Sociobiology 37: 169–178.

49. Turesson H, Brönmark C (2007) Predator-prey encounter rate in freshwater piscivores: effects of prey density and water transparency. Oecologia 153: 281–291.

50. Dobler E (1977) Correlation between the feeding time of the Pike (*Esox lucius*) and the dispersion of a school of Leucaspius delineatus. Oecologia 27: 93–96.

51. Winfield IJ, Nelson JS (1991) Cyprinid fishes: systematics, biology and exploitation. London: Chapman & Hall. 667 p.

52. Hemmings CC (1966) Olfaction and vision in fish schooling. Journal of Experimental Biology 45: 449-&.

53. Ryer CH, Olla BL (1998) Effect of light on juvenile walleye pollock shoaling and their interaction with predators. Marine Ecology-Progress Series 167: 215–226.

54. Miyazaki T, Shiozawa S, Kogane T, Masuda R, Maruyama K, et al. (2000) Developmental changes of the light intensity threshold for school formation in the striped jack *Pseudocaranx dentex*. Marine Ecology-Progress Series 192: 267–275.

55. Ali MA, Ryder RA, Anctil M (1977) Photoreceptors and visual pigments as related to behavioural-responses and prefferred habitat of perches (*Perca* Spp) and pikeperches (*Stizostedion* Spp). Journal of the Fisheries Research Board of Canada 34: 1475–1480.

56. Luchiari AC, Freire FAD, Koskela J, Pirhonen J (2006) Light intensity preference of juvenile pikeperch *Sander lucioperca* (L.). Aquaculture Research 37: 1572–1577.

57. Ljunggren L, Sandstöm A (2007) Influence of visual condition on foraging and growth of juvenile fishes with dissimilar sensory physiology. Journal of Fish Biology 70: 1319–1334.

58. Janssen J (1997) Comparison of response distance to prey via the lateral line in the ruffe and yellow perch. Journal of Fish Biology 51: 921–930.

59. Popova OA, Sytina LA (1977) Food and feeding relations of Eurasian perch (*Perca fluviatilis*) and pikeperch (*Stizostedion lucioperca*) in various waters of the USSR. Journal of the Fisheries Research Board of Canada 34: 1559–1570.

# 2

# Biogeochemical Typing of Paddy Field by a Data-Driven Approach Revealing Sub-Systems within a Complex Environment - A Pipeline to Filtrate, Organize and Frame Massive Dataset from Multi-Omics Analyses

**Diogo M. O. Ogawa**[1,2,3,4]**, Shigeharu Moriya**[4,5,6]**, Yuuri Tsuboi**[4]**, Yasuhiro Date**[4,6]**, Álvaro R. B. Prieto-da-Silva**[1,3,7]**, Gandhi Rádis-Baptista**[1,2,3]**, Tetsuo Yamane**[1,3,8]**, Jun Kikuchi**[4,6,9]*

1 Biotechnology and Natural Resources Program, University of the State of the Amazonas, Manaus, AM, Brazil, 2 Laboratory of Biochemistry and Biotechnology, Institute for Marine Sciences, Federal University of Ceara, Fortaleza, CE, Brazil, 3 Center for Environment and Biodiversity Studies, University of the State of the Amazonas, Manaus, AM, Brazil, 4 RIKEN Center for Sustainable Resource Science, and Biomass Engineering Corporation Division, Yokohama, Japan, 5 RIKEN Antibiotics Laboratory, Yokohama, Japan, 6 Graduate School of Medical Life Science, Yokohama City University, Suehiro-cho, Tsurumi-ku, Yokohama, Japan, 7 Laboratory of Genetics, Butantan Institute, Sao Paulo, SP, Brazil, 8 Center of Biotechnology of Amazon, Manaus, AM, Brazil, 9 Graduate School of Bioagricultural Sciences, Nagoya University, Nagoya, Japan

## Abstract

We propose the technique of biogeochemical typing (BGC typing) as a novel methodology to set forth the sub-systems of organismal communities associated to the correlated chemical profiles working within a larger complex environment. Given the intricate characteristic of both organismal and chemical consortia inherent to the nature, many environmental studies employ the holistic approach of multi-omics analyses undermining as much information as possible. Due to the massive amount of data produced applying multi-omics analyses, the results are hard to visualize and to process. The BGC typing analysis is a pipeline built using integrative statistical analysis that can treat such huge datasets filtering, organizing and framing the information based on the strength of the various mutual trends of the organismal and chemical fluctuations occurring simultaneously in the environment. To test our technique of BGC typing, we choose a rich environment abounding in chemical nutrients and organismal diversity: the surficial freshwater from Japanese paddy fields and surrounding waters. To identify the community consortia profile we employed metagenomics as high throughput sequencing (HTS) for the fragments amplified from Archaea rRNA, universal 16S rRNA and 18S rRNA; to assess the elemental content we employed ionomics by inductively coupled plasma optical emission spectroscopy (ICP-OES); and for the organic chemical profile, metabolomics employing both Fourier transformed infrared (FT-IR) spectroscopy and proton nuclear magnetic resonance ($^1$H-NMR) all these analyses comprised our multi-omics dataset. The similar trends between the community consortia against the chemical profiles were connected through correlation. The result was then filtered, organized and framed according to correlation strengths and peculiarities. The output gave us four BGC types displaying uniqueness in community and chemical distribution, diversity and richness. We conclude therefore that the BGC typing is a successful technique for elucidating the sub-systems of organismal communities with associated chemical profiles in complex ecosystems.

**Editor:** Gabriel Moreno-Hagelsieb, Wilfrid Laurier University, Canada

**Funding:** This research was supported in part by Grants-in-Aid for Scientific Research (C) (No. 25513012 to J.K.), Grants-in-Aid for Scientific Research (A) (No.21247010 to S.M. and J.K.) and Advanced Low Carbon Technology Research and Developmental Program (ALCA to J.K.) from the Ministry of Education, Culture, Sports, Science, and Technology, Japan. This study was also financially supported by RIKEN Yokohama Institute under the International Program Associate (IPA) program. The funders had no role in study design, data collection and analysis, decision to publish, or preparation of the manuscript.

**Competing Interests:** The authors have declared that no competing interests exist.

* Email: jun.kikuchi@riken.jp

## Introduction

Unravelling trends that rule complex aquatic environments is a puzzling task due to the myriad of possibilities of interactions presented between and within the hosted organismal consortia with organic and inorganic compounds.

As explaining the totality of the interactions is a goal hard to achieve, if not plainly impossible considering the never ending development on science, therefore, here we intend to frame sub-systems co-existing within a larger system using a data-driven approach [1]. To comprehend such interactions we gave rise to the biogeochemical typing (BGC typing), a flexible tool to bring forth and individualize a subset of structures underlying in the studied environment based on the correlation between the community and chemical profiles analysed.

The BGC typing analysis is a pipeline built using integrative statistical analysis and can treat massive datasets as used here produced by multi-omics analysis [2] which would otherwise be hard to visualize [3] and process [4]. It filters, organizes and frames the data based on the strength of the mutual trends working within the environment.

The multi-omics analyses here was composed by metagenomics which gave the community consortia profile, ionomics showing the elemental content a metabolomics for the organic chemical profile. Here, we regard metagenomics as applying solely to the characterization of small-subunit ribosomal RNA. Therefore, the multi-omics analysis provided who is there and what is there as explained as following.

In this study we researched on the aquatic environment of three distinct paddy fields and surrounding water located in Saitama Prefecture, Japan. The paddy field is the source of one of the most important staple foods in the world and a rich environment comparable to a natural wetland: more than merely the ability to sustain crops, it harbours an intricate net of life, and it is able to support even higher-trophic level organisms such as fish [5].

In a complex environment, one can find thousands of different organisms thriving. To answer who is there, we performed the metagenomics to identify the organismal consortia. The identification was expressed as operational taxonomic units (OTUs) [6] retrieved by high throughput sequencing (HTS) the polymerase chain reaction (PCR) products of Archaea-specific and universal 16S (Archaeal genes excluded) and universal 18S small-subunit ribosomal RNA primers.

To assess what is there we joined the pieces of information from ionomics and metabolomics.

The ionomics is the elemental analysis evaluating its variation over a set of samples in an approach as the one applied to plant assay [7]. The ionomic analysis was assessed by the use of inductively coupled plasma optical emission spectroscopy (ICP-OES).

For the metabolomics we used two techniques, the attenuated total reflectance Fourier transformed infrared (FT-IR) and the proton nuclear magnetic resonance ($^1$H-NMR).

The FT-IR is a technique easy to be employed by request little preparation to the sample and give us information about its organic chemical profile regarding the rotational-vibrational frequency from the chemical bonds present in the molecules being a useful tool in metabolomics [8].

The $^1$H-NMR has been proved for long to be also a powerful tool in metabolomics [9,10], assessing the information related to the structure from the molecules in our sample that contain hydrogen as the large majority of organic compounds.

Such multi-omics dataset was split in two groups of data: one derived from metagenomics aggregating the OTUs from Archaea, 16S rRNA and 18S rRNA forming our organismal community matrix (matrix community) and the second group of data formed by the ionomic information acquired by ICP-OES joint to the metabolomic information represented by the integration of FT-IR and $^1$H-NMR spectra, thus being fused to one matrix of chemical profile (matrix chemicals).

As in our method we are not able to differentiate whether the cells were dead or alive in the exact time of sampling and some organisms may feed on dead cells it was assumed that all matter including the cells constituents took part of the environmental condition, therefore, regarded on the chemical profile.

The integrated statistical analysis is the tool to connect and frame the various trends acting underneath the broader complexity of the totality of the environment. Our group has being developing statistical tools to grasp the explainable features present on diverse environments [11,12].

Here, we filter, organize and frame the data applying the pipeline of the BGC typing to expose the links amongst the organismal community and the chemical profile. It optimizes the set of information retrieved by filtering the data according to the strength of the correlation and individualizes sub-systems of organismal consortia along its chemical features framing our BGC types. Each BGC type thus comprises a small universe statistically isolated working within the environment, helping the understanding of the whole system.

The BGC typing pipeline is based on integrative statistical analysis: namely, Spearman correlation [13,14], least-squares structuring [15] and k-means clustering [16,17]. The set formed by the groups of organisms and the chemical profiles associated by this pipeline we call the BGC types which are meaningful a priori only in the specific study, nevertheless we expect to find similar BGC types spread on similar environments under similar conditions and analyses. The BGC typing then would improve and develop itself as more and more studies are done following this approach.

The description of four singular BGC types found in this study shows that we successfully established the technique of BGC typing as a tool to characterize sub-systems composed by the community distributions and structures associated to chemical profiles on a complex environment such the Japanese paddy field and surrounding waters (Fig. 1).

## Materials and Methods

### Sampling

We designed the sampling method to encompass what we regarded as three unities of sampling sites which comprised three samples from the water of a chosen paddy field plus its collector stream. We added two sampling points from the river that boundaries the paddy fields, the Ara river – one sampling was taken from the river right before it meets the paddy field area and another right after such paddy area ends.

The sampling site lies on the plains of the Hiki District of Saitama Prefecture (Japan) over a large agricultural area following the course of the Ara River for approximately 13 km. Samples were collected on August 23, 2011, a few weeks prior to harvest; the paddy fields had been flooded all summer to raise the crop (rice). Using sterile tubes, four 50 mL aliquots of water were collected per sample from each sampling point. The points were located in three individual paddy fields, their respective collector streams, and the Ara River itself, for a total of 14 sampling points (Fig. 2). Specifically, there were two samples from the Ara River, one upstream of the paddy field areas and the other downstream of the paddy fields (ara1 - 36°2′32″N 139°30′8″E; ara2 - 35°56′54″N 139°32′41″E); three samples from different points of paddy field 1 (p1f1–p1f3 - 36°2′23″N 139°29′51″E) and its collector stream (p1s - 36°2′25″N 139°29′46″E); three samples from different points of paddy field 2 (p2f1–p2f3 - 35°59′37″N 139°30′5″E) and its collector stream (p2s - 35°59′37″N 139°30′5″E); three samples from different points of paddy field 3 (p3f1–p3f3 - 35°58′29″N 139°30′33″E) and its collector stream (p3s - 35°58′29″N 139°30′32″E). The samples were stored at 4°C in a cooler box and returned to the laboratory immediately, where they were stored at −80°C. Before each analysis, a 60 h freeze-drying pre-processing step was performed. All freeze-dried samples were weighed. One 50-mL aliquot was used for DNA extraction and community analysis, another aliquot for ICP-OES, another aliquot for FT-IR, and another one for $^1$H-NMR.

## HTS of the PCR products from the ribosomal RNA gene from environmental samples

High throughput sequencing is a powerful tool to identify the constituent organisms in environmental studies [18].

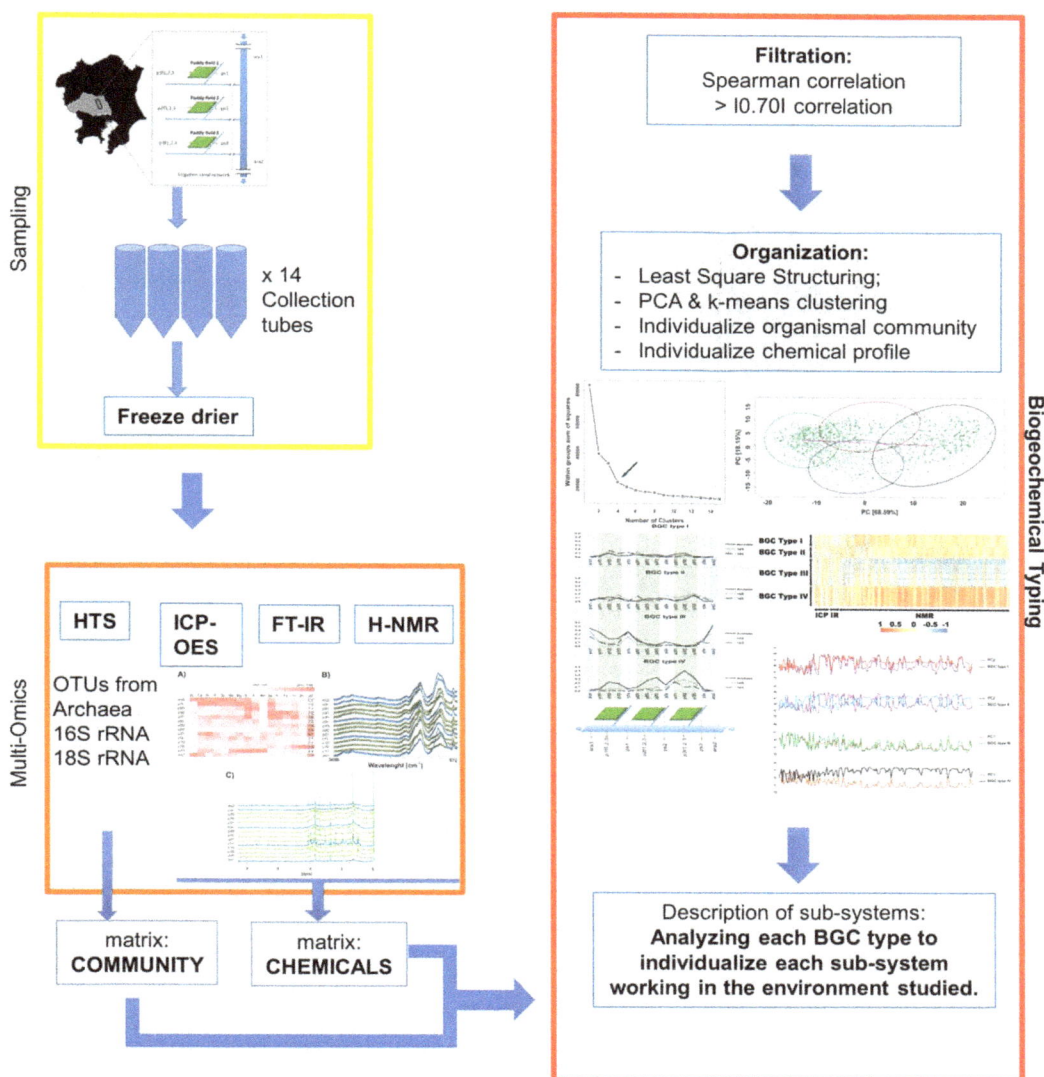

**Figure 1. Schematic representation for the Biogeochemical Typing (BGC typing).** Yellow box: steps for collection and pre-processing the samples. Orange box: steps for data acquisition and formatting for BGC typing. Red box: steps for BGC typing as the integrated statistical analyses.

The DNA was extracted from the freeze-dried samples by using the Power Soil DNA extraction kit (MoBio, CA, USA); the concentration of extracted nucleic acids was measured using a CLUBIO Micro Spectrophotometer.

We amplified by PCR the small sub-unit ribosomal RNA sequences from the extracted DNA. Fragments from 16S rRNA, 18S rRNA, and Archaeal rRNA were separately amplified. The hipervariable regions from V1 to V3 for the 16S rRNA were amplified using modified Ba27F (5′-AGAGTTT-GATCCTGGCTCAG-3′) as the forward primer [19] and PRUN518 (5′- ATTACCGCGGCTGCTGG-3′) as the reverse primer [20]. The hipervariable regions from V1 to V3 for the 18S rRNA were amplified using Euk1A (5′-CTGGTTGATCCTGC-CAG-3′) as the forward primer and Euk516R (5′- ACGGGGG-GACCAGACTTGCCCTCC-3′) as the reverse primer [21]. The hipervariable regions from V4 to V6 for the Archaeal rRNA were amplified using the 16S Archaea-specific rRNA pair of S-D-Arch-0519-a-S-15 (5′-CAGCMGCCGCGGTAA-3′) as the forward primer and S-D-Arch-1041-a-A-18 (5′-GGCCATG-

CACCWCCTCTC-3′) as the reverse primer [22]. The PCR products were subjected to agarose gel electrophoresis. Correctly sized fragments were retrieved from the gel and DNA was extracted using the Wizard SV Gel and PCR Clean-Up System (Promega, WI, USA). The final DNA concentration was measured using Invitrogen Quant-iT PicoGreen sDNA Reagent and Kits (Invitrogen, CA, USA). Correct dilutions were performed using Milli-Q water. The sequencing library for HTS was prepared using the GS Junior Titanium emPCR kit (Lib-L) (454 Life Sciences, CT, USA) by following the provided protocol. The library was read by a GS Junior sequencer following standard operating procedures.

The obtained reads from each GS Junior run were treated using QIIME software [23]. We followed the "454 Overview Tutorial: de novo OTU picking and diversity analyses using 454 data" (http://qiime.org/tutorials/tutorial.html) using default settings, with the following exceptions: de novo chimera detection and Trie pre-filtering in the OTU picking step [24]; uclust_ref as the clustering method [25]; SILVA 108 of the SILVA rRNA database

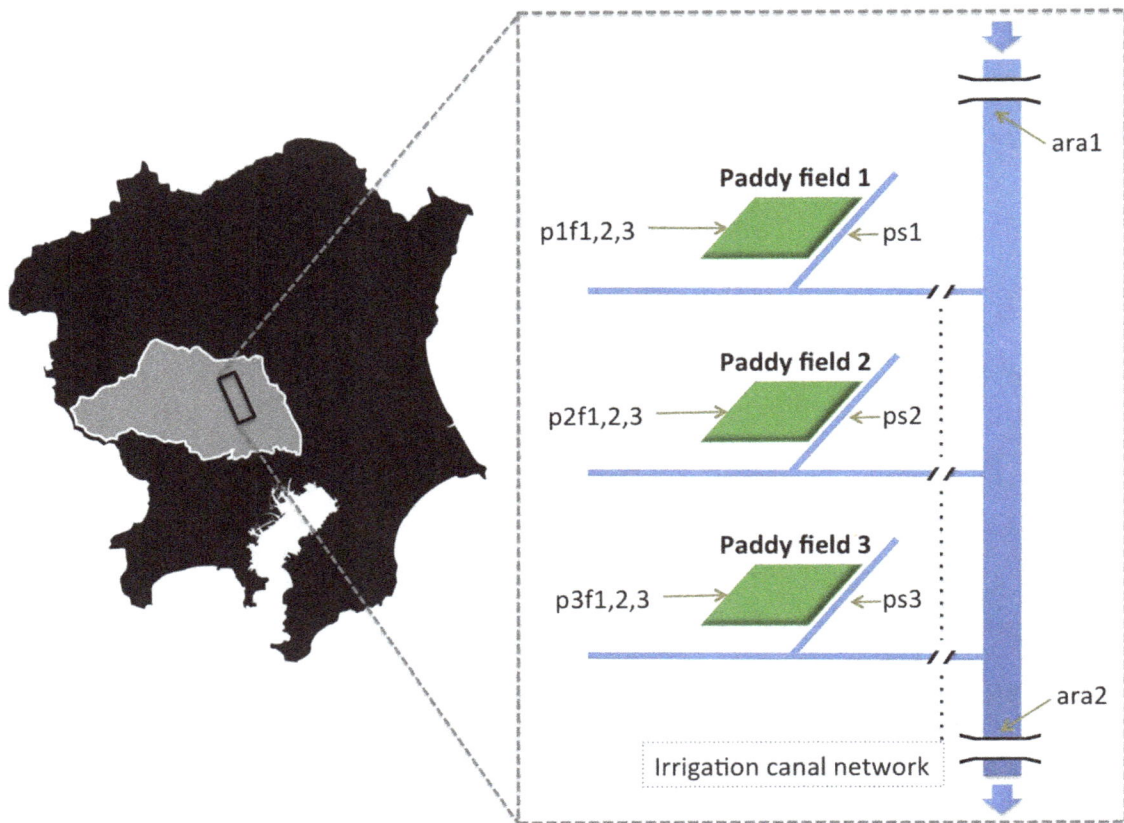

**Figure 2. Schematic representation for the sampling location.** Shadow map showing Kanto region (Japan). Grey area is Saitama prefecture. The rectangle indicates where the samples were taken. Magnified area is schematic map for the sampling site. At right side, schematic figure: Ara River was sampled in two points (ara1 and ara2) as well as paddy fields located within the area between these two river sampling points. Three independent paddy fields (paddy field 1, 2, 3) were selected and three samples were taken from each paddy field (p1f1-3, p2f1-3 and p3f1-3). These paddy fields were connected with Ara River through independent collector streams which were also sampled (p1s, p2s, p3s), respectively for each paddy field. Blue arrows indicate flow direction of Ara river and light brown arrows indicate each sampling points. Gaps in Ara River indicate bridges.

as a sequence reference [26]. The aligned sequences were assigned against the SILVA rRNA database. OTUs represented by a single read over all sampling points were filtered out to decrease computational demand, since our correlation method used would not be able to show any trend for a sequence read only once. OTUs generated by one set of primers (e.g., Archaea) that were aligned to other domains (e.g., Eukaryota) were also filtered out to prevent overrepresentation of organisms. To compare quantified information of each OTU amongst the sampling sites, the number of reads for each OTU was divided by the total number of reads for its sampling point in order to find the relative abundance for each OTU. This step was separately performed to the three domains studied. The resulting tables of OTUs against sampling points for Archaea rRNA, 16S rRNA, and 18S rRNA were fused to one matrix (matrix community). Although the aim of this study is not building an ultimate phylogenetic tree, neither this dataset allows such task, an emulation of a phylogenetic tree was built in QIIME [23,27] and exported to R to plot a more visually informative tree [28]. Observing the resultant tree with mixed Bacteria and Archaea domains, we opted to reassign the 16S rRNA and Archaeal OTUs according to data from the Ribosomal Database Project (RDP) [29]. The trees were built once again after the BGC typing for comparison and the 16S rRNA, 18S rRNA, and Archaeal OTUs appeared fairly distinguished.

## Inductively Coupled Plasma Optical Emission Spectrometry (ICP-OES)

For the preparation to the analysis, we suspended in Milli-Q water each freeze-dried sample to recreate the same 50 mL initial volume. Three dilutions were prepared–1:1, 1:10, and 1:100-for 6-mL aliquots in 15-mL plastic tubes. For the dilutions, elemental analysis was performed using SII model SPS 5510 CCD simultaneous ICP-OES (SII NanoTechnology Inc., Chiba, Japan) equipped with an SPS-3 auto-sampler (SII NanoTechnology Inc.) and using ICP Expert software (SII NanoTechnology Inc.). We used the Multi-Element Calibration Standards 3, 4, and 5 acquired from PerkinElmer (PerkinElmer Japan Co., Ltd., Yokohama, Japan) to calibrate the machine. A concentration of $1 \ mg \ L^{-1}$ of each standard dilution was used for this step. We quantified 27 chemical elements: Al, B, Ba, Be, Ca, Cd, Cr, Cs, Cu, Fe, Hg, K, Li, Mg, Mn, Na, Ni, P, Pb, Rb, S, Sb, Se, Si, Sn, Sr and Zn. Three emission wavelengths of each element were chosen to satisfy both the achievement of maximum intensities and the elimination or minimization of the interference effect for discrimination of each element in the samples. The ICP-OES operating conditions were as follows: power 1.2 kW, plasma gas flow 15 $L \ min^{-1}$, auxiliary gas flow 1.5 $L \ min^{-1}$, nebulizer gas flow 0.75 $L \ min^{-1}$, and peristaltic pump speed 15 rpm. From the obtained data, a matrix was built using the concentration in ppm for each element against the sampling points from the average result of the optimal dilution with the optimal wavelengths.

## Attenuated Total Reflectance Fourier Transformed Infrared (FT-IR) Spectroscopy

The freeze-dried samples were pressed directly on the crystal of Nicolet 6700 FT-IR spectrometer using the ATR smart iTR accessory with a high-pressure clamp (Thermo Scientific) to measure the absorbance from 4,000 to 650 cm$^{-1}$ at a resolution of 8 cm$^{-1}$. The peaks were annotated according to the absorbance wavelength. Distinguishable peaks consisting of regions with overlapping chemical bond signals were annotated by more than one chemical bond (i.e., as many as needed). The region of interest (ROI) was integrated for each assigned peak using an interval with no observed overlap. From 1,200 to 849 cm$^{-1}$ the signals for the chemical bonds were permitted to overlap, such that the peaks were assigned to more than one chemical bond candidate. The region over the interval from 847 to 650 cm$^{-1}$ was not used for annotation due to visually present but indistinguishable peaks. The spectra were integrated using Thermo OMIC software (USA). A matrix was built by assigning the sum of the integration values to a distinct part for each assigned peak as its relative amount.

## Proton Nuclear Magnetic Resonance ($^1$H-NMR)

Each freeze-dried sample was dissolved individually in a proportion of 1/9 (m/v) in 100 mM 95% deuterated phosphate buffer (100 mM $KH_2PO_4$ in 99% $D_2O$, pH 7.0) with 1 mM sodium 2,2-dimethyl-2-silapentane-5-sulfonate (DSS) as the internal standard. The solution was sonicated for 5 min at room temperature in a Bioruptor Diagenode (USA). Following sonication, centrifugation was performed at 8 krpm and supernatant was transferred to a 5-mm ø NMR tube.

All spectra were recorded at 298 K on a Bruker DRU-700 spectrometer (Germany) equipped with a $^1$H inverse cryogenic probe with triple-axis gradients operating at 700.15 MHz.

The $^1$H-NMR spectra were recorded at 32,768 points over 256 scans using the Watergate pulse sequence [30]. The J-resolved spectra were recorded in 32 scans per f1 increment with a total of 32 complex f1 and 16,384 complex f2 points.

The spectra were manually phased and calibrated in the Bruker Top Spin program. Integration of spectra was performed in advanced bucketing mode in Bruker AMIX 3.5 software on manually picked peaks using bucket widths equal to 0.02 ppm to find the integration value for each peak. Two broad peaks from 3.48 to 0.78 ppm were integrated by the sum of the integration for each bucket with no visible sharp peak for the region.

The peaks of the J-resolved spectra were assigned according to the Birmingham Metabolite Library [31]. The peaks for $^1$H-NMR spectra were assigned with the help of SpinAssign [32,33,34]. Both assigned tables were transposed to the $^1$H-NMR bucketed table (table for the integrated peaks) based on the chemical shifts. Broad peaks were assigned to proteins, whose presence was supported by positive Bradford protein assay results. 1D-STOCSY also was performed to the annotated bins correlated to the BGC types [35]. A column for the broad peak was added to the $^1$H-NMR assigned table to complete the $^1$H-NMR matrix.

## Results and Discussion

### HTS

HTS generated 100,641 sequences for the small-subunit Archaea rRNA primers, 302,974 sequences for the small-subunit 16S rRNA general primers, and 314,632 sequences for the small-subunit 18S rRNA general primers. The sequences repeated were collapsed and then assigned with similarity $\geq$97%, generating 5,688 OTUs for Archaea rRNA, 24,633 OTUs for 16S rRNA (no Archaeal OTU was found among universal 16S rRNA amplicons), and 9,355 OTUs for 18S rRNA. The rarefaction analysis showed a good coverage for the organismal community (Table S1, S2, S3, Fig. S1, S2, S3).

After filtering out the reads to OTUs appearing only once over the 14 sampling points, the respective totals for retrieved OTUs were 4,019, 14,942, and 6,391. Performing BGC typing as correlation filtering process, we retrieved 854 OTUs for Archaea rRNA, 1,743 OTUs for 16S rRNA and 815 OTUs for 18S rRNA. For each BGC type, we used QIIME to build an emulation of a phylogenetic tree and software R to plot and compare how the sequences were separated amongst the products retrieved from Archaea rRNA, 16S rRNA and 18S rRNA. The result displayed a fairly distinguished distribution of the OTUs (Fig. S4, S5, S6, S7).

### ICP-OES

We tested for 27 elements and detected 15 in our samples. Aluminium was detected in all samples with a lower concentration in paddy field waters than in the river. It is known that pH affects the concentration of elements; however, no clear trend between elemental concentration and pH was observed in the current study (Table S4).

### FT-IR

We retrieved 10 distinguishable peaks and annotated accordingly [36]. The peaks annotated to $NH_2$ and N-C = O follow the same pattern, suggesting that they are intrinsically linked as moieties of amino acids (Fig. 3). These peaks displayed the highest intensities at the two sampling points from the Ara River (ara1; ara2) and the one from the collector stream for paddy field 3 (p3s).

Also, we annotated on the spectra peaks as C-H and as O-H, suggesting organic energy available since molecules with aliphatic and alcohol bonds are prone to be oxidized by many organisms (Table S5).

### $^1$H-NMR

e integrated 161 individual peaks for $^1$H-NMR plus an integration for the two broad peaks, for a total of 162 $^1$H-NMR variables. The soluble organic compounds detected by this technique exhibited a clear pattern of samples from the Ara River having poorer concentrations (Fig. 3). Assignment of all peaks on the $^1$H-NMR spectra was difficult due to low concentrations of extractable compounds and to limited sensitivity insufficient to extend analysis to $^1$H-$^{13}$C correlation experiments. However, J-resolved $^1$H-NMR analysis allowed us to annotate a total of 60 buckets (Fig. S8).

These annotations constituted molecular residue information from the chemical compounds, amino acids, or organic products present in our samples. A fraction of the peaks correlated to the BGC types was evaluated by one-dimensional statistical total correlation spectroscopy (1D-STOCSY) to verify the degree of support for the annotation [37]. The tallest peak in the spectra was assigned to lactate with good support from 1D-STOCSY plot and direct comparison against the spectrum for the pure compound provided on the Bruker AMIX database (Table S6, Fig. S9, S10).

The two broad peaks observed in the spectra had the sum of their approximated area integrated discounting the buckets with sharp peaks. These broad peaks possessed characteristic patterns of proteins [38,39]. The Bradford assay [40] was performed on the samples with the largest broad peaks for each paddy field – p1f3, p2f3, and p3f2- with respective results of 18.1 (sd = 1.2), 29.8 (0.4), and 43.0 (0.2) µg mL$^{-1}$ of protein. Additionally, 1D-STOCSY

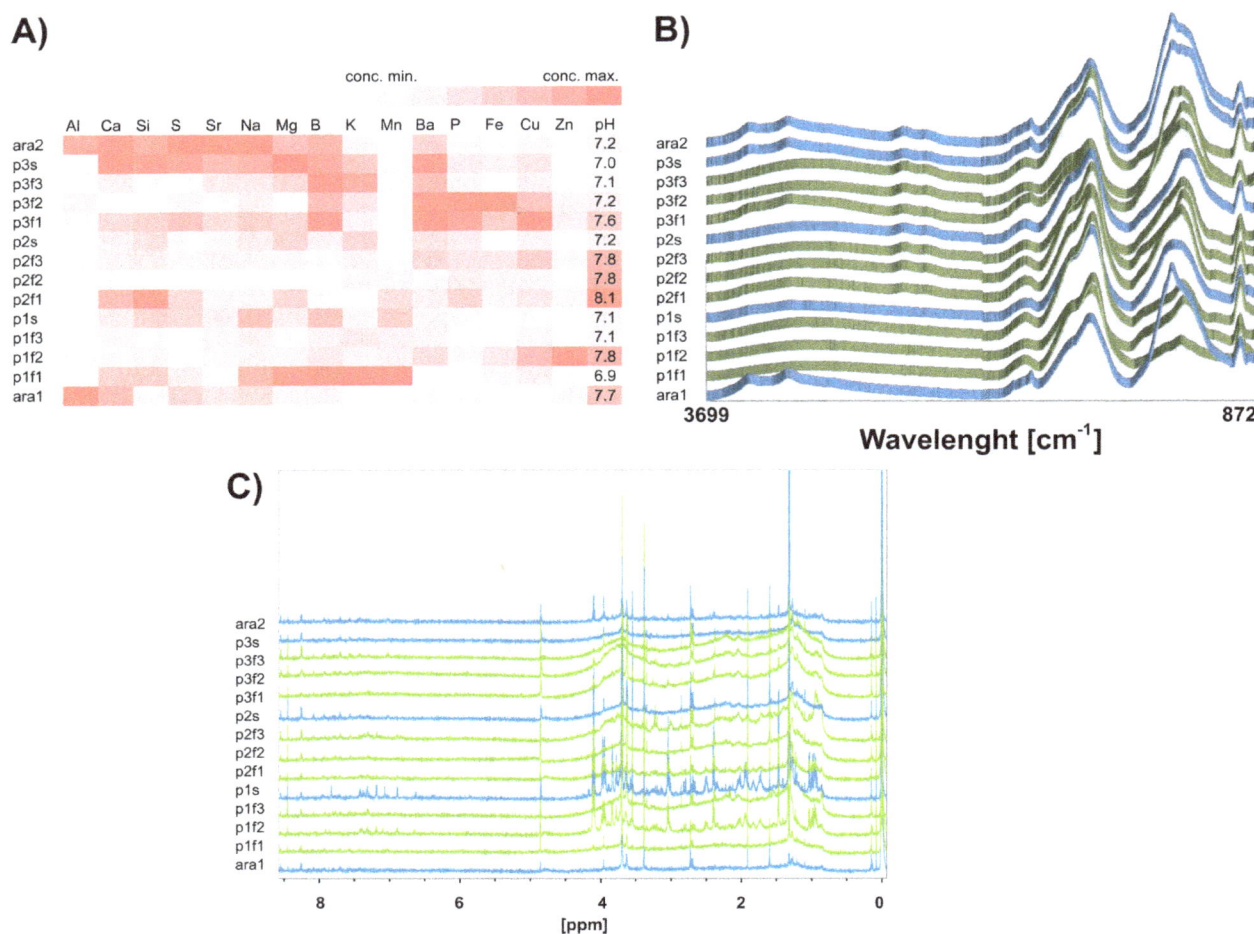

**Figure 3. Chemical profiles for the sampling points.** A) ICP-OES heatmap. X-axis: elements. Y-axis: sampling points. Red colour intensity corresponds to elemental concentration normalized by element. B) FT-IR spectra. X-axis: wavelength number. Y-axis: absorbance intensity. C) ¹H-NMR spectra. X-axis: chemical shift. Y-axis: intensity.

showed a correlation between protein concentrations against the large broad peaks in the $^1$H-NMR spectra, suggesting that soluble protein produced these peaks (Fig. S11).

## Biogeochemical typing (BGC Typing)

We constructed two independent matrices arranged by our sampling points, the matrix community and the matrix chemicals. We characterized four different groups composed by the correlated organismal and chemical profiles. We propose this statistical treatment as BGC typing.

Once the dataset is built, in this case using multi-omics approach, we applied the statistical process of BGC typing which can be divided into three main steps: 1) Filtration, 2) Organization and 3) Description.

**1) Filtration.** The matrices formed by the OTUs retrieved from Archaea rRNA, 16S rRNA and 18S rRNA sequencing were fused in one table as being the matrix community. The matrices from ICP-OES, FT-IR, and $^1$H-NMR were fused into a single matrix termed chemicals. Using R software [41], Spearman correlation was performed between the community and chemicals matrices. The resulting correlation matrix was then imported to Microsoft Excel to extract all OTUs with coefficients equal to or higher than |0.70|.

**2) Organization.** To evaluate the appropriate number of BGC types representing the sub-systems, we used the least square structuring testing up to 15 groups [42,43]. The curve inflexion indicated four as the optimal number of BGC types (Fig. 4) [15].

The correlation matrix was then divided within the principal component analysis (PCA) into four BGC types by the k-means clustering method [16] and plotted [44]. The BGC types were distributed in a cross-like fashion: two oriented horizontally according to the axis of principal component 1 (PC1) and two oriented vertically according to the axis of principal component 2 (PC2) (Fig. 5).

A table of the OTUs from each BGC type was built. For the community distribution analysis, we plotted the sum of all relative abundances along the sampling points for each BGC type (Fig. 6). To analyse the community structure, the OTUs were collapsed to the class level or beyond according to the next divergence on the taxon presented. The resulting matrices were used to find the percentage of abundance for the groups of organisms in each BGC type. To visualize the differences among chemical profiles presented by the different BGC types, we built a table and plotted (Fig. 7).

A table of the chemical profile for each BGC type was built by using the average correlation index for all the chemical variables for each chemical variable from the BGC type and the loadings for

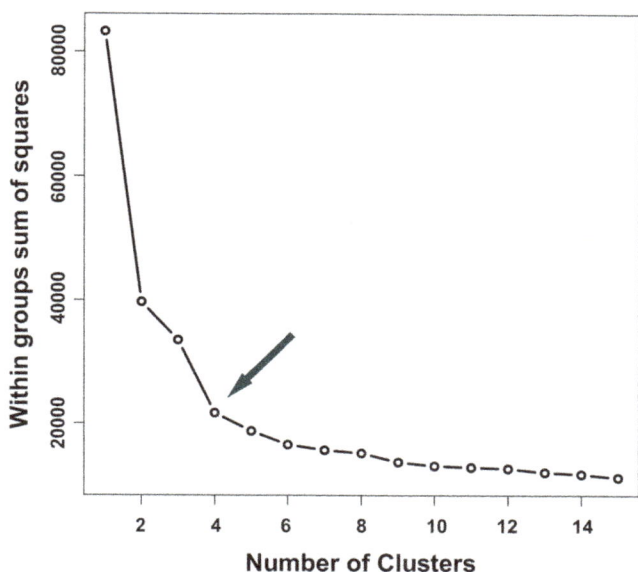

**Figure 4. Finding the optimal number of BGC types.** X-axis: number of clusters. Y-axis: sum of squared distances from each variable to the centroid within the BGC type.

**3) Description.** According the structure presented by each BGC type regarding the organismal community and chemical profiles along the sampling points, we analysed the features that individualize them and searched for the literature that could bring sense to the sub-systems exposed.

## Organismal community distribution

Organisms of BGC type I had a scattered distribution over the sampling points, with a tendency to be less present in the lotic waters ara1, p2s, p3s and ara2 with exception of p1s. BGC type II presented a scattered distribution over the paddy field and collector streams. BGC types I and II appeared to be minimally present in the Ara River (ara1, ara2).

Organisms in BGC type III were distributed predominantly on the lotic waters and had minimum presence in the lentic waters on the paddy fields. The organisms for BGC type IV were distributed predominantly on the lentic waters from the paddy fields. Conversely, BGC type IV had a minimal distribution on the river and collector streams, the lotic waters. When we summed the three domains for each paddy field data, a gradual increase was observed from paddy 1 to paddy 3, although none of the paddies was directly connected. The organismal community distribution over the sampling points can be visualized in Fig. 6.

## Organismal community structures

The retrieved Archaeal community profile was dominated by the phylum *Euryarchaeota*, mostly composed of the classes *Methanomicrobia*, *Thermoplasmata* and *Methanobacteria*. In addition, *Euryarchaeota* exhibited the largest abundance in three out of the four BGC types, being the dominant phylum on the paddy floodwaters and the collection streams for BGC types I, II and IV. The mentioned classes tend to be directly involved in methane production [45,46,47]. BGC type III, which was predominantly distributed over the lotic waters of the Ara River and the collector streams, had the phylum *Crenarchaeota*, class *Thermoprotei* as the most abundant kind of Archaea. *Crenarchaeota* is suggested to take part in primary production by active involvement in the

PC1 or PC2 according to the one that better explains the BGC type. In order to facilitate the comparison, the tables were scaled by unity of variance without centring (Fig. 8).

The PC which better explains each BGC type had the loadings compared against the average correlations from each chemical variable using the standard deviation of a population (STDEVP). Chemical variables with STDEVP of 0.20 and over were extracted to build the table of the chemical profile for each BGC type (Table 1).

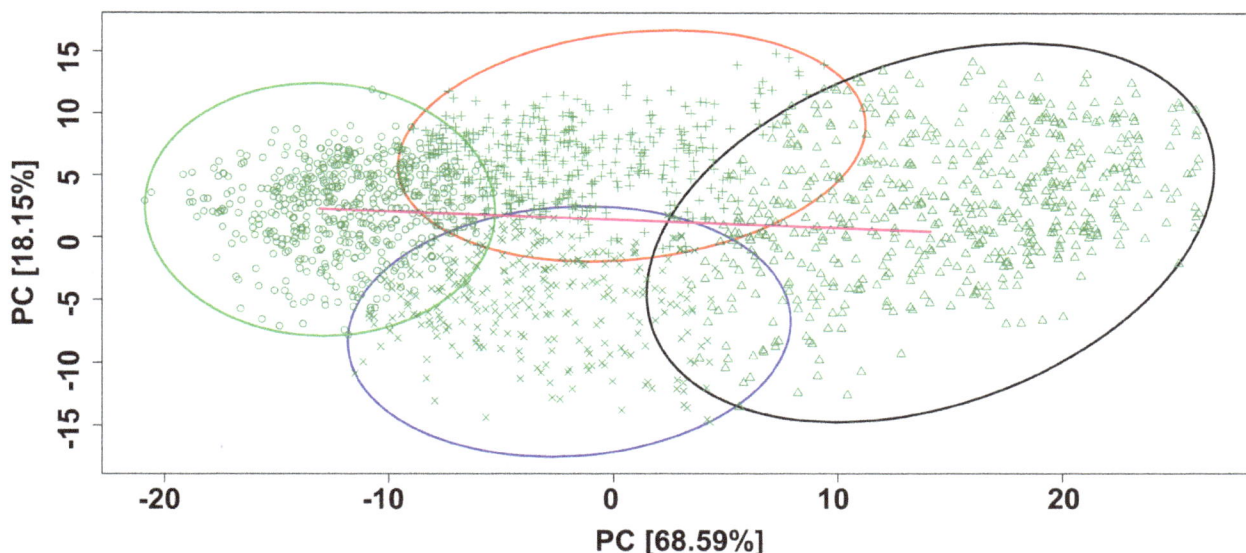

**Figure 5. Delimiting BGC types.** Plot of PCA scores for the extracted OTU matrix correlated with chemical profile. Four BGC types were delimited by k-means clustering. X-axis: PC1. Y-axis: PC2. BGC I: area enclosed in red with cross symbols, BGC II: area enclosed in blue with x symbols, BGC III: area enclosed in green with circular signals, BGC IV: area enclosed in pink with triangular signals. Arrows indicate the axes separating the BGC types; the quasi-horizontal arrow separates BGC III from BGC IV along the PC1 axis; the quasi-vertical arrow separates BGC I from BGC II along the PC2.

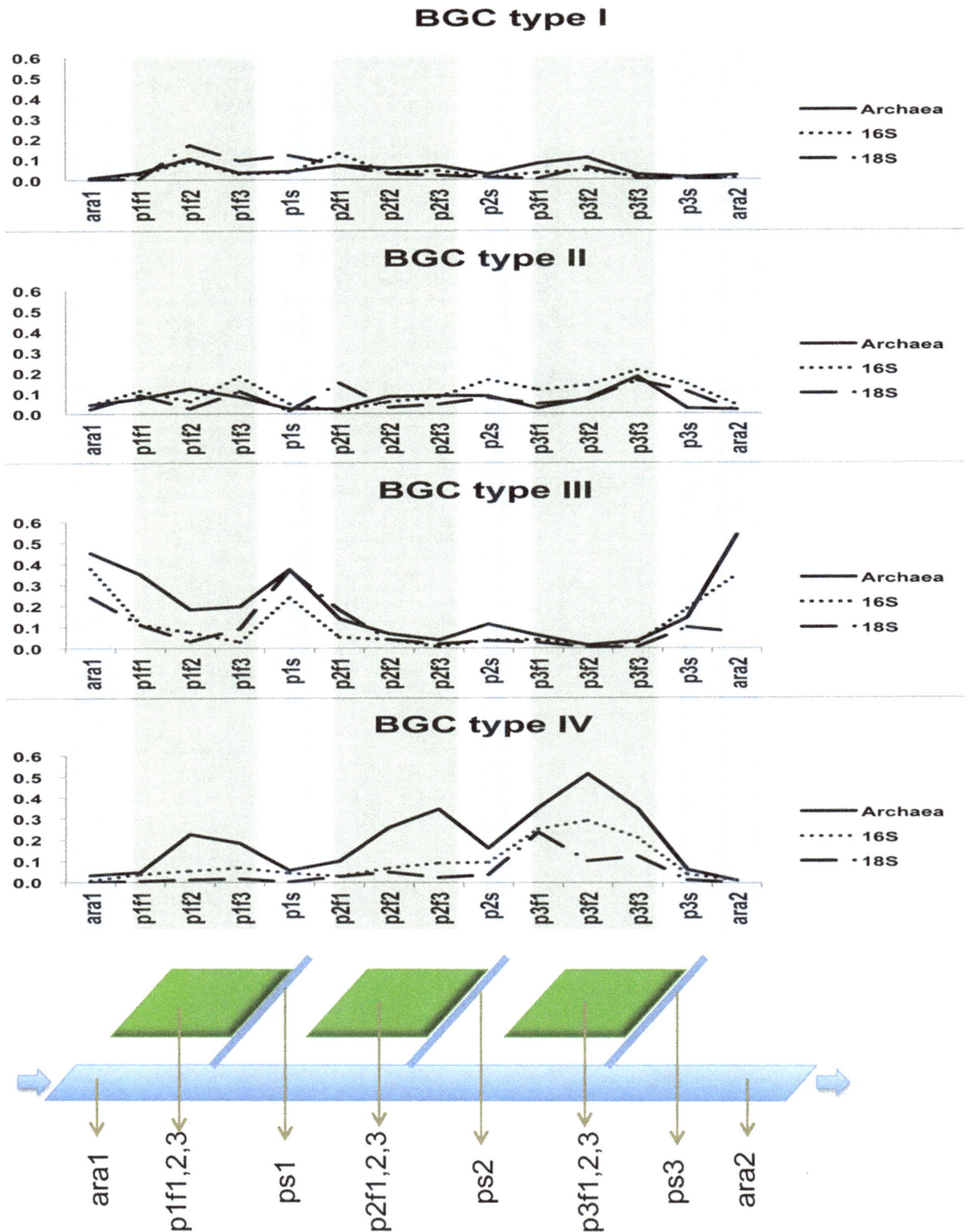

**Figure 6. Community distributions of the BGC types over the sampling points.** Distributions of relative abundances of communities identified as Archaea, 16S rRNA and 18S rRNA for the BGC types along the sampling points. X-axis: sampling points, Y-axis: relative abundance. Green shadows indicate sampling points on lentic waters and blue shadows indicate the sampling points over lotic waters. Below: a schematic drawing for the sampling points.

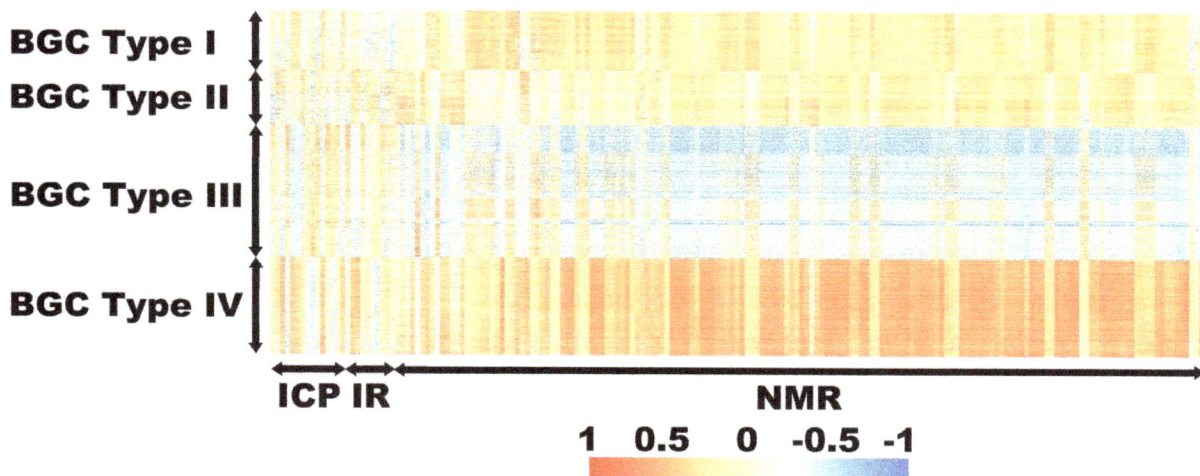

**Figure 7. Heatmap for the Spearman correlations of extracted OTUs against chemical profiles.** X-axis, in order: chemical elements (ICP-OES), wavelength number (FT-IR), chemical shifts (H-NMR). Y-axis: OTUs arranged from BGC type I (upper) to IV (bottom). BGC types are indicated on the left. The meaning of the colours is indicated in the legend at the bottom.

ammonia-oxidizing process of the nitrogen cycle and in autotrophic carbon assimilation [48,49,50]. This phylum has also been suggested to be dominant in fresh-water systems [51], which is in accordance with our findings for the distribution of BGC type III. However, such a pattern was clearly not mirrored in the paddy floodwaters, where the profile resembled that of soil [52] or a suboxic freshwater pond [53]. The BGC types distributed over the paddy fields and collector streams encompassed methanogenic candidates along with organisms that thrive in a wide range of aerobic conditions. Some of the associations might be possible by the formation of anoxic microzones in aggregates [54], although Archaea methanogens can survive in aerobic environments [55] (Fig. S12, S13, S14, S15).

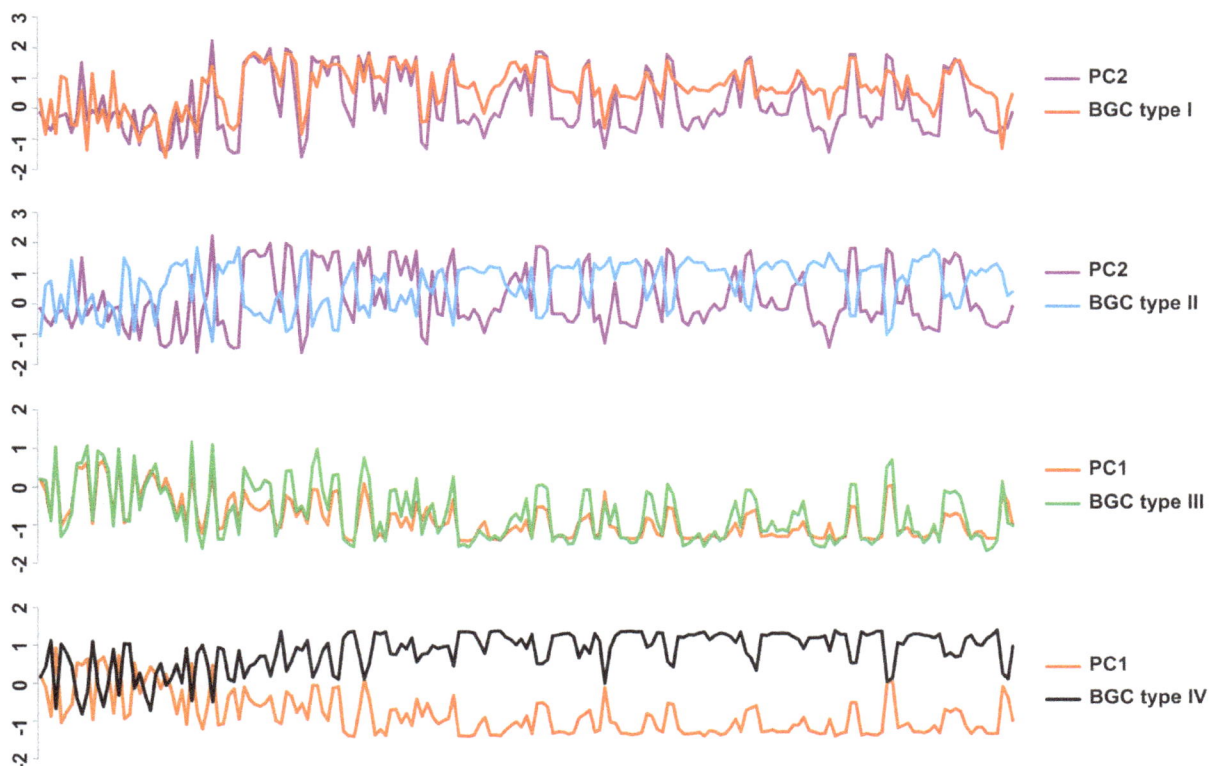

**Figure 8. Comparison of the average correlations for each BGC type against PCA loadings.** X-axis in order: chemical elements (ICP-OES), wavelength number (FT-IR), chemical shifts (H-NMR) (legend omitted). Y-axis: scaled scores. Average correlations from top to bottom are cluster 1 against PC2, cluster 2 against PC2, cluster 3 against PC1, and cluster 4 against PC1. Values are scaled by unity of variance.

**Table 1.** Chemical profiles.

| | BGC type I | BGC type II | BGC type III | BGC type IV |
|---|---|---|---|---|
| ICP-OES | | K | Si | Ba |
| | | | Mn | Cu |
| | | | Ca | Fe |
| | | | | P |
| | | | | Zn |
| FT-IR | | RC=O (1722–1647) | | C-H (3037–2864) |
| | | C=C or C-H or RS=O (879–849) | | O-H (3356–3055) |
| $^1$H-NMR | Phe (7.41, 7.39, 7.37, 7.35, 7.33) | 7.94 | 7.96 | protein |
| | Tyr 7.19 | 7.70 | 7.84 | formate 8.45 |
| | Ser (4.00, 3.94, 3.83) | 7.56 | | Phe 7.31 |
| | lactate (4.12, 4.10, 1.32, 1.30) | 7.04 | | Ser (4.03, 3.92, 3.80) |
| | Lys or Ala 3.72 | | | pyroglutamate (4.16, 2.50, 2.47, 2.44, 2.34, 2.09, 2.07, 2.04, 2.02, 2.00) |
| | Gly 3.55 | | | Ala or Lys 3.69 Lys (3.75, 3.09) |
| | Lys (3.06, 3.04, 3.01, 1.94) | | | Ala (3.67, 1.44) |
| | pyroglutamate 2.41 | | | Gly (3.58, 3.51) |
| | Ala (1.49, 1.46) | | | ketone (1.59, 1.57) |
| | 7.84 | | | amine or formaldehyel or methanol (3.38, 3.34) |
| | 7.09 | | | Val (1.10, 1.08) |
| | 6.65 | | | 8.08 7.94 7.91 7.87 7.81 7.79 7.56 |
| | | | | 7.22 7.07 7.04 7.02 4.08 4.06 3.90 |
| | | | | 3.87 3.78 3.65 3.63 3.60 3.48 3.45 |
| | | | | 3.42 3.40 3.32 3.28 3.26 3.22 3.20 |
| | | | | 3.17 3.13 2.99 2.96 2.93 2.91 2.88 |
| | | | | 2.85 2.81 2.77 2.75 2.69 2.67 2.64 |
| | | | | 2.62 2.60 2.57 2.54 2.30 2.28 2.26 |
| | | | | 2.22 2.20 2.16 2.12 1.89 1.86 1.84 |
| | | | | 1.82 1.80 1.78 1.76 1.74 1.72 1.68 |
| | | | | 1.66 1.64 1.62 1.53 1.51 1.41 1.39 |
| | | | | 1.37 1.35 1.28 1.26 1.23 1.21 1.18 |
| | | | | 1.16 1.13 0.92 0.89 0.87 0.84 0.82 |
| | | | | 0.80 0.78 |

The chemical profile for the BGC types were divided by ICP-OES, FT-IR and $^1$H-NMR variables (omitted groups of variables are those with no high positive statistical dependence for the BGC type). Variables for ICP-OES are elements. Variables for FT-IR are the integrated area corresponding to the chemical bond in interval of wavelength (cm$^{-1}$) showed in parentheses. Variables for $^1$H-NMR are the integrated area for the buckets (chemical shifts) in ppm; values in parentheses are chemical shifts assigned to the same compound or organic function.

The community identified by amplifying the 16S rRNA showed the ubiquitous class *Actinobacteria* to present the largest abundance (~38%) and indeed as one of the four most abundant classes across all BGC types. It prevailed in abundance even over the ~26% of the higher-taxon *Proteobacteria* phylum. *Actinobacteria* possess versatile metabolism, exemplified by their ability to decompose lignocellulose [56,57] or to produce bioactive metabolites as antibiotics. The high abundance of *Actinobacteria* in BGC types II and IV in particular mirrored studies of both soil and aquatic ecosystems where this phylum often appears as an

important component of the community alongside *Proteobacteria* [58,59,60]. Besides *Actinobacteria*, BGC types I and III shared *Proteobacteria*, *Flavobacteria* and *Sphingobacteria* as significant elements of the community. The bacteria detected in this study were largely from aerobic or facultative anaerobic taxa, suggesting aerobic or a micro-aerobic conditions (Fig. S16,S17, S18, S19).

The community identified amplifying the 18S rRNA, the *Eukaryota* fraction was dominated by chemoheterotrophic phyla such as *Metazoa* (with the classes *Gastrotricha* and *Rotifera* being first- and second-largest, respectively) and *Alveolata* (with the classes *Ciliophora* and *Apicomplexa* being first- and second-largest, respectively) that may serve as links in energy transfer from lower to higher trophic levels. The BGC types I, II and IV that were distributed over the paddy fields all shared this pattern. Jurgens and Gude [61] suggested that the presence of protozoans and metazoans might shape the characteristics of the bacterial community, not just regarding structure and diversity, but also triggering phenotype changes in response to predation. Similar to our findings for the Archaea community, BGC type III showed a different structure from the other BGC types, presenting the phylum *Viridiplantae* as the most abundant. *Viridiplantae* is a taxon formed by a wide range of terrestrial and aquatic primary producers that rely on chloroplasts to perform photosynthesis, which plants are part of this clade [62] (Fig. S20, S21, S22, S23).

## Chemical profiles and organismal community structures

The differences in the community distributions for BGC types I and II were not entirely clear. However, the differences in the chemical profiles revealed that all four BGC types are indeed unique. On the PCA score plot, the k-clustering method segregated four clusters as the BGC types in a cross-like fashion. This structure translates how each BGC type has its chemical profile distinguished from the others or, alternatively, its own environmental condition. In accordance with the results, BGC types III and IV were more distinct from each other than BGC types I and II were from each other, as the former were split along the PC1 axis and the latter along the PC2 axis as seen in Fig. 5. So that, BGC types III and IV had the average correlations for the chemical variables compared against PC1 while BGC I and II had the average correlations for the chemical variables compared against PC2 as seen in Fig. 8.

The chemical variables with positive correlations were extracted and assigned to the related BGC type. We listed only variables with positive correlations for each BGC type, since positivity implies that the organisms living under those specific environmental conditions are at least tolerant to those conditions. A negative correlation would not necessarily imply suppression arising from either side, biological or chemical, since the samples are from an open system with many variables that were not measured, such as geographical configuration or even weather conditions; however, it does imply the absence of a given chemical variable in the presence of a given organism, and vice-versa.

BGC type I was statistically independent from the elements tested and from the variation on the FT-IR profile. However, it presented a positive correlation to buckets on the aromatic side of the $^1$H-NMR spectra, including those annotated as phenylalanine, some annotated as other amino acids, and, most remarkably, to the largest peak on the $^1$H-NMR spectra, which was assigned as lactate. The presence of lactate suggests anoxic conditions where this molecule could be derived from pyruvate fermentation [63] as an adaptation of Metazoan organisms to thrive under suboxic conditions [64]. Despite a suggestion of anoxic conditions for this BGC type, we found that 20% of the 16S rRNA community were from chloroplasts. Therefore, we hypothesize that anoxic microzones might buffer the chloroplast-produced oxygen [65]. Another possibility is that the chloroplasts might even simply constitute debris, just dead organic material serving as a substrate to other organisms [66].

BGC type II presented the highest negative correlation to the buckets assigned to lactate on $^1$H-NMR. Since BGC types I and II inhabit the same set of samples to different degrees of colonization, this finding may suggest either competitive interactions between them or chemical heterogeneity in the occurrence of microzones

displaying different chemical compositions, and hence a resulting difference in the kinds of life supported.

Organisms from BGC type III showed a positive correlation with three elements tested using ICP-OES and with two non-annotated buckets in the aromatic region of the $^1$H-NMR. One of the elements was manganese and it is a key element to, among other biological functions, photosynthesis [67,68]. The other two elements were silica and calcium that can be linked to, but not restricted to, silification and calcification: both are processes related to increasing physical resistance in organisms as cytoskeletons and shells, features which are important in many groups within algae and microalgae, for example [69]. Indeed, for this BGC type, the phylum *Alveolata* within Eukaryota had the class *Dinophyceae* as its dominant class, instead of *Ciliophora*, which dominated the same phylum in the other BGC types. BGC type III encompassed the organisms thriving in the environments poorest in the organic compounds detected by $^1$H-NMR and FT-IR. The phyla *Viridiplantae* and *Crenarchaeota* would be the primary sources of organic carbon, with the latter also potentially serving as a nitrogen source. As the waters from paddy fields can drain into the river but the opposite is not possible, BGC type III could not act as a source of organic matter for the other BGC types. The richness in *Crenarchaeota* despite poor organic load is consistent with the findings of Ochsenreiter et al. [70], where the phylum was found throughout many kinds of environmental soil and freshwater. Despite poor organic matter content, BGC type III had the highest abundance of the three domains compared to all other BGC types, suggesting efficient cycling of photosynthesized compounds.

BGC type IV presented the group of organisms statistically correlated to more chemical variables than any of the previous ones. There are correlations to five elements within ICP-OES – barium, copper, iron, zinc and phosphorous. Sanchez-Moral et al. [71] suggested that barium precipitation can be bio-induced in pure cultures of *Actinobacteria* which was the most abundant class of bacteria present in this BGC type. Copper, iron and zinc are some of the trace elements essential for enzymatic activity in methanogenic systems, and deficiencies are suggested to diminish such activity [72]. Phosphorous is abundant in several metabolic pathways, being involved in structural biomolecules and the energy currency ATP. BGC type IV was also correlated to those bands of absorbance in FT-IR assigned to be C-H and O-H bonds: the former may suggest long carbon chains from lipids, and the latter sugar- or alcohol-related molecules. Both of them are intrinsically linked to high levels of chemical energy, either implicated in catabolism or anabolism. Within $^1$H-NMR, BGC type IV was correlated to a vast part of the spectra, suggesting a chemically rich organic environment, with relationships to many buckets assigned to amino acids and to the broad peaks assigned as protein. This would suggest nitrogen recycling, meaning the group presents an organismal community that may be both source and consumer of the organic compound assigned as protein.

## Conclusions

After the evaluation done for each BGC type, we can classify them roughly to the remarkable features that individualize them. According to the findings, BGC type I would represent the subsystem represented by the anoxic/near anoxic microzones by presenting lactate as an important energy source intermediate (aka the "BGC type Anoxic"). The BGC type II would be the aerophilic counter part of BGC type I, eventually presenting a relation energy transfer between them (aka the "BGC type Counter Part"). The BGC type III would represent a sub-system of photosynthetic organisms and the related chemical profile (aka the "BGC type Photosynthetic"). The BGC type IV would represent the most active in the cycling of organic compounds (aka the "BGC type Glutton").

Hence, we successfully established the BGC typing analysis pipeline technique and applied to the environment of Japanese paddy fields bringing forth four unique subset of organismal and chemical assembles. The technique is flexible and can accept any biogeochemical or omics measurements, since it operates on numerical tables of values (e.g., intensity, concentration, number of reads, etc.) and can contribute to further insights into biogeochemical cycles in other environments.

This holistic technique will broaden the understanding of "hidden" sub-systems working within the totality of the environments.

## Ethics Statement

There is no specific permission required for all of following sampling points as they are public places. Also the field does not host endangered or protected species. The exact location for the sampling points from Ara River (ara1, ara2) are 36°2'32"N 139°30'8"E and 35°56"54"N 139°32'41"E respectively, from paddy field 1 (p1f1–p1f3) are: 36°2'23"N 139°29'51"E, from paddy field 2 (p2f1–p2f3) are: 35°59'37"N 139°30'5"E, from paddy field 3 (p3f1–p3f3) are: 35°58'29"N 139°30'33"E, from collector stream from paddy field 1 (p1s) is: 36°2'25"N 139°29'46"E, from collector stream from paddy field 2 (p2s) is: 35°59'37"N 139°30'5"E and from collector stream from paddy field 3 (p3s) is: 35°58'29"N 139°30'32"E.

The whole of the sequences retrieved in our study are available in the DDBJ Sequenced Read Archive under accession number DRA002437.

## Supporting Information

**Figure S1  Rarefaction curve to total Archaea OTUs.** Rarefaction curve for Archaea rRNA OTUs for all samples. X-axis: number of readings. Y-axis: number of species (log). (PDF)

**Figure S2  Rarefaction curve to toatal 16S rRNA OTUs.** Rarefaction curve for 16S rRNA OTUs for all samples. X-axis: number of readings. Y-axis: number of species (log). (PDF)

**Figure S3  Rarefaction curve to total 18S rRNA OTUs.** Rarefaction curve for 18S rRNA OTUs for all samples. X-axis: number of readings. Y-axis: number of species (log). (PDF)

**Figure S4  Emulation of a phylogenetic tree for BGC type I.** Abundance range indicated by the size of the symbol indicated at upper right. Symbols indicative of the domain indicated at middle right. Colour codes for the most abundant classes indicated at lower right. The number of symbols for each branch is related to the number of sampling points within the OTUs present. (PDF)

**Figure S5  Emulation of a phylogenetic tree for BGC type II.** Abundance range indicated by the size of the symbol indicated at upper right. Symbols indicative of the domain indicated at middle right. Colour codes for the most abundant classes indicated at lower right. The number of symbols for each branch is related to the number of sampling points within the OTUs present. (PDF)

**Figure S6  Emulation of a phylogenetic tree for BGC type III.** Abundance range indicated by the size of the symbol indicated at upper right. Symbols indicative of the domain indicated at middle right. Colour codes for the most abundant classes indicated at lower right. The number of symbols for each branch is related to the number of sampling points within the OTUs present. (PDF)

**Figure S7  Emulation of a phylogenetic tree for BGC type IV.** Abundance range indicated by the size of the symbol indicated at upper right. Symbols indicative of the domain indicated at middle right. Colour codes for the most abundant classes indicated at lower right. The number of symbols for each branch is related to the number of sampling points within the OTUs present. (PDF)

**Figure S8  J-resolved $^1$H-NMR with annotations.** J-resolved $^1$H-NMR analysis for aromatic and non-aromatic regions of the spectra with annotations. (PDF)

**Figure S9  1D-STOCSY correlation for chemical shift 1.32 ppm.** 1D-STOCSY with centroid at 1.32 ppm showing high correlation with the other chemical shifts from the lactate assignment. X-axis: chemical shifts. Y-axis: degree of correlation. Colours simplify visualization: cold colours indicate negative correlations and hot colours indicate positive correlations. (TIF)

**Figure S10**  Comparison between the $^1$H-NMR spectra against the Bruker AMIX database for the pure compound lactate (lactic acid). The $^1$HNMR spectra with lower baselines are from this study. The upward-shifted baseline spectrum in orange is the lactate spectrum provided in Bruker AMIX software. Numbers indicate chemical shift. (PDF)

**Figure S11  1D-STOCSY correlation for the integration annotated as protein.** 1D-STOCSY with centroid at the region of interest integrated and annotated as protein showing high correlation with most of the chemical shifts from the spectra. X-axis: chemical shifts. Y-axis: degree of correlation. Colours simplify visualization: cold colours indicate negative correlations and hot colours indicate positive correlations. (TIF)

**Figure S12  Percentage of Archaea OTUs for BGC type I.** Archaeal OTUs for BGC type I collapsed to the class level or beyond according to the next divergence on the taxon presented. The four most abundant taxa are shown, with others collapsed. (PDF)

**Figure S13  Percentage of Archaea OTUs for BGC type II.** Archaeal OTUs for BGC type II collapsed to the class level or beyond according to the next divergence on the taxon presented. The four most abundant taxa are shown, with others collapsed. (PDF)

**Figure S14  Percentage of Archaea OTUs for BGC type III.** Archaeal OTUs for BGC type III collapsed to the class level or beyond according to the next divergence on the taxon presented. The four most abundant taxa are shown, with others collapsed. (PDF)

**Figure S15  Percentage of Archaea OTUs for BGC type IV.** Archaeal OTUs for BGC type IV collapsed to the class level or beyond according to the next divergence on the taxon presented. The four most abundant taxa are shown, with others collapsed. (PDF)

**Figure S16   Percentage of 16S rRNA OTUs for BGC type I.** 16S OTUs for BGC type I collapsed to the class level or beyond according to the next divergence on the taxon presented. The four most abundant taxa are shown, with others collapsed. (PDF)

**Figure S17   Percentage of 16S rRNA OTUs for BGC type II.** 16S OTUs for BGC type II collapsed to the class level or beyond according to the next divergence on the taxon presented. The four most abundant taxa are shown, with others collapsed. (PDF)

**Figure S18   Percentage of 16S rRNA OTUs for BGC type III.** 16S rRNA OTUs for BGC type III collapsed to the class level or beyond according to the next divergence on the taxon presented. The four most abundant taxa are shown, with others collapsed. (PDF)

**Figure S19   Percentage of 16S rRNA OTUs for BGC type IV.** 16S rRNA OTUs for BGC type IV collapsed to the class level or beyond according to the next divergence on the taxon presented. The four most abundant taxa are shown, with others collapsed. (PDF)

**Figure S20   Percentage of 18S rRNA OTUs for BGC type I.** 18S OTUs for BGC type I collapsed to the class level or beyond according to the next divergence on the taxon presented. The four most abundant taxa are shown, with others collapsed. (PDF)

**Figure S21   Percentage of 18S rRNA OTUs for BGC type II.** 18S OTUs for BGC type II collapsed to the class level or beyond according to the next divergence on the taxon presented. The four most abundant taxa are shown, with others collapsed. (PDF)

**Figure S22   Percentage of 18S rRNA OTUs for BGC type III.** 18S rRNA OTUs for BGC type III collapsed to the class level or beyond according to the next divergence on the taxon presented. The four most abundant taxa are shown, with others collapsed. (PDF)

**Figure S23   Percentage of 18S rRNA OTUs for BGC type IV.** 18S rRNA OTUs for BGC type IV collapsed to the class level or beyond according to the next divergence on the taxon presented. The four most abundant taxa are shown, with others collapsed. (PDF)

**Table S1   Archaea OTUs in number of reads along sampling points.** Rows: OTUs. Columns: sampling points. (XLSX)

**Table S2   16S OTUs in number of reads along sampling points.** Rows: OTUs. Columns: sampling points. (XLSX)

**Table S3   18S OTUs in number of reads along sampling points.** Rows: OTUs. Columns: sampling points. (XLSX)

**Table S4   Elemental concentrations along sampling points.** Rows: sampling points. Columns: elements and pH. (XLSX)

**Table S5   Infrared absorbance along sampling points.** Rows: sampling points. Columns: annotated regions of interest in intervals of wavelength ($cm^{-1}$) indicated in parentheses. (XLSX)

**Table S6   $^1$H-NMR along sampling points.** Rows: sampling points. Columns: bins for chemical shifts (ppm) with annotation where applicable and integration for the broad peaks annotated as protein. (XLS)

## Acknowledgments

We are grateful to Craig Everroad and Fredd Vergara for their valuable discussions and assistance. We also thank Masato Otagiri, Amiu Shino, and Taiji Watanabe for their significant technical assistance.

## Author Contributions

Conceived and designed the experiments: DMOO ARBPS GR-B TY SM. Performed the experiments: DMOO SM YT. Analyzed the data: DMOO SM GR-B YT JK. Contributed reagents/materials/analysis tools: SM JK. Wrote the paper: DMOO YT YD JK.

## References

1. Asakura T, Date Y, Kikuchi J (2014) Comparative analysis of chemical and microbial profiles in estuarine sediments sampled from Kanto and Tohoku regions in Japan. Anal Chem 86: 5425–5432.
2. Joyce AR, Palsson B (2006) The model organism as a system: integrating 'omics' data sets. Nat Rev Mol Cell Biol 7: 198–210.
3. Enjalbert B, Jourdan F, Portais JC (2011) Intuitive Visualization and Analysis of Multi-Omics Data and Application to Escherichia coli Carbon Metabolism. Plos One 6.
4. Castell WZ, Ernst D (2012) Experimental 'omics' data in tree research: facing complexity. Trees-Structure and Function 26: 1723–1735.
5. Xie J, Hu L, Tang J, Wu X, Li N, et al. (2011) Ecological mechanisms underlying the sustainability of the agricultural heritage rice-fish coculture system. Proceedings of the National Academy of Sciences of the United States of America 108: E1381–E1387.
6. Caporaso JG, Lauber CL, Walters WA, Berg-Lyons D, Lozupone CA, et al. (2011) Global patterns of 16S rRNA diversity at a depth of millions of sequences per sample. Proc Natl Acad Sci U S A 108 Suppl 1: 4516–4522.
7. Hirschi KD (2003) Strike while the ionome is hot: making the most of plant genomic advances. Trends Biotechnol 21: 520–521.
8. Sitole L, Steffens F, Krüger TP, Meyer D (2014) Mid-ATR-FTIR Spectroscopic Profiling of HIV/AIDS Sera for Novel Systems Diagnostics in Global Health. OMICS 18: 513–523.
9. Kikuchi J, Shinozaki K, Hirayama T (2004) Stable isotope labeling of Arabidopsis thaliana for an NMR-based metabolomics approach. Plant Cell Physiol 45: 1099–1104.
10. Everroad RC, Yoshida S, Tsuboi Y, Date Y, Kikuchi J, et al. (2012) Concentration of metabolites from low-density planktonic communities for environmental metabolomics using nuclear magnetic resonance spectroscopy. J Vis Exp: e3163.
11. Ogata Y, Chikayama E, Morioka Y, Everroad RC, Shino A, et al. (2012) ECOMICS: a web-based toolkit for investigating the biomolecular web in ecosystems using a trans-omics approach. PLoS One 7: e30263.
12. Nakanishi Y, Fukuda S, Chikayama E, Kimura Y, Ohno H, et al. (2011) Dynamic omics approach identifies nutrition-mediated microbial interactions. J Proteome Res 10: 824–836.
13. Gottel NR, Castro HF, Kerley M, Yang Z, Pelletier DA, et al. (2011) Distinct microbial communities within the endosphere and rhizosphere of Populus deltoides roots across contrasting soil types. Appl Environ Microbiol 77: 5934–5944.
14. Wubet T, Christ S, Schöning I, Boch S, Gawlich M, et al. (2012) Differences in soil fungal communities between European beech (Fagus sylvatica L.) dominated forests are related to soil and understory vegetation. PLoS One 7: e47500.
15. Mirkin B (1998) Least-Squares Structuring, Clustering and Data Processing Issues. The Computer Journal 41: 518–536.
16. Maechler M (2013) cluster: Cluster Analysis Basics and Extensions. R package version 1.14.4. In: Rousseeuw P, Hubert M, Hornik K, editors.

17. Dardenne F, Van Dongen S, Nobels I, Smolders R, De Coen W, et al. (2008) Mode of action clustering of chemicals and environmental samples on the bases of bacterial stress gene inductions. Toxicological Sciences 101: 206–214.

18. Margulies M, Egholm M, Altman WE, Attiya S, Bader JS, et al. (2005) Genome sequencing in microfabricated high-density picolitre reactors. Nature 437: 376–380.

19. Weisburg WG, Barns SM, Pelletier DA, Lane DJ (1991) 16S Ribosomal DNA Amplification for Phylogenetic Study. Journal of Bacteriology 173: 697–703.

20. Muyzer G, Dewaal EC, Uitterlinden AG (1993) Profiling of Complex Microbial-Populations by Denaturing Gradient Gel-Electrophoresis Analysis of Polymerase Chain Reaction-Amplified Genes-Coding for 16s Ribosomal-RNA. Applied and Environmental Microbiology 59: 695–700.

21. Diez B, Pedros-Alio C, Marsh TL, Massana R (2001) Application of denaturing gradient gel electrophoresis (DGGE) to study the diversity of marine picoeukaryotic assemblages and comparison of DGGE with other molecular techniques. Applied and Environmental Microbiology 67: 2942–2951.

22. Klindworth A, Pruesse E, Schweer T, Peplies J, Quast C, et al. (2013) Evaluation of general 16S ribosomal RNA gene PCR primers for classical and next-generation sequencing-based diversity studies. Nucleic Acids Research 41.

23. Caporaso JG, Kuczynski J, Stombaugh J, Bittinger K, Bushman FD, et al. (2010) QIIME allows analysis of high-throughput community sequencing data. Nature Methods 7: 335–336.

24. Bonder MJ, Abeln S, Zaura E, Brandt BW (2012) Comparing clustering and pre-processing in taxonomy analysis. Bioinformatics 28: 2891–2897.

25. Edgar RC (2010) Search and clustering orders of magnitude faster than BLAST. Bioinformatics 26: 2460–2461.

26. Quast C, Pruesse E, Yilmaz P, Gerken J, Schweer T, et al. (2013) The SILVA ribosomal RNA gene database project: improved data processing and web-based tools. Nucleic Acids Research 41: D590–D596.

27. Izquierdo-Carrasco F, Smith SA, Stamatakis A (2011) Algorithms, data structures, and numerics for likelihood-based phylogenetic inference of huge trees. Bmc Bioinformatics 12.

28. McMurdie PJ, Holmes S (2013) phyloseq: An R Package for Reproducible Interactive Analysis and Graphics of Microbiome Census Data. Plos One 8.

29. Wang Q, Garrity GM, Tiedje JM, Cole JR (2007) Naive Bayesian classifier for rapid assignment of rRNA sequences into the new bacterial taxonomy. Applied and Environmental Microbiology 73: 5261–5267.

30. Date Y, Iikura T, Yamazawa A, Moriya S, Kikuchi J (2012) Metabolic sequences of anaerobic fermentation on glucose-based feeding substrates based on correlation analyses of microbial and metabolite profiling. J Proteome Res 11: 5602–5610.

31. Ludwig C, Easton JM, Lodi A, Tiziani S, Manzoor SE, et al. (2012) Birmingham Metabolite Library: a publicly accessible database of 1-D H-1 and 2-D H-1 J-resolved NMR spectra of authentic metabolite standards (BML-NMR). Metabolomics 8: 8–18.

32. Akiyama K, Chikayama E, Yuasa H, Shimada Y, Tohge T, et al. (2008) PRIMe: a Web site that assembles tools for metabolomics and transcriptomics. In Silico Biol 8: 339–345.

33. Chikayama E, Sekiyama Y, Okamoto M, Nakanishi Y, Tsuboi Y, et al. (2010) Statistical indices for simultaneous large-scale metabolite detections for a single NMR spectrum. Anal Chem 82: 1653–1658.

34. Chikayama E, Suto M, Nishihara T, Shinozaki K, Kikuchi J (2008) Systematic NMR analysis of stable isotope labeled metabolite mixtures in plant and animal systems: coarse grained views of metabolic pathways. PLoS One 3: e3805.

35. Edoardo G (2012) muma: Metabolomics Univariate and Multivariate Analysis. R package version 1.4. In: Francesca C, Silvia M, Andrea S, Michela G, editors.

36. Ogura T, Date Y, Kikuchi J (2013) Differences in cellulosic supramolecular structure of compositionally similar rice straw affect biomass metabolism by paddy soil microbiota. PLoS ONE 8: e66919.

37. Cloarec O, Dumas ME, Craig A, Barton RH, Trygg J, et al. (2005) Statistical total correlation spectroscopy: An exploratory approach for latent biomarker identification from metabolic H-1 NMR data sets. Analytical Chemistry 77: 1282–1289.

38. Kikuchi J, Asakura T, Loach PA, Parkes-Loach PS, Shimada K, et al. (1999) A light-harvesting antenna protein retains its folded conformation in the absence of protein-lipid and protein-pigment interactions. Biopolymers 49: 361–372.

39. Kraft BJ, Masuda S, Kikuchi J, Dragnea V, Tollin G, et al. (2003) Spectroscopic and mutational analysis of the blue-light photoreceptor AppA: A novel photocycle involving flavin stacking with an aromatic amino acid. Biochemistry 42: 6726–6734.

40. Bradford MM (1976) Rapid and Sensitive Method for Quantitation of Microgram Quantities of Protein Utilizing Principle of Protein-Dye Binding. Analytical Biochemistry 72: 248–254.

41. Team RDC (2008) R: A Language and environment for statistical computing.: R Foundation for Statistical Computing.

42. Peeples MA (2011) R Script for K-Means Cluster Analysis.

43. Pollard D (1982) a Central Limit-Theorem for K-Means Clustering. Annals of Probability 10: 919–926.

44. Kolde R (2012) pheatmap: Pretty Heatmaps. R package version 0.7.4.

45. Sakai S, Imachi H, Hanada S, Ohashi A, Harada H, et al. (2008) Methanocella paludicola gen. nov., sp nov., a methane-producing archaeon, the first isolate of the lineage 'Rice Cluster I', and proposal of the new archaeal order Methanocellales ord. nov. International Journal of Systematic and Evolutionary Microbiology 58: 929–936.

46. Poulsen M, Schwab C, Jensen BB, Engberg RM, Spang A, et al. (2013) Methylotrophic methanogenic Thermoplasmata implicated in reduced methane emissions from bovine rumen. Nature Communications 4.

47. Whitford MF, Teather RM, Forster RJ (2001) Phylogenetic analysis of methanogens from the bovine rumen. BMC Microbiology.

48. Herrmann M, Saunders AM, Schramm A (2008) Archaea dominate the ammonia-oxidizing community in the rhizosphere of the freshwater macrophyte Littorella uniflora. Applied and Environmental Microbiology 74: 3279–3283.

49. Wang S, Wang Y, Feng X, Zhai L, Zhu G (2011) Quantitative analyses of ammonia-oxidizing Archaea and bacteria in the sediments of four nitrogen-rich wetlands in China. Applied Microbiology and Biotechnology 90: 779–787.

50. Pratscher J, Dumont MG, Conrad R (2011) Ammonia oxidation coupled to CO2 fixation by archaea and bacteria in an agricultural soil. Proceedings of the National Academy of Sciences of the United States of America 108: 4170–4175.

51. Ghai R, Rodriguez-Valera F, McMahon KD, Toyama D, Rinke R, et al. (2011) Metagenomics of the Water Column in the Pristine Upper Course of the Amazon River. Plos One 6.

52. Kudo Y, Nakajima T, Miyaki T, Oyaizu H (1997) Methanogen flora of paddy soils in Japan. Fems Microbiology Ecology 22: 39–48.

53. Briee C, Moreira D, Lopez-Garcia P (2007) Archaeal and bacterial community composition of sediment and plankton from a suboxic freshwater pond. Research in Microbiology 158: 213–227.

54. Ploug H, Kuhl M, BuchholzCleven B, Jorgensen BB (1997) Anoxic aggregates - an ephemeral phenomenon in the pelagic environment? Aquatic Microbial Ecology 13: 285–294.

55. Grossart H-P, Frindte K, Dziallas C, Eckert W, Tang KW (2011) Microbial methane production in oxygenated water column of an oligotrophic lake. Proceedings of the National Academy of Sciences of the United States of America 108: 19657–19661.

56. Crawford DL (1978) Lignocellulose Decomposition by Selected Streptomyces Strains. Applied and Environmental Microbiology 35: 1041–1045.

57. Ball AS, Godden B, Helvenstein P, Penninckx MJ, McCarthy AJ (1990) Lignocarbohydrate Solubilization From Straw by Actinomycetes. Applied and Environmental Microbiology 56: 3017–3022.

58. Peiffer JA, Spor A, Koren O, Jin Z, Tringe SG, et al. (2013) Diversity and heritability of the maize rhizosphere microbiome under field conditions. Proceedings of the National Academy of Sciences of the United States of America 110: 6548–6553.

59. Andreote FD, Javier Jimenez D, Chaves D, Franco Dias AC, Luvizotto DM, et al. (2012) The Microbiome of Brazilian Mangrove Sediments as Revealed by Metagenomics. Plos One 7.

60. Glockner FO, Zaichikov E, Belkova N, Denissova L, Pernthaler J, et al. (2000) Comparative 16S rRNA analysis of lake bacterioplankton reveals globally distributed phylogenetic clusters including an abundant group of actinobacteria. Applied and Environmental Microbiology 66: 5053–+.

61. Jurgens K, Gude H (1994) the Potential Importance of Grazing-Resistant Bacteria in Planktonic Systems. Marine Ecology Progress Series 112: 169–188.

62. Wodniok S, Brinkmann H, Gloeckner G, Heidel AJ, Philippe H, et al. (2011) Origin of land plants: Do conjugating green algae hold the key? Bmc Evolutionary Biology 11.

63. Semenza GL (2012) Hypoxia-Inducible Factors in Physiology and Medicine. Cell 148: 399–408.

64. Braeckman U, Vanaverbeke J, Vincx M, van Oevelen D, Soetaert K (2013) Meiofauna Metabolism in Suboxic Sediments: Currently Overestimated. Plos One 8.

65. Paerl HW, Bebout BM (1988) Direct Measurement of O-2-Depleted Microzones in Marine Oscillatoria - Relation to N-2 Fixation. Science 241: 442–445.

66. Kiorboe T, Tang K, Grossart HP, Ploug H (2003) Dynamics of microbial communities on marine snow aggregates: Colonization, growth, detachment, and grazing mortality of attached bacteria. Applied and Environmental Microbiology 69: 3036–3047.

67. Kanehisa M, Goto S, Sato Y, Furumichi M, Tanabe M (2012) KEGG for integration and interpretation of large-scale molecular data sets. Nucleic Acids Research 40: D109–D114.

68. Kanehisa M, Goto S (2000) KEGG: Kyoto Encyclopedia of Genes and Genomes. Nucleic Acids Research 28: 27–30.

69. Brownlee C, Taylor AR (2002) Algal Calcification and Silification. Encyclopedia of Life Sciences: Macmillan Publishers Ltd, Nature Publishing Group.

70. Ochsenreiter T, Selezi D, Quaiser A, Bonch-Osmolovskaya L, Schleper C (2003) Diversity and abundance of Crenarchaeota in terrestrial habitats studied by 16S RNA surveys and real time PCR. Environmental Microbiology 5: 787–797.

71. Sanchez-Moral S, Luque L, Canaveras JC, Laiz L, Jurado V, et al. (2004) Bioinduced barium precipitation in St. Callixtus and domitilla catacombs. Annals of Microbiology 54: 1–12.

72. Unal B, Perry VR, Sheth M, Gomez-Alvarez V, Chin K-J, et al. (2012) Trace elements affect methanogenic activity and diversity in enrichments from subsurface coal bed produced water. Frontiers in Microbiology.

# 3

# Getting What Is Served? Feeding Ecology Influencing Parasite-Host Interactions in Invasive Round Goby *Neogobius melanostomus*

**Sebastian Emde**[1], **Judith Kochmann**[2]*, **Thomas Kuhn**[1], **Martin Plath**[3], **Sven Klimpel**[1,2]

**1** Institute for Ecology, Evolution and Diversity, Goethe-University, Frankfurt am Main, Hesse, Germany, **2** Senckenberg Gesellschaft für Naturforschung, Biodiversity and Climate Research Centre, Frankfurt am Main, Hesse, Germany, **3** College of Animal Science and Technology, Northwest Agriculture & Forestry University, Yangling, Shaanxi Province, P. R. China

## Abstract

Freshwater ecosystems are increasingly impacted by alien invasive species which have the potential to alter various ecological interactions like predator-prey and host-parasite relationships. Here, we simultaneously examined predator-prey interactions and parasitization patterns of the highly invasive round goby (*Neogobius melanostomus*) in the rivers Rhine and Main in Germany. A total of 350 *N. melanostomus* were sampled between June and October 2011. Gut content analysis revealed a broad prey spectrum, partly reflecting temporal and local differences in prey availability. For the major food type (amphipods), species compositions were determined. Amphipod fauna consisted entirely of non-native species and was dominated by *Dikerogammarus villosus* in the Main and *Echinogammarus trichiatus* in the Rhine. However, the availability of amphipod species in the field did not reflect their relative abundance in gut contents of *N. melanostomus*. Only two metazoan parasites, the nematode *Raphidascaris acus* and the acanthocephalan *Pomphorhynchus* sp., were isolated from *N. melanostomus* in all months, whereas unionid glochidia were only detected in June and October in fish from the Main. To analyse infection pathways, we examined 17,356 amphipods and found *Pomphorhynchus* sp. larvae only in *D. villosus* in the river Rhine at a prevalence of 0.15%. *Dikerogammarus villosus* represented the most important amphipod prey for *N. melanostomus* in both rivers but parasite intensities differed between rivers, suggesting that final hosts (large predatory fishes) may influence host-parasite dynamics of *N. melanostomus* in its introduced range.

**Editor:** Raul Narciso C. Guedes, Federal University of Viçosa, Brazil

**Funding:** The present study was financially supported by the research funding programme "LOEWE –Landes-Offensive zur Entwicklung Wissenschaftlich-ökonomischer Exzellenz" of Hesse's Ministry of Higher Education, Research, and the Arts. The funders had no role in study design, data collection and analysis, decision to publish, or preparation of the manuscript.

**Competing Interests:** The authors have declared that no competing interests exist.

* Email: judith.kochmann@senckenberg.de

## Introduction

Biological invasions have increased exponentially in recent years due to human activities, especially shipping, along with the adverse effects of environmental changes such as global warming [1–3]. Although brackish waters have the highest risk for species introductions, freshwater ecosystems are also strongly affected, especially by the introduction of non-indigenous fishes [4,5]. Once established in their new environment, invasive non-indigenous species can have tremendous effects on local populations of indigenous species, e.g., through competitive [6], predator-prey [7–9], or host-parasite interactions [10,11], all of which have the potential to result in altered ecosystem functioning (see review by Strayer [12]).

To date, several studies in aquatic ecosystems have considered the question of how invasive predators can affect native prey populations [13–15], or how invasive prey populations can alter indigenous prey communities [16,17], and whether or not non-indigenous prey species become integrated into the prey spectrum of indigenous predators [18]. Furthermore, studies have started to concentrate on parasitization patterns of native and invasive species, and several different scenarios are possible: (i) invasive hosts may lose their original parasite load ('enemy release hypothesis'), providing invasive species with an initial benefit in their novel range [10,19,20]. (ii) Introduced hosts may carry new parasite species (parasite spill-over), which may adversely affect native host species [21]. (iii) Invasive hosts may serve as intermediate hosts or vectors for local parasites or diseases (parasite spillback) [21]. (iv) Finally, shift and/or loss of local parasite species would be predicted if the invader is replacing local host species but cannot function as intermediate or definitive host in the parasite life cycles (dilution effect) [22,23]. Few studies, however, have simultaneously considered predator-prey interactions and parasitization patterns of different trophic levels in ecosystems that are heavily influenced by invasive species [24–26]. This is surprising, given that many parasites with indirect life-cycles rely on the ingestion of their intermediate hosts by further (intermediate or final) host species to successfully complete their life cycles [27,28]. Biological invasions could provide large numbers of host specimens within a very short time-span (e.g.

[29]) that could affect parasite transmission patterns in entire fish communities.

The round goby *Neogobius melanostomus* (Pallas, 1814) is a frequent invader of brackish and freshwater habitats worldwide, reaching enormous population densities and causing changes of food web dynamics at different trophic levels, e.g., in the North American Great Lakes [30] and in large European rivers, e.g. the Danube [29]. Round gobies nowadays make up app. 80% of fish catches in the Rhine [31], and so an alteration of ecological interactions is also expected for the Rhine. For example, it is known that round gobies act as competitors of spawning or foraging sites with native species [30]. Feeding patterns of *N. melanostomus* vary in different distribution areas. While dreissenid mussels play an important role in the feeding ecology of *N. melanostomus* in the Great Lakes and in the Baltic Sea [32,33], amphipods seem to be their main forage in German rivers [24,25,31]. In the Rhine, the Ponto-Caspian amphipod *Dikerogammarus villosus* (Sowinsky, 1894) has been described as dominating communities of macroinvertebrates and as an important prey species of *N. melanostomus* [24,25,31,34,35]. Both species, *D. villosus* and *N. melanostomus* function as intermediate hosts for different parasites (e.g., *Pomphorhynchus* spp. and *Raphidascaris* spp.) and may be responsible for the spread of these parasites, which could increasingly affect native vertebrate and invertebrate hosts as well [24,36].

Studies on *N. melanostomus* that combine the analysis of their feeding habits with parasitological analyses are rare and have focused on the Danube [25,29] and Rhine [24,31]. To analyse the role of different amphipod species for metazoan fish parasite transmission as well as temporal variation of diet compositions in invasive *N. melanostomus*, samples from the rivers Main and Rhine were compared in this study. We hypothesized that (a) *N. melanostomus* will mainly feed on amphipods throughout the course of our repeated monthly sampling and in both rivers, and accordingly, (b) the availability of amphipod species in a given river will reflect their relative contribution to gut contents of *N. melanostomus*. Moreover, we expected that (c) monthly infestation rates of amphipods with parasite species and monthly feeding rates of amphipods by *N. melanostomus* should reflect parasite infestation rates in *N. melanostomus*. Finally, a detailed description of parasite fauna for two sampling locations in the rivers Main and Rhine was intended to complement current parasite diversity estimates of *N. melanostomus* in its introduced range.

## Materials and Methods

### Sampling

A total of $n = 350$ *N. melanostomus* were collected from June to October 2011 in the rivers Rhine (49°51'54.7"N 8°21'40.2"E) and Main (50°04'48.9"N 8°31'19.6"E) in Germany. Both sites were similar in habitat structure with rip-rap embanked shorelines (technolithal) that led into bottom substrate of sand and gravel. In contrast to the Rhine, the river bank of the Main had little more vegetation with roots partly reaching into the water.

35 *N. melanostomus* specimens per site were caught randomly on top of and around rip-raps (depths of ~40–200 cm) during one day at the end of each month (between ~9 am–2 pm) using a hook and line technique. Since standardized angling is known to yield an equilibrated sex ratio and homogenously distributed, relatively large-sized specimens in *N. melanostomus* [37], a fishing rod equipped with an anti-tangle bottom rig consisting of a special sinker (Tiroler Hölzl, 80 g) was used to avoid entanglement between rip-rap interstices. A small, round hook (Owner, barb special, size 14, FRL-044) was baited with 1–3 fly maggots. All hooked fish were used for subsequent examination in the laboratory without any size or sex selection. Each fish was carefully hooked off with a special hook removal tool and was humanely killed inside a plastic bag in order to avoid losing gut contents or parasites. To prevent further digestion or migration of parasites to other organs, fishes were kept separately in plastic bags in a cooling box filled with ice and stored afterwards at $-20°C$ for later examination.

Amphipods were also collected monthly at the same sampling sites turning around large stones and using the 'kick-sampling' method after Storey et al. [38]. A small fishing net (15×20 cm, mesh size ~1 mm) was used to catch as many amphipods as possible within 30 minutes along a 10 m stretch at a depth of up to 50 cm. Amphipods were kept together with organic material and some stones in plastic bags. Entire samples were frozen at $-20°C$ and later separated from sediment to identify amphipods to species level.

### Parasitological examination and feeding ecology of *N. melanostomus*

Gobies were measured for total length (cm) and weight (g), condition factors (CF) were calculated according to Schäperclaus [39]. These measures are key parameters in studies on fish biology and were reported in (Text S1, Table S1) to facilitate comparisons with other studies.

Fish were then examined for their metazoan parasite fauna and stomach content using a stereomicroscope (Olympus SZ 61, magnification x 6.7–45). At first, skin, fins and gills were inspected for ectoparasites. Afterwards, the body cavity was opened to separate the inner organs. Body cavity, rinsed with 0.9% NaCl, gastrointestinal tract, gonads, kidney, liver, mesenteries, spleen and eyes were dissected and examined for endoparasites. Isolated parasites were freed from host tissue and preserved in 70% ethanol (with 4% glycerol) for morphological identification. To this end, glycerine preparations were made according to Riemann [40]. Determination under a microscope (Leitz Dialux 22, magnification x 15.75–630) was aided by original descriptions and descriptions of Golvan [41] and Špakulová et al. [42] for acanthocephalans, and Moravec [43] for nematodes. Subsamples were stored in 100% ethanol for genetic analysis (see Text S2).

Since gobies have no clearly separated stomach and a very short gut, the entire gastrointestinal tract was carefully cut lengthwise with a small pair of scissors. The weights of full and empty stomachs and the weights of each food item were recorded to the nearest 0.001 g after pat-drying on absorbent paper. Very small, as well as almost digested and defragmented parts of one prey group that could not be identified to species level were referred to as 'not determined' (indet.) and weighted as a pooled subsample. Only specimens that could clearly be identified, e.g. using assignable parts like eyes or telson, were identified and counted. Other components, mainly mucus and sand, but also undeterminable items were neglected. Prey organisms were sorted and identified to the lowest possible taxon and grouped into the following categories: amphipods, molluscs, insects and 'others' (plants, vertebrates, Acari). Isolated food organisms and parasites were preserved in 70% ethanol (with 4% glycerol) for morphological identification.

Amphipods were identified to species-level following Eggers & Martens [44,45] and preserved in 70% ethanol. For parasitological examination, all amphipods were dissected and carefully screened under a stereomicroscope. Isolated parasites were stored in 100% ethanol. From each monthly sampling, fifty amphipods of each species were randomly taken to determine sex, body size and weight using an ocular micrometer and a micro-balance. Size was measured from the anterior rostrum to the base of the telson while animals were stretched in a straight position [46]. Data are reported in (Text S3, Figure S1).

## Statistical analyses

We first tested if the relative abundance of amphipods on site (covariate, arcsine(square root)-transformed percentages relative to the highest monthly abundance value observed for the respective site) determines the proportion of amphipods in *N. melanostomus* gut contents (monthly mean values were treated as the dependent variable) using analysis of covariance (ANCOVA using SPSS vs. 22), in which 'site' was a fixed factor. A Chi$^2$ goodness-of-fit test (using R; R Development Core Team [47]) was then applied to test whether amphipod species compositions as encountered on site are reflected in gut contents.

Gut content analyses comprised calculations of the numerical percentage of prey (N%), the weight percentage of prey (W%), and the frequency of occurrence of prey (F%) [48,49]. On the basis of these three indices, the index of relative importance (IRI) of different food items was calculated [50]. Differences in gut content assemblage structure between months and rivers were also assessed using two-factorial permutation ANOVA (PERMANOVA; 999 permutations) on Bray-Curtis dissimilarities of 4$^{th}$-root transformed weights (mg) of the different species in each fish gut using the PRIMER v6 and PERMANOVA+ add-on package (PRIMER-e, Plymouth, UK). The SIMPER procedure [51] was used for post hoc identification of the source of variation.

Parasitological analyses comprised calculations of standard parameters: the prevalence (P), mean intensity (mI), intensity (I) and mean abundance (mA) for each parasite species according to Bush et al. [52]. High mean intensities of *Pomphorhynchus* sp. infections were found (see results), and previous studies suggested transmission pathways into *N. melanostomus* via amphipods, especially *D. villosus* [24]. Therefore, we used a repeated-measures General Linear Model (rmGLM using SPSS vs. 22) to test if mean intensities of *Pomphorhynchus* sp. in round gobies (dependent variable) differed between sexes (rm) and sites (fixed factor), and if the proportion of amphipods in the gut contents (arcsine(square root)-transformed numerical percentages, covariate) had an effect. The nematode *R. acus* was also relatively abundant in fish samples, but we restricted our analysis to non-parametric Wilcoxon signed-rank test (using SPSS vs. 22) to test whether differences in infection rates existed between the two rivers.

## Results

### Amphipod communities

717 to 3,758 amphipods were collected during the monthly samplings, with a total of $n = 9,820$ in the Rhine and $n = 7,536$ in the Main (see Table S2). Five invasive but no native amphipod species were found in both rivers, namely *D. villosus*, *Echinogammarus trichiatus* (Martynov, 1932), *Echinogammarus ischnus* (Stebbing, 1899), *Chelicorophium curvispinum* (Sars, 1895) and *Chelicorophium robustum* (Sars, 1895). *Cryptorchestia cavimana* (Heller, 1865) occurred only in samples from the Main. *Dikerogammarus villosus* was dominating in all samples from the Main (total $n = 5,346$; 69%), except for September (Figure 1). In contrast, *E. trichiatus* was the dominant species in all samples from the Rhine (total $n = 8,463$; 86%; Figure 1). In both rivers a more balanced sex ratio was found for *D. villosus* (males:females, Rhine: 1:1.03; Main: 1:1.29) than for *E. trichiatus* (Rhine: 1:2.36; Main: 1:3.10).

## General feeding ecology of *N. melanostomus*

18 (Rhine) and 16 (Main) different prey items were identified in *N. melanostomus* guts (Table S3, Table S4). The index of relative importance (IRI) found amphipods to be the main diet component of *N. melanostomus*, with an overall contribution of 71% in the Rhine and 46% in the Main (Figure 2). In the Rhine, amphipods contributed with at least 30% in each monthly sample (Figure 2). The second most important group was molluscs, which contributed with 7–38% to the overall gut content. The widespread and common species *Bithynia tentaculata*, *Potamopyrgus antipodarum* and *P. antipodarum f. carinata* were distinguishable, but, due to a high degree of fragmentation, were combined into 'Gastropoda indet.'. Insects were rarely consumed, except for July where the IRI for Chironomidae rose to 2,288.83 (Table S3) when very little gut content was found overall. In the Main, highest proportions of amphipods (over 80%) occurred in September and October (Figure 2). Insects were consumed more often than in the Rhine, especially in June (79%) and August (36%). Fish diet was based on molluscs with 50% and 45% in July and August, respectively. Fishes, plants and Acari were rarely consumed in both rivers.

Gut content assemblage structure showed strong fluctuations between months and rivers. They differed significantly between June and July and June and August in the Rhine, whereas June and August were different from all other months in the Main (PERMANOVA: pseudo-$F = 8.64$, $df = 4$, $p = 0.001$ for the interaction 'river × month'; for post hoc results see Table 1). Amphipods mostly accounted for the highest average dissimilarity between different monthly samples in the Rhine, whereas amphipods and insects accounted for the highest average dissimilarity between months in the Main (SIMPER procedure).

## Amphipod prey preference of *N. melanostomus*

Few individuals of *C. curvispinum* were found in *N. melanostomus* guts, and the dominating amphipod species was *D. villosus*, especially in the Main, but to a lesser degree also in the Rhine. This was reflected in the ANCOVA, which detected a significant interaction between 'site' and 'relative abundance of amphipods on site' (Table 2).

*Dikerogammarus villosus* was disproportionally frequent in gut contents given its availability relative to that of other amphipod species on site (Chi$^2$ goodness-of-fit tests, $p<0.001$; except for the July sampling in the Main when *D. villosus* overall was highly abundant in the field; Figure 1). Therefore, an additional ANCOVA with similar model structure was run using percentages of *D. villosus* in the gut content of *N. melanostomus* as the dependent variable (Table 3). Whereas a decrease (not increase) of numerical percentages of *D. villosus* in the gut content of *N. melanostomus* with increasing availability of *D. villosus* on site was found in the Main (driving a significant main effect of the covariate; Table 3), this pattern was not observed in the river Rhine (see significant interaction effect in Table 3; Figure 3).

## Fish parasites: species identity and general biology

In total, three metazoan parasite species, two in the Rhine and three in the Main, could be isolated from *N. melanostomus*. The following taxa were identified morphologically: *Pomphorhynchus* sp., *Raphidascaris acus*, and Glochidia indet. (Table 4). As noted by Špakulová et al. [42] and Emde et al. [24], morphological identification of species within the acanthocephalan genus *Pomphorhynchus* can be difficult. Therefore, molecular barcoding was conducted on a subset of $n = 3$ specimens that were morphologically identified as *P. tereticollis*. Sequence data for ITS-1/5.8S/ITS-2 (Genbank accession numbers KJ756498–KJ756500) were almost identical (99.0% similarity, e-value: 0.00)

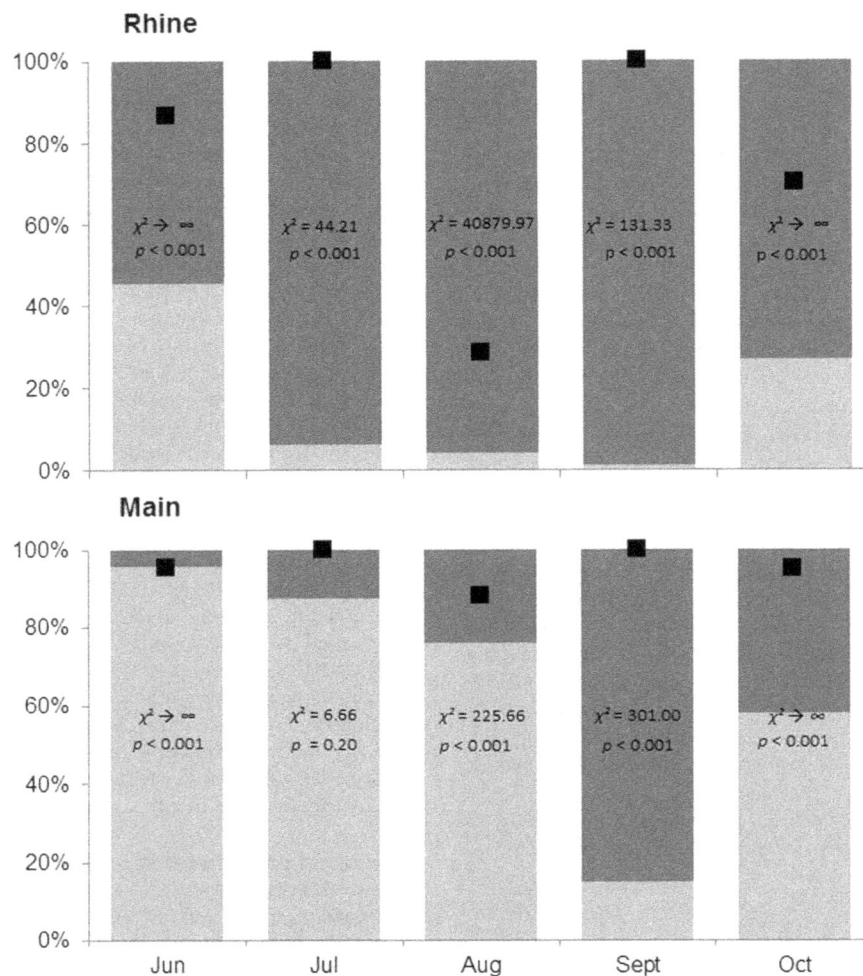

**Figure 1. Dominant amphipod species.** Fraction of the two dominant amphipod species (*D. villosus* = light grey, *E. trichiatus* = dark grey) in samples collected at our two study sites and numerical percentages of *D. villosus* in gut contents of *N. melanostomus* (black squares). Chi$^2$ goodness-of-fit tests were used to compare the availability of different amphipod species on site (expected values) with observed compositions in gut contents. For total numbers of individuals and amphipod species see Table S2.

to a sequence from *P. laevis* isolated from the cyprinid *Leuciscus cephalus* from the Czech Republic (Genbank accession number AY135415), suggesting that all acanthocephalan individuals in this study may belong to the same species. Due to a mismatch between the morphological identification characteristics and genetic information, acanthocephalan specimens were referred to as *Pomphorhynchus* sp. in this study.

All parasites were larval stages (Table 4). *Pomphorhynchus* sp. occurred only in the cystacanth stage. In the Rhine 91% of specimens were encysted in the mesenteries and liver and 9% were living freely in the body cavity. A similar pattern was found in the Main with 96% encysted in mesenteries and liver and 4% freely in the body cavity. The body cavity also harboured encysted *R. acus*, which occurred predominantly as $L_2$-larvae (91% in the Main, 88% in the Rhine), and $L_3$-larvae.

### Fish parasites: faunal composition

The most prevalent metazoan parasite type was *Pomphorhynchus* sp. with 100% prevalence in August and September in fish caught in the Rhine (Table 4). Maximum intensity reached 118 specimens per fish. Highest prevalence of *Pomphorhynchus* sp.

in the Main was recorded in June with 74.3%. Mean intensity of *Pomphorhynchus* sp. was an order of magnitude larger in fishes sampled from the Rhine (maximum mI = 34.6) than from the Main (maximum mI = 3.48) and always greater in female than in male *N. melanostomus* (rmGLM, significant interaction of 'sex × site'; Table 5; Figure 4). The nematode *R. acus* occurred with significantly lower prevalence in the Rhine (min. 28.57%, max. 57.14%) than in the Main (74.29% and 91.43%; Wilcoxon signed-rank test, $z = -2.023$, $p = 0.043$; Table 4). A maximum intensity of specimens of *R. acus* per fish was detected. Undetermined glochidia, i.e., parasitic larvae of unionid bivalves were detected on fish gills only in June (P = 54.3%) and October (P = 38.1%) in the Main.

### Parasites retrieved from amphipods

*Pomphorhynchus* sp. was the only parasite species that could be detected in amphipod samples. Two individuals were retrieved from *D. villosus* in the Rhine; the first was detected in samples from August (157 amphipods screened, P = 0.64%), the second in samples from October (671 amphipods screened, P = 0.15%). Overall, *Pomphorhynchus* sp. occurred at a prevalence of 0.15% in

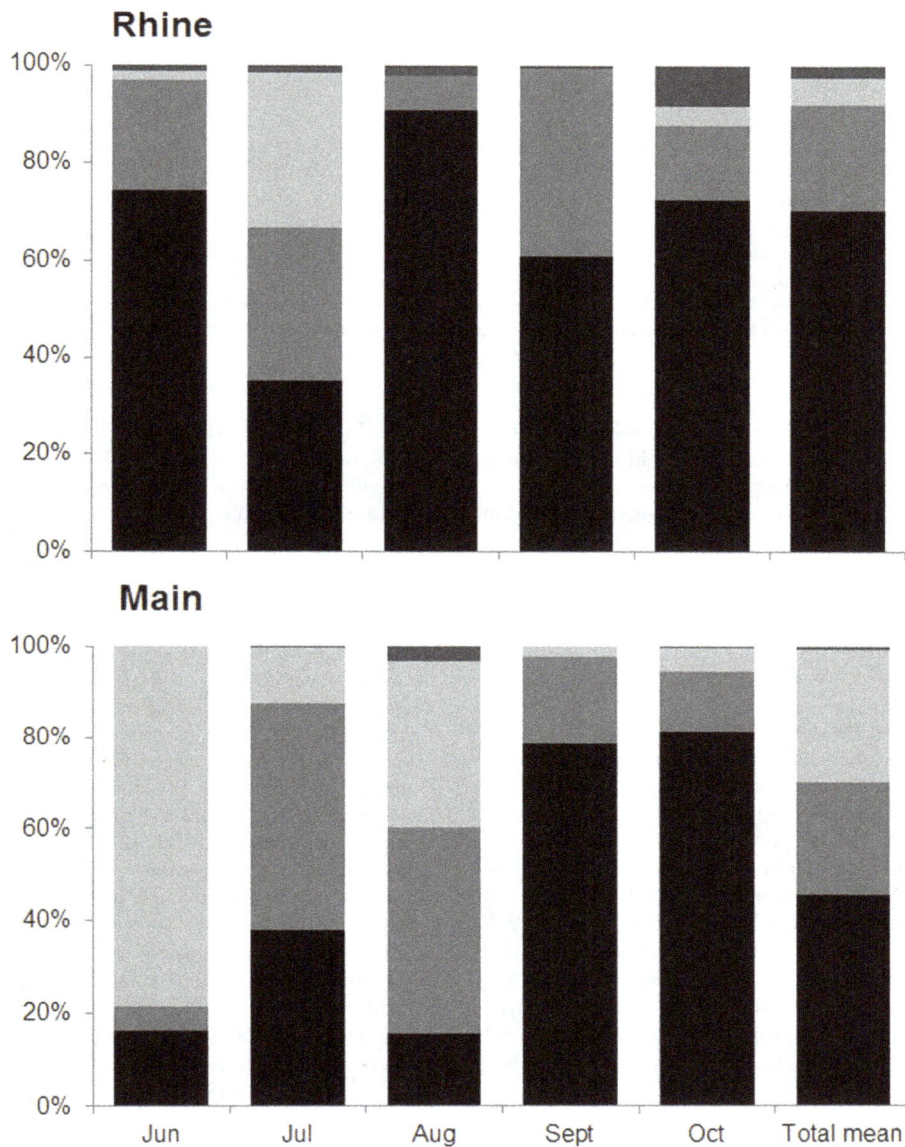

**Figure 2. Gut contents of *Neogobius melanostomus*.** Relative compositions (index of relative importance, IRI) of gut contents of *N. melanostomus* in two rivers from June until October 2011 as well as the total mean. Bar plot, from bottom to top: Amphipoda (black), Mollusca (medium grey), Insecta (light grey), others (dark grey).

**Table 1.** PERMANOVA results (post hoc procedure).

| River Rhine | $t$ | $p$ (perm) | River Main | $T$ | $p$ (perm) |
|---|---|---|---|---|---|
| Jun, Jul | 2.842 | 0.001 | Jun, Jul | 5.006 | 0.001 |
| Jul, Aug | 3.310 | 0.001 | Jun, Aug | 3.155 | 0.001 |
| | | | Jun, Sept | 5.943 | 0.001 |
| | | | Jun, Oct | 3.758 | 0.001 |
| | | | Jul, Aug | 2.835 | 0.001 |
| | | | Aug, Sept | 4.269 | 0.001 |
| | | | Aug, Oct | 2.447 | 0.003 |

Differences in gut contents assemblage structure (based on species' weights) of *N. melanostomus* between months. Significant results (permutation $p < 0.05$) after Bonferroni correction for multiple comparisons are shown.

**Table 2.** ANCOVA results – all amphipods.

| | Source | df | MS | F | p | Partial Eta squared |
|---|---|---|---|---|---|---|
| All amphipods | Site | 1 | 0.317 | 9.330 | **0.022** | 0.609 |
| | Rel. abundance | 1 | 0.003 | 0.091 | 0.773 | 0.015 |
| | Site × rel. abundance | 1 | 0.549 | 16.183 | **0.007** | 0.730 |
| | Residuals | 6 | 0.034 | | | |

Numerical percentages of all amphipods in the gut content of N. melanostomus in relation to the relative abundance of amphipods on site. Significant effects are in bold.

D. villosus in the river Rhine (two out of 1,350 specimens). The total number of D. villosus was four times larger in the Main than in the Rhine (i.e., n = 5,346), still, no parasites were detected. Low overall abundance precluded an analysis of potential temporal fluctuation in parasite infections of amphipods. Numerical percentages of amphipods in fish gut contents did not predict mean intensities of acanthocephalan parasites in N. melanostomus (Table 5).

## Discussion

### Feeding ecology of N. melanostomus

Co-evolved trophic relationships can facilitate biological invasions, as exemplified by communities of coexisting invasive N. melanostomus, dreissenid mussels and E. ischnus in the North American Great Lakes [53,54]. Presence of co-evolved prey, however, appears not to be a prerequisite for N. melanostomus in German rivers, since N. melanostomus was characterized by an opportunistic and broad feeding strategy [see also 30,31]. Opportunistic feeding might also provide a plausible explanation for why we detected no positive correlation between the abundance of D. villosus in the field (generally a preferred type among amphipod prey) and their proportional contribution to gut contents. This was obvious especially during early summer, when prey species other than amphipods became more relevant (higher index of relative importance), especially in the Main, where insects and molluscs became the main food sources. Similarly, the importance of amphipod prey (D. villosus and others) for N. melanostomus in the Danube increased from early to late summer while the importance of chironomid larvae decreased [25]. Ingested insects in our present study were mostly nematoceran larvae, which are generally abundant in slow-flowing waterways like the Main. Non-biting midges (Chironomidae) no longer dominate the invertebrate community of the navigable main

channel of the upper Rhine [55], which may explain why insects, overall, were barely ingested. While N. melanostomus is commonly regarded as a predator of fish eggs and fry (e.g. [56]), these were only rarely retrieved from gut contents.

An ontogenetic size dependent diet shift from amphipods and insects to a diet dominated mainly by molluscs is known for round gobies (e.g. [25]), however, fish lengths where shifts seem to occur vary substantially between study regions and most likely depend on availability and abundance of prey organisms [57,58] as well as on time since invasion [29]. In our present study, the genus Dreissena seems to play a subordinate role compared to the Great Lakes and the Baltic Sea, which may be attributable to more readily available food sources, like insect larvae and amphipods. Generally, a tendency of increasing absolute numbers with increasing fish size was observed for D. villosus and nematocerans. In this context, Emde et al. [24] already demonstrated a size-dependent increase in acanthocephalan infections, which was inter alia explained by a correlation between goby and amphipod (D. villosus) prey body size, as it seems likely that the development of acanthocephalan larvae might only grow in amphipods above a certain size threshold. Thus, smaller gobies, feeding on smaller D. villosus, are less infected by acanthocephalans.

All amphipods found during monthly sampling were non-indigenous species from the Ponto-Caspian region (i.e., Black and Caspian Seas), corroborating studies in several European watersheds [24,44,59]. The most common non-indigenous amphipod species were D. villosus und E. trichiatus. Dikerogammarus villosus was dominant in samples from the Main, whereas E. trichiatus was dominant in Rhine samples, suggesting that faunal compositions of invasive amphipods may be more stable temporally and to a lesser degree spatially within the Rhine drainage (see also [24,60]). Dikerogammarus villosus was detected six years earlier than E. trichiatus in the Rhine and is known for its strong predation on other gammarids [7,61]. However, the total number

**Table 3.** ANCOVA results – D. villosus.

| | Source | df | MS | F | p | Partial Eta squared |
|---|---|---|---|---|---|---|
| D. villosus | Site | 1 | 0.155 | 31.261 | **0.001** | 0.839 |
| | Rel. abundance | 1 | 0.053 | 10.644 | **0.017** | 0.640 |
| | Site × rel. abundance | 1 | 0.127 | 25.487 | **0.002** | 0.809 |
| | Residuals | 6 | 0.005 | | | |

Numerical percentages of D. villosus in the gut content of N. melanostomus in relation to the relative abundance of D. villosus on site. Significant effects are in bold.

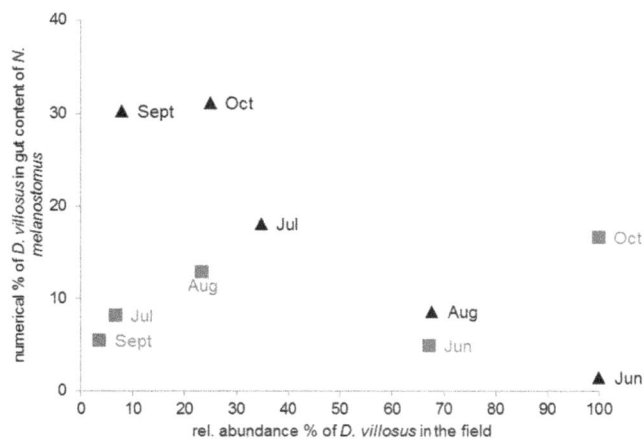

**Figure 3. *Dikerogammarus villosus* in fish guts and in the field.** Numerical percentage of *D. villosus* in gut contents of *N. melanostomus* in relation to the relative abundance of *D. villosus* at the Main (black) and Rhine (grey) between June and October 2011.

of individuals caught in the Rhine was an order of magnitude lower than that of *E. trichiatus*. Whether higher predation on *D. villosus* by *N. melanostomus* in the Rhine compared to the Main could explain this pattern remains uncertain, since no fish densities at both sites were recorded herein.

Regardless of the high numbers of *E. trichiatus* in the Rhine, *N. melanostomus* primarily fed on *D. villosus*. How can this pattern be explained? Sih and Christensen [62] argued that variation in prey behaviour is more likely to affect the direction of predator-prey interactions than active prey choice of predators. Qualitatively, we noted that *E. trichiatus* at our study sites occurred closer to riverbanks, while *D. villosus* were found in both shallow and deeper waters, and so *E. trichiatus* could avoid fish predation in shallower habitats or by hiding between rip-rap interstices. Spatial niche segregation between *E. trichiatus* and *D. villosus* was previously observed in the Netherlands where the former seems to occur on soft substrates whereas the latter is most abundant on hard substrates [63]. Thus, different microhabitat use or different activity patterns in *D. villosus* are likely explanations for their dominance among amphipod prey in *N. melanostomus*.

Parasites can manipulate the predator avoidance of freshwater amphipods, rendering them more vulnerable to their fish predators (for *Gammarus pulex* see [64,65]). Whether infections by *Pomphorhynchus* sp. affect the predator avoidance of *D. villosus* is currently not known, but if infected individuals were indeed more prone to predation, this would provide a striking explanation for our finding that gobies were highly infected by *Pomphorhynchus* sp., yet infectious stages were barely found in their amphipod prey (i.e., *D. villosus*), and were even completely absent in the Main. It seems reasonable to argue that infected *D. villosus* were ingested at an accelerated rate compared to uninfected specimens. Generally, infection rates of invertebrate intermediate hosts, especially crustaceans, tend to be low, often ranging between 0.01 and 0.1% prevalence [23,28]. A possible reason for the higher infestation rates of *D. villosus* in the Rhine might be the presence of more final hosts (like common barbel *Barbus barbus* and European chub *Squalius cephalus*, however this assumption is not based on quantitative data but on personal observations only (S. Emde, personal observation).

*Pomphorhynchus* sp. is known to include a variety of different first intermediate hosts in its life cycle, such as *D. villosus* [24], *G. pulex* [64] and *C. curvispinum* [65]. *Gammarus pulex* seems to be

completely displaced by invasive species in the Rhine and Main [24] and was not part of the gobies' diet at both sampling sites. Following a massive decrease since 1995, *C. curvispinum* currently also plays a negligible role in the gobies' diet [65]. In the light of the decrease of other amphipod species and the observed dominance of *D. villosus* in the gobies' diet, we suggest that *D. villosus* currently represents the most relevant intermediate host for *Pomphorhynchus* sp. Still, future studies could investigate additional invertebrate groups and might uncover additional first intermediate hosts for the opportunistic parasites of the genus *Pomphorhynchus*.

## Parasite fauna of *N. melanostomus*

More than 94 parasites of *N. melanostomus* have been recorded worldwide [66], and in its introduced range in Europe, 35 metazoan parasite species have been detected so far (e.g., [66–69]). *Neogobius melanostomus* usually carries more than ten different parasite species per population in its native range [70]. Herein, only three parasite species could be detected in 350 round gobies examined, suggesting that the diversity of *N. melanostomus* parasites in the Rhine did not change over the past four years ([24], S. Emde personal observation). In other regions, the parasite fauna of invasive *N. melanostomus* increased rapidly, e.g., in the Gulf of Dansk, where numbers rose from 4 to 12 parasite species within two years [68]. Only 6 to 10 years have passed since round gobies were first recorded in German inland waterways, while the first report of round gobies at our sampling sites was in 2007 [71,72]. Our results support the 'enemy release hypothesis' [19], and release from natural parasites could be one reason promoting the fast spread of round gobies worldwide. This advantage over indigenous fishes, however, will likely be lost if the diversity of the parasite fauna of *N. melanostomus* increases with time [73]. Whether or not such an increase of parasite diversity will occur in the future requires further monitoring.

All parasites detected in *N. melanostomus* were larval stages, and so we tentatively argue that currently no native parasite species is able to use *N. melanostomus* as its final but only as a paratenic host. A higher parasitization of *N. melanostomus* was observed in the Rhine, where fishes were also smaller and had a lower condition factor than in the Main (Text S1, Table S1). A high parasite load can lead to decreased growth in their fish hosts [74], however, infection studies in controlled environments would be needed to further address this hypothesis.

*Pomphorhynchus* sp. (Acanthocephala) and *Raphidascaris acus* (Nematoda) have been detected before in *N. melanostomus* caught in the Rhine, with similar infection rates for *Pomphorhynchus* sp. [24]. Latest data of the Danube River also described high abundances of this parasite but detected highest abundances in more recently invaded areas [29]. Similarly high prevalences of *R. acus* as found in our current study (up to 91.43%) are known from studies in other sections of the Rhine (56%) [75] and the Danube (P = 57%) [67]. Generally, differences in infection rates (prevalence/intensities) among studies could be related to the presence/absence as well as abundance of the parasites' final hosts. For adult *R. acus* the European pike (*Esox lucius*) and brown trout (*Salmo trutta fario*) are known as principal final hosts [43], whereas it is barbel (*Barbus barbus*) and chub (*Squalius cephalus*) for *Pomphorhynchus* sp. [24]. However, *N. melanostomus* seems to represent a new, additional intermediate host for these parasites and thus, bridges the trophic level towards new potential, predatory final hosts. Other potential definitive hosts in the rivers Rhine and Main are trout (*Salmo trutta*) and catfish (*Silurus glanis*) for *Pomphorhynchus* [76,77] and the European eel (*Anguilla anguilla*), European perch (*Perca fluviatilis*) and pike-

**Table 4.** Parasitological parameters for the parasite fauna of *N. melanostomus*.

| | P (%) | | | | | $I_{min}$–$I_{max}$/Mi | | | | | mA | | | | |
|---|---|---|---|---|---|---|---|---|---|---|---|---|---|---|---|
| | Jun | Jul | Aug | Sept | Oct | Jun | Jul | Aug | Sept | Oct | Jun | Jul | Aug | Sept | Oct |
| *Rhine* | | | | | | | | | | | | | | | |
| **Nematoda** | | | | | | | | | | | | | | | |
| *Raphidascaris acus* (Cyst, L) in body cavity/mesentery/liver | 50.00 | 28.57 | 57.14 | 55.56 | 34.29 | 1–6/2.44 | 1–5/2.10 | 1–18/3.80 | 1–17/3.67 | 1–7/10.25 | 1.22 | 0.60 | 2.17 | 1.89 | 3.51 |
| **Acanthocephala** | | | | | | | | | | | | | | | |
| *Pomphorhynchus* sp. (L) in body cavity/mesentery/liver | 86.11 | 71.43 | 100 | 100 | 74.29 | 1–118/22.65 | 1–35/9.04 | 1–101/24.26 | 2–95/32.71 | 1–20/8.46 | 19.50 | 6.46 | 24.26 | 32.71 | 6.29 |
| **Total** | 100 | 77.14 | 100 | 100 | 88.57 | 1–118/20.72 | 1–39/9.15 | 1–105/26.43 | 2–95/34.60 | 1–23/11.06 | 20.72 | 7.06 | 26.43 | 34.60 | 9.80 |
| *Main* | | | | | | | | | | | | | | | |
| **Nematoda** | | | | | | | | | | | | | | | |
| *Raphidascaris acus* (Cyst, L) in body cavity/mesentery/liver | 74.29 | 85.71 | 82.86 | 88.57 | 91.43 | 1–20/3.85 | 1–29/6.17 | 1–24/6.55 | 1–9/3.39 | 1–15/5.63 | 2.86 | 5.29 | 5.43 | 3.00 | 5.14 |
| **Acanthocephala** | | | | | | | | | | | | | | | |
| *Pomphorhynchus* sp. (L) in body cavity/mesentery/liver | 74.29 | 60.00 | 37.14 | 37.14 | 25.71 | 1–16/2.77 | 1–15/3.48 | 1–6/2.62 | 1–9/3.00 | 1–16/3.22 | 2.06 | 2.09 | 0.97 | 1.11 | 0.83 |
| **Bivalvia** | | | | | | | | | | | | | | | |
| Glochidia indet. (L) in gills | 54.29 | – | – | – | 38.10 | 1–8/3.68 | – | – | – | 1–23/9.88 | 2.00 | – | – | – | 3.76 |
| **Total** | 97.14 | 85.71 | 82.86 | 94.29 | 94.29 | 1–30/7.12 | 1–44/8.60 | 1–29/7.72 | 1–17/4.36 | 1–35/8.73 | 6.91 | 7.37 | 6.40 | 4.11 | 8.23 |
| **Wilcoxon signed–rank test** | | | | | | | | | | | | | | | |
| *Raphidascaris acus* | z=−2.023 p=**0.043** (Rhine/Main) | | | | | z=−0.405 p=0.686 (Rhine/Main) | | | | | z=−2.023 p=**0.043** (Rhine/Main) | | | | |
| *Pomphorhynchus* sp. | z=−2.023 p=**0.042** (Rhine/Main) | | | | | z=−2.023 p=**0.043** (Rhine/Main) | | | | | z=−2.023 p=**0.043** (Rhine/Main) | | | | |

I = Intensity, L = larvae, mA = mean abundance, mI = mean intensity and P = prevalence.

**Table 5.** Repeated–measures GLM results.

| | Source | df | MS | F | p | Partial Eta squared |
|---|---|---|---|---|---|---|
| **Within–subjects effects** | Sex | 1 | 35.935 | 3.689 | 0.096 | 0.345 |
| | Sex × amphipods in gut | 1 | 3.973 | 0.408 | 0.543 | 0.055 |
| | Sex × site | 1 | 129.212 | 13.263 | **0.008** | 0.655 |
| | Residuals (Sex) | 7 | 9.742 | | | |
| **Between–subjects effects** | Intercept | 1 | 510.007 | 3.202 | 0.117 | 0.314 |
| | Amphipods in gut | 1 | 46.853 | 0.294 | 0.604 | 0.040 |
| | Site | 1 | 1459.710 | 9.164 | **0.019** | 0.567 |
| | Residuals | 7 | 159.282 | | | |

Repeated–measures GLM on mean intensities of *Pomphorhynchus* sp. in round gobies in relation to fish sex, numerical percentages of amphipods (*D. villosus* and Amphipoda indet.) in the gut content and site. Significant effects are in bold.

perch (*Sander lucioperca*) for *R. acus* [78]. Infection studies need to show whether the female parasite attains gravidity in the potential definitive host or whether these predatory fishes may only act as para-definitive hosts in which the parasite matures but is unable to produce eggs [78]. If they do not act as definitive hosts, the large number of parasite larvae in *N. melanostomus* will

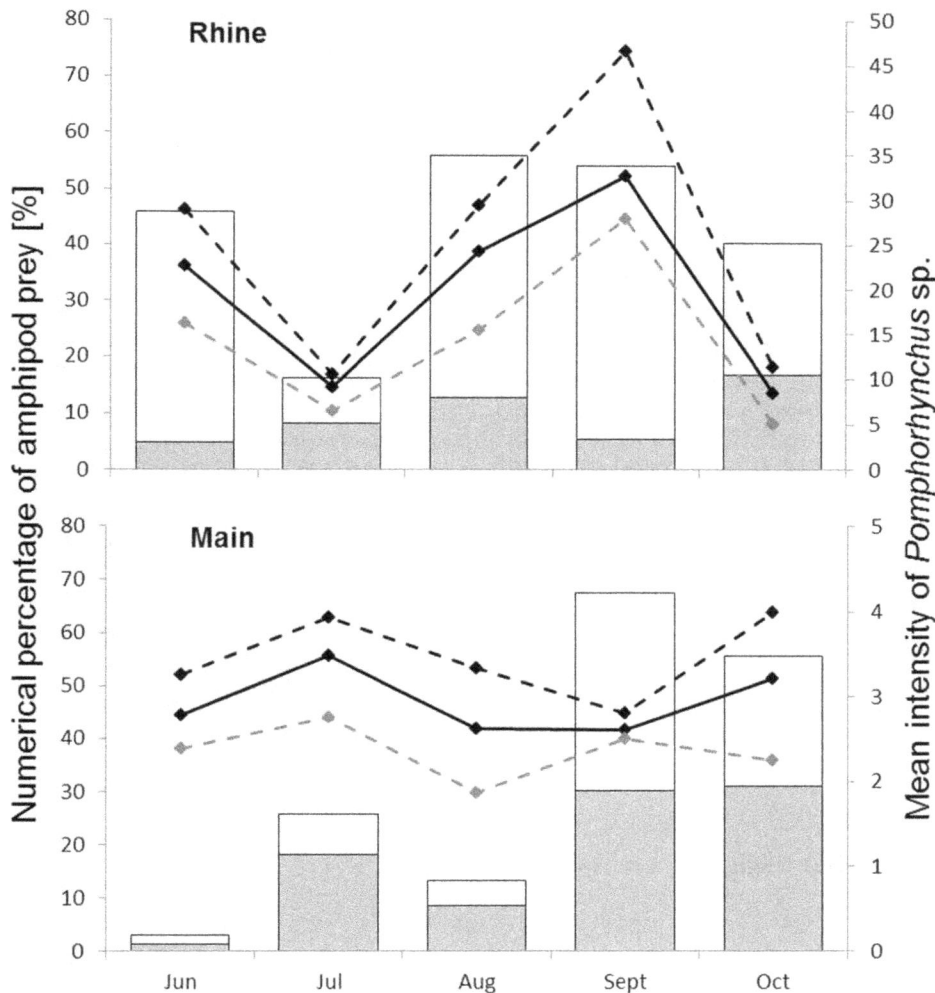

**Figure 4. Amphipod prey and infections with *Pomphorhynchus* sp.** Relationship between numerical percentages of *D. villosus* (grey) and Amphipoda indet. (white) in the gut content of *N. melanostomus* and mean intensities (ml, black line) of *Pomphorhynchus* sp. in male (grey dashed line) and female (black dashed line) *N. melanostomus*. For numbers of individuals please refer to Table S3 and Table S4.

be transmitted to these predatory fishes, however, not be able to complete their life cycle. This would lead to a dilution effect, resulting in a continued loss of infection within the system as has been described for different parasite-host communities before [79,80] and would therefore be an alternative plausible explanation for the lower infection rates in the Main than in the Rhine.

Parasitic larval stages (Glochidia) of freshwater mussels of the family Unionidae were found in samples from the river Main, which confirms a former report of *N. melanostomus* serving as a host for unionid glochidia in the Danube [67]. Glochidia could be detected only during some months, because river mussels (*Unio* sp.) spawn in early summer and swan mussels (*Anodonta* sp.) in late summer, and glochidia attach to fish gills for only a few weeks [81]. Although unionid mussels are known to occur in the Rhine [82], no glochidia were detected on the gills of *N. melanostomus*, which could suggest an abundance-correlated effect. Alternatively, *N. melanostomus* might be a bad host for unionids [83]. Authors infected gobies with Glochidia of which 98% were lost within 16 days. Based on that study, our findings of Glochidia attached to gills of *N. melanostomus* could therefore be a finding that was the result of a very recent infection.

We initially hypothesized monthly infestation rates of *D. villosus* with *Pomphorhynchus* sp. potentially reflecting infestation rates in *N. melanostomus*. Due to overall low abundances of *Pomphorhynchus* sp. in *D. villosus* a statistical analysis in this direction was not possible. We also tested whether the numerical percentage of *D. villosus* in gut contents predicts mean intensities of *Pomphorhynchus* sp. but found no such effect. The timing of the parasite's life cycle, however, has not yet been examined, and so our analysis (that was based on monthly sampling) may not have been appropriate to capture such potential effect.

Sex-related differences in parasite infections are common and can be ascribed to sex-specific behavioural, physiological or morphological differences [84,85]. In this study, mean intensity of *Pomphorhynchus* sp. was significantly higher in females than males in the Rhine, supporting the finding of Brandner et al. [29] from the Danube River. No significant sex differences were observed in the Main, but *Pomphorhynchus* sp. mean intensities were low in the Main overall. Males can allocate much less time to feeding than females (for poeciliid fishes see [86,87]) lowering their risk to take up parasites from food. Indeed, Charlebois et al. [88] found *N. melanostomus* males to cease feeding during brood care, while females producing eggs should have increased energy demands.

Our study confirmed that *D. villosus* functions as the main amphipod prey species for *N. melanostomus* in German rivers, however, parasite intensities in *N. melanostomus* differed between sampling locations of Rhine and Main independently of amphipod abundances. We suggest that a characterization of new final fish hosts, especially for *Pomphorhynchus* sp., at the sites investigated herein could provide important new insight into the ecological causes of variation in parasitization patterns of *N. melanostomus* in its introduced range.

## Supporting Information

**Figure S1 Box–plots of total length and total weight of two amphipod species.**

## References

1. Carlton JT, Geller JB (1993) Ecological roulette: the global transport of nonindigenous marine organisms. Science 261: 78–82.
2. Walther G-R, Post E, Convey P, Menzel A, Parmesan C, et al. (2002) Ecological responses to recent climate change. Nature 416: 389–395.

(TIF)

**Table S1 Biological parameters of *Neogobius melanostomus*.**
(DOCX)

**Table S2 Amphipod fauna.**
(DOCX)

**Table S3 Gut contents and parameters of *Neogobius melanostomus* for the river Rhine.**
(DOC)

**Table S4 Gut contents and parameters of *Neogobius melanostomus* for the river Main.**
(DOC)

**Text S1 Size measurements and condition factors of *N. melanostomus*.**
(DOCX)

**Text S2 Genetic identification of parasites.**
(DOCX)

**Text S3 Size measurements of *D. villosus* and *E. trichiatus*.**
(DOCX)

## Acknowledgments

We thank J. Schneider (Office for fish ecological studies – BFS, Frankfurt), S. Gallus and S. Schierz (Goethe University, Frankfurt) for their support with data assessment. We further wish to thank C. Koehler (Dezernat V 51.1 Landwirtschaft-Landschaftspflege-Fischerei, Regierungspräsidium Darmstadt) for providing a fishing license to catch gobies. We are grateful to D. Green who helped with statistics. Finally, we thank the reviewers of this article for their helpful suggestions. The authors declare no conflict of interest.

### Ethics Statement

Approval of our present study by a review board institution or ethics committee was not necessary because all fish were caught by a person (S. Emde) holding a valid local fishing license (No. 06258) for the river Main, issued by the 'Höchster Fischereigenossenschaft', 65830 Kriftel, Germany. For the river Rhine a special permit (F4/Di-Zi) was issued by the 'Hessische Landesgesellschaft mbH', 34121 Kassel, Germany. No living animals were used. In Germany, the fishing license permits the holder to capture and sacrifice the fish, which can be used not only for consumption but also for research purposes. All fish were stunned by a blow on the head and expertly killed immediately by cervical dislocation and a cardiac stab according to the German Animal Protection Law (§ 4) and the ordinance of slaughter and killing of animals (*Tierschlachtverordnung* § 13). Because of public accessibility no permissions were required to enter the sampling sites.

## Author Contributions

Conceived and designed the experiments: SE SK. Performed the experiments: SE TK. Analyzed the data: SE JK MP SK. Contributed reagents/materials/analysis tools: SK MP. Contributed to the writing of the manuscript: SE JK TK MP SK.

3. Leprieur F, Beauchard O, Blanchet S, Oberdorff T, Brosse S (2008) Fish invasions in the world's river systems: when natural processes are blurred by human activities. Plos Biol 6: e28.
4. Vitule JRS, Freire CA, Simberloff D (2009) Introduction of non-native freshwater fish can certainly be bad. Fish Fish 10: 98–108.

5. Ricciardi A, MacIsaac HJ (2011) Impacts of biological invasions on freshwater ecosystems. In: Fifty Years of Invasion Ecology: The Legacy of Charles Elton (Ed. Richardson D. M.), 211–224, Blackwell Publishing.

6. Martin CW, Valentine MM, Valentine JF (2010) Competitive interactions between invasive Nile Tilapia and native fish: the potential for altered trophic exchange and modification of food webs. Plos One 5: e14395. doi:10.1371/journal.pone.0014395.

7. Dick JTA, Platvoet D (2000) Invading predatory crustacean Dikerogammarus villosus eliminates both native and exotic species. P Roy Soc Lond B Bio 267: 977–983.

8. Salo P, Korpimaki E, Banks PB, Nordstrom M, Dickman CR (2007) Alien predators are more dangerous than native predators to prey populations. P Roy Soc B-Biol Sci 274: 1237–1243.

9. Paolucci EM, MacIsaac HJ, Ricciardi A (2013) Origin matters: alien consumers inflict greater damage on prey populations than do native consumers. Divers and Distrib 19: 988–995.

10. Prenter J, MacNeil C, Dick JTA, Dunn AM (2004) Roles of parasites in animal invasions. Trends Ecol Evol 19: 385–390.

11. Douda K, Lopes-Lima M, Hinzmann M, Machado J, Varandas S, et al. (2013) Biotic homogenization as a threat to native affiliate species: fish introductions dilute freshwater mussel's host resources. Divers Distrib 19: 933–943.

12. Strayer DL (2012) Eight questions about invasions and ecosystem functioning. Ecol Lett 15: 1199–1210.

13. Witte F, Goldschmidt T, Wanink J, van Oijen M, Goudswaard K, et al. (1992) The destruction of an endemic species flock: quantitative data on the decline of the haplochromine cichlids of Lake Victoria. Environ Biol Fish 34: 1–28.

14. Kats LB, Ferrer RP (2003) Alien predators and amphibian declines Review of two decades of science and the transition to conservation. Divers and Distrib 9: 99–110.

15. Machida Y, Akiyama YB (2013) Impacts of invasive crayfish (Pacifastacus leniusculus) on endangered freshwater pearl mussels (Margaritifera laevis and M. togakushiensis) in Japan. Hydrobiologia 720: 145–151.

16. Ricciardi A, Whoriskey FG, Rasmussen JB (1996) Impact of Dreissena polymorpha on native unionid bivalves in the upper St. Lawrence River. Can J Fish Aquat Sci 53: 1434–1444.

17. Dick JTA, Platvoet D, Kelly DW (2002) Predatory impact of the freshwater invader Dikerogammarus villosus (Crustacea: Amphipoda). Can J Fish Aquat Sci 59: 1078–1084.

18. Carlsson NOL, Sarnelle O, Strayer DL (2009) Native predators and exotic prey – an acquired taste? Front Ecol Environ 7: 525–532.

19. Crawley MJ (1987) What makes a community invasible? In: Colonization, succession, and stability (Eds. Gray A.J., Crawley M.J., Edwards P.J.) 429–453, Blackwell, Oxford.

20. Torchin ME, Lafferty KD, Dobson AP, McKenzie VJ, Kuris AM (2003) Introduced species and their missing parasites. Nature 421: 628–630.

21. Kelly DW, Paterson RA, Townsend CR, Poulin R, Tompkins DM (2009) Parasite spillback: A neglected concept in invasion ecology? Ecology 90: 2047–2056.

22. Kopp K, Jokela J (2007) Resistant invaders can convey benefits to native species. Oikos 116: 295–301.

23. Paterson RA, Townsend CR, Poulin R, Tompkins DM (2011) Introduced brown trout alternative acanthocephalan infections in native fish. J Anim Ecol 80: 990–998.

24. Emde S, Rueckert S, Palm HW, Klimpel S (2012) Invasive Ponto-Caspian amphipods and fish increase the distribution range of the acanthocephalan Pomphorhynchus tereticollis in the river Rhine. Plos One 7: e53218 doi:10.1371/journal.pone.0053218.

25. Brandner J, Auerswald K, Cerwenka AF, Schliewen UK, Geist J (2013) Comparative feeding ecology of invasive Ponto-Caspian gobies. Hydrobiologia 703: 113–131.

26. Locke SA, Bulté G, Marcogliese DJ, Forbes MR (2014) Altered trophic pathway and parasitism in a native predator (Lepomis gibbosus) feeding on introduced prey (Dreissena polymorpha). Oecologia 175: 315–24.

27. Rohde K (2005) Marine parasitology. CSIRO Publishing.

28. Busch MW, Kuhn T, Münster J, Klimpel S (2012) Marine crustaceans as potential hosts and vectors for metazoan parasites. In: Arthropods as vectors of emerging diseases (Ed. H. Mehlhorn), 329–360, Parasitol Res Monographs 3, Springer, Berlin Heidelberg.

29. Brandner J, Cerwenka AF, Schliewen UK, Geist J (2013) Bigger is better: Characteristics of round gobies forming an invasion front in the Danube River. PLoS ONE 8(9): e73036.

30. Kornis MS, Mercado-Silva N, van der Zanden MJ (2012) Twenty years of invasion: a review of round goby Neogobius melanostomus biology, spread and ecological implications. J Fish Biol 80: 235–285.

31. Borcherding J, Dolina M, Heermann L, Knutzen P, Krüger S, et al. (2013) Feeding and niche differentiation in three invasive gobies in the Lower Rhine, Germany. Limnologica 43: 49–58.

32. Skóra KE, Rzeznik J (2001) Observations on diet composition of Neogobius melanostomus Pallas 1811 (Gobiidae, Pisces) in the Gulf of Gdansk (Baltic Sea). J Great Lakes Res 27: 290–299.

33. Rakauskas V, Bacevičius E, Pūtys Ž, Ložys L, Arbačiauskas K (2008) Expansion, feeding and parasites of the round goby, Neogobius melanostomus (Pallas, 1811), a recent invader in the Curonian Lagoon, Lithuania. Acta Zoologica Lituanica 18, 3: 180–190.

34. Haas G, Brunke M, Streit B (2002) Fast turnover in dominance of exotic species in the Rhine river determines biodiversity and ecosystem function: an affair between amphipods and mussels. In: Invasive aquatic species of Europe: distribution, impacts, and management (eds. Leppäkoski E, Gollasch S, Olenin S), 426–432, Dordrecht.

35. Bernauer D, Jansen W (2006) Recent invasions of alien macroinvertebrates and loss of native species in the upper Rhine River, Germany. Aquatic Invasions 1, 2: 55–71. Available: http://aquaticinvasions.net/2006/AI_2006_1_2_Bernauer_Jansen.pdf. Accessed 2012 June 5.

36. Ondračková M, Francová K, Dávidová M, Poláčik M, Jurajda P (2010) Condition status and parasite infection of Neogobius kessleri and N. melanostomus (Gobiidae) in their native and non-native area of distribution of the Danube River. Ecol Res 25: 857–866.

37. Brandner J, Pander J, Mueller M, Cerwenka AF, Geist J (2013) Effects of sampling techniques on population assessment of invasive round goby Neogobius melanostomus. J Fish Biol 82: 2063–2079.

38. Storey AW, Edward DHD, Gazey P (1991) Surber and kick sampling: a comparison for the assessment of macroinvertebrate community structure in streams of south-western Australia. Hydrobiologia 211: 111–121.

39. Schäperclaus W (1991) Fish Diseases. Volume 1. (Ed. Kothekar V.S.), Akademie-Verlag, Berlin.

40. Riemann F (1988) Introduction to the study of meiofauna. Higgins RP und Thiel H (Eds.). Smithsonian Institution Press: 293–301.

41. Golvan YJ (1969) Systematique des acanthocephales (Acanthocephala, Rudolphi 1801). L'ordre des Palaeacanthocephala Meyer 1931. La superfamille des Echinorhynchoidea (Cobbold 1876) Golvan et Houin, 1963. Mémoires du Museum National d'Histoire Naturelle, Série A, Zoologie Band 57, Paris: 373 p.

42. Špakulová M, Perrot-Minnot M-J, Neuhaus B (2011) Resurrection of Pomphorhynchus tereticollis (Rudolphi, 1809) (Acanthocephala: Pomphorhynchidae) based on new morphological and molecular data. Helminthologia 48, 3: 268–277.

43. Moravec F (1994) Parasitic nematodes of freshwater fishes of Europe. Academy of Sciences of the Czech Republic, Academia.

44. Eggers TO, Martens A (2001) Bestimmungsschlüssel der Süßwasser–Amphipoda (Crustacea) Deutschlands. Lauterbornia 42: 1–68.

45. Eggers TO, Martens A (2004) Ergänzungen und Korrekturen zum "Bestimmungsschlüssel der Süßwasser-Amphipoda (Crustacea) Deutschlands". Lauterbornia 50: 1–13.

46. Quigley MA, Lang GA (1989) Measurement of amphipod body length using a digitizer. Hydrobiologia 171: 255–258.

47. R Development Core Team (2010) R: A language and environment for statistical computing. Foundation for Statistical Computing, Vienna, Austria.

48. Hyslop EJ (1980) Stomach content analysis - a review of methods and their application. J Fish Biol 17: 411–429.

49. Amundsen PA, Gabler HM, Staldvik FJ (1996) A new approach to graphical analysis of feeding strategy from stomach contents data – modification of the Costello (1990) method. J Fish Biol 48: 607–614.

50. Pinkas L, Oliphant MD, Iverson ILK (1971) Food habits of albacore, bluefin tuna and bonito in Californian waters. Calif Fish Game 152: 1–105.

51. Clarke KR (1993) Non-parametric multivariate analyses of changes in community structure. Aust J Ecol 18: 117–143.

52. Bush O, Lafferty AD, Lotz JM, Shostak AW (1997) Parasitology meets ecology on his own terms: Margolis, et al. revisited. J Parasitol 83: 575–583.

53. McKinney ML, Lockwood JL (1999) Biotic homogenizaton: a few winners replacing many losers in the next mass extinction. Trends Ecol Evol 14: 450–453.

54. Vanderploeg HA, Nalepa TF, Jude DJ, Mills EL, Holeck KT, et al. (2002) Dispersal and emerging ecological impacts of Ponto-Caspian species in the Laurentian Great Lakes. Can J Fish Aquat Sci 59: 1209–1228.

55. IKSR (2002) Das Makrozoobenthos des Rheins 2000, Internationale Kommission zum Schutz des Rheins (IKSR), Bericht Nr. 128-d.doc; Koblenz.

56. Corkum LD, Sapota MR, Skora KE (2004) The round goby, Neogobius melanostomus, a fish invader on both sides of the Atlantic Ocean. Biol Invasions 6: 173–181.

57. Karlson AML, Almqvist G, Skóra KE, Appelberg M (2007) Indications of competition between non-indigenous round goby and native flounder in the Baltic Sea. ICES J Mar Sci 64: 479–486.

58. Campbell LM, Thacker R, Barton D, Muir DCG, Greenwood D, et al. (2009) Re-engineering the eastern Lake Erie littoral food web: the trophic function of non-indigenous Ponto-Caspian species. J Great Lakes Res 35: 224–231.

59. Grabowski M, Jażdżewski K, Konopacka A (2007) Alien Crustacea in Polish waters. Aquatic Invasions 2: 25–38.

60. Chen W, Bierbach D, Plath M, Streit B, Klaus S (2012) Distribution of amphipod communities in the Middle to Upper Rhine and five tributaries. BioInvasions Rec 1: 263–271.

61. Podraza P, Ehlert T, Roos P (2001) Erstnachweis von Echinogammarus trichiatus (Crustacea: Amphipoda) im Rhein. Lauterbornia: 41: 129–133.

62. Sih A, Christensen B (2001) Optimal diet theory: when does it work, and when and why does it fail? Anim Behav 61: 379–390.

63. Boets P, Lock K, Tempelman D, van Haaren T, Platvoet D, et al. (2012) First occurrence of the Ponto-Caspian amphipod Echinogammarus trichiatus (Martynov, 1932) (Crustacea: Gammaridae) in Belgium. BioInvasions Rec 1: 115–120.

64. Baldauf SA, Thünken T, Frommen JG, Bakker TC, Heupel O, et al. (2007) Infection with an acanthocephalan manipulates an amphipod's reaction to a fish predator's odours. Int J Parasitol 37: 61–65.

65. Van Riel MC, Van der Velde G, and Bij de Vaate A (2003): *Pomphorhynchus* spec. (Acanthocephala) uses the invasive amphipod *Chelicorophium curvispinum* (G.O. Sars, 1895) as an intermediate host in the river Rhine. Crustaceana 7: 241–246.

66. Kvach J, Stepien CA (2008) Metazoan parasites of introduced Round and Tubenose Gobies in the Great Lakes: Support for the "Enemy Release Hypothesis". J Great Lakes Res 34: 23–35.

67. Ondračková M, Dávidová M, Pečínková M, Blažek R, Gelnar M, et al. (2005) Metazoan parasites of *Neogobius* fishes in the Slovak section of the River Danube. J Appl Ichthyol 21: 345–349.

68. Kvach J, Skóra KE (2007) Metazoa parasites of the invasive round goby *Apollonia melanostoma* (*Neogobius melanostomus*) (Pallas) (Gobiidae: Osteichthyes) in the Gulf of Gdańsk, Baltic Sea, Poland: a comparison with the Black Sea. Parasitol Res 100: 767–774.

69. Francová K, Ondračová M, Polačik M, Jurajda P (2011) Parasite fauna of native and non-native populations of *Neogobius melanostomus* (Pallas, 1814) (Gobiidae) in the longitudinal profile of the Danube River. J Appl Ichthyol 27: 879–886.

70. Kvach Y (2005) A comparative analysis of helminth faunas and infection of ten species of gobiid fishes (Actinopterigii: Gobiidae) from the North-Western Black Sea. Acta Ichthyol Piscat 35: 103–110.

71. Borcherding J, Staas S, Krüger S, Ondračková M, Ślapansky L, et al. (2011) Non-native Gobiid species in the lower River Rhine (Germany): recent range extensions and densities. A review of Gobiid expansion along the Danube-Rhine corridor – geopolitical change as a driver for invasion. J Appl Ichthyol 27: 153–155.

72. Roche KF, Janač M, Jurajda P (2013) A review of Gobiid expansion along the Danube-Rhine corridor – geopolitical change as a driver for invasion. Knowl Manag Aquat Ec 411: 01.

73. Gendron AD, Marcogliese DJ, Thomas M (2012) Invasive species are less parasitized than native competitors, but for how long? The case of the round goby in the Great Lakes-St. Lawrence Basin. Biol Invasions 14: 367–384.

74. Woo PTK, Buchmann K (2012). Fish Parasites. Pathobiology and Protection. CABI.

75. Nachev M, Ondračková M, Severin S, Ercan F, Sures B (2010) The impact of invasive gobies on the local parasite fauna of the family percidae and the gudgeon (*Gobio gobio*) in the Rhine River. In: Tagungsband der Deutschen Gesellschaft für Protozoologie und Parasitologie 2010.

76. Hine PM, Kennedy CR (1974) Observations on the distribution, specificity and pathogenicity of the acanthocephalan *Pomphorhynchus laevis* (Müller). J Fish Biol 6: 521–535.

77. Dezfuli BS, Castaldelly G, Bo T, Lorenzoni M, Giari L (2011) Intestinal immune response of *Silurus glanis* and *Barbus barbus* naturally infected with *Pomphorhynchus laevis* (Acanthocephala). Parasite Immunol 33: 116–123.

78. Moravec F (2013) Parasitic Nematodes of Freshwater fishes of Europe. Academia Praha, 264–284.

79. Kopp K, Jokela J (2007) Resistant invaders can convey benefits to native species. Oikos 116: 295–301.

80. Telfer S, Bown KJ, Sekules R, Begon M, Hayden T, et al. (2005) Disruption of a host-parasite system following the introduction of an exotic host species. Parasitology 130: 661–668.

81. Brodniewicz I (1968) On glochidia of the genera Unio and Anodonta from the quaternary fresh-water sediments of Poland. Acta Palaeontol Pol XIII: 619–631.

82. Zieritz A, Gum B, Kuehn R, Geist J (2012) Identifying freshwater mussels (Unionoida) and parasitic glochidia larvae from host fish gills: a molecular key to the North and Central European species. Ecol Evol 2: 740–750.

83. Taeubert JE, Gum B, Geist J (2012) Host-specificity of the endangered thick-shelled river mussel (*Unio crassus*, Philipsson 1788) and implications for conservation. Aquat Conserv 22: 36–46.

84. Poulin R. (1996) Helminth growth in vertebrate hosts: Does host sex matter? Int J Parasitol, 2: 1311–1315.

85. Robinson SA, Forbes MR, Hebert CE, McLauglin JD (2010) Male biased parasitism in cormorants and relationships with foraging ecology on Lake Erie, Canada. Waterbirds, 33: 307–313.

86. Koehler A, Hildenbrand P, Schleucher E, Riesch R, Arias-Rodriguez L, et al. (2011) Effects of male sexual harassment on female time budgets, feeding behavior, and metabolic rates in a tropical livebearing fish (*Poecilia mexicana*). Behav Ecol Sociobiol 65: 1513–1523.

87. Scharnweber K, Plath M, Tobler M (2011) Trophic niche segregation between the sexes in two species of livebearing fishes (Poeciliidae). Bull Fish Biol 13: 11–20.

88. Charlebois PM, Marsden JE, Goettel RG, Wolfe RK, Jude DJ, et al. (1997) The round goby, *Neogobius melanostomus* (Pallas): a review of European and North American literature. Illinois-Indiana Sea Grant Program and Illinois Natural History Survey. INHS Special Publication No.20.

# 4

# Sequential Cross-Species Chromosome Painting among River Buffalo, Cattle, Sheep and Goat: A Useful Tool for Chromosome Abnormalities Diagnosis within the Family Bovidae

**Alfredo Pauciullo**[1]*, **Angela Perucatti**[1], **Gianfranco Cosenza**[2], **Alessandra Iannuzzi**[1], **Domenico Incarnato**[1], **Viviana Genualdo**[1], **Dino Di Berardino**[2], **Leopoldo Iannuzzi**[1]

**1** Institute for Animal Production System in Mediterranean Environment, National Research Council, Naples, Italy, **2** Department of Agriculture, University of Naples Federico II, Portici, Italy

## Abstract

The main goal of this study was to develop a comparative multi-colour Zoo-FISH on domestic ruminants metaphases using a combination of whole chromosome and sub-chromosomal painting probes obtained from the river buffalo species (*Bubalus bubalis*, 2n = 50,XY). A total of 13 DNA probes were obtained through chromosome microdissection and DOP-PCR amplification, labelled with two fluorochromes and sequentially hybridized on river buffalo, cattle (*Bos taurus*, 2n = 60,XY), sheep (*Ovis aries*, 2n = 54,XY) and goat (*Capra hircus*, 2n = 60,XY) metaphases. The same set of paintings were then hybridized on bovine secondary oocytes to test their potential use for aneuploidy detection during *in vitro* maturation. FISH showed excellent specificity on metaphases and interphase nuclei of all the investigated species. Eight pairs of chromosomes were simultaneously identified in buffalo, whereas the same set of probes covered 13 out 30 chromosome pairs in the bovine and goat karyotypes and 40% of the sheep karyotype (11 out of 27 chromosome pairs). This result allowed development of the first comparative M-FISH karyotype within the domestic ruminants. The molecular resolution of complex karyotypes by FISH is particularly useful for the small chromosomes, whose similarity in the banding patterns makes their identification very difficult. The M-FISH karyotype also represents a practical tool for structural and numerical chromosome abnormalities diagnosis. In this regard, the successful hybridization on bovine secondary oocytes confirmed the potential use of this set of probes for the simultaneous identification on the same germ cell of 12 chromosome aneuploidies. This is a fundamental result for monitoring the reproductive health of the domestic animals in relation to management errors and/or environmental hazards.

**Editor:** Qinghua Shi, University of Science and Technology of China, China

**Funding:** This research was financially supported by CISIA-VARIGEAV project, National Research Council (CNR) of Italy. The funders had no role in study design, data collection and analysis, decision to publish, or preparation of the manuscript.

**Competing Interests:** All authors have declared that no competing interests exist.

* Email: alfredo.pauciullo@cnr.it

## Introduction

One of the main goals of cytogeneticists is the characterization of chromosomes by simple, rapid and reliable approaches. In this regard, the classical banding techniques are still the most used procedures since they represent standard and well established karyotyping methods. This is particularly true for the farm animal populations, whose routine cytogenetic analysis has been performed mainly by the application of classical methods [1].

Despite their wide application, several technical restrictions characterize the classical banding techniques [2] among which the size variations in a chromosomal band or the chromosome itself require a deep knowledge of the banding pattern to resolve complex karyotypes. However, in the last decade, the development of molecular cytogenetic techniques based on fluorescence *in situ* hybridization (FISH) led to a significant improvement in the accuracy of cytogenetic investigation, representing a valid alternative to the standard methods. In humans, the achievement of 24 colour FISH-based karyotyping (M-FISH, SKY, COBRA) [3,4,5] was the culmination of this technological progress.

Further advancements were reached in humans with chromosome arm-specific [4,6], region-specific [7,8], centromeric [9] and sub-telomeric probe sets [10,11], until arriving to the recent karyomapping [12] which offers a very fine clinical investigation for chromosome imbalances and miscarriage detections.

The applications of FISH techniques in farm animals and humans are very similar and approximately the same level of advancement was reached from animal cytogeneticists in the last two decades. The use of chromosome paintings and DNA probes in domestic animals allowed several important questions to be resolved, including a) detection of chromosome aberrations and complex karyotypes; b) gene mapping and comparative mapping; c) identification of conserved syntenic blocks between species and d) description of chromosome evolution [13,14].

Contrary to humans, the use of multi-colour FISH is still very limited in animal cytogenetics. This is mainly due to the lack of existing commercial probes which are essentially limited to sex chromosomes for most of the domestic species and to two autosomes in cattle for the detection of rob (1; 29) translocation. In addition, only some laboratories of excellence in the world have probes availability [14], thus limiting the application of the method to few research groups. For instance, Kubickova et al. [15] proposed a tri-colour FISH to resolve rob (14; 20) and rob (16; 20) translocations in cattle, later seven-colour FISH using a specific paint pool was used in camels to identify smaller chromosomes [16], whereas recently a pool of 13 chromosome-specific painting probes were used to develop a sequential multi-colour FISH in river buffalo to quickly identify submetacentric chromosomes and gonosomes [17].

The aim of this investigation was to use a combination of whole chromosome and sub-chromosomal painting probes derived from river buffalo (*Bubalus bubalis*, river type, 2n = 50,XY) in a comparative multi-colour Zoo-FISH study on domestic ruminants such as cattle (*Bos taurus* L., 2n = 60,XY), sheep (*Ovis aries* L., 2n = 54,XY) and goats (*Capra hircus* L., 2n = 60,XY) to develop the first comparative M-FISH karyotype. In addition, we report on the application of this pool of probes on bovine secondary oocytes as potential tool for aneuploidy detection during *in vitro* maturation.

## Materials and Methods

### Ethics statements

Procedures were in accordance with the ethical standards of the national ethics committee on research on animal science of 7th June 2011. All institutional and national guidelines for the care and use of laboratory animals were followed. The protocol was approved by the Committee on the Ethics of Animal experiments of the CNR-ISPAAM (Permit Number: 0000391-18/03/2014).

### Cell cultures

Peripheral blood cultures from eight clinically healthy adult males (two river buffalo bulls, two cattle bulls, two goat rams and two sheep rams) reared in southern Italy were performed following the method described by Iannuzzi and Di Berardino [1]. Four replicates for each sample were prepared according to the conventional cultures protocol and subjected to 20 min of colcemid (0.05 µg/ml) treatment, followed by centrifugation steps, hypotonic (KCl 75 mM) and fixative methanol/glacial acetic acid (3:1) treatments.

### In vitro maturation of COCs and oocyte fixation

Ovaries were collected from two slaughtered bovine cows and transported to the laboratory within two hours. Cumulus-oocyte complexes (COCs) were collected through aspiration with 21-gauge needles, washed in TCM-199 (Sigma, USA), and examined on Petri dishes under a stereomicroscope. Only oocytes with

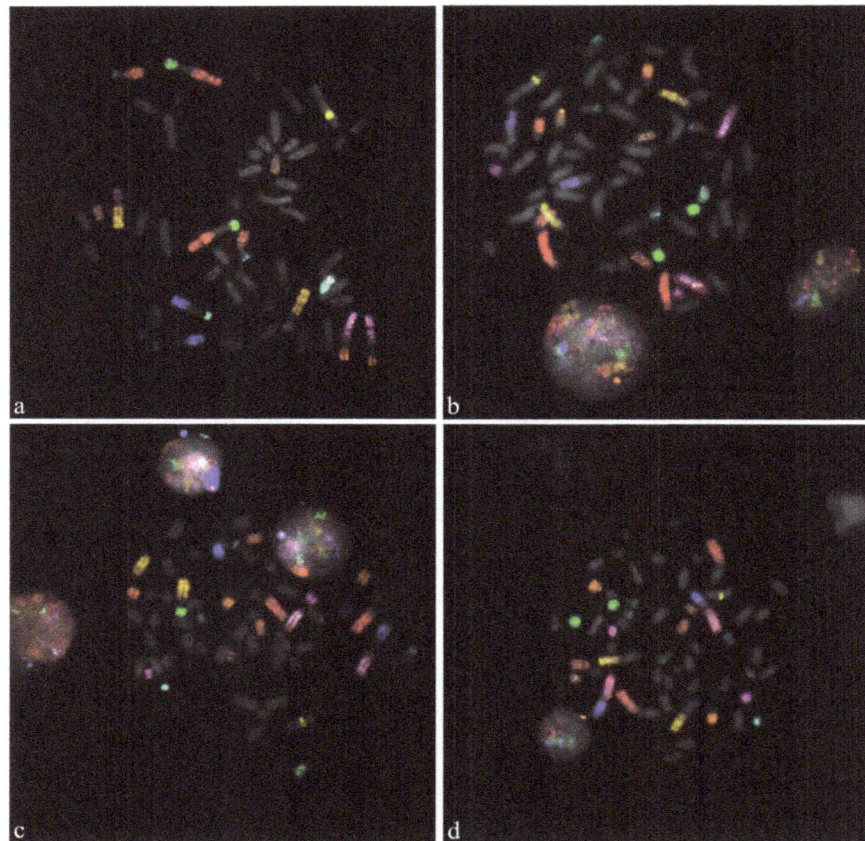

**Figure 1. Sequential multicolour hybridization with 13 river buffalo DNA probes on normal males metaphase spreads.** Specific signals were clearly detected on: a) river buffalo (*Bubalus bubalis*, 2n = 50) mitosis used as control; b) cattle (*Bos taurus*, 2n = 60); c) goat (*Capra hircus*, 2n = 60) and d) sheep (*Ovis aries*, 2n = 54) mitosis in Zoo-FISH experiments.

| FISH | River buffalo (2n=50) | Cattle (2n=60) | Sheep (2n=54) | Goat (2n=60) | Imposed colour |
|------|------|------|------|------|------|
| 1 | 1p | 27 | 26 | 27 | |
|   | 1q | 1 | 1q | 1 | |
| 2 | 2p | 23 | 20 | 23 | |
|   | 2q | 2 | 2q | 2 | |
|   | 18 | 18 | 14 | 18 | |
| 3 | 3p | 19 | 11 | 19 | |
|   | 3q | 8 | 2p | 8 | |
| 4 | 4p | 28 | 25 | 28 | |
|   | 4q | 5 | 3q | 5 | |
| 5 | 5p | 29 | 21 | 29 | |
|   | 5q | 16 | 12 | 16 | |
| 6 | X | X | X | X | |
|   | Y | Y | Y | Y | |

**Figure 2. Round of FISH, corresponding homologous chromosomes in river buffalo, cattle, sheep and goat (from ISCNDB, 2000) and super imposed colour for the 13 chromosome/arm specific painting probes.**

several compact cumulus cell layers and good morphology were selected for the maturation procedure. Groups of oocytes selected from each donor were transferred into 50-mL droplets of maturation medium consisting of TCM-199+10% foetal bovine serum (Gibco), supplemented with 0.5 mg/mL follicle-stimulating hormone (FSH) (Sigma), 5 mg/mL luteinizing hormone (LH) (Sigma), covered with sterile mineral oil (Sigma) and allocated in a humidified atmosphere containing 5% $CO_2$ in air at 39°C for 24 h.

After 24 h maturation, the COCs were incubated for a few minutes in a 1 mg/mL hyaluronidase solution (Sigma) to remove the cumulus cells, washed in Phosphate Buffered Saline (PBS), and exposed to a hypotonic sodium citrate solution (0.8% w/v) for 3 min, followed by KCl (75 mM) treatment for 3 min. The fixation was carried out using cold methanol/glacial acetic acid (1:1) solution. Oocytes were individually fixed at the center of a pre-cleaned slide, air-dried, and kept at −20°C until analysis.

## Chromosome microdissection and painting probes preparations

For the production of probes via chromosome microdissection, the river buffalo fixed lymphocyte suspension was spread onto a pre cleaned 24×60 mm coverslip, air dried and then treated for GTG-banding. According to Pauciullo et al. [17], the probes corresponding to the biarmed pairs (from 1 to 5) were produced by dissecting out the centromeric area, to avoid unspecific repetitive amplification of the centromeric regions. The probe corresponding

to the X chromosome was produced by dissecting the region Xq21–25, analogous to the Xcen region of the bovine chromosome [18]. The probes corresponding to chromosomes 18 and Y were produced by scraping the entire chromosomes.

Briefly, each micro-needle used for microdissection was broken in a 0.2 ml tube containing a collection buffer made of 5X Sequenase reaction buffer (Affimetrix, USA) and water in a final volume of 3.4 µl. On average 15 copies of the same chromosome were collected in the each tube. All tubes underwent to topoisomerase I (10 U/µl) treatment at 37°C for 30 min followed by enzyme inactivation at 95°C per 10 min. Highly processive chromosomal amplification was accomplished by degenerate oligonucleotide primer and sequenase ver. 2.0 DNA polymerase (Affimetrix) through a primary DOP-PCR reaction carried out at 94°C for 1 min, 30°C for 1 min and 37°C for 2 min. The enzyme was diluted according to the manufacture's guidelines and added during the annealing step at every cycle of the reaction for the first 8 cycles. Further 40 cycles of PCR amplification were performed at 94°C for 1 min, 56°C for 1 min and 72°C for 2 min in a reaction volume of 50 µl made of 1X AmpliTaq buffer, 3.5 mM of $MgCl_2$, 1 pmol of primer, dNTPs each at 200 µM, 2.5 U of AmpliTaq DNA Polymerase (Applied Biosystem).

Each probe was labelled separately by using a secondary DOP-PCR using 2 µL of products from the first reaction as template. Labelling scheme was performed according to Pauciullo et al. [17], with Spectrum Orange-dUTP and Spectrum Green-dUTP (Abbott, USA).

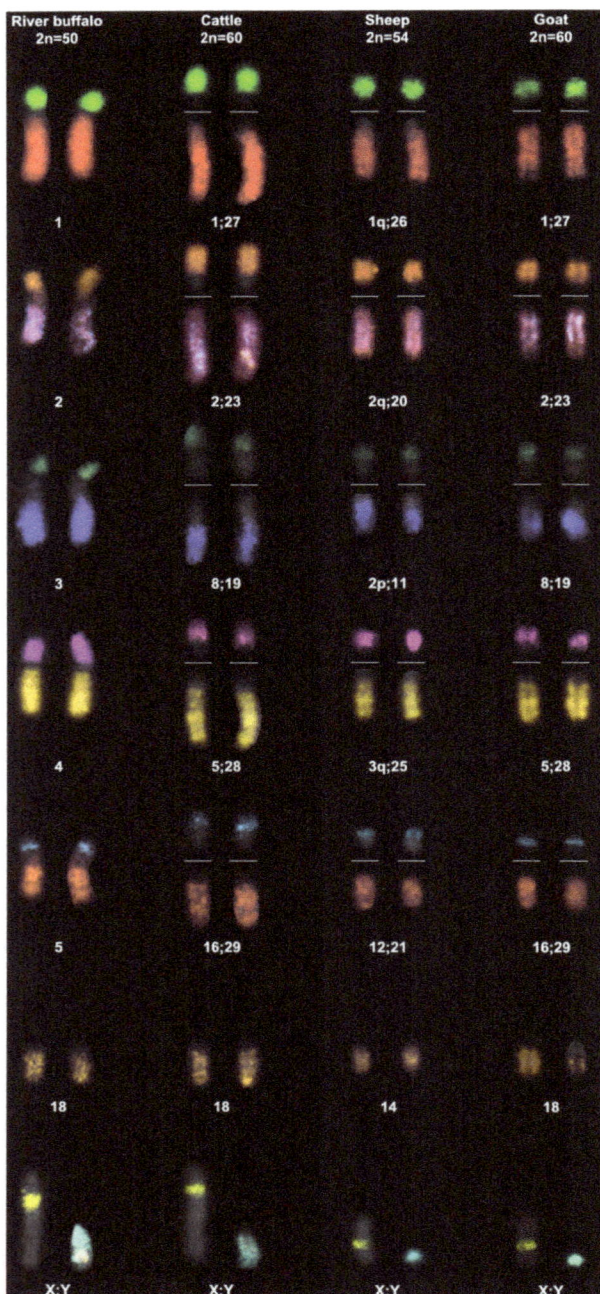

**Figure 3. Comparative M-FISH karyotypes generated from the metaphases of the figure 1.** River buffalo was taken as reference to build the partial karyotype limited to 13 DNA probes. Homologous chromosomes show the same colour among the investigated species. The buffalo Y-chromosome shows a hybrid signal (white) as result of the cross-hybridization of two sex painting probes. This chromosomal band corresponds to the pseudo-autosomal region.

## Fluorescent in situ hybridization (FISH)

Six sequential rounds of FISH were performed on the same slide. Each round was realized by using two probes simultaneously hybridized on the metaphase plate according to Pauciullo et al. [17], with the exception of the second FISH round in which 3 probes (2p, 2q and 18) were used simultaneously. The labeled probes were mixed and each precipitated in absolute ethanol

together with 10 µg salmon sperm DNA and 10 µg calf thymus DNA (both from Sigma). The pellets were vacuum-dried and then resuspended in 15 µl hybridization solution (50% formamide in 2X SSC+10% dextran sulfate) for 1 h at 37°C. The probe solutions were denatured for 10 min at 75°C and pre-hybridized for 60 min at 37°C.

Metaphase preparations were denatured for 3 min in a solution of 70% formamide in 2X SSC (pH 7.0) at 75°C. The slides were hybridized in a moist chamber at 37°C overnight. After hybridization, coverslips were removed by a gentle washing step in 2X SCC. The slides were then washed 3×4 min in washing solution (50% formamide in 2X SSC) at 42°C, followed by 3 additional washing steps for 4 min in 2X SSC at 42°C. Slides were counterstained with DAPI (4,6-diamidino-2-phenylindole) solution (0.24 µg/ml; Sigma) in Antifade (Vector Lab).

The slides were observed at 100x magnification with a Leica DM5500 fluorescence microscope equipped with DAPI, FITC, Spectrum orange specific filters, the FITC/Spectrum orange double filter, and provided with a Cytovision MB 8 image-analysis system (Leica Microsystems, Wetzlar, Germany). Digital images were captured in gray-scale, whereas false colours were created by the image-analyzing system for a reliable evaluation of the painting probes. Approximately 25–30 metaphases were acquired for each slide.

At the end of each round of FISH, the oil for microscope observation was removed from the coverslips and the slides were washed 2×15 min in PBST in a gently shacking, then air dried and immediately reused in the denaturation step for the next round of FISH.

## Results and Discussion

Thirteen chromosome-specific painting probes, generated from river buffalo metaphases via chromosome microdissection and the DOP-PCR procedure were sequentially hybridized on river buffalo, cattle, sheep and goat metaphases in a multi-colour zoo FISH experiment.

Typically 25–30 metaphase spreads per species were analysed. The DNA probes showed excellent specificity on buffalo mitosis, and the cross hybridization revealed very clearly FISH painting signals in the metaphases and interphase nuclei of all the investigated species (Figure 1). Eight pairs of chromosomes corresponding to the 5 sub-metacentric, two gonosomes and chromosome 18 were simultaneously identified in river buffalo. The same set of probes in the cross-species hybridization experiments covered nearly half of the bovine and goat karyotypes (13 out 30 chromosome pairs), and 40% of the sheep karyotype (11 out 27 chromosome pairs). These results are summarized in Figure 2. Although this set of probes only partially covers the chromosomal make-up of the investigated species, it allowed the first comparative M-FISH karyotype within the domestic ruminants to be developed (Figure 3).

The chromosomal comparison of different species and the detection of similarities between them is not new. For example, extensive comparative studies have taken place between human and cattle [19], whereas painting probes prepared from flow-sorted chromosomes and made available from the laboratory of Ferguson-Smith have been used in comparative studies in a number of species including human, mouse, pig, cattle, dog, lemurs, Indian and Chinese muntjacs, brown brocket deer, chicken, etc…[13]. Correlation between cytogenetic and gene mapping data is amply shown within the Bovidae (cattle, sheep, goat and buffalo), where similarities in banding patterns are a strong indication of homology at the DNA level [20]. However,

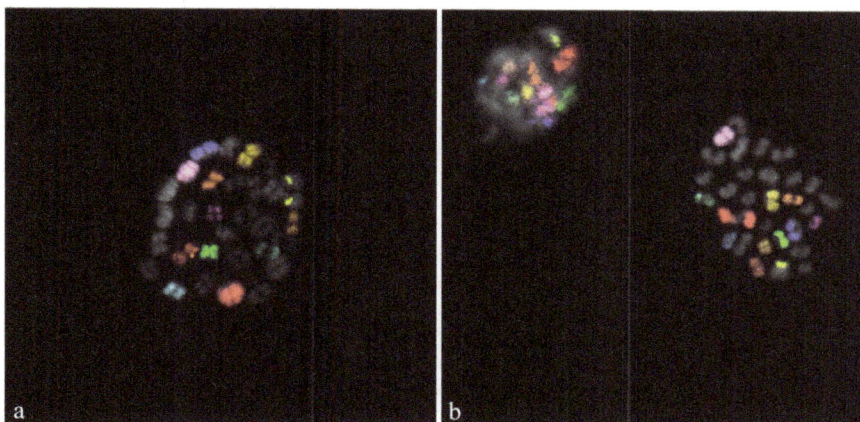

**Figure 4. M-FISH carried out on bovine *in vitro* matured secondary oocytes.** Specific fluorescent signals were identified on: a) oocyte at the diakinesis/metaphase I stage of the meiosis; b) oocyte at MII and corresponding PB I. Correct chromosomal segregation can be clearly indicated for 11 autosomes and X chromosome.

sometimes banding patterns can be of little consequence even though the species may belong to the same family (e.g. the species in the family Equidae). The comparative chromosome painting has proved to be an ideal alternative to bypass these problems [13] and several misleading conclusions from earlier Giemsa banding have been refuted by cross-species painting. For example, the conclusion that nucleolus-organizing chromosomes were shared between lesser apes and Old World monkeys was found to be incorrect [21]. The molecular resolution of complex karyotypes by the use of FISH is very helpful within the domestic bovids. This is largely true for the small chromosomes, whose similarity in the banding patterns makes their identification difficult, like those pointed out by the changes in chromosome nomenclature [22,23,24] before the approval of the International standards [25].

The use of the FISH with chromosome-specific probes removes any ambiguity in chromosome identification and improves the accuracy of diagnosis for Robertsonian translocations, fusions and more difficult structural rearrangements like reciprocal translocations and inversions. For example, the Robertsonian translocations in the cattle karyotype ($2n = 60$ with all acrocentric autosomes) might be easily detected by the classical banding methods, in a similar manner to the discovery of the first rob (1; 29) [26]. However, the use of painting probes, BAC probes and molecular markers resolved more complicated cases like the revision of rob (6; 8) and rob (26; 29) [27], or the recent identification of two new rob(14; 17) and rob(21; 23) translocations [28,29].

The use of classical banding techniques also complicates the identification of chromosomes involved in the fusions (especially in the case of small acrocentric chromosomes), as well as the detection of reciprocal translocations. Bovine whole chromosome paintings were instead successfully used for the identification of a rcp (2; 5) in a mosaic pattern in the Blonde d'Aquitaine breed [30] and recently a rcp (2; 4) (q45; q34) was detected in an Ayrshire bull by Switonski et al. [31]. Inversions are generally complicated to be detected by whole chromosome paintings. Although pericentric inversions could be revealed if appropriate arm specific probes are used, the combination with direct banding would be preferable to maximize the cytogenetic information obtainable [4]. Furthermore, the utility of M-FISH can be also extended to the detection of cryptic aberrations, which is routinely problematic for the animal cytogeneticists, such as telomeric translocations, interstitial deletions, duplications, etc...

The chromosomal identification by classical cytogenetic methods is not suitable also in the analysis of the meiotic metaphases, whose arrangement is usually evaluated in relation of numerical abnormalities. In this regard, the use of whole chromosome paintings in multicolour experiments became extremely useful in the application of preconception genetic diagnosis procedure for the prediction of chromosomal aneuploidies in human secondary oocytes [32].

With the same rationale, the estimation of aneuploidy in the oocytes of the various domestic species and breeds can be considered as an essential step for improving the *in vitro* production of embryos destined for the embryo transfer industry, as well as for monitoring future trends of the reproductive health of domestic animals in relation to management errors and/or environmental hazards. In this perspective, the complete set of probes herein produced (with the obvious exception of the Y-probe) was hybridized on 20 bovine secondary oocytes matured *in vitro* to test their potential use for aneuploidy detection. The interpretation of the results is based on the consideration that the first polar body (PB I) is the mirror image of the secondary oocyte metaphase (MII), therefore the lack of any chromosome in the MII (nullisomy) has its counterpart in the corresponding PB, which therefore results disomic and viceversa.

Specific fluorescent signals were clearly identify for each chromosome in all investigated oocytes. Two out of 20 oocytes showed the presence of bivalents (Figure 4a). Although the pairing of the autosomal bivalents and the sex chromosomes is normal, the occurrence of tetrads and the absence of the corresponding PB I reveals the interruption of the *in vitro* maturation at the diakinesis/metaphase I stage of the meiosis (Figure 4a).

The remaining 18 oocytes underwent a normal meiotic division. Specific fluorescent signals were visible on both MII and corresponding PB I (Figure 4b), thus evidencing the correct chromosomal segregation and therefore the lack of abnormalities for the investigated oocytes.

The scarce availability of commercially available chromosome-specific probes -in domestic ruminants- is a limiting factor for the investigation of aneuploidy rates by FISH. This is particularly evident in cattle where only few chromosomes were investigated and few studies are available so far [33,34]. In addition, the application of other molecular methods like the comparative genomic hybridization (CGH) is not suitable for the analysis of bovine aneuploidies. In fact, differently to what observed in pig [35], the acrocentric nature of cattle autosomes hampers the chromosomal identification after CGH hybridization. As consequence, in this species the identification of specific gains/losses of chromosomal DNA can be detected only by FISH.

Although no abnormalities were detected in the investigated oocytes, these data confirm the potential use of the river buffalo probes for aneuploidy detection in germ cells, thus opening further opportunity of investigation for clinical cytogenetic applications also in the other species with difficult karyotype.

## Conclusions

A DNA collection made of 13 probes generated by chromosome microdissection and DOP-PCR was sequentially hybridised on river buffalo, cattle, sheep and goat metaphase spreads in cross-species hybridization experiments. Nearly half of the bovine and goat karyotypes (13 out 30 chromosome pairs), and 40% of the sheep karyotype (11 out of 27 chromosome pairs) were covered. This allowed the development -for the first time- a comparative M-FISH karyotype for the domestic bovids, which represents a fundamental step for the future achievement of: a) health screening programs of the breeds (highly productive, endangered, indigenous, etc...) related to these species on a molecular cytogenetic basis; b) rapid identification of simple and complex chromosomal

rearrangements; c) cross-species hybridization experiments within the family Bovidae and more generally, for comparative evolutionary studies with species of other families; d) resolution of complex karyotypes with particular regard to the detection of hybrid animals; e) evaluation of the aneuploidy level in germ cells as tool for the monitoring the reproductive health of animals in relation to management errors (hormonal imbalances, nutritional and diet mistakes) and/or environmental hazards (mutagens, mitotic poisons) which are known to damage the mitotic/meiotic machinery of the cell.

## Acknowledgments

We thank the owners of the farms and Dr. Giuseppe Auriemma and Giuseppe Grazioli for the possibility of sampling and for providing information on animals. We also thank M.Sc. Lorenzo Pucciarelli for the technical support in the lymphocyte cell cultures.

## Author Contributions

Conceived and designed the experiments: A. Pauciullo DD LI. Performed the experiments: A. Pauciullo A. Perucatti VG DI. Analyzed the data: A. Pauciullo A. Perucatti AI VG. Contributed reagents/materials/analysis tools: A. Pauciullo GC LI. Contributed to the writing of the manuscript: A. Pauciullo A. Perucatti. Revised the article critically for important intellectual content: GC DD LI.

## References

1. Iannuzzi L, Di Berardino D (2008) Tools of the trade: diagnostic and research applied to domestic animal cytogenetics. J Appl Genet 49: 357–366.
2. Claussen U, Michel S, Mühlig P, Westermann M, Grummt UW, et al. (2002) Demystifying chromosome preparation and the implications for the concept of chromosome condensation during mitosis. Cytogenet Genome Res 98: 136–146.
3. Schröck E, du Manoir S, Veldman T, Schoell B, Wienberg J, et al. (1996) Multicolour spectral karyotyping of human chromosomes. Science 273: 494–497.
4. Speicher MR, Gwyn Ballard S, Ward DC (1996) Karyotyping human chromosomes by combinatorial multi-fluor FISH. Nat Genet 12: 368–375.
5. Tanke HJ, Wiegant J, van Gijlswijk RP, Bezrookove V, Pattenier H, et al. (1999) New strategy for multi-colour fluorescence in situ hybridisation: COBRA: COmbined Binary RAtio labelling. Eur J Hum Genet 7: 2–11.
6. Karhu R, Ahlstedt-Soini M, Bittner M, Meltzer P, Trent JM, et al. (2001) Chromosome arm-specific multicolour FISH. Genes Chromosom Cancer 30: 105–109.
7. Muller S, O'Brien PC, Ferguson-Smith MA, Wienberg J (1998) Cross-species colour segmenting: a novel tool in human karyotype analysis. Cytometry 33: 445–452.
8. Chudoba I, Plesch A, Lorch T, Lemke J, Claussen U, et al. (1999) High resolution multicolour-banding: a new technique for refined FISH analysis of human chromosomes. Cytogenet Cell Genet 84: 156–160.
9. Henegariu O, Bray-Ward P, Artan S, Vance GH, Qumsyieh M, et al. (2001) Small marker chromosome identification in metaphase and interphase using centromeric multiplex fish (CMFISH). Lab Invest 81: 475–481.
10. Brown J, Saracoglu K, Uhrig S, Speicher MR, Eils R, et al. (2001) Subtelomeric chromosome rearrangements are detected using an innovative 12-colour FISH assay (M-TEL). Nat Med 7: 497–501.
11. Fauth C, Zhang H, Harabacz S, Brown J, Saracoglu K, et al. (2001) A new strategy for the detection of subtelomeric rearrangements. Hum Genet 109: 576–583.
12. Handyside AH, Harton GL, Mariani B, Thornhill AR, Affara N, et al. (2010) Karyomapping: a universal method for genome wide analysis of genetic disease based on mapping crossovers between parental haplotypes. J Med Genet 47: 651–658.
13. Ferguson-Smith MA, Yang F, O'Brien PCM (1998) Comparative mapping using chromosome sorting and painting. ILAR J 39: 68–76.
14. Rubes J, Pinton A, Bonnet-Garnier A, Fillon V, Musilova P, et al. (2009) Fluorescence in situ hybridization applied to domestic animal cytogenetics. Cytogenet Genome Res 126: 34–48.
15. Kubickova S, Cernohorska H, Musilova P, Rubes J (2002) The use of laser microdissection for the preparation of chromosome-specific painting probes in farm animals. Chromosome Res 10: 571–577.
16. Balmus G, Trifonov VA, Biltueva LS, O'Brien PCM, Alkalaeva ES, et al. (2007) Cross-species chromosome painting among camel, cattle, pig and human: further insights into the putative Cetartiodactyla ancestral karyotype. Chromosome Res 15: 499–514.

17. Pauciullo A, Perucatti A, Iannuzzi A, Incarnato D, Genualdo V, et al. (2014) Development of a sequential multicolor-FISH approach with 13 chromosome-specific painting probes for the rapid identification of river buffalo (Bubalus bubalis, 2n = 50) chromosomes. J Appl Genetics 55: 397–401.
18. Nicodemo D, Pauciullo A, Castello A, Roldan E, Gomendio M, et al. (2009) X-Y Sperm aneuploidy in 2 cattle (Bos taurus) breeds as determined by dual colour fluorescent in situ hybridization (FISH). Cytogenet Genome Res 126: 217–225.
19. Solinas-Toldo S, Lengauer C, Fries R (1995) Comparative genome map of human and cattle. Genomics 33: 214–219.
20. Di Meo GP, Goldammer T, Perucatti A, Genualdo V, Iannuzzi A, et al. (2011) Extended cytogenetic maps of sheep chromosome 1 and their cattle and river buffalo homoeologues: comparison with the OAR1 RH map and human chromosomes 2, 3, 21 and 1q. Cytogenet Genome Res 133: 16–24.
21. Stanyon R, Arnold N, Koehler U, Bigoni F, Wienberg J (1995) Chromosomal painting shows that "marked chromosomes" in lesser apes and Old World monkeys are not homologous and evolved by convergence. Cytogenet Cell Genet 68: 74–78.
22. Evans HJ, Buckland RA, Sumner AT (1973) Chromosome homology and heterochromatin in goat, sheep and ox studied by banding techniques. Chromosoma 42: 383–402.
23. Di Berardino D, Hayes H, Friers R, Long S (1990) ISCNDA 1989 - International System for Cytogenetic Nomenclature of Domestic Animals. Cytogenet Cell Genet 53: 65–79.
24. Iannuzzi L (1994) Standard karyotype of the river buffalo (Bubalus bubalis L., 2n = 50). Cytogenet Cell Genet 67: 102–113.
25. Di Berardino D, Di Meo GP, Gallagher DS, Hayes H, Iannuzzi L (2001) ISCNDB 2000 - International System for Chromosome Nomenclature of Domestic Bovids. CytogenetCell Genet 92: 283–299.
26. Gustavsson I, Rockborn G (1964) Chromosome abnormality in three cases of lymphatic leukaemia in cattle. Nature 203: 990.
27. Di Meo GP, Molteni L, Perucatti A, De Giovanni A, Incarnato D, et al. (2000) Chromosomal characterization of three centric fusion translocations in cattle using G-, R- and C-banding and FISH technique. Caryologia 53: 213–218.
28. De Lorenzi L, Molteni L, De Giovanni A, Parma P (2008) A new case of rob(14; 17) in cattle. Cytogenet Genome Res 120: 144–146.
29. De Lorenzi L, Molteni L, Denis C, Eggen A, Parma P (2008) A new case of centric fusion in cattle: rob(21; 23). Anim Genet 39: 454–455.
30. Pinton A, Ducos A, Yerle M (2003) Chromosomal rearrangements in cattle and pigs revealed by chromosome microdissection and chromosome painting. Genet Sel Evol 35: 685–696.
31. Switonski M, Andersson M, Nowacka-Woszuk J, Szczerbal I, Sosnowski J, et al. (2008) Identification of a new reciprocal translocation in an AI bull by synaptonemal complex analysis, followed by chromosome painting. Cytogenet Genome Res 121: 245–248.
32. Gianaroli L, Magli MC, Cavallini G, Crippa A, Capoti A, et al. (2010) Predicting aneuploidy in human oocytes: key factors which affect the meiotic process. Hum Reprod. 25: 2374–2386.

33. Nicodemo D, Pauciullo A, Cosenza G, Peretti V, Perucatti A, et al. (2010). Frequency of aneuploidy in in vitro-matured MII oocytes and corresponding first polar bodies in two dairy cattle (Bos taurus) breeds as determined by dual-colour fluorescent in situ hybridization. Theriogenology, 73: 523–529.

34. Pauciullo A, Nicodemo D, Cosenza G, Peretti V, Iannuzzi A, et al. (2012). Similar rate of chromosomal aberrant secondary oocytes in two indigenous cattle (*Bos taurus*) breeds as determined by dual-colour FISH. Theriogenology 77: 675–683.

35. Hornak M, Jeseta M, Musilova P, Pavlok A, Kubelka M, et al. (2011) Frequency of aneuploidy related to age in porcine oocytes. PLoSOne 6:e18892.

# 5

# Shifting Regimes and Changing Interactions in the Lake Washington, U.S.A., Plankton Community from 1962–1994

Tessa B. Francis[1]*, Elizabeth M. Wolkovich[2,3¤a], Mark D. Scheuerell[4], Stephen L. Katz[5¤b], Elizabeth E. Holmes[6], Stephanie E. Hampton[2¤c]

1 University of Washington Tacoma, Puget Sound Institute, Tacoma, Washington, United States of America, 2 National Center for Ecological Analysis and Synthesis, University of California Santa Barbara, Santa Barbara, California, United States of America, 3 The Biodiversity Research Centre, University of British Columbia, Vancouver, British Columbia, Canada, 4 Fish Ecology Division, Northwest Fisheries Science Center, National Marine Fisheries Service, National Oceanic and Atmospheric Administration, Seattle, Washington, United States of America, 5 Channel Islands National Marine Sanctuary, National Ocean Service, National Oceanic and Atmospheric Administration, Santa Barbara, California, United States of America, 6 Conservation Biology Division, Northwest Fisheries Science Center, National Marine Fisheries Service, National Oceanic and Atmospheric Administration, Seattle, Washington, United States of America

## Abstract

Understanding how changing climate, nutrient regimes, and invasive species shift food web structure is critically important in ecology. Most analytical approaches, however, assume static species interactions and environmental effects across time. Therefore, we applied multivariate autoregressive (MAR) models in a moving window context to test for shifting plankton community interactions and effects of environmental variables on plankton abundance in Lake Washington, U.S.A. from 1962–1994, following reduced nutrient loading in the 1960s and the rise of *Daphnia* in the 1970s. The moving-window MAR (mwMAR) approach showed shifts in the strengths of interactions between *Daphnia*, a dominant grazer, and other plankton taxa between a high nutrient, *Oscillatoria*-dominated regime and a low nutrient, *Daphnia*-dominated regime. The approach also highlighted the inhibiting influence of the cyanobacterium *Oscillatoria* on other plankton taxa in the community. Overall community stability was lowest during the period of elevated nutrient loading and *Oscillatoria* dominance. Despite recent warming of the lake, we found no evidence that anomalous temperatures impacted plankton abundance. Our results suggest mwMAR modeling is a useful approach that can be applied across diverse ecosystems, when questions involve shifting relationships within food webs, and among species and abiotic drivers.

**Editor:** Elliott Lee Hazen, UC Santa Cruz Department of Ecology and Evolutionary Biology, United States of America

**Funding:** This research was initiated while TBF held a National Research Council Research Associateship Award at NOAA's Northwest Fisheries Science Center, and conducted in part while TBF was employed by the Puget Sound Institute, funded by the Environmental Protection Agency (Grant #PC-00J303-01). This work was conducted in part while EMW was a postdoctoral associate at the National Center for Ecological Analysis and Synthesis, a Center funded by the National Science Foundation (Grant #EF-0553768), the University of California, Santa Barbara, and the State of California, and in part while she was a National Science Foundation Postdoctoral Research Fellow in Biology (DBI-0905806), and also while she was supported by the NSERC CREATE training program in biodiversity research. The funders had no role in study design, data collection and analysis, decision to publish, or preparation of the manuscript.

**Competing Interests:** The authors have declared that no competing interests exist.

\* Email: tessa@uw.edu

¤a Current address: Organismic and Evolutionary Biology, Harvard University, Cambridge, Massachusetts, United States of America
¤b Current address: School of the Environment, Washington State University, Pullman, Washington, United States of America
¤c Current address: Center for Environmental Research, Education and Outreach, Washington State University, Pullman, Washington, United States of America

## Introduction

One of the most important challenges facing ecologists is specifying how global change will affect community stability and the production of associated critical ecosystem services. Community stability is mediated by species interactions, which are sensitive to changing environmental conditions [1,2], and therefore estimating the effects of environmental drivers on food web dynamics is critical for understanding how anthropogenic forces have altered ecosystems and for anticipating further change [3,4]. Analyzing food web dynamics is complicated in part because the communities we observe are likely not in "equilibrium" as we might have once expected [5]. There is increasing evidence that

the structure of communities and the nature of species' responses to each other and to their environments are not static, but rather shift over time. In particular, anthropogenic pressures may be pushing communities further from equilibrium [6], with communities exhibiting a variety of non-equilibrium dynamics from smooth trends to abrupt step changes [7]. Changes in abiotic conditions of ecosystems can directly and indirectly affect food web structure [8]. Thus, food web models must account for diverse temporal changes in community dynamics. In some systems, while we may have a good understanding of average species interactions or effects of the environment on food web dynamics over key time periods, we may still lack important information about whether and how such dynamics changed over time in response to large shifts in the ecosystem.

Lake Washington, U.S.A., is an example of an aquatic ecosystem that experienced a series of well-described dramatic changes in its environmental conditions and plankton community in the mid-20th Century. This time period included a regime shift from one of high nutrient loading from sewage inputs to one of increased water clarity, as well as temperature and species abundance changes [9–11]. The lake also experienced shifting regimes in terms of plankton community dominance. During the era of high sewage inputs, the lake experienced extensive nuisance algal blooms, especially of the cyanobacterium, *Oscillatoria rubescens*. Following sewage diversion, water clarity increased substantially [12]; subsequently, the influential grazer *Daphnia* established in the lake [11] and *Oscillatoria* effectively disappeared from the record. In more recent years, warming temperatures have caused phenological changes in phytoplankton and zooplankton [10,13,14]. What is unclear is how these changes in the plankton community and abiotic conditions affected interactions within the food web concomitant with the changing environment. Such shifts in plankton community interactions – such as weakening of grazer effects on phytoplankton, or increased competition among grazing zooplankton guilds – would have consequences for higher trophic levels in lakes, as plankton provides an important component of the energetic support for some lacustrine fish [15], including in Lake Washington [16]. Moreover, plankton community structure and indirect effects of herbivore-plant interactions can influence fundamental lake characteristics such as light, temperature and water clarity [17,18]. In this paper, we introduce an extension of a well-used static food web model – a multivariate autoregressive (MAR) model [19–21] – to study Lake Washington's dynamically changing food web and ecosystem.

Over the last several decades, multivariate time-series methods have been used to estimate the strength and pattern of species interactions and the effect of abiotic drivers on communities [20,22]. MAR models provide a locally linear approximation of non-linear stochastic multispecies processes. They have been particularly useful in aquatic ecosystems and for understanding plankton dynamics in part because of the tight coupling between plankton and their environment. MAR models have also become useful in broader aquatic food web analyses [23,24], as they can incorporate multiple trophic levels and environmental drivers.

Prior implementations of MAR models have assumed that the interactions in the study system were unchanging over the time period encompassed by the data. This approach maximized the performance of parameter estimation given the properties of monitoring data, but only estimated the average interaction strengths over a time series. In contrast, if food web dynamics shift in response to changing drivers [25], then a better analytical approach would accommodate and capture this non-stationarity in modeling the food web. A suite of statistical methods can be applied to ecological time series to examine non-stationarity – such as shifts in abiotic conditions or periodicities – through time. Methods such as wavelets [26,27], single-spectrum [28] and breakpoint analyses have been used in climatology and paleoclimatology, and have also recently been applied to ecological data [29,30]. Such methods allow ecologists to see how abundances may be shifting [30] or how interactions among species may change over time in simple lab systems [29], but they do not provide a cohesive ecosystem approach to examining how integrated abiotic and biotic forces may change through time. In particular, food web responses to changes in the strength or nature of abiotic drivers would be predicted to cause cascading shifts in

the interactions among many members of a food web, and may also feed back to how community members respond to other environmental drivers. Examining such a suite of interactions and drivers, however, would require a model that analyzes all the variables at once, and that allows estimation of such shifts through time.

A running or moving window approach is another tool that has long been used in other disciplines, such as finance, to examine non-stationarity in time series. In this approach, consecutive and overlapping subsets of time series – or windows – are analyzed individually to detect changes through time in a historical record [31,32]. This approach has recently been used with univariate autoregressive models to develop leading indicators of regime change [33–35]. Here we offer an extension of the MAR model, which we term "moving-window MAR" (mwMAR), and we use it to examine a case of shifting species interactions and environmental effects on species through time. Our approach blends the community focus of the MAR model with the moving window approach of detecting historical changes in time-series data. We describe the mwMAR model and then apply the model to long-term monitoring data from Lake Washington, U.S.A., to show how interactions among dominant taxa of the plankton community shifted following sewage diversion. Because food webs show sensitivity to changes in their abiotic environment [6–8], we hypothesize that changes in the nutrient status, clarity, and dominant plankton taxa of the lake would cascade throughout the plankton food web, resulting in shifts in the direction and strength of community interactions, which would in turn affect community stability.

## Materials and Methods

### Model configuration

We estimated interaction strengths among phytoplankton and zooplankton guilds, environmental effects on phytoplankton and zooplankton abundance, plankton intrinsic growth rates, and plankton community stability in Lake Washington from 1962–1994 using multivariate autoregressive (MAR) models. MAR models are stochastic models describing changes in species abundance through time as a function of species interactions and environmental influences, while accounting for temporal autocorrelation in species abundances [20,36,37]. MAR models can also be used to estimate various metrics of community stability, such as return time to a stationary state following an ecosystem perturbation, or the distance away from a stationary state that an ecosystem can be pushed by a perturbation. Previous work has used MAR models to describe environmental effects on, and interactions among, lake phytoplankton and zooplankton [20,22,38,39], effects of climate regime shifts on interactions among marine plankton [40], causes of estuarine fish declines [24], and effects of fishing on marine food webs [23]. Extended descriptions of MAR approaches to time-series data have been given previously [19,20,37], so we provide only a brief review of the model structure here.

MAR models are written in matrix form as:

$$\mathbf{X}_t = \mathbf{A} + \mathbf{B}\mathbf{X}_{t-1} + \mathbf{C}\mathbf{U}_t + \mathbf{E}_t \qquad (\text{Eq.1})$$

where, for $p$ interacting species and $q$ environmental covariates, $\mathbf{X}_t$ is a $p \times 1$ vector of species abundances (here, natural log-transformed) at time $t$; $\mathbf{A}$ is a $p \times 1$ vector of constants, representing intrinsic per-capita growth rates; $\mathbf{B}$ is a $p \times p$ species interaction matrix, with off-diagonal elements describing inter-specific interactions, and diagonal elements describing intra-specific interactions (i.e., density-dependence); $\mathbf{C}$ is a $p \times q$ matrix with elements describing environmental effects on species abundance; $\mathbf{U}_t$ is a $q \times 1$ vector of environmental covariates at time $t$; and $\mathbf{E}_t$ is a $p \times 1$ vector of process errors at time $t$, representing environmental variation not otherwise accounted for in the model. $\mathbf{E}_t$ is distributed as a multivariate Normal with mean $\mathbf{0}$ and a diagonal variance matrix $\sum$. Elements of $\mathbf{B}$ and $\mathbf{C}$ typically range from -1 to 1, with distance from 0 representing increasing negative or positive interaction strength. The diagonal elements of $\mathbf{B}$ typically range from 0 to 1, with values closer to 0 representing higher density dependence.

We also used MAR models to estimate community stability. Specifically, we estimated the rate at which the system returns to its stationary distribution following a disturbance by the maximum eigenvalue of the $\mathbf{B}$ matrix (that maximum eigenvalue is henceforth referred to as lambda, $\lambda$). Systems with values of $\lambda$ closer to 0 are considered to be more stable because they tend to return to equilibrium conditions faster than systems with values of $\lambda$ farther from 0 [21].

MAR models estimate mean intrinsic growth rates (captured by the $\mathbf{A}$ vector), community interactions (captured by the $\mathbf{B}$ matrix), environmental effects (captured by the $\mathbf{C}$ matrix), and community stability (captured by $\lambda$) across a given time series [20]. Here we use MAR models to quantify changes in interactions through time, by modeling community interactions for overlapping subsets of a time series, or moving "windows" of time, thereby estimating trends in MAR parameters. For a $p \times n$ matrix $\mathbf{X}$ of time series observations consisting of successive $p \times 1$ vectors $\mathbf{X}_1$, $\mathbf{X}_2$,..., $\mathbf{X}_n$, and a moving window of size $W < n$, we estimated MAR parameters within $n-W-1$ successive windows. These windows contained data from $\mathbf{X}_2 : \mathbf{X}_{W+1}$, $\mathbf{X}_3 : \mathbf{X}_{W+2}$,..., $\mathbf{X}_{n-W+1} : \mathbf{X}_n$. Note that the time series starts at $t = 2$ to allow for the lag-1 effect in Eq. 1. The output of the mwMAR analysis is a new time series of estimated MAR parameters.

## Lake Washington data and analysis

To investigate changes in interactions among zooplankton and phytoplankton guilds and the effects of environmental covariates in Lake Washington through time, we implemented the mwMAR approach using monthly plankton and environmental data from Lake Washington (Washington, U.S.A.) spanning 1962 to 1994 (396 timesteps; see Figure S1 for plankton time series). Our 33-year time series begins in the year of maximum sewage input (1962) when the lake experienced extensive nuisance algal blooms, especially of the cyanobacterium, *Oscillatoria rubescens*. Sewage diversion began the following year (1963), and was completed in 1968. Water clarity increased substantially by 1971 [12] and continued to improve through 1976, when the influential grazer *Daphnia* established in the lake [11] and *Oscillatoria* abundance decreased dramatically. Despite low abundances at times, and periods when they were not observe in samples, neither *Daphnia* nor *Oscillatoria* ever technically went extinct in Lake Washington. Before they begun to be observed at high abundances in 1973, *Daphnia* were observed every year but one (1971). Likewise, after their period of dominance ended in 1980, *Oscillatoria* continued (and continue) to appear in plankton samples, appearing in all but 3 years between 1980–1994.

The lake has additionally undergone significant warming throughout the historical record [10], which has altered the timing of zooplankton abundance cycles [14,41]. Recent work, however, suggests species and nutrient (phosphorus) shifts related to the sewage effluent have had a stronger influence on the lake than shifts associated with warming [42]. These well-documented shifts in environmental drivers and plankton dynamics make Lake

Washington an ideal ecosystem for evaluating the mwMAR model's sensitivity to non-stationary process. Indeed, the dominant environmental drivers and species interactions in Lake Washington are well-studied via observational [9,12], experimental [43,44] and traditional MAR approaches [39,45], offering the necessary background to build informed community and environmental interaction matrices ($\mathbf{B}$ and $\mathbf{C}$ matrices, respectively).

For our analyses we aggregated physical, chemical and plankton community data, which were collected at various intervals, into monthly means. Previous analyses of the Lake Washington plankton community interactions identified a simplified food web containing species that demonstrated strong roles in structuring the community [39,45]. We targeted the most strongly-interacting taxa of this simplified food web with the present analysis, to determine how the dominant interactions changed through time. While weak species interactions can be important in structuring food webs, we chose to focus on the dominant taxa and interactions as a first test of this new method. These taxa were pooled into four taxonomic groups: diatoms and green algae – "DG," both palatable food for grazing zooplankton; *Oscillatoria* – known to suppress *Daphnia* [44]; *Daphnia;* and non-daphnid and non-cladoceran crustaceans – "NDC," comprised of non-daphnid cladocerans, *Cyclops*, and *Diaptomus*. Group abundance data were log-transformed to better capture non-linearities [20]. A more complete description of the data is available in Hampton et al. [39], and the raw data are available in Appendix S1.

We included as covariates in the mwMAR model surface temperature and total phosphorus, because they were previously identified as the strongest environmental drivers of plankton abundance in the lake [39,45]. However, rather than simply use temperature as a covariate by itself, we instead used the data to estimate (1) a mean monthly signal indicative of long-term seasonal forcing, and (2) monthly deviations from the mean to capture short-term anomalies (e.g., a particularly warm July) or long-term trends (e.g., an overall increase). To ease comparison of effect sizes across all environmental covariates, we standardized all covariate data to a mean of 0 and a standard deviation of 1.

For our environmental covariate matrix ($\mathbf{C}$) we included *a priori* only biologically meaningful interactions based on established environmental relationships: we assumed total phosphorus could not directly affect *Daphnia* or other zooplankton taxa. We expected shifts in mwMAR coefficients to lag behind known dates of change in the biotic community, sewage diversion and water clarity because our moving window size (7 years) is much larger than the timescale of most known changes. We graphically present all data at the end year of the moving window; thus, in our figures, results based on data from 1963–1970 would appear on the x-axis at year 1970.

## Sensitivity analysis

The accuracy and precision of parameter estimates by the mwMAR model, as with other statistical methods, are sensitive to and affected by multiple factors, including food web configuration (i.e., the number of interacting species and covariates), window size, the variance structure of the process errors, and outliers in the data (see Appendix S2 for discussion and additional model validation). We conducted several sensitivity analyses to ensure such factors were not influencing the mwMAR model estimates. For example, the Lake Washington dataset is of high quality, and our outlier inspection showed no influence of outliers on the final results. In addition, because there is a tradeoff between precision of parameter estimates and accuracy of those estimates that is defined by window size, we conducted tests using simulated time series based on the Lake Washington food web configuration, to determine the appropriate window size for analysis of the Lake Washington dataset. In those simulations, the parameter estimation accuracy decreased sharply at window sizes smaller than 75 time steps (see Figure S1), and therefore we use a window size of 84 (the next factor of 12 larger than 75, given the monthly time step in the Lake Washington data). We also conducted simulations to determine what bias, if any, exist in parameter estimates during periods when a system is undergoing transition between states, for example between a eutrophic and clear-water state as was the case with Lake Washington. Last, to ensure that the mwMAR model was capable of capturing shifts in species interactions and environmental conditions outside of the Lake Washington case study, we fit mwMAR models to simulated time series with known interactions (see Appendix S2).

## Statistical programming

MAR and mwMAR modeling was done in MATLAB (2007, The MathWorks), using the open-source program LAMBDA ([46]; freely available from http://conserver.iugo-cafe.org/user/e2holmes/LAMBDA) with additional programming by the authors. The coefficients of the **A**, **B** and **C** matrices were estimated using conditional least squares (CLS), and confidence intervals around each coefficient were established using 2,000 bootstrapped data sets. Each bootstrapped data set was generated by creating random **E** matrices and fitting the rest of the parameters using CLS (see [21] for details).

## Results

The mwMAR approach revealed changes in interaction strengths in the Lake Washington plankton community between 1962 and 1994 (Figures 1–5; Figures S2–S3). For example, there were changes in the effects of *Oscillatoria* on *Daphnia* and diatoms and green algae (DG) coincident with the community composition shift during which *Oscillatoria* abundance decreased and *Daphnia* appeared (Figure 1, Table 1). In the period following the first appearance of *Daphnia* in Lake Washington, the effect of *Oscillatoria* on *Daphnia* became increasingly negative and was strongest in 1976 (Figure 1A). Following the decrease in *Oscillatoria* abundance, the negative effect of *Oscillatoria* on *Daphnia* weakened, and there was no significant effect of *Oscillatoria* on *Daphnia* from late in 1982 until the end of the time series. There was no effect of *Daphnia* on *Oscillatoria* (Figure 1B) until after the decline in *Oscillatoria* and increase in *Daphnia*. By 1980, the interaction coefficient became negative, weakened in the late 1980s, and returned to neutral after 1990. *Oscillatoria* also had a negative effect on DG in the beginning of the time series, and this effect disappeared by the mid-1970s (Figure 1C).

The effects of *Daphnia* on other plankton groups in Lake Washington also varied through time (Figure 2). *Daphnia* had a negative effect on its main food source, DG, starting in the early 1980s, and the effect strengthened until the mid-1980s (Figure 2A). The effect remained negative, though slightly weaker, until the end of the time series. The effect of *Daphnia* on other zooplankton (NDC) also varied through time (Figure 2B). Similar to the *Daphnia*-DG interaction, after *Daphnia* established in Lake Washington, the effect of *Daphnia* on NDC became increasingly negative, reached its peak in the mid-1980s, then remained negative but weakened to the end of the time series.

Density dependence also varied through time for all plankton groups. Density dependence in DG decreased (i.e., the diagonal **B** matrix coefficient associated with DG increased) until after *Daphnia* established in the lake, after which density dependence

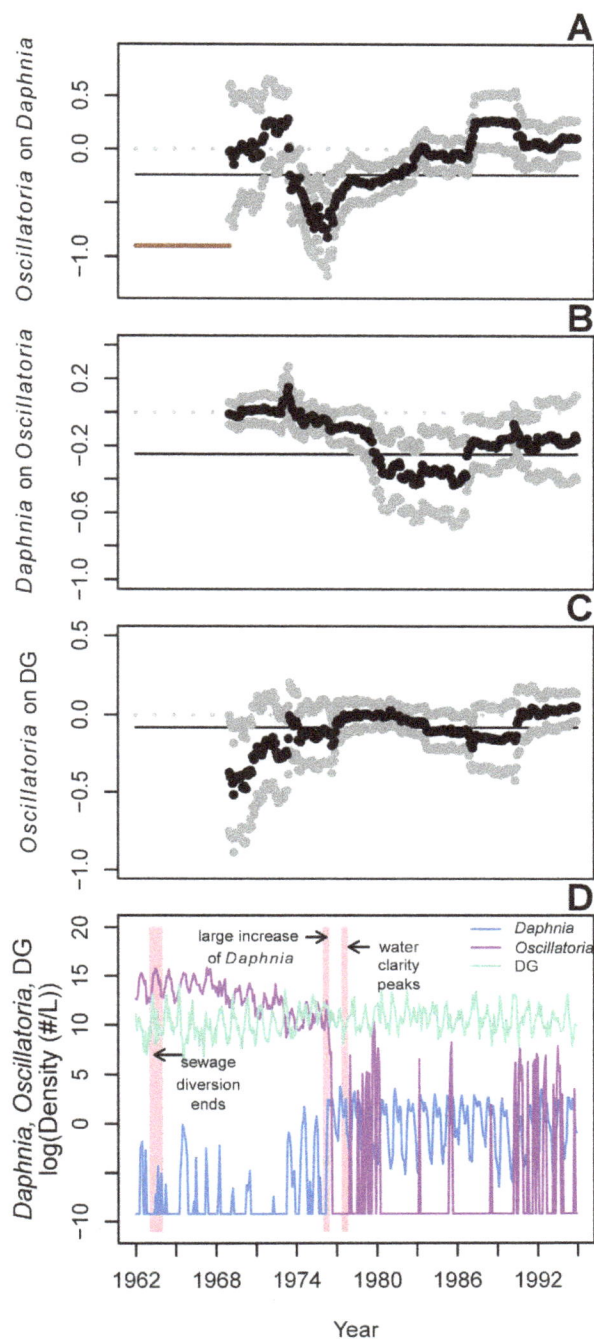

**Figure 1. Shifting impacts of *Oscillatoria* on the Lake Washington plankton community.** Effects of *Oscillatoria* on *Daphnia* (A); *Daphnia* on *Oscillatoria* (B); and *Oscillatoria* on diatoms/green algae, DG, (C) estimated by a mwMAR model using an 84-timestep window (indicated by solid red horizontal line shown in A). The mwMAR-estimated effect of *Oscillatoria* on NDC was non-significant. Estimates are shown with 95% upper and lower CIs. Grey dotted lines indicate a neutral interaction; solid black lines indicate the average interaction across the full time series, as estimated by a traditional MAR model. The raw time-series data are given in (D), with years of significant known changes shown in shaded vertical bars. All results are presented at the end year of the moving window.

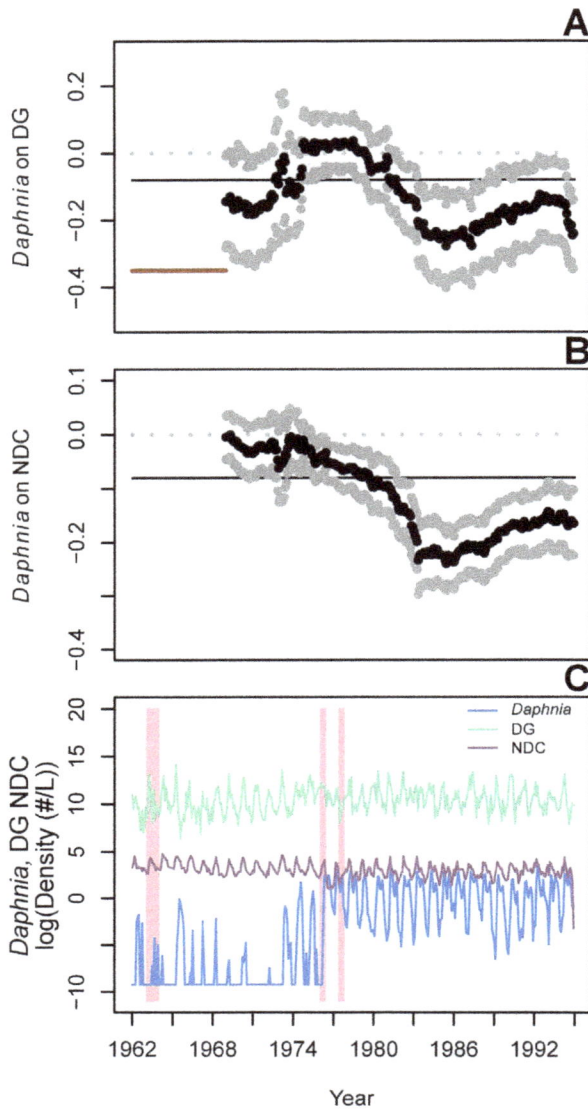

**Figure 2. Shifting effects of *Daphnia* on the Lake Washington plankton community.** Effects of *Daphnia* on diatoms and green algae, DG, (A) and non-daphnid cladocerans and non-cladoceran crustaceans, NDC, (B) through time as estimated by a mwMAR model with an 84-timestep window (indicated by solid red horizontal line in A). Estimates are shown with 95% upper and lower CIs. Grey dotted lines indicate coefficient values of 0; solid black lines indicate the average interaction across the full time series, as estimated by a traditional MAR model. The raw time-series data are given in (C), with years of influential known changes shown in shaded vertical bars. All results are presented at the end year of the moving window.

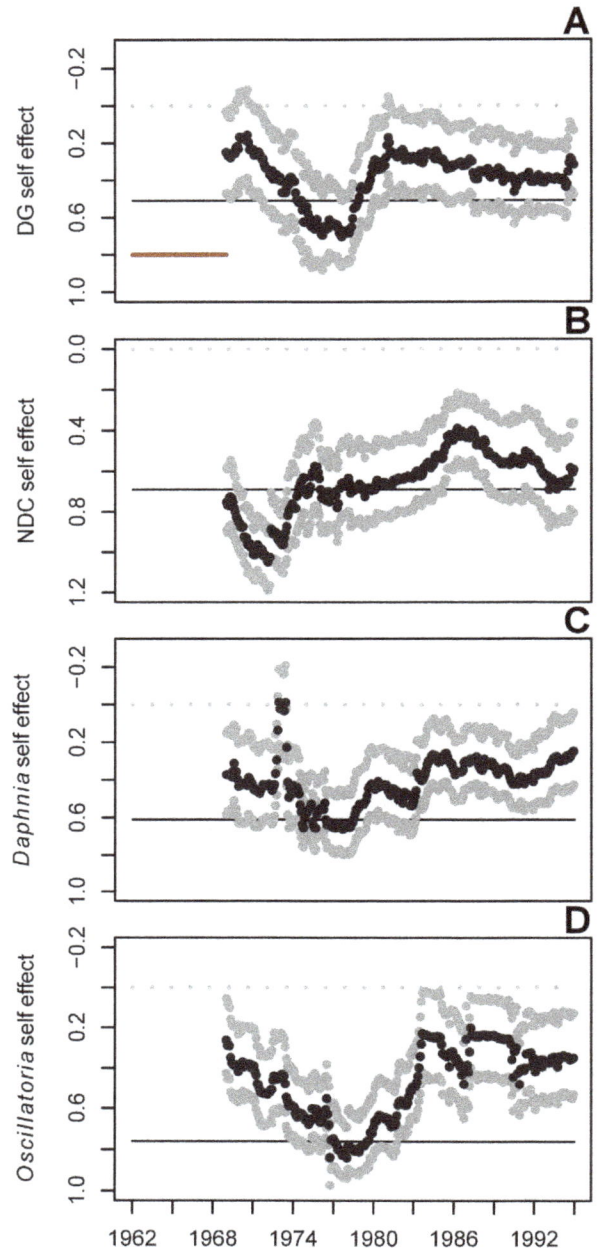

**Figure 3. Shifting density-dependent effects of all plankton groups.** Coefficients are estimated by a mwMAR model with an 84-timestep window (indicated by solid red horizontal line in A). Estimates are shown with 95% upper and lower CIs. DG = diatoms and green algae; NDC = non-daphnid cladocerans and non-cladoceran crustaceans. Grey dotted lines indicate coefficient values of 0; solid black lines indicate the average effect across the full time series, as estimated by a traditional MAR model. All results are presented at the end year of the moving window.

increased (Figure 3A). Density dependence in the grazer group NDC increased steadily from the early 1970s until the late 1980s, after which it weakened until the end of the time series (Figure 3B). *Daphnia* density dependence increased from the time it established in Lake Washington until the end of the time series (Figure 3C). *Oscillatoria* density dependence weakened until its decline in abundance in the late 1970s, at which point it increased and held more or less steady from the early 1980s until the end of the time series (Figure 3D).

The MAR model also estimates density-independent intrinsic population growth (the **A** vector), and while many of the confidence intervals surrounding the **A** estimates overlapped zero for a portion of the time series, there were consistent trends in the estimates among different plankton groups (Figure 4). For all four plankton groups, there were three distinct periods of intrinsic growth rate estimates: (1) before regular appearances of *Daphnia*

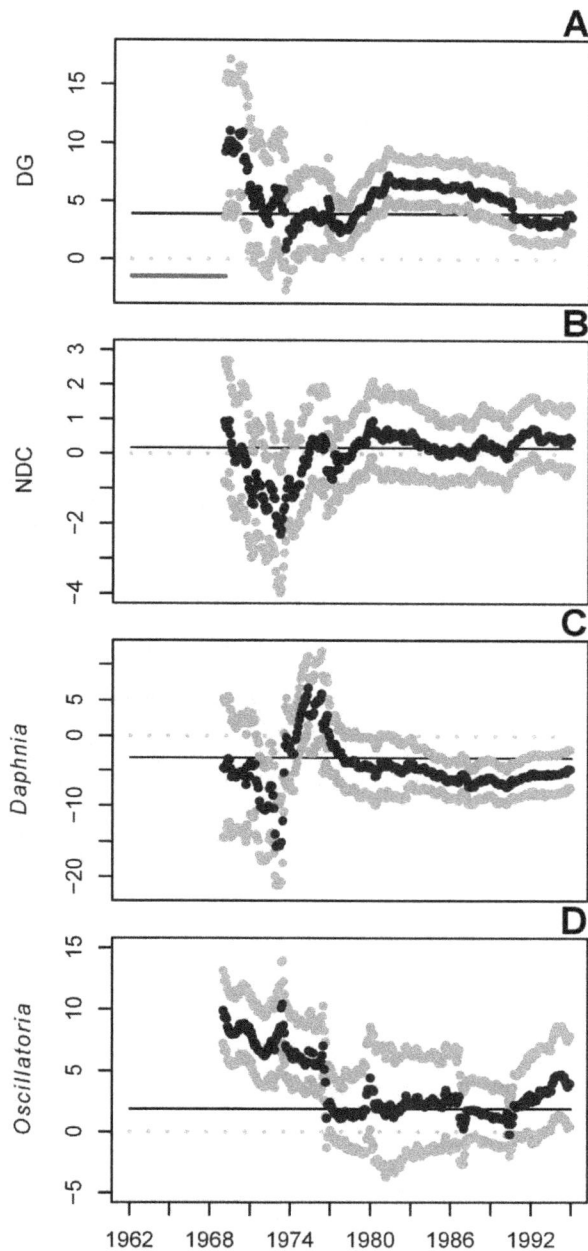

**Figure 4. Intrinsic growth rates of Lake Washington plankton groups.** Growth rates are estimated by a mwMAR model with an 84-timestep window (indicated by solid red horizontal line in A). Estimates are shown with 95% upper and lower CIs. Grey dotted lines indicate coefficient values of 0; solid black lines indicate the average rate across the full time series, as estimated by a traditional MAR model. DG = diatoms and green algae; NDC = non-daphnid cladocerans and non-cladoceran crustaceans. All results are presented at the end year of the moving window.

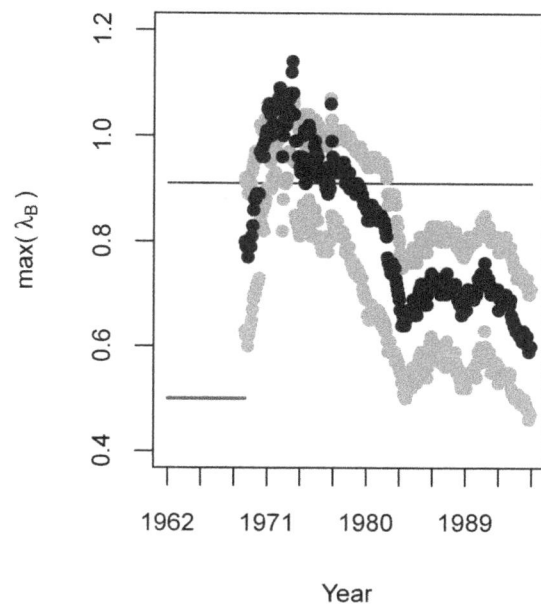

**Figure 5. Shifting community stability.** Stability is given by $\lambda$, the maximum eigenvalue of the community interaction matrix, as estimated by a mwMAR model using an 84-timestep window (indicated by solid red horizontal line). Estimates are shown with 95% upper and lower CIs. The grey dotted line indicates coefficient value of 0; the solid black line indicates the average stability across the full time series, as estimated by a traditional MAR model. Results are presented at the end year of the moving window.

in Lake Washington (pre-summer 1973); (2) between when *Daphnia* began to make regular appearances, and when *Daphnia* established in the lake and *Oscillatoria* declined dramatically (summer 1973 – spring 1976); and (3) after the rise of *Daphnia* and decline of *Oscillatoria* (summer 1976 onward). During the first period, there was high variability and negative trends in all **A** estimates. During the second period, the DG growth rate was

mostly constant (Figure 4A), both grazer groups growth rates increased (though some CIs overlapped 0; Figure 4B, C), and the *Oscillatoria* growth rate decreased (Figure 4D). During the final period, from the mid-1970s to the end of the time series, the growth rates of most groups were constant, except for an increase in the DG growth rate. Both *Oscillatoria* and NDC had growth rates equal to zero during this period.

Stability ($\lambda$) decreased sharply from the beginning of the time series, and the system was least stable (i.e., $\lambda$ was at its maximum value) in the early 1970s (Figure 5). Following this nadir, community stability increased and reached maximum stability (i.e., the lowest $\lambda$ value) at the end of the time series. Following bootstrapping, mean temperature had significant effects on all plankton groups. In contrast, very few effects of temperature anomalies or total phosphorus on plankton groups in Lake Washington were retained in the final mwMAR model (Table 1; Figure S3).

We assessed the fit of the best mwMAR model to the Lake Washington data, and found that fewer than 1% of correlations between model residuals and data were significant. We also tested the model assumption of normally-distributed errors by applying the Shapiro-Wilk test [47] to the residuals of the MAR fit to each data window (**E**, from Equation 1), with a Bonferroni-corrected alpha [48] to account for multiple null hypotheses. We rejected the null hypothesis of normally distributed errors in 65/312 windows for *Daphnia*, and in 217/312 windows for *Oscillatoria* (and in 0 windows for DG and NDC; Figure S4). These data windows for which the null hypothesis was rejected corresponded to periods in the time series when the abundance of each species was zero, i.e., the long one-sided tails in the data.

**Table 1.** Interaction ranges estimated by the mwMAR model for Lake Washington across all 84-timestep windows of data in the historical time series.

| | Community Interactions | | | | Covariate Effects | | |
| --- | --- | --- | --- | --- | --- | --- | --- |
| | DG | NDC | Daphnia | Oscillatoria | Season | Temperature | Phosphorus |
| DG | 0.15-0.70 (-0.09, 0.89) | | -0.28-0.04 (-0.40, 0.18) | -0.52-0.05 (-0.89, 0.20) | -0.92-1.21 (-2.75, 2.74) | -1.25-0.10 (-2.33, 1.23) | -2.09-0.61 (-3.99, 1.49) |
| NDC | 0.00-0.18 (-0.12, 0.28) | 0.39-1.05 (0.22, 1.19) | -0.24-0.00 (-0.30, 0.05) | -0.08-0.09 (-0.17, 0.18) | -1.53-1.70 (-0.77, 0.80) | -0.59-0.73 (-1.23, 1.25) | n/a |
| Daphnia | -0.32-0.46 (-0.67, 0.91) | 0.08-1.46 (-0.65, 2.10) | -0.01-0.66 (-0.21, 0.80) | -0.82-0.29 (-1.18, 0.66) | -1.38-2.72 (-4.40, 5.39) | -1.88-2.14 (-3.61, 4.53) | n/a |
| Oscillatoria | -0.44-0.21 (-0.91, 0.48) | -0.64-0.69 (-1.16, 1.43) | -0.43-0.15 (-0.68, 0.27) | 0.20-0.84 (0.02, 0.97) | -0.63-3.60 (-4.18, 6.74) | -2.02-1.45 (-4.37, 4.22) | -2.12-1.66 (-4.96, 3.96) |

Empty cells reflect interactions not retained by the model. Parenthetical values represent ranges of 95% confidence intervals estimated for 2,000 bootstrapped data sets. Coefficients represent effects of variables in columns on variables in rows. DG: Diatoms/Green algae; NDC: Non-daphnid cladocerans and non-cladoceran crustaceans. n/a indicates a coefficient that was a priori excluded from the model.

## Discussion

### Shifting plankton dynamics in Lake Washington

We hypothesized that the mwMAR would show shifts in the interactions among the major taxa corresponding roughly with known periods of change in Lake Washington (e.g., years of and around 1968–1971, and 1976). For example, it has long been hypothesized that the highly-abundant *Oscillatoria*, owing to its low palatability, inhibited *Daphnia* before *Daphnia*'s increase in Lake Washington in 1976 [11], during the period of time when the two species overlapped but *Oscillatoria* abundance was decreasing. These dynamics have been demonstrated experimentally [44], but our results are the first to corroborate this hypothesis using historical data. During the period of time between the peak in water quality (1971) and the dramatic increase in *Daphnia* abundance (1976) – the period of overlap between *Oscillatoria* and *Daphnia* and hypothesized inhibition of *Daphnia* by *Oscillatoria* – we found an increasingly negative effect of *Oscillatoria* on *Daphnia*. Once the mwMAR window included only dates following the large increase in *Daphnia* (i.e., 1976 and later), there was no detectable effect of *Oscillatoria* on *Daphnia*. The long, filamentous shape of *Oscillatoria* generally makes it inedible for *Daphnia*, which is one likely source of the negative per-capita effect estimated here during their period of overlap.

*Oscillatoria* also had a negative effect on diatoms and edible green algae, the main food source for *Daphnia* and other grazers in the lake. High intrinsic growth rates in edible phytoplankton estimated at the start of the time series decreased during the period when *Oscillatoria* was dominant. At the same time, density dependence in diatoms and green algae also decreased, suggesting inhibition in growth, possibly resulting from competition for limiting nutrients, or physical shading or toxic effects of excretions by *Oscillatoria*. Such inhibition of algae by *Oscillatoria* has also been demonstrated experimentally [44]. This apparent inhibition of phytoplankton by *Oscillatoria* rapidly decreased following an abrupt transition in the mid-1970s when the negative effect of *Oscillatoria* on DG decreased and disappeared (Figure 1). Coincident with these dynamics, the effect of *Oscillatoria* on *Daphnia* also weakened and the intrinsic growth rate of *Daphnia* increased from its minimum in 1972 to its peak in 1976 (Figure 4C). After 1976, *Daphnia*'s intrinsic growth rate decreased and density dependence increased (Figure 3C) as the *Daphnia* population increased in abundance. In addition, while the result was not significant (95% CIs overlapped zero), DG may have had a bottom-up positive effect on *Daphnia* after being freed from inhibition by *Oscillatoria*, in the latter half of the time series (Figure S2). Taken together, these results corroborate the hypothesis that the establishment of *Daphnia* following the improvement of water quality in Lake Washington was impeded directly and indirectly by the cyanobacterium *Oscillatoria*.

Grazers are known to inhibit cyanobacteria under some environmental conditions [49], and our analysis found a negative effect of *Daphnia* on *Oscillatoria* coincident with *Oscillatoria*'s decrease in abundance. In general, the frequency of cyanobacteria blooms is associated with the relationships between grazers and edible phytoplankton, such that when grazers and edible phytoplankton dynamics are stable (i.e., abundances do not undergo large, intrinsic oscillations), cyanobacteria are controlled by grazers [49]. These dynamics are often associated with phosphorus inputs to a lake. We observed a similar pattern in Lake Washington. As phosphorus inputs decreased, the grazing effect of *Daphnia* on edible phytoplankton increased concomitant with the inhibiting effect of *Daphnia* on *Oscillatoria* (Figures 1 and 2).

MAR coefficients have been shown previously to reflect changes in community dominance, when an increase in the abundance of one species or group coincides with a decrease in another [40], and therefore the negative effect of *Daphnia* on *Oscillatoria* may represent shifting dominance between the two taxa. The transition from *Oscillatoria* to *Daphnia* dominance was reflected in interactions among other plankton groups in the community. As the negative effect of *Oscillatoria* on *Daphnia* declined in the mid-1970s through to the early 1980s, and as *Daphnia* increased in abundance, *Daphnia* had stronger impacts on their main food source (DG) and competitors (NDC; Figure 2). At the same time, the strength of density-dependence (Figure 3) and density-independent growth rates increased for the grazing zooplankton groups (Figure 4), suggesting the release of the grazer community from inhibition by *Oscillatoria*. No previous work has shown an effect of *Oscillatoria* on other grazer groups beyond *Daphnia*, but the increase in population growth rates (**A** matrix elements) of the NDC group following *Oscillatoria*'s decline suggests a possible negative interaction.

The negative effects of *Daphnia* on their food and competitors weakened towards the end of the time series, apart from an intensified grazing effect of *Daphnia* on DG at the very end. One potential explanation for the weakened grazing effect at the end of the time series relates indirectly to the warming of the lake during this time. Between 1962 and 2002, the lake surface temperature increased by 1.4°C during the stratified months, and associated with this warming was an advance in the spring phytoplankton bloom by 19 days [10]. Most of the warming and spring bloom advance occurred in the period 1962-1994. The weakening of the effect of total *Daphnia* on the phytoplankton group during that period, in the present analysis, could be a reflection of shifts in species-specific phenology and grazing characteristics [14,41].

The results presented here highlight opportunities to learn more from time series data about how species interactions shift with changes in the environment across ecosystem types, and how those changing food web dynamics are liable to affect community stability and resilience to further disturbance. Ecosystem-based approaches to management often include a focus on food web dynamics, but quantifying changes in species interactions, and how those changes map onto the environmental template, proves difficult. Linking shifts in species interactions to specific environmental drivers opens opportunities to focus efforts aimed at retaining resilience as ecosystems undergo rapid change.

## Community stability and environmental covariates

The Lake Washington system has undergone major shifts in chemistry and ecology that are reflected in community stability. The peak of instability occurred in April 1973, (a window that included data from May 1966 – April 1973; Figure 5). Values of λ greater than 1 indicate an unstable system [21], and here λ exceeded 1 for windows ending in November 1970 – November 1974, representing the period of time between December 1963 – November 1974, inclusive. This is the time period that included major ecosystem shifts: high nutrient levels, sewage diversion, and nutrient reduction; high *Oscillatoria* abundance followed by its decline; and the first rare appearances of *Daphnia*. By the time of *Oscillatoria*'s disappearance, maximum water clarity, and establishment of *Daphnia* in 1976, the community stability was increasing, and continued to increase until the end of the time series. Thus, the period of time the lake was undergoing the most substantial and dramatic shifts throughout the ecosystem, and before *Daphnia* gained a foothold, was the least stable period in the community as well.

We observed effects of monthly mean temperature on the abundance of all plankton guilds (Figure S3), which agrees with previous MAR analyses [39,45], and with the MAR model estimated here from the whole Lake Washington time series (Table S1). Previous work has suggested that Lake Washington plankton phenology also responds to lake warming [10,50], and that the relationships between temperature and plankton taxa are evidence of the potential influence of a warming lake on food web dynamics [39]. However, we found no significant effects of deviations from the long-term seasonal temperature patterns, suggesting that lake-warming effects are not detectable in the abundance of these plankton guilds.

## Caveats and considerations

Our results suggest moving-window MAR models may be useful in systems with sufficient time-series data for understanding shifting abiotic and biotic dynamics. As with all statistical methods, however, practitioners must consider possible caveats and issues in advance of and throughout analyses. The data and ecosystem considerations applicable to prior MAR model applications also extend to our moving-window approach. Users must have sufficient time-series data for valid parameter estimation, which varies depending on the time scale of interactions in the system and frequency of observations. The moving-window MAR model imposes the further consideration of having sufficient time-series data for multiple windows and surrounding the event(s) of interest. Importantly, bias in model estimates shrinks as the ratio between window size and system transition period increases, and users are cautioned to interpret model estimates during system transitions with consideration of such bias. However, the window could be configured for different purposes: made smaller to detect changes before they occur, or sized to optimize detection of a change in a particular state variable.

Applications of this method will benefit from *a priori* knowledge of ecological interactions and drivers in the modeled system to build a robust MAR model. In our analysis of the Lake Washington plankton community, we simplified the plankton community based on previous work that highlighted the strongest food web interactions and key environmental covariates [39]. However, Hampton et al. [39] also pointed out the importance of other plankton taxa in driving the dynamics of the dominant species in Lake Washington, such as *Cryptomonas*, picoplankton and non-colonial rotifers. Therefore, it is possible that additional food web dynamics contribute to the interaction coefficients observed here, which could be highlighted by future analyses. Furthermore, if the model failed to include an influential environmental driver of Lake Washington plankton dynamics, the model results might be erroneously interpreted: if one plankton guild responds negatively to an unmeasured environmental variable, and another guild responds positively, this might incorrectly be interpreted as a negative interaction between the two guilds. In the Lake Washington case, years of experimental work and field observations have identified environmental variables that are robust driving signals. In addition, preliminary, exploratory MAR model runs were performed to screen a broad suite of potential drivers on plankton time series data. The analyses here rely heavily on those two approaches to validation, and potential users are advised to similarly behave as ecological detectives.

Additionally, as with prior MAR approaches, users must invest time in simulation modeling that allows them to test how the approach is likely to work with data similar to theirs. Simulation of data from a model with similar parameters to the study ecosystem helps identify the appropriate moving window size and, thus, estimate the precision associated with future predictions of system change. Because much of the MAR approach is based on iterative fitting approaches, creating and testing simulation data sets from known parameter values with similar lengths, covariate and taxa numbers, and variance, is critical to interpreting knowledge gained from MAR models. For the moving-window approach, users should carefully examine the effect of window size on their simulation datasets (see Appendix S2 for an example analysis using simulated datasets). *A priori* knowledge or hypotheses related to the resolution of data and interactions as well as the strength and timing of the predicted shift should be considered during the process of simulation modeling. Comparison of the mwMAR output with whole time-series MAR estimates is useful in assessing when the broad confidence intervals estimated with the mwMAR model are potentially masking significant interactions.

## Conclusions

Ecologists have recently gained an appreciation for the need to develop methods based on the underlying hypothesis that many systems are rarely, if ever, stationary. Here we present a method that allows researchers and managers alike to examine long-term monitoring data and develop a dynamic record of shifting interactions and drivers. By calculating indirect and direct effects over time, and their changes, mwMAR allows researchers to understand how species invasions and extinctions, shifts in temperature and nutrient loadings, and other anthropogenic perturbations may cascade and feedback through food webs and ecosystems.

## Supporting Information

**Figure S1 Lake Washington plankton densities from 1962–1994.** Monthly means of densities for aggregated plankton groups used in mwMAR analyses. NDC = non-daphnid cladocerans; DG = diatoms and green algae.
(DOCX)

**Figure S2 Time series of all community interactions.** Interaction coefficients estimated for the Lake Washington time series with a mwMAR model, using an 84-month window. Figures show per-capita effects of plankton guilds in columns on plankton guilds in rows. Diagonal figures represent self-effects, or density-dependent effects on abundance.
(DOCX)

**Figure S3 Time series of environmental covariate effects.** Interaction coefficients estimated for the Lake Washington time series with a mwMAR model, using an 84-month window. Figures show the effects of covariates in columns on plankton guilds in rows.
(DOCX)

**Figure S4 Quantile-quantile plots of residuals for the *Daphnia* and *Oscillatoria* time series.** Shown are theoretical versus observed distributions of mwMAR model residuals for all windows where the Shapiro-Wilk test statistic was below the alpha value required to reject the null hypothesis of normally-distributed errors (61/1248 for *Daphnia*, 175/1248 for *Oscillatoria*, 0 for DG and 0 for NDC).
(DOCX)

**Table S1 Community and covariate matrix coefficients estimated by a MAR model for the full Lake Washington time series.**
(DOCX)

**Appendix S1 Lake Washington plankton and covariate data, 1962–1994.**
(CSV)

**Appendix S2 Moving-window MAR Model Testing.** Validation of the moving window MAR model approach, including accuracy of parameter estimation and estimation of bias during system transition.
(DOC)

## Acknowledgments

Many thanks to J. Regetz for assistance with coding. We thank D.E. Schindler for generous access to the Lake Washington data, the numerous people who have contributed to the data set, including A. Litt and S. Abella, and the organizations that have financially supported it, including the Mellon Foundation. The manuscript was improved by comments from D.E. Schindler, E.J. Ward and two anonymous reviewers.

## Author Contributions

Analyzed the data: TF EW MS SK. Contributed reagents/materials/analysis tools: TF EW MS SK EH. Wrote the paper: TF EW MS SK EH SH.

## References

1. Burkle LA, Marlin JC, Knight TM (2013) Plant-Pollinator Interactions over 120 Years: Loss of Species, Co-Occurrence, and Function. Science 339: 1611–1615.
2. Tylianakis JM, Didham RK, Bascompte J, Wardle DA (2008) Global change and species interactions in terrestrial ecosystems. Ecol Lett 11: 1351–1363.
3. Holyoak M (2000) Habitat subdivision causes changes in food web structure. Ecol Lett 3: 509–515.
4. Schlesinger MD, Manley PN, Holyoak M (2008) Distinguishing stressors acting on land bird communities in an urbanizing environment. Ecology 89: 2302–2314.
5. Hastings A (2004) Transients: the key to long-term ecological understanding? Trends Ecol Evol 19: 39–45.
6. Harley CDG, Hughes AR, Hultgren KM, Miner BG, Sorte CJB, et al. (2006) The impacts of climate change in coastal marine systems. Ecol Lett 9: 228–241.
7. Walther GR (2010) Community and ecosystem responses to recent climate change. Philos Trans R Soc B-Biol Sci 365: 2019–2024.
8. Tunney TD, McCann KS, Lester NG, Shuter BJ (2014) Effects of differential habitat warming on complex communities. Proceedings of the National Academies of Science 111: 8077–8082.
9. Edmondson WT (1970) Phosphorus, nitrogen, and algae in Lake Washington after diversion of sewage. Science 169: 690–691.
10. Winder M, Schindler DE (2004) Climatic effects on the phenology of lake processes. Glob Change Biol 10: 1844–1856.
11. Edmondson WT, Litt AH (1982) Daphnia in Lake Washington. Limnol Oceanogr 27: 272–293.
12. Edmondson WT (1991) The Uses of Ecology. Lake Washington and Beyond: University of Washington Press. 329 p.
13. Winder M, Schindler DE, Essington TE, Litt AH (2009) Disrupted seasonal clockwork in the population dynamics of a freshwater copepod by climate warming. Limnol Oceanogr 54: 2493–2505.
14. Hampton SE, Romare P, Seiler DE (2006) Environmentally controlled Daphnia spring increase with implications for sockeye salmon fry in Lake Washington, USA. J Plankton Res 28: 399–406.
15. Francis TB, Schindler DE (2009) Shoreline urbanization reduces terrestrial insect subsidies to fishes in North American lakes. Oikos 118: 1872–1882.
16. Beauchamp DA, Sergeant CJ, Mazur MM, Scheuerell MD, Scheuerell JM, et al. (2004) Spatial-temporal dynamics of early feeding demand and food supply for sockeye fry in Lake Washington. Transactions of the American Fisheries Society 133: 1014–1032.
17. Mazumder A, Taylor WD, McQueen DJ, Lean DRS (1990) Effects of fish and plankton on lake temperature and mixing depth. Science 247: 312–315.
18. Sarnelle O (1993) Herbivore effects on phytoplankton succession in a eutrophic lake. Ecol Monogr 63: 129–149.
19. Ives AR (1995) Measuring resilience in stochastic systems. Ecol Monogr 65: 217–233.
20. Ives AR, Dennis B, Cottingham KL, Carpenter SR (2003) Estimating community stability and ecological interactions from time-series data. Ecol Monogr 73: 301–330.

21. Ives AR, Gross K, Klug JL (1999) Stability and variability in competitive communities. Science 286: 542–544.
22. Hampton SE, Holmes EE, Scheef LP, Scheuerell MD, Katz SL, et al. (2013) Quantifying effects of abiotic and biotic drivers on community dynamics with multivariate autoregressive (MAR) models. Ecology 94: 2663–2669.
23. Lindegren M, Mollmann C, Nielsen A, Stenseth NC (2009) Preventing the collapse of the Baltic cod stock through an ecosystem-based management approach. Proc Natl Acad Sci U S A 106: 14722–14727.
24. Mac Nally R, Thomson JR, Kimmerer WJ, Feyrer F, Newman KB, et al. (2010) Analysis of pelagic species decline in the upper San Francisco Estuary using multivariate autoregressive modeling (MAR). Ecol Appl 20: 1417–1430.
25. Kordas RL, Dudgeon S (2011) Dynamics of species interaction strength in space, time and with developmental stage. Proc R Soc B-Biol Sci 278: 1804–1813.
26. Torrence C, Webster PJ (1999) Interdecadal changes in the ENSO-monsoon system. Journal of Climate 12: 2679–2690.
27. Menard F, Marsac F, Bellier E, Cazelles B (2007) Climatic oscillations and tuna catch rates in the Indian Ocean: a wavelet approach to time series analysis. Fisheries Oceanography 16: 95–104.
28. Jevrejeva S, Moore JC, Grinsted A (2004) Oceanic and atmospheric transport of multiyear El Nino-Southern Oscillation (ENSO) signatures to the polar regions. Geophys Res Lett 31.
29. Beninca E, Johnk KD, Heerkloss R, Huisman J (2009) Coupled predator-prey oscillations in a chaotic food web. Ecol Lett 12: 1367–1378.
30. Winder M, Cloern JE (2010) The annual cycles of phytoplankton biomass. Philosophical Transactions of the Royal Society B: Biological Sciences 365: 3215–3226.
31. Hsieh CH, Chen CS, Chiu TS, Lee KT, Shieh FJ, et al. (2009) Time series analyses reveal transient relationships between abundance of larval anchovy and environmental variables in the coastal waters southwest of Taiwan. Fisheries Oceanography 18: 102–117.
32. Breaker LC (2006) Nonlinear aspects of sea surface temperature in Monterey Bay. Prog Oceanogr 69: 61–89.
33. Carpenter SR, Cole JJ, Pace ML, Batt R, Brock WA, et al. (2011) Early warnings of regime shifts: A whole-ecosystem experiment. Science 332: 1079–1082.
34. Dakos V, Carpenter SR, Brock WA, Ellison AM, Guttal V, et al. (2012) Methods for detecting early warnings of critical transitions in time series illustrated using simulated ecological data. PLoS One 7: e41010.
35. Seekell DA, Carpenter SR, Pace ML (2011) Conditional heteroscedasticity as a leading indicator of ecological regime shifts. American Naturalist 178: 442–451.
36. Elkinton JS, Healy WM, Buonaccorsi JP, Boettner GH, Hazzard AM, et al. (1996) Interactions among gypsy moths, white-footed mice, and acorns. Ecology 77: 2332–2342.
37. Ives AR (1995) Predicting the response of populations to environmental change. Ecology 76: 926–941.
38. Hampton SE, Izmest'eva LR, Moore MV, Katz SL, Dennis B, et al. (2008) Sixty years of environmental change in the world's largest freshwater lake - Lake Baikal, Siberia. Glob Change Biol 14: 1947–1958.
39. Hampton SE, Scheuerell MD, Schindler DE (2006) Coalescence in the Lake Washington story: interaction strengths in a planktonic food web. Limnol Oceanogr 51: 2042–2051.
40. Francis TB, Scheuerell MD, Brodeur RD, Levin PS, Ruzicka JJ, et al. (2012) Climate shifts the interaction web of a marine plankton community. Glob Change Biol 18: 2498-2508.
41. Winder M, Schindler DE (2004) Climate change uncouples trophic interactions in an aquatic ecosystem. Ecology 85: 2100–2106.
42. Law T, Zhang WT, Zhao JY, Arhonditsis GB (2009) Structural changes in lake functioning induced from nutrient loading and climate variability. Ecological Modelling 220: 979–997.
43. Edmondson WT (1991) Sedimentary record of changes in the condition of Lake Washington. Limnol Oceanogr 36: 1031–1044.
44. Infante A, Abella SEB (1985) Inhibition of *Daphnia* by *Oscillatoria* in Lake Washington. Limnol Oceanogr 30: 1046–1052.
45. Hampton SE, Schindler DE (2006) Empirical evaluation of observation scale effects in community time series. Oikos 113: 424–439.
46. Viscido SV, Holmes EE (2010) Statistical modelling of communities and ecosystems using the LAMBDA software tool. Environmental Modelling and Software 25: 1905–1908.
47. Shapiro SS, Wilk MB (1965) An analysis of variance test for normality (complete samples). Biometrika 52: 591–611.
48. Hochberg Y (1988) A sharper Bonferroni procedure for multiple tests of significance. Biometrika 75: 800–802.
49. Carpenter SR (1992) Destabilization of planktonic ecosystems and blooms of blue-green algae. In: Kitchell JF, editor.Food Web Management, A Case Study of Lake Mendota.New York: Springer New York. pp. 461–481.
50. Hampton SE (2005) Increased niche differentiation between two Conochilus species over 33 years of climate change and food web alteration. Limnol Oceanogr 50: 421–426.

# 6

# Invasibility of Mediterranean-Climate Rivers by Non-Native Fish: The Importance of Environmental Drivers and Human Pressures

**Maria Ilhéu[1,2]\*, Paula Matono[1,2], João Manuel Bernardo[1]**

**1** Departamento de Paisagem Ambiente e Ordenamento, Escola de Ciências e Tecnologia, Universidade de Évora, Évora, Portugal, **2** Instituto de Ciências Agrárias e Ambientais Mediterrânicas, Universidade de Évora, Évora, Portugal

## Abstract

Invasive species are regarded as a biological pressure to natural aquatic communities. Understanding the factors promoting successful invasions is of great conceptual and practical importance. From a practical point of view, it should help to prevent future invasions and to mitigate the effects of recent invaders through early detection and prioritization of management measures. This study aims to identify the environmental determinants of fish invasions in Mediterranean-climate rivers and evaluate the relative importance of natural and human drivers. Fish communities were sampled in 182 undisturbed and 198 disturbed sites by human activities, belonging to 12 river types defined for continental Portugal within the implementation of the European Union's Water Framework Directive. Pumpkinseed sunfish, *Lepomis gibbosus* (L.), and mosquitofish, *Gambusia holbrooki* (Girard), were the most abundant non-native species (NNS) in the southern river types whereas the Iberian gudgeon, *Gobio lozanoi* Doadrio and Madeira, was the dominant NNS in the north/centre. Small northern mountain streams showed null or low frequency of occurrence and abundance of NNS, while southern lowland river types with medium and large drainage areas presented the highest values. The occurrence of NNS was significantly lower in undisturbed sites and the highest density of NNS was associated with high human pressure. Results from variance partitioning showed that natural environmental factors determine the distribution of the most abundant NNS while the increase in their abundance and success is explained mainly by human-induced disturbance factors. This study stresses the high vulnerability of the warm water lowland river types to non-native fish invasions, which is amplified by human-induced degradation.

**Editor:** Maura (Gee) Geraldine Chapman, University of Sydney, Australia

**Funding:** This study was partially funded by INAG (Instituto da Água) as part of a larger programme on the implementation of the Water Framework Directive and by the Foundation for Science and Technology (FCT) through the research project (PTDC/AAC-AMB/102541/2008). Additional funding was provided by FEDER under the programme COMPETE and by National Funds through FCT under the Strategic Project PEst-OE/AGR/UI0115/2014. P. Matono was supported by a PhD grant from FCT (SFRH/BD/23435/2005). The funders had no role in study design, data collection and analysis, decision to publish, or preparation of the manuscript.

**Competing Interests:** The authors have declared that no competing interests exist.

\* Email: milheu@uevora.pt

## Introduction

The rate and extent of invasions in freshwater ecosystems are particularly alarming in the Mediterranean region, which is among the most heavily invaded ecosystems in the world [1–3]. The Iberian freshwater habitats are one of the most paradigmatic examples where there are constant reports of new invading fish species and colonization of new areas (see [4–10] as examples).

The success of biological invasions can be explained by several factors, with environmental drivers being among the most important ones [11–12]. One of the most frequently stated hypotheses in the biological invasion literature is that species should have a greater chance of success if they are introduced to an area with a climate that closely matches that of their original range [13–15]. Other environmental drivers, such as spatial heterogeneity and environmental variability, may also be important [16–18]. The success of an invading fish may be predicted with reference to environmental conditions at different scales, ranging from habitat to river types or ecoregions. Some of the features of the recipient ecosystem and the success of an introduced species may be determined by human-induced pressure. The "human activity" hypothesis argues that human activities facilitate the establishment of non-native species (NNS) by disturbing natural landscapes and by increasing propagule pressure [2,8]. In Mediterranean-climate rivers, both landscape and human disturbance factors are expected to play a major role in the biological invasions, as these systems are largely governed by stochastic processes and have suffered a long history of human-induced pressure [19].

Although non-native species are not mentioned specifically in the European Union's Water Framework Directive (WFD) [20], in the context of the Directive's objectives NNS represent an important pressure (listed in the WFD annexes), since they can modify the structure of native biota and the ecological functioning of aquatic systems. NNS affect the ecological quality of natural environments in multiple ways and may represent a serious threat to native communities. Potential effects include genetic alterations within populations, spreading of pathogens and parasites, competition with and replacement of native species, and habitat deterioration or modification (see [21] for Iberian freshwater ichthyofauna). All the combined effects may result in changes in ecosystem function and global homogenization [1,3,22–23], with profound impact at ecological, evolutionary, genetic, and economic levels [24–25].

The apparently strong relationship between introduced fish species and degraded river conditions and their potential impact on native species and ecosystem suggest that NNS may be useful indicators of biological integrity and river health [26–28] and they are therefore incorporated in river monitoring schemes under the WFD and elsewhere. The incorporation of NNS into the ecological assessment will require knowledge about the density, distribution and potential risk of each species for each water body. According to the WFD, with the exception of particular cases, the ecological assessment involves the previous derivation of typologies for the aquatic systems. The ecological assessment is performed within groups of similar ecosystems and is expressed as a deviation from the reference conditions, i.e. conditions found in sites without human pressures. In fact, differences in climate, hydrology, geomorphology, geology, soil and vegetation make comparison of river communities difficult if not impossible. The use of river typology, by stratifying the spatial variability, makes the ecological assessment more practical, transparent and understandable by decision-takers, managers, technicians and public, and for these reasons it was the chosen approach on the WFD implementation. Moreover, approaching invasibility at a regional framework is useful for management purposes because it allows the identification of water bodies particularly vulnerable to NNS invasion and helps to regionalize monitoring schemes [29–31].

Understanding the factors promoting successful invasions is of great conceptual and practical importance. From a practical point of view, it should help to prevent future invasions and to mitigate the effects of recent invaders through early detection and prioritization of management measures. Thus, the identification of the environmental determinants of fish invasions is relevant in forecasting the overall impact of invasions on a global scale and prerequisite management authorities to adopt sound conservation policies.

The objectives of the present study were to determine: i) the patterns of non-native fish richness and abundance in Mediterranean-climate river types of Portugal; ii) the environmental drivers that favour the invasiveness of non-native fish within the morpho-climatic gradients; iii) the relationship between human-induced disturbance and non-native fish abundance and iv) the relative importance of environmental variables and human pressures for the occurrence and abundance of non-native fish species.

## Methods

### Study Area

The study considered 12 river types out of 15 defined for continental Portugal under the implementation of the WFD [32] (Fig. 1), as the 3 very large and highly modified river types were not considered.

The climate of Portugal presents high intra- and inter-annual precipitation and discharge variation, with severe and unpredictable floods between autumn and spring and persistent summer droughts [33], although the influence of factors such as topography and proximity to the Atlantic Ocean causes significant climatic contrasts. The general conditions of the atmospheric circulation cause a decrease in precipitation from north to south and from west (coast) to east (inland), enhanced by orographic asymmetry. Indeed, the mountain barrier in the north and center causes less rainfall in the interior regions. The temperature shows

an opposite pattern, increasing from north to south. In a general way, the altitude causes a decrease in temperature and an increase in rainfall.

Portuguese river types reflect two main geographical and climatic gradients: north–south and west–east. The north–south gradient is associated with a decrease in altitude, precipitation, annual discharge, and increasing temperature. The west–east gradient is related to a continental effect, with a precipitation decrease and an increase in temperature extremes. Owing to these features, most rivers are permanent in the north and intermittent in the south. The most relevant climatic and morphological characteristics of the basins of the river-types are presented in Table 1 (see also [32]). The river types may be described as:

M - Mountain rivers of the North: trout rivers with small drainage area located in the North-West and North-Center regions with low mean temperature and high precipitation; N1< $100 \text{ km}^2$ - North rivers of small drainage area: low mean temperature, altitude mainly in the 200–600 m range; N1> $100 \text{ km}^2$ - North rivers of medium and large drainage area: characteristics similar to the previous type; N2 - Medium-large Alto Douro rivers: drainage area larger than $100 \text{ km}^2$, high temperature (similar to the south types), lower rainfall; N3 - Small Alto Douro rivers: drainage area smaller than $100 \text{ km}^2$ and characteristics similar to the previous type; N4 - North-South transition type: located at the center region with intermediate characteristics; L - Littoral type: coastal streams of small-medium dimensions located at the littoral center region, low altitude; S1< $100 \text{ km}^2$ - Small South rivers: a large group of low altitude temporary rivers, in a region of high temperature and low precipitation; S1>$100 \text{ km}^2$ - South rivers of medium-large dimension: larger drainage temporary rivers with similar characteristics to the previous type; S2 - South mountain rivers: small streams in the mountain regions of the south with lower temperature and higher precipitation than in the other south river types; S3 - Rivers of Tagus and Sado sedimentary basins: small and medium dimension, low altitude and high mineralization, smaller rivers with temporary regime; S4 - Small chalk rivers of Algarve: low altitude, low precipitation and high temperature.

Native freshwater fish fauna of Portuguese rivers presents relatively low species richness per site and most fish species are endemic cyprinids with high conservation value, particularly in the south. Many of those species are threatened with extinction [34–35].

### Sampling and data collection

Sampling was carried out between 2004 and 2006 during spring at 380 sites in the main Portuguese river basins (Fig. 1). The number of sampled sites in each river type was proportional to its basin area: M = 28 sites; N1<$100 \text{ km}^2$ = 60 sites; N1> $100 \text{ km}^2$ = 68 sites; N2 = 20 sites; N3 = 28 sites; N4 = 16 sites; L = 33 sites; S1<$100 \text{ km}^2$ = 26 sites; S1>$100 \text{ km}^2$ = 33 sites; S2 = 22 sites; S3 = 37 sites; S4 = 9 sites. In each river type, sampling sites included undisturbed or least disturbed sites and disturbed ones. For the site selection, a preliminary human-induced pressure screening using GIS and information on pollution loads was followed. The final selection was based on the human disturbance level regarding ten variables [36–37]: land use, urban area, riparian zone, connectivity of the river segment, sediment load, flow regime, morphological condition, presence of artificial lentic water bodies, toxic and acidification levels, and nutrient/organic load. The spatial scale of the assessment depended on the character of each pressure; some were made at local scale, others at the fluvial segment, and some others at the basin scale. Each variable was scored from 1 (minimum

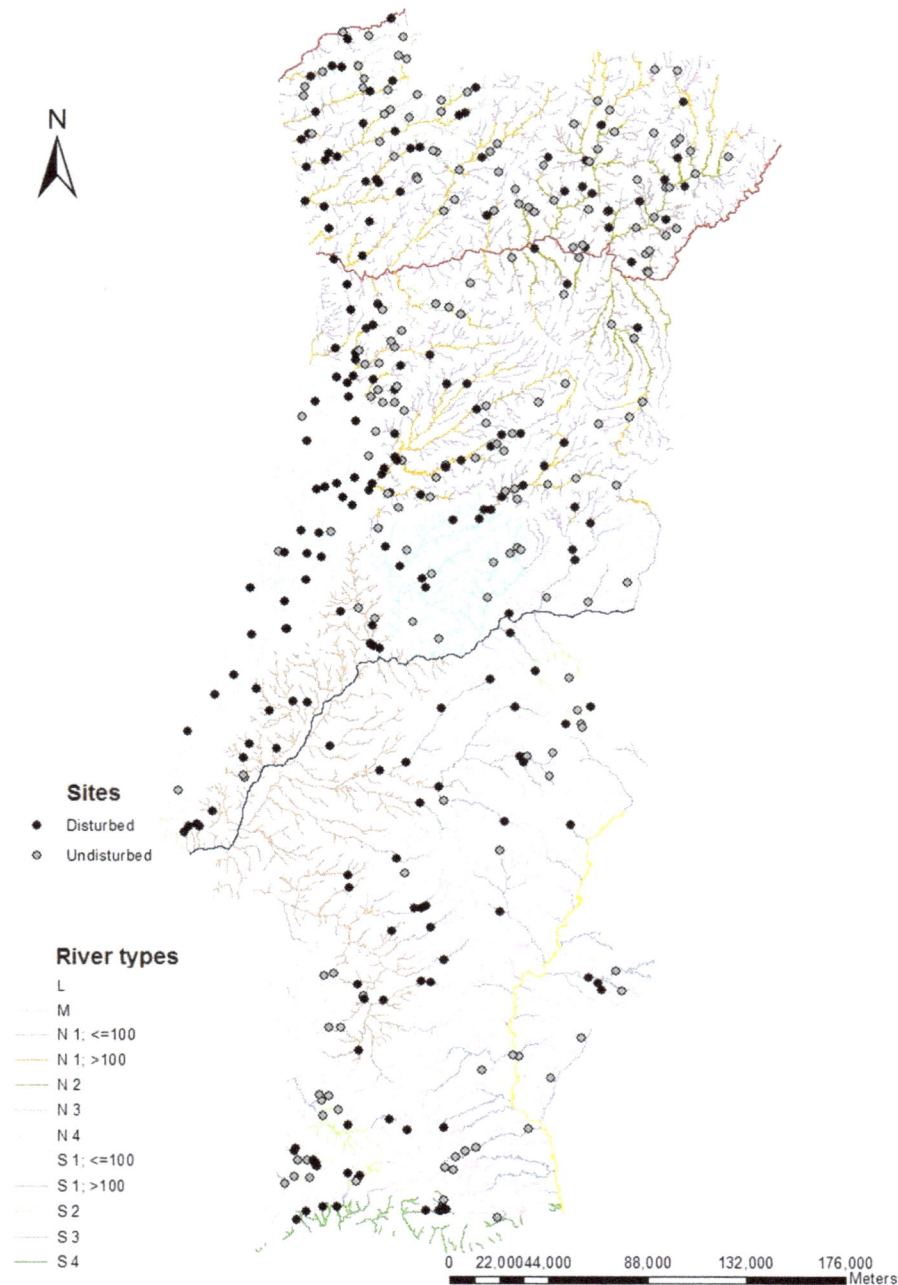

**Figure 1. Map of the river types defined to continental Portugal showing undisturbed (grey dots) and disturbed (black dots) sampling sites.**

disturbance) to 5 (maximum disturbance) (Tables 2 and 3) and only sites with scores of 1 and/or 2 and only one variable scored with a 3 were considered as undisturbed or least disturbed (i.e., reference sites). The sum of these scores represents the total human pressure. Several water quality variables complemented the evaluation of human pressure in each site, following Standard Methods for the Examination of Water and Wastewater [38]: Biochemical Oxygen Demand ($BOD_5$, mg $L^{-1}$), Chemical Oxygen Demand (COD, mg $L^{-1}$), Total Suspended Solids (TSS, mg $L^{-1}$), Soluble Reactive Phosphorous (SR-P, mg $L^{-1}$), nitrite ($NO_2^-$, mg $L^{-1}$), nitrate ($NO_3^-$, mg $L^{-1}$), ammonium

($NH_4^+$, mg $L^{-1}$), and total inorganic dissolved nitrogen (N, mg $L^{-1}$).

Environmental characterization of sites was based on regional and local variables. Regional variables were obtained from digital cartography with free Internet access and included the drainage area of the basin upstream the site ($km^2$), distance from source (km), altitude (m), slope (%), mean annual discharge (mm), mean annual air temperature (°C), and mean annual rainfall (mm). Rainfall, temperature, and flow variables were described from 30-year data series. Topographical variables were derived from a Digital Elevation Model (DEM), with a 90-m grid cell resolution

**Table 1.** Main morpho-climatic characteristics of river types in continental Portugal (mean and SD).

| River-Types | Mean Annual Temperature (°C) | Mean Annual Precipitation (mm) | Altitude (m) | Drainage Area (km²) |
|---|---|---|---|---|
| **M** | 11.0 | 1944 | 506 | 24.8 |
| northern mountain streams | (1.5) | (379) | (300) | (17) |
| **N1<100 km²** | 12.4 | 1190 | 413 | 33 |
| north streams with small drainage area | (1.3) | (358) | (242) | (23) |
| **N1>100 km²** | 12.6 | 1196 | 274 | 549 |
| north streams with large drainage area | (1.2) | (374) | (205) | (65) |
| **N2** | 13.1 | 596 | 300 | 960 |
| streams from Alto Douro with large drainage area | (1.0) | (81) | (141) | (1115) |
| **N3** | 13.0 | 671 | 432 | 32 |
| streams from Alto Douro with small drainage area | (0.8) | (134) | (160) | (23) |
| **N4** | 14.1 | 1065 | 280 | 151 |
| transition streams between north and south | (0.7) | (168) | (122) | (361) |
| **L** | 14.8 | 941 | 44 | 180 |
| littoral streams from west/centre region | (0.3) | (118) | (44) | (671) |
| **S1<100 km²** | 15.7 | 628 | 183 | 30 |
| south streams with small drainage area | (0.9) | (86) | (75) | (21) |
| **S1>100 km²** | 15.8 | 587 | 137 | 439 |
| south streams with large drainage area | (0.9) | (84) | (68) | (579) |
| **S2** | 15.4 | 743 | 175 | 60 |
| southern mountain streams | (0.3) | (85) | (147) | (87) |
| **S3** | 15.6 | 730 | 54 | 388 |
| streams of sedimentary deposits in Tagus and Sado basins | (0.4) | (118) | (46) | (1081) |
| **S4** | 16.9 | 632 | 54 | 67 |
| southern carsick streams of Algarve | (0.5) | (60) | (57) | (89) |

(CGIAR-CSI 2005), using ArcMap 9.1. Local variables were assessed *in situ* at each sampling occasion: water temperature (°C), conductivity ($\mu$S cm$^{-1}$), pH, dissolved oxygen (mg L$^{-1}$), mean stream wetted width (m), maximum and mean water depth (m), mean current velocity (m s$^{-1}$), dominant substrate class, adapted from the Wentworth scale [39]: 1 – mud and sand; 2 – gravel; 3 – pebble; 4 – cobble; 5 – boulders; 6 – boulders larger than 0.50 m, riparian vegetation (%), shadow (%), and proportion of different habitat types (pool, run, riffle).

Human disturbance level was evaluated *in situ* regarding the same ten variables described above for site selection. A total of 182 undisturbed and 198 disturbed sites were sampled covering the full human pressure gradient in each river type. The proportion of sites along the degradation gradient was sampled according to their availability within each river type.

Fish were collected by electrofishing following the WFD-compliant and CEN sampling protocol [40–41]. A backpack battery-powered electrofishing equipment was used, wading in shallow reaches (<1.2 m) or from a boat in deeper areas. The power and pulse frequency of the equipment were adjusted to ensure capture efficiency but also that fish were only stunned for a short period of time. Captured specimens were carefully handled, ensuring the minimum possible stress, and were quickly returned to the river. No specimens of endangered or protected species were injured or sacrificed.

The present study was part of a large programme coordinated by the National Water Institute (INAG, Instituto da Água, now part of APA, Agência Portuguesa do Ambiente). The National Institute for the Nature Conservation and Forestry (ICNF, Instituto de Conservação da Natureza e das Florestas (http://www.proforbiomed.eu/project/partners/institute-conservation-nature-and-forest-icnf) provided the necessary fishing permits. All sampling took place in national public rivers, which were under the jurisdiction of the National Water Institute and ICNF. As all the fish manipulation was restricted to the sampling procedures and animal care was taken into account, it was not necessary to obtain permission from the National *Animal Care* and Use *Committee* (Comissão de Ética, Bioética e Bem-estar Animal).

Captures were quantified as density (individuals/100 m²) and biomass (grams/100 m²). The degree of invasibility was measured using NNS richness and abundance.

A nonparametric ANCOVA performed using the rank transformation analysis of covariance evaluated the existence of significant differences in the density and number of NNS between river types [42,43]. The total human pressure was included as a covariable, in order to exclude its effects from the analysis.

Mann-Whitney test was used to identify significant differences in the number and density of NNS between undisturbed and disturbed sites. The Z test of proportions [44] evaluated the existence of significant differences in the frequency of occurrence of NNS between undisturbed and disturbed sites.

**Table 2.** Description, assessment scale and methods, and scoring criteria of the variables; land use, urban area, riparian vegetation, morphological condition and sediment load - used to evaluate the level of anthropogenic disturbance in sampled sites.

| Variables | Description | Assessment scale | Score | Criteria | Methods |
|---|---|---|---|---|---|
| **Land use** | Impact of farming/ forestry practices | River segment | 5 | >40% Agricultural use (intensive agriculture), very severe impact (rice field) | Local expert assessment complemented with Corine Land Cover (2000, 2006)* |
| | | | 4 | >40% Strong impact (area with strong forestry, including clearcuts) | |
| | | | 3 | <40% Moderate impact (subsistence gardens, pastures) | |
| | | | 2 | <40% Small impact (cork and holm oaks, high-growth forest) | |
| | | | 1 | <10% No significant impacts (natural forest and bush) | |
| | Land cover and bankface characterization | Local | 5 | Irrigated crops and/or high stocking | Local expert assessment complemented with Corine Land Cover (2000, 2006)* |
| | | | 4 | Horticultural crops, semi-intensive grazing | |
| | | | 3 | Extensive cultures (e.g. pastures, cereal crops, pine, eucalyptus), extensive grazing | |
| | | | 2 | Cork and holm oaks | |
| | | | 1 | Natural | |
| **Urban area** | Impact of urban areas | River segment | 5 | Very severe (location near a city with basic sanitation needs) | Local expert assessment complemented with Corine Land Cover (2000, 2006) * |
| | | | 4 | Town | |
| | | | 3 | Village | |
| | | | 2 | Hamlet | |
| | | | 1 | Negligible (isolated dwellings) | |
| **Riparian vegetation** | Deviation from the natural state of the riparian zone | River segment | 5 | Lack of riparian shrubs and trees (only the presence of annual plants) | Local expert assessment |
| | | | 4 | Fragmented vegetation with bushes and/or the presence of reed | |
| | | | 3 | Second replacement step (dominance of dense brushwood) | |
| | | | 2 | First replacement step (presence of shrub or tree strata with some level of preservation). | |
| | | | 1 | Potential vegetation (presence of shrub and tree strata according to the geo-series) | |
| **Morphological condition** | Deviation from the natural state of the stream bed and banks | Local | 5 | Transverse and longitudinal profile of the channel completely changed, with very few habitats | Local expert assessment |
| | | | 4 | Channelized sector, missing most of the natural habitats | |
| | | | 3 | Channelized sector, missing some types of natural habitats, but maintaining much of the shape of the natural channel | |
| | | | 2 | Poorly changed sector, close to the natural mosaic of habitats. | |
| | | | 1 | Morphological changes absent or negligible | |

**Table 2.** Cont.

| Variables | Description | Assessment scale | Score | Criteria | Methods |
|---|---|---|---|---|---|
| **Sediment load** | Deviation from the natural sediment load (both carried in the water column and deposited on the riverbed) | River segment and local | 5 | >75% of coarse particles of the stream bed are covered with fine sediments (sand, silt, clay) | Local expert assessment |
| | | | 4 | 50–75% of coarse particles of the stream bed are covered with fine sediments (sand, silt, clay) | |
| | | | 3 | 25–50% of coarse particles of the stream bed are covered with fine sediments (sand, silt, clay) | |
| | | | 2 | 5–25% of coarse particles of the bed are covered with fine sediments (sand, silt, clay) | |
| | | | 1 | <5% of coarse particles of the stream bed are covered with fine sediments (sand, silt, clay) | |

*information available from http://sniamb.apambiente.pt/clc/frm/.

The response pattern of the most abundant NNS to the human-induced pressure gradient was tested with a nonparametric ANCOVA (detailed above) [42,43] of NNS density between five human pressure classes. The establishment of these five quality classes (High, Good, Moderate, Poor and Bad) for the stressor gradient (total human pressure) to which sites were assigned, followed the approach used in the REFCOND Guidance Document [37]. To account for the possible influence of the environmental variables, a Principal Components Analysis (outputs not shown) was performed in order to extract the most relevant environmental gradient (PCA1), which was then included as a covariable in this analysis. For each species, only data from river types with occurrences were considered.

Univariate Generalized Linear Models (GLM) were used for partitioning the variance [45] among three sets of explanatory variables - environmental, human-induced pressure and spatial trends on the occurrence and density of the most abundant NNS. For each species, only data from river types with NNS occurrences were considered. The set of environmental variables included altitude, mean annual discharge, drainage area of the basin, slope, mean annual temperature, mean water depth, mean stream width, mean current velocity, dominant substrate class, proportion of each habitat type, pH, dissolved oxygen and conductivity. The set of human pressure variables included $BOD_5$, COD, TSS, N, SRP, and the 10 human pressure variables evaluated at each site (Table 2). The spatial structure of data was incorporated into the analyses to prevent misinterpretation of relations between and within the spatially arranged datasets [46]. Owing to spatial autocorrelation, values of particular variables in neighbouring sites are more or less similar than they would be in a random set of observations [47]. Autocorrelation is a frequently observed feature in spatially sampled biological data that may make the identification of plausible relationships between biota and the environment difficult [48]. The spatial structure of data was explored, including geographical coordinates of sites and their higher and cross product terms, in the modelling procedure ($x$, $y$, $xy$, $x^2$, $y^2$, $x^2y$, $xy^2$, $x^3$ and $y^3$) [47]. The $x$ and $y$ coordinates were centred to zero mean before computing the matrix of spatial descriptors, to reduce collinearity between successive terms when fitting the polynomial. GLMs were performed using a forward selection procedure of the explanatory variables. The best models (minimal adequate) were estimated according to the lowest Akaike Information Criterion (AIC). Plots of residuals (not shown) were examined to complement AIC values and to confirm goodness-of-fit (see [49]). For the occurrence of species (presence–absence data), GLMs were performed using the binomial distribution and logit link function. In order to overcome the problems of distribution fitting resulting from the high number of absences in the density matrix, this continuous response variable was standardized to the maximum value for the species and converted into four classes: 0 (0% of individuals), 1 (0% - 10% of individuals), 2 (10% - 50%) and 3 (>50%). For this transformed density response variable (count data), GLMs were performed using the Poisson distribution and log link function. The existence of over-dispersion in data (variance higher than the mean; dispersion parameter>1) was checked during the analysis. If the value observed was higher than the threshold limit then quasi-binomial and quasi-Poisson distributions should be used, respectively. The total variation within the response variables was decomposed among the three sets of explanatory variables and the percentage of explained deviance calculated for eight different components [50]: (i) pure effect of environmental drivers, (ii) pure effect of anthropogenic disturbance, (iii) pure effect of spatial trends, (iv) combined variation due to the joint effect of environmental and anthropogenic components, (v) combined variation due to the joint effect of environmental and spatial components, (vi) combined variation due to the joint effect of anthropogenic and spatial components, (vii) combined variation due to the joint effect of the three components and (viii) variation not explained by the independent variables included in the analysis.

Multicollinearity among explanatory variables may result in the exclusion of ecologically meaningful variables from the models if another intercorrelated variable or variables happen to explain the variation better in statistical terms. Therefore, some of the most clearly intercorrelated variables (Spearman's rank correlation IrI> 0.75; $P<0.05$) were initially excluded. The exclusion decision took into account the potential relevance of the variable in the occurrence and distribution of NNS. Furthermore variables were maintained in the models only if their addition did not cause any Variation Inflation Factor (VIF) to exceed 3.0, therefore ensuring that no covariation exists between the selected variables in the models.

**Table 3.** Description, assessment scale and methods, and scoring criteria of the variables: hydrological regime, toxic and acidification levels, organic and nutrient loads, artificial lentic waters - used to evaluate the level of anthropogenic disturbance in sampled sites.

| Variables | Description | Assessment scale | Score | Criteria | Methods |
|---|---|---|---|---|---|
| **Hydrological regime** | Deviation from the natural hydrological regime (flow pattern and/or quantity). Includes all sources of hydrologic alteration, such as significant water abstraction | Local | 5 | <50% and strong deviation from the natural variability of the flow regime | Local expert assessment complemented with SNIRH |
| | | | 4 | <50% and moderate deviation from the natural variability of the flow regime | |
| | | | 3 | >50% and duration of flood periods close to the natural | |
| | | | 2 | >75% and duration of flood periods close to the natural | |
| | | | 1 | >90% and normal duration of natural flood periods | |
| | | Local | 5 | <10% of mean annual discharge | Local expert assessment complemented with SNIRH |
| | | | 4 | <15% of mean annual discharge | |
| | | | 3 | >15% of mean annual discharge | |
| | | | 2 | >30% of mean annual discharge | |
| | | | 1 | >90% of mean annual discharge | |
| **Toxic and acidification levels** | Deviation from the natural state of toxicity conditions, including acidification and oxygen levels | Local | 5 | Constant for long periods (months) or frequent occurrence of strong deviations from natural conditions (e.g. pH <5.0, DO <30%) | Local expert assessment complemented with SNIRH |
| | | | 4 | Constant for long periods (months) or frequent occurrence of strong deviations from natural conditions (e.g. pH <5.5, DO <30–50%) | |
| | | | 3 | Occasional deviations (single measurements or episodic) in relation to natural conditions (e.g. pH <5.5, DO <30–50%) | |
| | | | 2 | Occasional deviations (single measurements or episodic) in relation to natural conditions (e.g. pH <6.0) | |
| | | | 1 | Conditions within the normal range of variation | |
| **Organic and nutrient loads** | Deviation from the normal values of BOD, COD, ammonium, nitrate and phosphate concentrations | Local | 5 | >20% of values in classes D or E | SNIRH (classification of water quality for multiple uses, according to the guidelines from the National Water Institute*), complemented with local expert assessment |
| | | | 4 | >10% of values in classes D or E | |
| | | | 3 | >10% of values in class C | |
| | | | 2 | No obvious or too small signs of eutrophication and organic loading | |
| | | | 1 | No signs of eutrophication and organic loading | |
| **Artificial lentic water bodies** | Impact related to the presence of artificial lentic water bodies upstream and/or downstream of the site (upstream change in thermal and flow regimes; downstream invasion by exotic species of lentic character) | Local | 5 | Local immediately downstream of a large reservoir or within the influence area of its backwater | SNIRH and available cartography |

**Table 3.** Cont.

| Variables | Description | Assessment scale | Score | Criteria | Methods |
|---|---|---|---|---|---|
| | | | 4 | Local immediately downstream of a mini-hydro or within the influence area of its backwater | |
| | | | 3 | Local downstream of a reservoir or within the influence area of the reservoir | |
| | | | 2 | Local downstream of a mini-hydro or within the influence area of its backwater | |
| | | | 1 | No influence of reservoirs | |
| Connectivity | Impact of artificial barriers to fish migration | River basin and segment | 5 | Permanent artificial barrier | SNIRH, available cartography, documental data and local expert assessment |
| | | | 4 | Occasional passage of some species | |
| | | | 3 | Passage of certain species or only in certain years | |
| | | | 2 | Passage of most species in most years | |
| | | | 1 | No barriers or existence of an effective pass-through device | |

*information available from http://snirh.pt/snirh/_dadossintese/qualidadeanuario/boletim/tabela_classes.php.

Prior to the analyses, all data were either log (x+1) (linear measurements) or arcsin [sqrt(x)] (percentages) transformed to improve normality. Statistical analyses were performed using CANOCO 4.5, STATISTICA 6, PRIMER 6, SPSS 21 and BRODGAR 2.6 software applications.

## Results

### Patterns of richness, distribution and abundance of non-native species

A total of 41 fish species were captured, including 6 diadromous species and 10 NNS: pumpkinseed sunfish *Lepomis gibbosus* (L.), Iberian gudgeon *Gobio lozanoi* Doadrio and Madeira, mosquito-fish *Gambusia holbrooki* (Girard), carp *Cyprinus carpio* L, goldfish *Carassius auratus* (L.), largemouth bass *Micropterus salmoides* (Lacépède), chameleon cichlid *Herichthys facetum* (Jenyns), black bullhead *Ameiurus melas* (Rafinesque), pike-perch *Sander lucio-perca* (L.), and bleak *Alburnus alburnus* (L.) (Table 4).

NNS occurred in 33% of the sites, representing nearly 11% of both the total mean density (3.58 ind./100 m², SD = 12.54) and number of species (0.54, SD = 0.94) per site. Although the absolute values may seem low, they represent an important percentage of the total density and number of species, as Mediterranean streams usually show low species richness and density per site. Biomass values were not shown or included in the analysis, as they followed the density pattern and were therefore redundant.

There were significant differences in the density (F $_{(11,380)}$ = 8.04; $P<0.001$) and proportion of non-native species (F $_{(11,380)}$ = 9.20; $P<0.001$) among river types considering the total human pressure as a covariable in the analysis. NNS did not occur in the mountain river type or in the southern chalk river type. The small northern stream types, N1<100 km², N3 and N4, registered very low densities, frequency of occurrence, and percentage of NNS (Table 4). N2 and S2 also showed low densities of NNS, but these species represented more than 15% of the fish assemblages in each site and registered high values of frequency of occurrence (f. oc.).

N1>100 km² and S1<100 km² presented high values of occurrence and density of NNS, also representing more than 10% of total density and species richness. Littoral and southern river types S1> 100 km² and S3 showed the highest occurrence and abundance of NNS; in Littoral type, NNS represented almost 12% of the total density and 15% of the total species richness, S1>100 km² and S3 registered percentages between 20% and 30%.

The most frequent and abundant NNS were *L. gibbosus* (mean density = 0.89 ind./100 m², SD = 3.3; f. oc. = 0.21), *G. lozanoi* (mean density = 1.79 ind./100 m², SD = 10.9; f. oc. = 0.14), and *G. holbrooki* (mean density = 0.54 ind./100 m², SD = 3.1; f. oc. = 0.11) (Table 4). The remaining species registered low occurrences and abundances. *L. gibbosus* showed a wide distribution in both northern and southern river types - N1> 100 km², N2, L, and mostly southern river types (Table 4), which are associated with high annual temperature and conductivity, large drainage area, and low altitude (Table 1). *G. holbrooki* were the most abundant species in the southern river types, showing a distribution almost limited to these river types, while *G. lozanoi* was the dominant species in north/central river types, exhibiting high occurrence in river types with large drainage area and low altitude, especially in the north/centre region - N1>100 km², N2, and L (Table 4). *C. carpio* and *M. salmoides* occurred in several river types, showing higher values in the south. *H. facetum*, *A. melas*, and *A. alburnus*, only occurred in the southern river types S1<100 km², S1>100 km², and S3 and *S. lucioperca* only occurred in N1>100 km², all presenting very low frequencies of occurrence and abundance. *C. auratus* registered nearly vestigial occurrence in N1>100 km², N2, and S3 (Table 4).

Overall, disturbed sites showed significantly higher occurrence (f. oc. = 0.46, $P<0.001$) and abundance (mean = 6.41, SD = 16.8, $U_{380}$ = 12155.00, $Z$ = -6.56, $P<0.001$) of NNS than undisturbed ones (f. oc. = 0.18; mean = 0.5, SD = 1.7). *M. salmoides*, *A. melas*, *S. lucioperca*, and *A. alburnus* only occurred in disturbed sites. All NNS but *H. facetum*, exhibited significantly higher occurrence ($P<0.01$) in disturbed sites. Fish density was significantly higher in

**Table 4.** Frequency of occurrence (f.oc.), number of species (%) and density (mean and SD) (ind/100 m²) of non-native fish species (NNS) for total data (a), undisturbed (b) and disturbed (c) sites in each river-type (see Table 1) with NNS occurrence.

| | | NNS f.oc. | Total | N1<100 km² | N1>100 km² | N2 | N3 | N4 | L | S1<100 km² | S1>100 km² | S2 | S3 |
|---|---|---|---|---|---|---|---|---|---|---|---|---|---|
| NNS f.oc. | a) | 0.30 | | 0.1 | 0.4 | 0.7 | 0.07 | 0.2 | 0.5 | 0.4 | 0.7 | 0.3 | 0.5 |
| | b) | 0.18 | | 0 | 0.23 | 0.67 | 0.12 | 0.11 | 0 | 0.13 | 0.50 | 0.25 | 0.17 |
| | c) | 0.46 | | 0.25 | 0.58 | 0.80 | 0 | 0.29 | 0.58 | 0.70 | 0.79 | 0.40 | 0.52 |
| Nbr. NNS | a) | | 12.0 | 3.0 | 13.1 | 19.3 | 2.9 | 4.6 | 14.8 | 13.1 | 31.1 | 16.3 | 19.6 |
| (%) | | | (21.0) | (9.2) | (19.3) | (14.2) | (11.2) | (10.0) | (19.6) | (23.2) | (30.1) | (31.1) | (25.9) |
| | b) | | 5.3 | 0 | 5.6 | 19.0 | 4.9 | 2.2 | 0 | 2.6 | 17.3 | 7.4 | 4.2 |
| | | | (12.0) | | (10.9) | (15.2) | (14.2) | (6.7) | | (7.2) | (19.5) | (15.6) | (10.2) |
| | c) | | 18.2 | 6.4 | 19.0 | 20.0 | 0 | 7.7 | 18.8 | 29.8 | 41.2 | 27.0 | 22.6 |
| | | | (25.2) | (12.7) | (22.4) | (12.6) | | (13.1) | (20.3) | (30.1) | (32.8) | (41.7) | (27.0) |
| NNS Mean density | a) | | 3.6 | 2.3 | 4.8 | 1.2 | 0.07 | 0.3 | 9.3 | 3.6 | 6.4 | 1.1 | 6.3 |
| (ind/100 m²) | | | (12.5) | (9.1) | (12.8) | (1.6) | (0.3) | (0.8) | (30.1) | (8.6) | (11.5) | (2.1) | (12.5) |
| | b) | | 0.5 | 0 | 0.4 | 1.2 | 0.1 | 0.3 | 0 | 1.6 | 1.4 | 0.7 | 0.2 |
| | | | (1.7) | | (1.2) | (1.8) | (0.5) | (1.0) | | (4.4) | (1.8) | (1.6) | (0.4) |
| | c) | | 6.4 | 4.8 | 8.2 | 1.2 | 0 | 0.2 | 11.8 | 6.9 | 10.1 | 1.5 | 7.5 |
| | | | (16.8) | (13.0) | (16.3) | (0.9) | | (0.4) | (33.6) | (12.4) | (14.1) | (2.7) | (13.3) |
| Lepomis gibbosus | | 0.21 | 0.9 | 0.1 | 1.1 | 0.6 | 0.01 | 0.3 | 0.8 | 2.3 | 2.1 | 0.06 | 2.3 |
| | | | (3.3) | (0.9) | (3.7) | (0.9) | (0.02) | (0.8) | (2.1) | (6.7) | (4.4) | (0.3) | (5.4) |
| Cyprinus carpio | | 0.03 | 0.1 | 0 | 0 | 0.01 | 0 | 0 | 0.09 | 0.01 | 0.9 | 0 | 0.08 |
| | | | (1.6) | | | (0.01) | | | (0.4) | (0.06) | (5.5) | | (0.4) |
| Carassius auratus | | 0.008 | 0.01 | 0 | 0.01 | 0.01 | 0 | 0 | 0 | 0 | 0 | 0 | 0.01 |
| | | | (0.03) | | (0.06) | (0.01) | | | | | | | (0.02) |
| Gobio lozanoi | | 0.14 | 1.8 | 2.2 | 3.4 | 0.6 | 0.07 | 0 | 7.5 | 0 | 0 | 0 | 1.6 |
| | | | (10.9) | (9.1) | (10.6) | (1.2) | (0.3) | | (29.9) | | | | (6.7) |
| Micropterus salmoides | | 0.02 | 0.07 | 0 | 0.2 | 0 | 0 | 0 | 0.2 | 0 | 0.1 | 0.07 | 0.08 |
| | | | (0.7) | | (1.5) | | | | (0.6) | | (0.6) | (0.3) | (0.5) |
| Herichthys facetum | | 0.01 | 0.03 | 0 | 0 | 0 | 0 | 0 | 0 | 0.3 | 0.1 | 0 | 0 |
| | | | (0.4) | | | | | | | (1.3) | (0.3) | | |
| Ameiurus melas | | 0.003 | 0.1 | 0 | 0 | 0 | 0 | 0 | 0 | 0 | 0 | 0 | 1.2 |
| | | | (2.2) | | | | | | | | | | (7.1) |
| Gambusia holbrooki | | 0.11 | 0.5 | 0 | 0.03 | 0 | 0 | 0.01 | 0.7 | 1.1 | 2.7 | 0.9 | 1.1 |
| | | | (3.1) | | (0.2) | | | (0.06) | (2.0) | (2.5) | (9.0) | (2.1) | (3.2) |
| Sander lucioperca | | 0.003 | 0.01 | 0 | 0.04 | 0 | 0 | 0 | 0 | 0 | 0 | 0 | 0 |
| | | | (0.1) | | (0.3) | | | | | | | | |
| Alburnus alburnus | | 0.005 | 0.03 | 0 | 0 | 0 | 0 | 0 | 0 | 0 | 0.4 | 0 | 0 |
| | | | (0.5) | | | | | | | | (1.6) | | |

disturbed sites only for the most abundant species, *L. gibbosus* ($U_{380} = 15029.50$, $Z = 3.94$, $P<0.001$), *G. lozanoi* ($U_{380} = 15594.50$, $Z = 3.73$, $P<0.001$), *G. holbrooki* ($U_{380} = 14676.50$, $Z = 5.68$, $P<0.001$), and *M. salmoides* ($U_{380} = 17199.00$, $Z = 2.91$, $P<0.01$). In all river types but N3, the occurrence of NNS was significantly higher in disturbed sites ($P<0.01$). The abundance and percentage of NNS were higher in human-disturbed sites in L, N1<100 km$^2$, N1>100 km$^2$, S1<100 km$^2$, and S1>100 km$^2$ (Table 4). The majority of river types that did not revealed significant differences between undisturbed and disturbed sites presented null or very low abundance of non-native fishes.

No significant correlations (Spearman's rank correlations IrI < 0.5) were observed between the density of the most abundant NNS and total human pressure, as this relationship was not linear. A marked increase in fish density along the human disturbance gradient considering the environmental gradient (PCA1) as a covariable was observed, but there was a decrease in fish abundance in the extreme of the gradient, when the pressure was maximum (Fig. 2). This was particularly evident for *G. lozanoi* ($F_{(4,246)} = 5.34$, $P<0.001$) and *G. holbrooki* ($F_{(4,235)} = 7.07$, $P<0.001$), as mean fish densities reached the highest values in the poor quality class. The increase in fish density from moderate to poor quality classes was particularly marked for *G. holbrooki*. This species showed a comparatively higher mean density in the bad quality class and a smooth decrease between the poor and bad quality classes. *L. gibbosus* densities showed a considerable dispersion in each quality class. Nevertheless, the highest mean density was observed in moderate human pressure conditions, decreasing in the poor quality class and presenting very low values in the extreme of the pressure gradient (Fig. 2). Overall, the response of this species along the disturbance gradient was less marked ($F_{(4,343)} = 0.30$, $P>0.05$) than that observed for the other species.

## Relative importance of pure environmental and human pressure variables

The results obtained above showed that both environmental and human pressure variables play a determinant role in the occurrence and abundance of non-native species. Moreover, the performed analyses revealed the existence of significant covariation ($P<0.001$) between these two groups of variables. Therefore, the analysis of the relationship between the occurrence and abundance of the most abundant NNS and the environmental and

human pressure gradients was complemented with a GLM approach in order to simultaneously evaluate the relative influence of environmental variables, human-induced pressure, and spatial trends. All GLMs showed over-dispersion parameters under the threshold limit of 1, and were therefore performed using binomial and Poisson distributions as previously described. For all the response variables, most of the explained variability was accounted for by unique, or "pure", components. Shared effects of explanatory variables were responsible for a comparatively small proportion of data variability (Fig. 3). Together, environmental variables and human pressure accounted for the majority of the explained variation in the occurrence and abundance of the three considered NNS (31.7% to 40.2%) (Fig. 3).

The partition of the variance showed some differences in the relative influence of each variable set among the three NNS. The occurrence of NNS was mainly determined by environmental variables, whereas abundance showed a proportionally higher influence of human pressure variables. *L. gibbosus* occurrence and abundance revealed a high dependence on environmental variables. *G. lozanoi* presented a balanced influence of environmental and pressure variables. *G. holbrooki* showed a strong influence of human pressure on both occurrence and abundance (Fig. 3). The results suggest that these three species have different levels of tolerance to human pressure. In all models, spatial variables (autocorrelation) represented the smaller fraction of the explained variation in both occurrence (0.8% to 4.8%) and abundance of the three NNS (1.4% to 2.5%) (Fig. 3). A higher influence of these variables on the occurrence of *G. lozanoi* and *G. holbrooki* was observed, which is in accordance with a more restricted and concentrated distribution of these two species.

Overall, both regional and local explanatory variables were present in the best ecological models in environmental and human pressure GLMs (Table 5). Amongst environmental GLMs, drainage area, in particular, showed a significant reduction in deviance, and positively shaped all the response variables. Slope (positive coefficient) and percentage of riffles (negative coefficient) were ranked as the second most important predictors respectively in G. lozanoi and L. gibbosus occurrence and abundance. For G. holbrooki only mean annual temperature (positive coefficient) revealed a higher positive influence than drainage area. Concerning human pressure variables, toxicity and acidification levels were the most important predictors and were negatively related to the occurrence and abundance of the three NNS. For L. gibbosus,

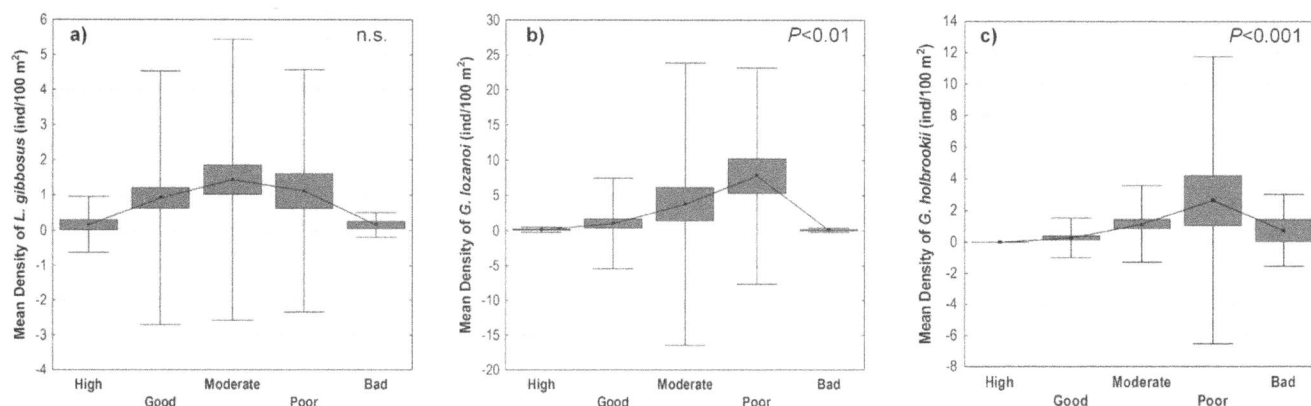

**Figure 2. Response pattern of the most abundant NNS to human-induced pressure gradient, established according to 5 environmental quality classes: a) Mean density of *L. gibbosus*; b) Mean density of *G. lozanoi*; c) Mean density of *G. holbrooki*. (■):** Mean; box:±SE; whisker:±SD. Significance of ANCOVA results with total human pressure as a covariable is shown.

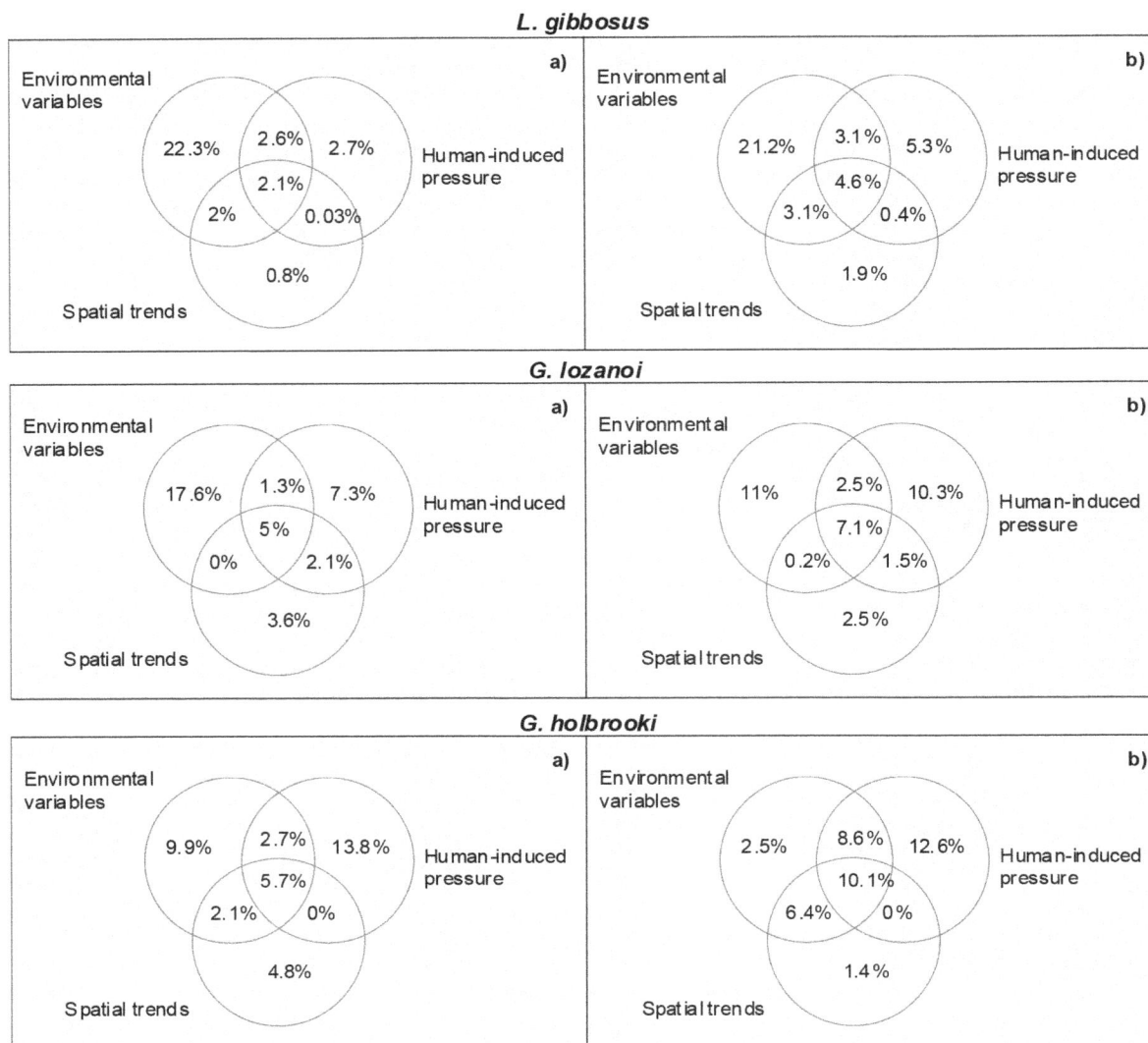

**Figure 3. Veen diagrams showing the partition of the variance in the occurrence (a) and mean density (b) of the most abundant NNS explained by environmental variables, human-induced pressure and spatial trends.**

TSS and COD also assumed a significant positive influence. Although nutrient/organic loads were retained in both L. gibbosus human pressure GLMs, these variables did not present a relevant decrease of deviance. Total dissolved nitrogen (N) and urban area showed an important relation with G. lozanoi occurrence and density. Sediment load revealed a particular influence on G. holbrooki occurrence, presenting the highest deviance reduction in this model.

For all the models produced, values of VIF were very low (Table 5), indicating the lack of significant multicollinearity between predictors, ensuring that no covariation exists among the variables included in the models, and therefore ensuring the reliability of the conclusions arising from the results.

## Discussion

Non-native fish present in Portuguese rivers have different histories. Although rapid spread is possible in a short period, the timing and location of NNS introductions is relevant to the expansion of species distribution and abundance, which explains the relatively narrow distribution and abundance of A. melas, A. alburnus, and S. lucioperca, all recently introduced [9–10,51] at the date of this study, compared with other NNS introduced earlier. The most frequent and abundant species were L. gibbosus, G. holbrooki and G. lozanoi. In southern rivers the most common and abundant NNS were the former ones while G. lozanoi dominated the northern rivers. As such, the discussion will mainly focus on these NNS, which presented the higher invasive potential and are totally established in Portuguese continental waters for more than forty years.

A striking fact regarding the distribution of NNS in this study was their absence from mountain and southern chalk river types. Mountain rivers and small highland streams, with high slope and energy and low productivity (such as M, N4, S4 and N3), seem to be much less vulnerable to invasion by the set of introduced species. The upstream reaches of most Mediterranean basins experience strong seasonal patterns in their environmental conditions. In winter and spring, these streams present high flows whereas in summer they are frequently reduced to small isolated pools. This hydrological regime may prevent the invasion of NNS

**Table 5.** Summary of the environmental and human pressure GLMs for *L. gibbosus*, *G. lozanoi* and *G. holbrooki* (response variables occurrence and density).

| Species | Response variables | Models features | Models AIC | Explanatory variables | Type | Deviance reduction | Coef. sign | VIF | P |
|---|---|---|---|---|---|---|---|---|---|
| *L. gibbosus* | Occurrence | Binomial distribution "logit" link function | 274.7 | Drainage area | Env | 76.31 | + | 2.14 | *** |
| | | | | % Riffles | Env | 21.95 | - | 1.5 | * |
| | | | | TSS | Pres | 21.93 | + | 1.09 | * |
| | | | 356.5 | Nutrient/organic loads | Pres | 3.06 | + | 2.09 | * |
| | Density | Poisson distribution "log" link function | | Drainage area | Env | 69.53 | + | 2.36 | *** |
| | | | | % Riffles | Env | 28.44 | - | 1.63 | ** |
| | | | 430.9 | Mean annual runoff | Env | 7.66 | - | 1.78 | * |
| | | | | Toxic/acid levels | Pres | 18.99 | - | 1.44 | * |
| | | | | COD | Pres | 13.08 | + | 1.31 | ** |
| | | | | TSS | Pres | 5.66 | + | 1.12 | * |
| | | | 496.1 | Nutrient/organic loads | Pres | 2.68 | + | 2.19 | * |
| *G. lozanoi* | Occurrence | Binomial distribution "logit" link function | 216.3 | Drainage area | Env | 38.40 | + | 2.21 | ** |
| | | | | Slope | Env | 11.22 | + | 1.61 | * |
| | | | | Dissolved N | Pres | 17.62 | + | 1.42 | ** |
| | | | | Toxic/acid levels | Pres | 13.53 | - | 1.92 | * |
| | | | | COD | Pres | 5.39 | - | 2.27 | * |
| | Density | Poisson distribution "log" link function | 236.1 | TSS | Pres | 4.31 | + | 1.29 | ** |
| | | | | Drainage area | Env | 26.26 | + | 2.73 | ** |
| | | | 286.4 | Slope | Env | 13.73 | + | 1.69 | * |
| | | | | COD | Pres | 15.06 | - | 2.17 | ** |
| | | | | Urban area | Pres | 11.41 | + | 2.62 | * |
| | | | | TSS | Pres | 4.94 | + | 1.22 | * |
| | | | | Toxic/acid levels | Pres | 6.16 | - | 1.69 | * |
| | | | 293.38 | Land use | Pres | 6.13 | - | 2.43 | * |
| *G. holbrooki* | Occurrence | Binomial distribution "logit" link function | 186.1 | Mean annual temperature | Env | 22.88 | + | 2.12 | * |
| | | | | Drainage area | Env | 14.87 | + | 2.38 | ** |
| | | | | % Riffles | Env | 7.86 | - | 1.44 | * |
| | | | | Sediment load | Pres | 22.06 | + | 2.31 | *** |
| | | | 186.7 | Toxic/acid levels | Pres | 13.21 | - | 1.98 | *** |
| | Density | Poisson distribution "log" link function | | Mean annual temperature | Env | 26.06 | + | 2.25 | * |
| | | | 247.3 | Drainage area | Env | 9.48 | + | 2.29 | * |
| | | | | Toxic/acid levels | Pres | 42.29 | - | 1.77 | ** |
| | | | | Sediment load | Pres | 9.70 | + | 1.36 | * |
| | | | 240.8 | Artif. lentic water bodies | Pres | 6.40 | + | 2.75 | * |

Explanatory variables type: environmental (Env) and human-induced pressure (Pres). Significance levels (*P*) at
*$P < 0.05$.
**$P < 0.01$.
***$P < 0.001$.

which are poorly adapted to high discharge events, including flash flows (e.g. [52–53]). Conversely, NNS occurrence and abundance were particularly meaningful in river types with medium and large drainage areas, both in the northern "cool–warm water" and in the south. The high proportion of non-native fishes in the middle and lower reaches of Mediterranean-climate rivers was also reported in other studies (e.g. [54–56]). In general, the pattern of NNS distribution and abundance among river types followed the pattern of human-induced disturbance. Although human-induced disturbance occurred among all river types, some areas exhibited higher degradation. Overall, the sites located in littoral regions, both in the centre-north and in the southern region, which are associated with higher human density, presented higher degraded conditions and higher occurrence and abundance of non-native fishes. The least disturbed river types were located in higher altitude regions and with small drainage areas, where NNS were absent or present in very low densities. Most of those streams are located in isolated areas with difficult human access, far from the main human pathways. Remote areas with low human disturbance receive much lower propagule pressure than areas hosting immense human settlement [57].

Sites with higher human pressure exhibited significantly higher occurrence, abundance, and percentage of NNS. This pattern was observed in all river types and for all species, excepting those with very low abundance of NNS and *H. facetum*. Several species (*A. alburnus*, *A. melas*, *S. lucioperca*) occurred exclusively in impaired sites. The response of the most abundant NNS to the human pressure gradient was shaped by a marked increase in their abundance along the quality classes. Though the same general pattern was observed for the three species, *G. lozanoi* and *G. holbrooki* were particularly responsive to degradation, reaching the highest abundances in the poor quality class.

Non-native freshwater fish have commonly been documented to succeed in degraded aquatic habitats in many areas of the world [28,58–61]. Disturbed systems and communities may attract biological invasions more than pristine ecosystems due to the redistribution of space and energy resources and may promote new vacant niches for the most adaptable and tolerant invaders [11,62]. One of the most important human pressures in Mediterranean-climate rivers is related to nutrient and organic loads, which under the highly favourable climatic conditions increase the aquatic productivity and food availability. The resource availability theory argues that invasibility is directly influenced by nutrient enrichment and resource availability [11,63], which may facilitate biological contamination by reducing resource limitation and therefore competition. The high proportion of NNS in medium and large lowland rivers where productivity is naturally higher corroborates this hypothesis. Moreover, these river types (Littoral, N1>100 km$^2$ and S1>100 km$^2$) are also the most affected by hydrological disturbances, flow regulation, connectivity loss, and dam construction. Dams typically create lentic conditions that favour non-native fishes, thus presenting higher abundance in regulated streams than in unregulated ones [64–65]. These river types are particularly vulnerable to non-native fish invasions, as some species, namely *L. gibbosus* and *G. holbrooki*, prefer enriched lentic habitats (e.g. [53,66]).

This study reveals that natural landscape/environmental drivers and human-induced disturbance are both important factors determining the distribution and invasiveness of NNS, and their relative importance vary from species to species.

The occurrence of *G. lozanoi* was mostly explained by environmental variables, while its abundance was equally determined by landscape/habitat and human-disturbance factors. In the Portuguese continental waters *G. lozanoi* populations were particularly abundant in N1>100 km$^2$ and littoral river types which is consistent with previous work associating this species to lowland rivers with slow flow [67]. Being absent from high altitude rivers and warm waters, this species occurrence was positively related to slope and drainage area (larger areas providing water flow all year long). For this reason, *G. lozanoi* does not occur in the warmer southern and central rivers, which present low or no flow during summer. The species abundance was positively correlated with human-induced hydrological disturbance, habitat modification, water quality degradation (TSS, COD, acidification) and agricultural areas. Other studies also evidenced this species response to anthropogenic hydrological disturbance [64,68].

*G. holbrooki* is mainly present in southern lowland river types (S1>100 km$^2$, S1<100 km$^2$ and S3) with the highest water temperature and conductivity. This species occurrence was positively correlated with temperature and drainage area. Although *G. holbrooki* is able to withstand wide temperature ranges, it prefers warm water temperatures [69–70]. Conversely, the lower temperatures of the northern rivers could limit the species proliferation, due to the effect of latitude on life-history traits, namely the reproductive potential [71]. At the habitat scale, *G. holbrooki* was negatively associated with riffles, that is, it displays a preference for standing or slow flowing waters rich in organic detritus and muddy sediments. High river discharges tend to displace individuals and eliminate populations [52,58,72], as this species has no swimming abilities and behavioural response to fast flowing waters [73]. The invasiveness of *G. holbrooki* seems to be mostly determined by human-induced disturbance factors. Its abundance was strongly related to indicators of disturbance describing local in-stream habitat modifications and organic and sediment loads. The response to degradation is partially related to the species tolerance to a wide range of environmental conditions, including pH (from 3.9 to 8.8, [74]) and dissolved oxygen (0.28 mg/L, [69–70]. The ability to tolerate low dissolved oxygen enables it to survive in anoxic eutrophic waters with high organic and nutrient loads, and it is often the only fish species present in these water bodies (Ilhéu pers. obs.). Moreover, this species is also tolerant to a wide range of pollutants, including organic wastes, phenols, pesticides, and heavy metals, due to the species' phenotypic plasticity [75–76].

*L. gibbosus* is one of the most widespread NNS in Portugal and a very successful invader. This species was particularly frequent and abundant in warm water rivers of southern and central Portugal – littoral and southern river types, with the exception of southern mountain rivers – while in the north it occurred mainly in larger rivers (N1>100 km$^2$ and N2). *L. gibbosus* responded positively to a certain degree of anthropogenic disturbance, namely hydrological and habitat modification, and nutrient and organic loads, probably because these conditions lead to an increase in suitable habitats and also offer feeding advantages [77]. The intermediate tolerance of *L. gibbosus* to degradation, also reported by other authors [78], supports the idea that human-induced disturbance is not a requisite for successful invasion by all introduced species [79–80], and that a particular site may contain introduced organisms simply because they were introduced there and natural environmental conditions were favourable to their establishment. Indeed, this species abundance and distribution was mostly explained by pure environmental variables; its density was positively related to drainage area and negatively related with runoff and percentage of riffles, which is consistent with its higher occurrences and abundance in the warm water lowland rivers. Sunfishes exhibit low ability to withstand high current velocities and are often reduced in number by flood events [73,81].

Differences in swimming ability or behavioural response to high current velocities between native and non-native species may be the mechanism responsible for the observed differential removals [65–66]. Furthermore, *L. gibbosus* is known to be invasive in some southern and central European locations but not in England or Norway [82–84], being the climatic conditions the limiting factor in the species invasiveness [85]. This illustrates the high vulnerability of the warm water rivers of southern Europe, where environmental conditions tend to favour the proliferation of this type of species, being high flows the major constraint.

The invasive potential of NNS seems to be determined by different environmental factors that may interact direct and indirectly. However, other factors may also explain the success of many NNS invasions, namely the spatial structure and the biota features, including species life-histories and interspecific interactions [86–88]. Thus further studies integrating all these factors should be performed in order to enhance the ability to predict the vulnerability of ecosystems to invasive species.

## Conclusions

This study reveals that natural landscape/environmental drivers are major factors determining the distribution of non-indigenous species in Portuguese rivers, while their abundance seems to be largely promoted by human-induced disturbance. The importance and role of different invasion factors are context dependent, because of the interaction between species traits and environment, regarding habitat, climate match, and human disturbance, and thus the river type approach seems to be useful to predict NNS

invasiveness. This study also stresses the high vulnerability of the warm water lowland river types to invasion by NNS where conditions are amplified by human-induced degradation. Although some warm water species may be present in most basins (e.g. *L. gibbosus*), they become successful invaders mainly in large lowland rivers, particularly in the south. In contrast, high energy and low productivity rivers, typically at higher altitudes, seem to resist better to the invasion of NNS even when subjected to anthropogenic disturbance. These are relevant issues for assessment of the ecological status of aquatic systems, river basin management and river rehabilitation programmes.

## Acknowledgments

We are grateful to several persons and their teams, who have taken part in data collection and entry: field colleagues from Universidade de Évora, M. T. Ferreira from Instituto Superior de Agronomia/Universidade de Lisboa, P. Raposo de Almeida from Instituto de Oceanografia/ Universidade de Évora, N. Formigo from Universidade do Porto and R. Cortes from Universidade de Trás-os-Montes e Alto Douro. We are also grateful to Dárcio Sousa for GIS help and A. Márcia Barbosa for statistical advice in GLM procedure. Valuable comments and critical review of an earlier draft of this manuscript by Thierry Oberdorff are deeply acknowledged, as well as the insightful comments of three anonymous referees, which have improved the manuscript.

## Author Contributions

Conceived and designed the experiments: MI PM JMB. Performed the experiments: MI PM JMB. Analyzed the data: PM MI. Contributed reagents/materials/analysis tools: MI JMB. Wrote the paper: MI PM JMB.

## References

1. Clavero M, Garcia-Berthou E (2006) Homogenization dynamics and introduction routes of invasive freshwater fish in the Iberian Peninsula. Ecol Appl 16: 2313–2324.
2. Leprieur F, Beauchard O, Blanchet S, Oberdorff T, Brosse S (2008) Fish invasions in the world's river systems: when natural processes are blurred by human activities. PloS Biol 6: 404–410.
3. Marr SM, Olden JD, Leprieur F, Arismendi I, Ćaleta M, et al. (2013) A global assessment of freshwater fish introductions in Mediterranean-climate regions. Hydrobiologia 719: 317–329.
4. Benejam L, Carol J, Benito J, Garcia-Berthou E (2007) On the spread of the European catfish (*Silurus glanis*) in the Iberian Peninsula: first record in the Llobregat river basin. Limnetica 26: 169–171.
5. Caiola N, de Sostoa A (2002) First record of the Asiatic cyprinid *Pseudorasbora parva* in the Iberian Peninsula. J Fish Biol 61: 1058–1060.
6. Comesaña J, Ayres C (2009) New data on the distribution of pumpkinseed *Lepomis gibbosus* and largemouth bass *Micropterus salmoides*, and of non-endemic Iberian gudgeon *Gobio lozanoi* in the Galicia region (NW Spain). Aquat Invasions 4: 425–427.
7. Miranda R, Leunda PM, Oscoz J, Vilches A, Tobes I, et al. (2010) Additional records of non-native freshwater fishes for the Ebro River basin (Northeast Spain). Aquat Invasions 5: 291–296.
8. Pérez-Bote JP, Romero RR (2009) First record of *Sander lucioperca* (Perciformes, Percidae) in the Alqueva reservoir, Guadiana basin (SW Iberian Peninsula). Limnetica 28: 225–228.
9. Ribeiro F, Chaves ML, Marques TA, Moreira da Costa L (2006) First records of *Ameiurus melas* (Siluriformes, Ictaluridae) in the Alqueva Reservoir, Portugal. Cybium 30: 283–284.
10. Ribeiro F, Gante HF, Sousa G, Filipe AF, Alves MJ, et al. (2009) New records, distribution and dispersal pathways of *Sander lucioperca* (Linnaeus, 1758) in Iberian freshwaters. Cybium 33: 255–256.
11. Davis M, Grime J, Thompson K (2000) Fluctuating resources in plant communities: a general theory of invasibility. J Ecol 88: 528–534.
12. Ruesink JL (2005) Global analysis of factors affecting the outcome of freshwater fish introductions. Conserv Biol 19: 1883–1893.
13. Bomford M, Barry SC, Lawrence E (2009) Predicting establishment success of introduced freshwater fishes: a role for climate matching. Biol Invasions 12: 2559–2571.
14. Duncan RP, Bomford M, Forsyth DM, Conibear L (2001) High predictability in introduction outcomes and the geographical range size of introduced Australian birds: a role for climate. J Anim Ecol 70: 621–32.
15. Williamson M (1996) Biological Invasions. London: Chapman and Hall. 244 p.

16. Fausch KD, Taniguchi Y, Nakano S, Grossman G, Townsend C (2001) Flood disturbance regimes influence rainbow trout invasion success among five holarctic regions. Ecol Appl 11: 1438–1455.
17. Davies KF, Chesson P, Hamson S, Inouye BD, Melbourne BA, et al. (2005) Spatial heterogeneity explains the scale dependence of the native-exotic diversity relationship. Ecology 86: 1602–1610.
18. Blanchet S, Leprieur F, Beauchard O, Staes J, Oberdorff T, et al. (2009) Broadscale determinants of non-native fish species richness are context-dependent. P Roy Soc Lond B Bio 276: 2385–2394.
19. Gasith A, Resh VH (1999) Streams in Mediterranean-climate regions: abiotic influences and biotic responses to predictable seasonal events. Annu Rev Ecol Syst 31: 51–58.
20. European Commission (2000) Directive 2000/60/EC of the European Parliament and of the Council of 23 October 2000. Establishing a framework for Community action in the field of water policy. Official Journal of the European Communities L327.
21. Leunda PM (2010) Impacts of non-native fishes on Iberian freshwater ichthyofauna: current knowledge and gaps. Aquat Invasions 5: 239–262.
22. Olden JD, Poff NL (2003) Towards a mechanistic understanding and prediction of biotic homogenization. Am Nat 162: 442–460.
23. Rahel FJ (2000) Homogenization of fish faunas across the United States. Science 288: 854–856.
24. Leprieur F, Olden J, Lek S, Brosse S (2009) Contrasting patterns and mechanisms of spatial turnover for native and exotic freshwater fish in Europe. J Biogeogr 36: 1899–1912.
25. Gozlan RE (2010) The cost of non-native aquatic species introductions in Spain: fact or fiction? Aquat Invasions 5: 231–238.
26. Harris JH, Silveira R (1999) Large-scale assessments of river health using an Index of Biotic Integrity with low-diversity fish communities. Freshwater Biol 41: 235–252.
27. Joy MK, Death RG (2004) Application of the index of biotic integrity methodology to New Zealand freshwater fish communities. Environ Manage 34: 415–428.
28. Kennard MJ, Arthington AH, Pusey BJ, Harch BD (2005) Are alien fish a reliable indicator of river health? Freshwater Biol 50: 174–193.
29. Angermeier PL, Smogor RA, Stauffer JR (2000) Regional frameworks and candidate metrics for assessing biotic integrity in Mid-Atlantic highland streams. Trans Am Fish Soc 129: 962–981.
30. Hughes RM, Larsen DP (1988) Ecoregions: an approach to surface water protection. J Water Pollut Control Fed 60: 486–493.
31. Schmutz S, Melcher A, Frangez C, Haidvogl G, Beier U, et al. (2007) Spatially based methods to assess the ecological status of riverine fish assemblages in European ecoregions. Fisheries Manag Ecol 14: 441–452.

32. INAG IP (2008) River Typology in Continental Portugal under the implementation of Water Framework Directive – Abiotic characterization. Lisboa: Instituto da Água, IP (in Portuguese).

33. Miranda R, Oscoz J, Leunda PM, García-Fresca C, Escala MC (2005) Effects of weir construction on fish population structure in the River Erro (North of Spain). Ann Limnol-Int J Lim 41: 7–13.

34. Cabral MJ, Almeida J, Almeida PR, Dellinger TR, Ferrand de Almeida N, et al. (2005) Livro Vermelho dos Vertebrados de Portugal. Lisboa: Instituto da Conservação da Natureza. 659 p.

35. Smith KG, Darwall WRT, editors (2006) The status and distribution of freshwater fish endemic to the Mediterranean basin. Gland, Switzerland, and Cambridge: IUCN. 34 p.

36. FAME (2004) Development, Evaluation and Implementation of a Standardised Fish-based Assessment Method for the Ecological Status of European Rivers-A Contribution to the Water Framework Directive. Final Report (Co-ordinator: Schmutz S). Institute for Hydrobiology and Aquatic Ecosystem Management, University of Natural Resources and Applied Life Sciences, Vienna. Available: http://fame.boku.ac.at. Accessed: 2007 Jul 23.

37. CIS-WFD (2003) Guidance on Establishing Reference Conditions and Ecological Status Class Boundaries for Inland Surface Waters. Final Version. Brussels: EU Common Implementation Strategy for the Water Framework Directive.

38. Clesceri LS, Greenberg AE, Eaton AD (1998) Standard methods for the examination of water and wastewater, 20th Edition. Washington: American Public Health Association, American Water Works Association, Water Environmental Federation.

39. Giller PS, Malmqvist B (1998) The biology of streams and rivers. New York: Oxford University Press. 296 p.

40. INAG IP (2008) Manual for biological assessment of water quality in rivers according to Water Framework Directive. Sampling and analysis protocol for fish fauna. Lisboa: Instituto da Água, I.P. (in Portuguese).

41. CEN (2003) Water Quality–Sampling of Fish with Electricity. European standard–EN 14011:2003. Brussels: European Committee for Standardization.

42. Quade D (1967) Rank analysis of covariance. J Am Stat Assoc 62: 1187–1200.

43. Conover WJ, Iman RL (1982) Analysis of covariance using the rank transformation. Biometrics 38: 715–724.

44. Daniel WW (1987) Biostatistics: A foundation for analysis in the health sciences. New York: Wiley. 734 p.

45. Borcard D, Legendre P, Drapeau P (1992) Partialling out the spatial component of ecological variation. Ecology 73: 1045–1055.

46. Hinch SG, Sommers KM, Collins NC (1994) Spatial autocorrelation and assessment of habitat-abundance relationships in littoral zone fish. Can J Fish Aquat Sci 51: 701–712.

47. Legendre P (1993) Spatial autocorrelation: Trouble or new paradigm. Ecology 74: 1659–1673.

48. Legendre P, Fortin MJ (1989) Spatial pattern and ecological analysis. Vegetation 80: 107–138.

49. Zuur AF, Ieno EN, Smith GM (2007) Analyzing Ecological Data. New York: Springer. 672 p.

50. Andersen MJ, Cribble NA (1998) Partitioning the variation among spatial, temporal and environmental components in a multivariate dataset. Aust J Ecol 23: 158–167.

51. Ribeiro F, Collares-Pereira MJ, Moyle PB (2009) Non-native fish in the fresh waters of Portugal, Azores and Madeira Islands: a growing threat to aquatic biodiversity. Fisheries Manag Ecol 16: 255–264.

52. Meffe GK (1984) Effects of abiotic disturbance on coexistence of predator-prey fish species. Ecology 65: 1525–1534.

53. Bernardo JM, Ilhéu M, Matono P, Costa AM (2003) Interannual variation of fish assemblage structure in a Mediterranean river: implications of streamflow on the dominance of native over exotic species. River Res Appl 19: 1–12.

54. Ilhéu M (2004) Patterns of habitat use by freshwater fish in Mediterranean streams. PhD Thesis, University of Évora.

55. Vila-Gispert A, Alcaraz C, García-Berthou E (2005) Life-history traits of invasive fish in small Mediterranean streams. Biol Invasions 7: 107–116.

56. Hermoso V, Clavero M, Blanco-Garrido F, Prenda J (2011) Invasive species and habitat degradation in Iberian streams: an analysis of their role in freshwater fish diversity loss. Ecol. Appl 21: 175–188.

57. Drake JA, Mooney HA, DiCastri F, Grove RH, Kruger FJ, et al. (1989) Biological Invasions: a Global Perspective. New York: Wiley. 525 p.

58. Arthington AH, Hamlet S, Bluhdorn DR (1990) The role of habitat disturbance in the establishment of introduced warm-water fishes in Australia. In: Pollard DA, editor.Introduced and translocated fishes and their ecological effect.Can-Canberra: Australian Government Publishing Service. pp. 61–66.

59. Gido KB, Brown JH (1999) Invasion of North American drainages by alien fish species. Freshwater Biol 42: 387–399.

60. Meador MR, Brown LR, Short T (2003) Relations between introduced fish and environmental Conditions at large geographic scales. Ecol Indic 3: 81–92.

61. Ross R, Lellis WA, Bennett RM, Johnson CS (2001) Landscape determinants of non indigenous fish invasions. Biol Invasions 3: 347–361.

62. Belote RT, Jones RH, Hood SM, Wender B (2008) Diversity-invasibility across an experimental disturbance gradient in Appalachian forests. Ecology 89: 183–192.

63. González AL, Kominoski JS, Danger M, Ishida S, Iwai N et al. (2010) Can ecological stoichiometry help explain patterns of biological invasions? Oikos 119: 779–790.

64. Alexandre CM, Almeida PR (2010) The impact of small physical obstacles on the structure of freshwater fish assemblages. River Res Appl 26: 977–994.

65. Power ME, Dietrich WE, Finlay JC (1996) Dams and downstream aquatic biodiversity: Potential food web consequences of hydrologic and geomorphic change. Environ Manage 20: 887–895.

66. Godinho FN, Ferreira MT (2000) Composition of endemic fish assemblages in relation to exotic species and river regulation in a temperate stream. Biol Invasions 2: 231–244.

67. Doadrio I, Madeira MJ (2004) A new species of the genus Gobio Cuvier, 1816 (Actynopterigii, Cyprinidae) from the Iberian Peninsula and Southwestern France. Graellsia 60: 107–116.

68. Lobón-Cervia J, Montanes C, de Sostoa A (1991) Influence of environment upon the life history of gudgeon, Gobio gobio (L.): a recent and successful colonizer of the Iberian Peninsula. J Fish Biol 39: 285–300.

69. Lloyd LN (1984) Exotic Fish: Useful Additions or "Animal Weeds"? J Austr New Guinea Fishes Assoc 1: 31–42.

70. Pyke GH (2005) A review of the biology of Gambusia affinis and G. holbrooki. Rev Fish Biol Fisher 15: 339–365.

71. Benejam L, Alcaraz C, Sasal P, Simon-Levert G, Garcia-Berthou E (2009) Life history and parasites of the invasive mosquitofish (Gambusia holbrooki) along a latitudinal gradient. Biol Invasions 1: 2265–2277.

72. Galat DL, Robertson B (1992) Response of endangered Poeciliopsis occidentalis sonoriensis in the Rio Yaqui drainage, Arizona, to introduced Gambusia affinis. Environ Biol Fish 33: 249–264.

73. Minckley WL, Meffe GK (1987) Differential selection by flooding in stream fish communities of the arid American Southwest. In: Matthews WJ, Heins DC, editors.Community and evolutionary ecology of North American stream fishes.Norman: University of Oklahoma Press. pp. 96–104.

74. Brown-Peterson N, Peterson MS (1990) Comparative life history of female mosquitofish, Gambusia affinis, in tidal freshwater and oligohaline habitats. Environ Biol Fish 27: 33–41.

75. Andreasen JK (1985) Insecticide resistance in mosquitofish of the lower Rio Grande Valley of Texas - an ecological hazard? Arch Environ Con Tox 14: 573–577.

76. Saiki MK, Martin BA, May TW (2004) Reproductive status of western mosquitofish inhabiting selenium-contaminated waters in the Grassland Water District, Merced County, California. Arch Environ Con Tox 47: 363–369.

77. Almeida D, Almódovar A, Nicola GG, Elvira B (2009) Feeding tactics and body condition of two introduced populations of pumpkinseed Lepomis gibbosus: taking advantages of human disturbances? Ecol Freshw Fish 18: 15–23.

78. Halliwell DB, Langdon RW, Daniels RA, Kurtenbach JP, Jacobson RA (1999) Classification of freshwater fish species of the northeastern United States for use in the development of indices of biological integrity, with regional applications. In: Simon TP, editor.Assessing the Sustainability and Biological Integrity of Water Resource Quality Using Fish Communities.Boca Raton: CRC Press. pp. 301–333.

79. Niemela J, Spence JR (1991) Distribution and abundance of an exotic groundbeetle (Carabidae): a test of community impact. Oikos 62: 351–335.

80. Townsend CR (1996) Invasion biology and ecological impacts of brown trout Salmo trutta in New Zealand. Biol Conserv 78: 13–22.

81. Schultz AA, Maughan OE, Bonar SA, Matter WJ (2003) Effects of flooding on abundance of native and nonnative fishes downstream from a small impoundment. N Am J Fish Manage 23: 503–511.

82. Copp GH, Fox MG, Przybylski M, Godinho FN, Vila-Gispert A (2004) Life-time growth patterns of pumpkinseed Lepomis gibbosus introduced to Europe, relative to native North American populations. Folia Zool 53: 237–254.

83. García-Berthou E, Alcaraz C, Pou-Rovira Q, Zamora L, Coenders G, et al. (2005) Introduction pathways and establishment rates of invasive aquatic species in Europe. Can J Fish Aquat Sci 62: 453–463.

84. Sterud E, Jorgensen A (2006) Pumpkinseed Lepomis gibbosus (Linnaeus, 1758) (Centrarchidae) and associated parasites introduced to Norway. Aquat Invasions 1: 278–280.

85. Villeneuve F, Copp GH, Fox MG, Stakénas S (2005) Interpopulation variation in growth and life-history traits of the introduced sunfish, pumpkinseed Lepomis gibbosus, in southern England. J Appl Ichthyol 21: 275–281.

86. Allen CR, Nemec KT, Wardwell DA, Hoffman JD, Brust M, et al. (2013) Predictors of regional establishment success and spread of introduced non-indigenous vertebrates. Glob Ecol Biogeogr 22: 889–899.

87. Filipe AF, Magalhães F, Collares-Pereira MJ (2010) Native and introduced fish species richness in Mediterranean streams: the role of multiple landscape influences. Divers Distrib 16: 773–785.

88. Olden JD, Poff LN, Bestgen KR (2006) Life-history strategies predict fish invasions and extirpations in the Colorado River Basin. Ecol Monograph 76: 25–40.

# 7

# Extensive Behavioural Divergence following Colonisation of the Freshwater Environment in Threespine Sticklebacks

**Carole Di-Poi**[1], **Jennyfer Lacasse**[1], **Sean M. Rogers**[2], **Nadia Aubin-Horth**[1]*

1 Département de Biologie & Institut de Biologie Intégrative et des Systèmes (IBIS), Université Laval, Québec, Québec, Canada, 2 Department of Biological Sciences, University of Calgary, Calgary, Alberta, Canada

## Abstract

Colonisation of novel environments means facing new ecological challenges often resulting in the evolution of striking divergence in phenotypes. However, little is known about behavioural divergence following colonisation, despite the predicted importance of the role of behavioural phenotype-environment associations in adaptive divergence. We studied the threespine stickleback (*Gasterosteus aculeatus*), a model system for postglacial colonisation of freshwater habitats largely differing in ecological conditions from the ones faced by the descendants of the marine ancestor. We found that common-environment reared freshwater juveniles were less social, more active and more aggressive than their marine counterparts. This behavioural divergence could represent the result of natural selection that acted on individuals following freshwater colonisation, with predation as a key selection agent. Alternatively, the behavioural profile of freshwater juveniles could represent the characteristics of individuals that preferentially invaded freshwater after the glacial retreat, drawn from the standing variation present in the marine population.

**Editor:** Walter Salzburger, University of Basel, Switzerland

**Funding:** Funding was provided to C. Di-Poi by the Government of Canada Post-Doctoral Research Fellowship, to N. Aubin-Horth and S. Rogers through Natural Sciences and Engineering Research Council of Canada Discovery grants and an Alberta Innovates Technology Futures New Faculty Award to S. Rogers. The funders had no role in study design, data collection and analysis, decision to publish, or preparation of the manuscript.

* E-mail: Nadia.Aubin-Horth@bio.ulaval.ca

## Introduction

The novel ecological challenges that individuals face when colonising new environments have been shown in many instances to result in adaptive trait divergence. Morphology has often been shown to diverge following colonisation as the result of natural selection [1]. Remarkable examples include the diversification of leg and tail lengths in anole lizards that use different microhabitats in trees and on land [2] and trophic morphology associated with different feeding ecology in African cichlid fishes [3]. Much less information is available on whether behaviours that could influence fitness in the new ecological context also evolve. This is true even for very well-known model systems of phenotypic divergence following colonisation of a novel environment. Yet, evolutionary changes in behaviour may be integral in initiating adaptive shifts [4]. It has been shown that behavioural divergence can be associated with morphological divergence in a new ecological context, whereby populations differ in the behavioural traits used to exploit different resources, such as in the association between beak shape and foraging behaviour in Darwin's finches [5]. The specific ecological constraints acting in the novel environment could directly affect the benefits associated with expressing a given behaviour. For example, when there are no substantial benefits associated with group life, such as in low-productivity freshwater habitats where predation pressure is weaker and intraspecific competition is the most relevant process

affecting fitness, the advantageous strategy might be to adopt the solitary lifestyle and to increase foraging activity, boldness and aggressiveness towards conspecifics [6–8]. Boldness, exploratory behaviours and sociability (schooling) have been shown to diverge when individuals face new ecological conditions (poeciliid *Brachyraphis episcopi* [9]; *Poecilia reticulata* [10,11]). Such behavioural divergence has been shown to enhance the probability of ecological speciation [12]. However, quantitative measurements of behavioural divergence between the ancestral population and the population in a new environment, potentially as the result of selection, remain scarce compared to other traits. Understanding the ecological context of behavioural divergence will help draw a complete picture of integrated phenotypes evolving in response to adaptive peak shifts [13].

To address this problem, we used a well-studied model system of colonisation of new environments, the threespine stickleback (*Gasterosteus aculeatus*). In the last 20,000 years, marine threespine sticklebacks have successfully colonised many freshwater systems with widely divergent ecological conditions in the Northern hemisphere [14]. A distinctive feature of this system is that the ancestral marine population is still extent, although it has also potentially diverged from the marine ancestor during this time period, allowing comparison with populations that have colonised new freshwater environments. Comparative studies between marine and freshwater populations have shown that colonisation of freshwater has resulted in rapid evolutionary changes in numerous traits in morphology (diminution of lateral armour and spine development, lower number of iridophores and presence of barred flanks [15,16]), life history (younger and smaller at reproduction [17]), and physiology (lower levels of thyroid hormones, lower critical swimming speed [18,19]). Furthermore, freshwater juveniles show higher territoriality [20] and a lower schooling tendency [21] than juvenile marine fish (representing the ancestral population). However, whether other ecologically relevant behaviours diverged following colonisation of the freshwater habitat by threespine sticklebacks has not been thoroughly studied.

We hypothesised that behaviour would have diverged between populations of sticklebacks that inhabit contrasting marine and freshwater environments [13,22] since the costs and benefits of expressing a particular behaviour may vary according to ecological demands [21,23]. We focused on genetically based differences in behaviour, rather than on phenotypic plasticity by studying common-environment reared juveniles originating from two populations of threespine sticklebacks representing the marine ancestor and the result of freshwater colonisation. We quantified boldness, aggressiveness, sociability, activity and exploratory behaviours, as they are relevant to fitness and ultimately to the ecology and evolution of populations [24]. Previous studies comparing two freshwater populations of sticklebacks differing in predation pressure have shown divergence in behaviour in sociability [25], activity [26,27], aggressiveness [27,28], and boldness [26,28,29]. Therefore, the potential behavioural divergence observed between marine and freshwater sticklebacks may be the result of natural selection, with predation as a key selection agent. Thus we predicted that sticklebacks from a freshwater lake population that experienced a presumed lower predation pressure based on their lower number of lateral armour plates would exhibit a lower sociability tendency, higher activity and exploratory behaviours, as well as a higher propensity to be aggressive towards a conspecific and to be bolder when foraging in a risky situation, compared to the individuals originating from the marine ancestor population that face a high-predation level environment.

## Methods

### Ethics statement

Wild animals were sampled under the Department of Fisheries and Ocean Permit #XR 315 2011 and the BC Ministry of Environment NA-SU-PE-10-63485 permit. The research adheres to the ASAB/ABS Guidelines for the Use of Animals in Research and was approved by the Comité de Protection des Animaux de l'Université Laval (permit #2010012-1).

### Sampling and rearing

Adult threespine sticklebacks from marine and freshwater populations were collected using minnow traps in British Columbia (Canada) in two breeding sites, which differ in fundamental ecological characteristics [13,22]. Marine adult fish were collected from a small and shallow inlet off Malaspina Strait (Oyster Lagoon, 49°36'48''N 124°01'47''W) where salinity ranges from 20 ppt in winter to 32 ppt in summer. The surface area of the lagoon is 1.8 ha and the maximum depth is 3 m. Marine sticklebacks in Oyster Lagoon face a wide variety of predators [22]. Birds, such as great blue herons, mergansers and kingfishers, are the primary piscivores in the lagoon. Mammals (river otters, racoons, mink), and reptiles (e.g. Pacific garter snakes) are also potential predators of sticklebacks. Sculpins also feed on adult and juvenile sticklebacks in the lagoon [22]. The marine fish from Oyster Lagoon have a high-plated phenotype [30], which

has been associated with high predation by vertebrate predators [31]. Freshwater adult fish were collected from Hoggan Lake on Gabriola Island (49°09'08''N 123°49''W). The lake is not connected to the ocean, and this population has been isolated from marine ancestors for approximately 10,000 years. It covers a surface area of 19.7 ha and has a perimeter of 2.2 km. The lake does not contain sculpin, a known predator of sticklebacks, but fish stocking records indicate that it was stocked with rainbow trout (*Oncorhynchus mykiss*) and coastal cutthroat trout (*Oncorhynchus clarkii clarkii*) in 1927 [32] with both of these species being intermittently observed in this environment, indicating that either stocked fish established a self-sustaining population, or that the lake has always contained a natural population. There is no known benthic-limnetic phenotypic divergence in this population. The fish from this population have a low number of lateral armour plates [13], which has been associated with low predation regimes [31].

Adult threespine sticklebacks were crossed to produce pure lines of F1 individuals, resulting in 14 full-sub F1 families, seven from each population. Juveniles were kept in 110 L-tanks under a 12L:12D cycle, with temperature at 18°C in the Life and Science Animal Research Centre at the University of Calgary and fed frozen bloodworms to satiation twice daily. Salinity was maintained at 5 ppt to promote health and maintain proper development in both marine and freshwater stickleback (see [33] for a similar rearing protocol of marine and freshwater populations). Juveniles (8-month old) were air-shipped to the "Laboratoire de Recherche en Sciences Aquatiques" at Université Laval. Juveniles were held in separate 80 L-tanks (stocking density: 0.11 g/L) under similar rearing conditions with the exception that all fish were acclimated to 15°C with a 8L:16D photoperiod (winter conditions). Juveniles from all families originating from a given environment (marine or freshwater) were kept together during air transport and later rearing, such that the information about the family of origin of individuals is not available. All fish were in a non-reproductive state (body length: average of 46.4±0.6 mm for fish from Hoggan Lake; 42.5±0.5 mm for fish from Oyster Lagoon) and sex was determined using a genetic sex marker [34].

### Behavioural series

Behavioural experiments began after at least two weeks of acclimatisation. Each individual fish (n = 36 fish/population) was exposed to five behavioural tests over a 5-day period (sociability, exploration, aggressiveness, activity and boldness, see Fig. 1A for a timeline). After the sociability assay, the focal individual was transferred to an individual 45 L-experimental tank with plants and shelter as enrichment. The focal individual stayed in this tank for the rest of the experiment. In total, there were 6 series of 5 days (6 fish per population tested per series; 12 fish per series). All tests were done at the same time of day for all fish. Repeatability of these behaviours has already been demonstrated in the threespine stickleback [35,36]. Each behavioural assay was filmed with a digital camera (JVC model GZ-MS120) mounted in front of the tank. The tanks were drained, cleaned and filled with water again between each series.

**Sociability.** We used an assay previously described in [21] that ensures repeatable stimulus conditions and the absence of confounding effects of the behaviour of the conspecifics in the school. A motorised mobile reproducing a school of five fish with the same colour as juvenile sticklebacks and of a smaller body length than the average size of the populations (36 mm) was constructed. The artificial school was immersed 10 cm from the water surface and from the tank bottom in an 80 L-tank. The focal fish was added in the tank in the presence of the motionless school for a 15-minute acclimatisation period. The rotating motor

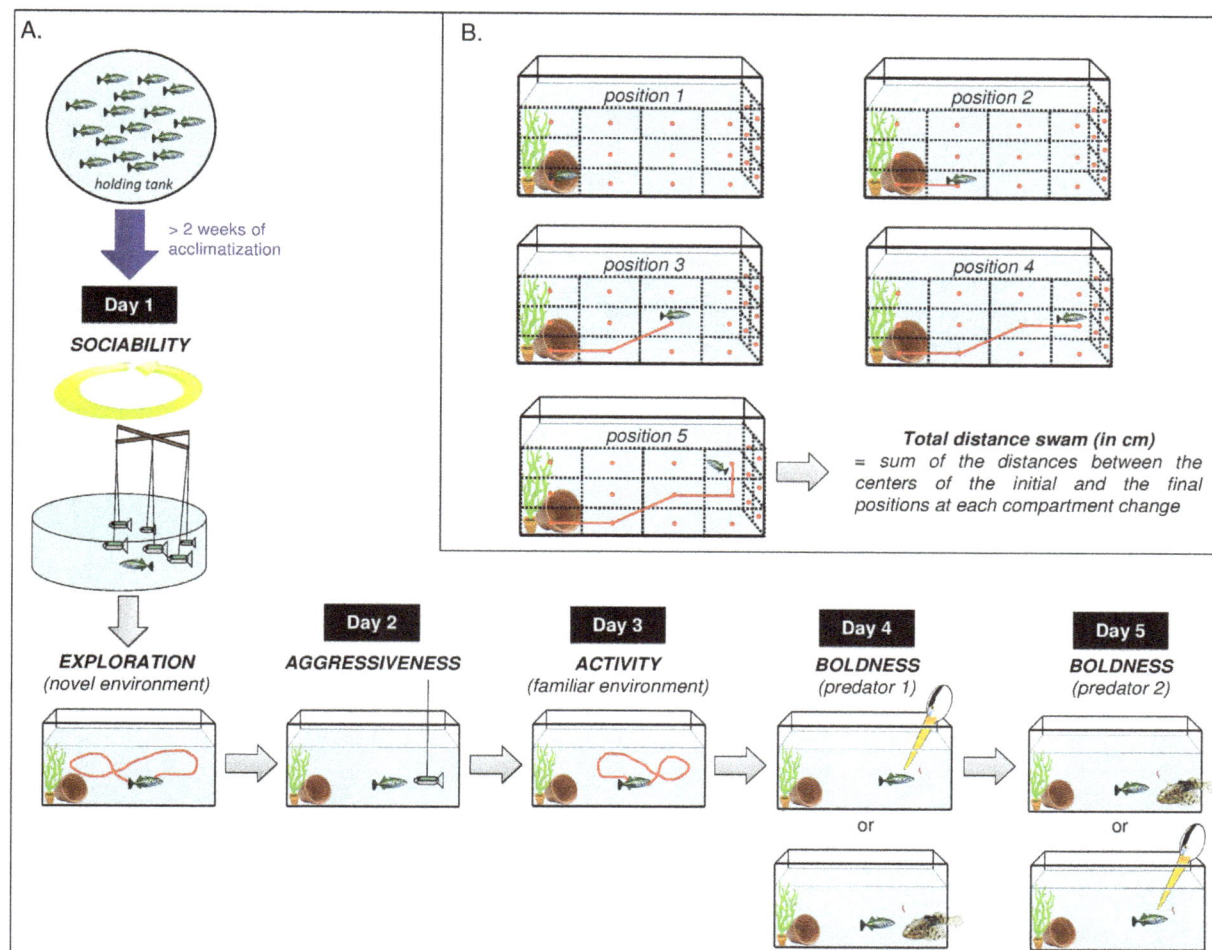

**Figure 1. Illustrations of (A) the set-ups and the timeline of the experiment, and (B) the method to calculate the total distance swam of each fish in the exploration and the activity trials.**

connected to the mobile then moved the fish school around the tank for 15 minutes. The focal fish's schooling behaviour was filmed from above. The time spent swimming within one body length of the group in the same direction and at the same speed (in seconds) was quantified as sociability.

**Exploration.** Immediately after the sociability test, the focal fish was moved into its novel individual experimental tank, where it remained throughout the rest of the behavioural experiment. The fish was placed in a 7 cm-diameter opaque cylinder with an open end, which was then introduced into the tank. The exploratory behaviour in this novel environment was quantified using three measures: (1) latency to exit the cylinder (in seconds) was measured (if the fish was still inside the cylinder after ten minutes following introduction, the cylinder was removed and it was given the maximum possible latency score, i.e. 600 sec), (2) after cylinder removal, time spent swimming (in seconds) in the new environment, and (3) total length swam (in cm) were measured for 5 minutes. The swimming activity of each fish was observed on the video recording and the total distance swam quantified manually using a 3-D grid divided into $4 \times 2 \times 3$ compartments. Each time the fish swam from a compartment to another, the distance between the centers of the initial and the final compartments was measured (Fig. 1B). All distances were summed-up over the 5-min period.

**Aggressiveness.** On the second day, a model juvenile stickleback fish (body length: 36 mm, colour: olive green and grey) was placed as an intruder in the observation tank and the resulting interactions were quantified for a period of 15 minutes. The number of bites made by the focal fish against the intruder was used as a measure of aggressiveness [37]. The percentage of individuals exhibiting at least one biting event was also quantified.

**Activity.** Activity is a measure of the general level of swimming activity in its familiar environment, in contrast to exploration, which is measured in a novel environment. On the third day when the fish was familiarised with its experimental tank, the time spent moving (in seconds), as well as the total length swam (in cm) were measured, as previously explained for the exploration assay, for each subject during a 5-minute period, at the same time of day as in the exploration test.

**Boldness.** The response of an individual in a risky situation was used as a measure of boldness. On the fourth and fifth day, during the feeding session, a simulated predator attack was manually performed by the experimenter on the focal fish using a replica of the head of a great blue heron or a flexible lure in the shape of a sculpin fish, two natural predators of sticklebacks. The feeding session started with provision of a single bloodworm, followed by a 30-second wait and a second bloodworm. The predator attack was simulated at the third bloodworm, when the

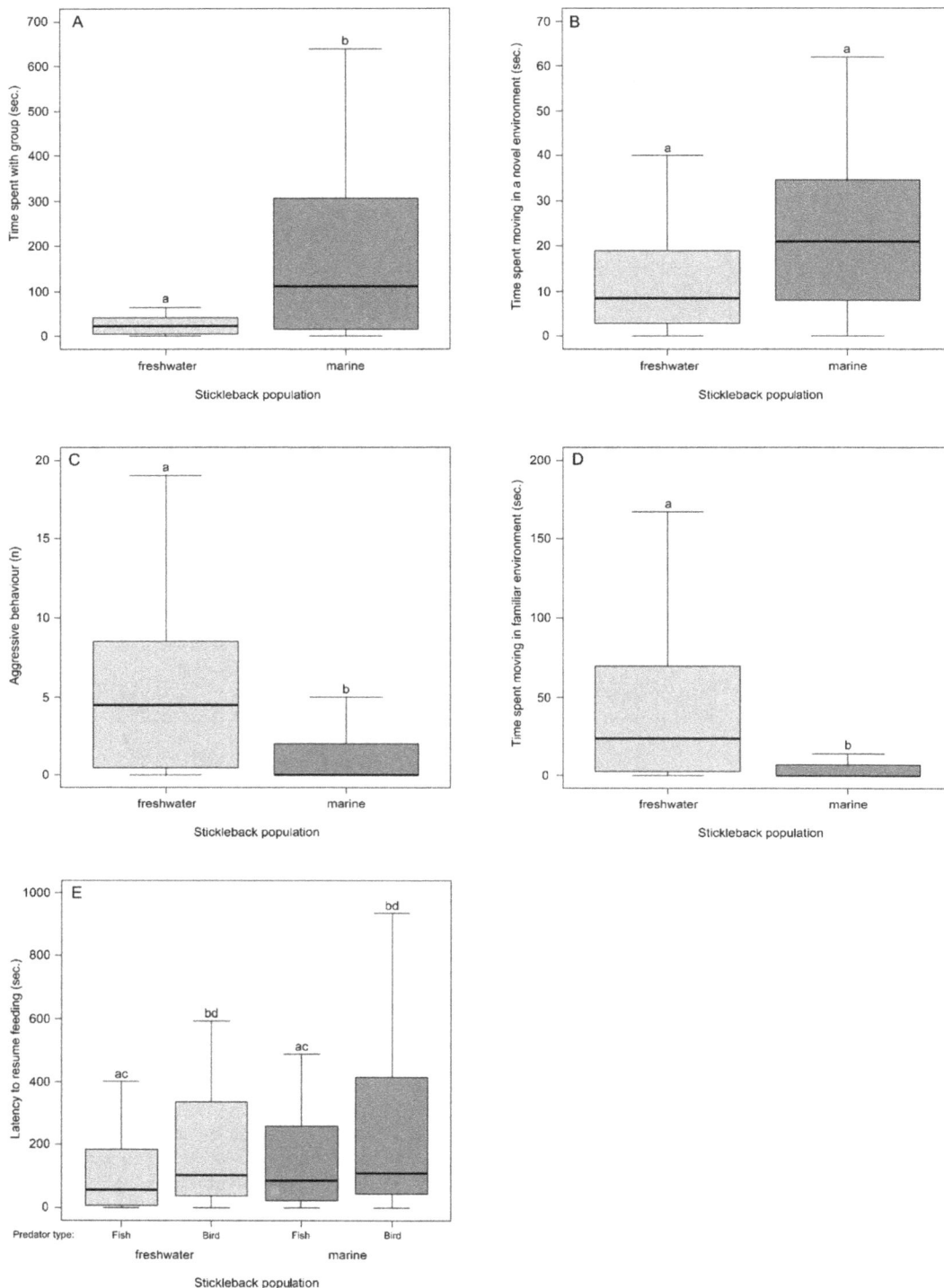

**Figure 2. Behaviour (median ± quartiles) of juvenile threespine sticklebacks from a marine and a freshwater population.** (A) Sociability, (B) exploratory behaviour in a novel environment, (C) aggressiveness, (D) activity in a familiar environment, and (E) boldness. Different letters indicate a significant difference between populations.

focal fish approached within at least one body length of the food. Striking the water with the beak directly over the area where food was distributed simulated the sudden overhead attack of a heron searching for prey. The beak was plunged about 5 cm below the water surface and moved in the water for 10 seconds. To simulate

a large sculpin attack, the sculpin replica was suddenly placed outside the front of the aquarium and moved along the length of the tank for 10 seconds. The experimenter was hidden behind a black curtain, such that no human movement was perceptible to the fish. Latency to resume feeding after the attack (in seconds) was used as a measure of boldness. Each focal fish was tested for each predator on two days (one predator per day). The order of appearance of the two predators was randomised for each fish to minimise carry over effects.

## Data availability

The entire dataset used is available as Table S1.

## Data analysis

All statistical analyses were performed using R software 3.0.2 (http://cran.r-project.org/bin/windows/base/). Normal distribution of data and homogeneity of variances were assessed using Shapiro-Wilk and Bartlett tests respectively; however, the data did not meet the assumption of normality. Because the data was not normally distributed, we used a linear model for each behaviour, with p-values obtained by permutation tests (lmPerm package, Permutation tests for linear models in R). These models were used to test for behavioural differences between the two populations including population, sex and length of the fish as fixed effects. For boldness, the significance of interactions between the effects of population, predator type, day of tests, as well as fish identification number (ID) were tested in order to correct for pseudo-replication (as individuals were tested twice on the fourth and fifth days). Only population effects are reported in the text as no effect of sex, fish length, day and fish ID was found on any behaviour tested. Median ± quartiles are presented and p<0.05 was used as the statistical significance threshold (see Table S1 for the dataset).

For each population separately, Spearman's rank correlations were calculated between body length of fish, and each of the eight behavioural parameters measured. No correlation was found with body length and these terms were subsequently removed from the analysis. Correlations between behaviours were also quantified using Spearman's rank correlations. As multiple correlations were performed, sequential Benjamini and Hochberg corrections [38] were applied to reduce type I errors by controlling for the false discovery rate. Correlations were found between different behaviour components within a population. However, after correcting for multiple testing, the only correlations that were significant were found between different measures of the same behaviour (see Table S2).

## Results

### Average behavioural divergence between populations

**Sociability.** The two populations exhibited different schooling tendencies under identical social circumstances (Table 1 & Fig. 2A). The model school assay elicited schooling behaviour for 85% of individuals in the freshwater population and for 90% of the individuals in the marine fish. During the entire 15 min-session, freshwater fish spent an average of 4.4% of their time schooling (mean = 39.4 sec; SEM: ±9.9) while marine fish spent 21% of their time grouping with the school (mean = 188.9 sec; SEM: ±35.5; Fig. 2A). Accordingly, freshwater sticklebacks showed a significantly lower schooling behaviour than the marine fish (Linear model: population term, p = 0.002, n = 35 marine fish, n = 35 freshwater fish).

**Exploration.** Exploration in a novel environment was not significantly different between populations for the three measures observed (Linear model: population term, p>0.05 for latency to exit a refuge, n = 32 marine fish, n = 34 freshwater fish; total distance swam, n = 33 marine fish, n = 31 freshwater fish; and time spent moving, n = 35 marine fish, n = 32 freshwater fish; Table 1 & Fig. 2B).

**Aggressiveness.** About 75% and 39% of the fish exhibited aggressiveness in the freshwater and marine population respectively, whereas the rest did not show a reaction to the intruder (i.e. no approach within one body length). When comparing all individuals tested, the number of bites was four times higher in the freshwater than in the marine population (Linear model: population term, p = 0.002, n = 26 marine fish, n = 32 freshwater fish; Table 1 & Fig. 2C).

**Activity.** Activity in a familiar environment measured as time spent moving was five times higher for freshwater individuals than marine ones (Linear model: population term, p = 0.0006 for time spent moving, n = 36 marine fish, n = 36 freshwater fish; Table 1 & Fig. 2D) and a significant difference was also found for total distance swam (p = 0.010, n = 36 marine fish, n = 36 freshwater fish; Table 1).

**Boldness.** In the boldness under predation risk test, 94% of freshwater fish responded to the heron predator attacks and 79% to the sculpin by hiding in the shelter, while 97% of marine fish showed this reaction for both predators. The two populations did not differ in boldness; all fish exhibited the same average latency to resume feeding after a predator stimulus (Linear model: population term, p = 0.961, n = 32 marine fish, n = 35 freshwater fish for the fish predator, n = 33 marine fish, n = 35 freshwater fish for the bird predator; Table 1 & Fig. 2E). Fish responded differently to the type of predator in both populations (Linear model: predator type term, p = 0.041; no significant population X predatory type interaction). The bird-shaped predator induced the longest latency in both populations (Fig. 2E).

## Discussion

We found that common-environment reared and predator-naïve juvenile threespine sticklebacks originating from a freshwater habitat were less social, more active in a familiar environment and more aggressive towards conspecifics than their marine conspecifics. Given that all adult fish were bred in the laboratory and that juveniles were raised in the same setting without parental care, our results suggest that a genetic difference may underlie the measured divergence in behaviour; although further investigations are needed to confirm this hypothesis. It has been shown that boldness [28,39], aggressiveness [27,28,39], exploration in an unfamiliar environment [28,39] and activity in a familiar environment [27,28,39] among others show genetic variation between freshwater populations in threespine stickleback, while schooling behaviour has been shown to differ between marine and freshwater populations of this species [21]. We thus uncover important behavioural differences between individuals belonging to the ancestral marine population and the freshwater one for two previously under- or unstudied ecologically-relevant behaviours, i.e. activity and aggressiveness, in addition to confirming divergence in sociability, in line with what has been found previously in another population pair [21].

Our results suggest that the behavioural divergence observed could represent the result of natural selection that acted on individuals after freshwater colonisation [40]. However, the hypothesis that this phenotypic divergence could result from other causes than selection cannot be excluded. Neutral processes acting during and after colonisation on genetic diversity of the population, e.g. genetic drift, could explain the divergence [41], although the findings that sociability diverges in the same direction as what was found previously for other stickleback populations [21] and other species [11] tends to support the hypothesis of the effects of natural selection. The evidence gathered demonstrating the parallel evolution of various traits (morphology, physiology)

**Table 1.** Summary of linear model statistical tests conducted to compare the two juvenile threespine stickleback populations for five behaviours.

| Behaviours | Variable | Estimate | P-value | Population trends |
|---|---|---|---|---|
| Sociability | population | −70.782 | 0.011* | marine > freshwater |
| | sex | 3.410 | 0.922 | |
| | length | −1.912 | 0.784 | |
| Exploration – latency to exit cylinder | population | 13.159 | 0.902 | marine = freshwater |
| | sex | 10.506 | 0.638 | |
| | length | 6.331 | 0.302 | |
| Exploration – time spent swimming | population | −3.626 | 0.745 | marine = freshwater |
| | sex | 3.285 | 0.510 | |
| | length | −0.878 | 0.317 | |
| Exploration – total length swam | population | 0.332 | 0.961 | marine = freshwater |
| | sex | −1.132 | 0.961 | |
| | length | 1.921 | 0.312 | |
| Aggressiveness | population | 2.297 | 0.002** | marine < freshwater |
| | sex | −0.970 | 0.270 | |
| | length | −0.220 | 0.265 | |
| Activity – time spent swimming | population | 19.573 | 0.002** | marine < freshwater |
| | sex | −3.572 | 0.388 | |
| | length | −0.522 | 0.638 | |
| Activity – total length swam | population | 40.187 | 0.013* | marine < freshwater |
| | sex | −18.725 | 0.171 | |
| | length | −0.002 | 1.000 | |
| Boldness | population | −46.004 | 0.961 | marine = freshwater |
| | sex | −0.838 | 1.00 | |
| | length | 1.965 | 1.00 | |
| | day | −31.193 | 0.863 | |
| | fish_ID | −0.419 | 0.863 | |
| | predator type | −53.608 | 0.041* | |

Estimate and p-value probability obtained using the lmp function of the lmPerm package in R. Asterisk (*) denotes significant difference.

occurring in hundreds of geographically and genetically isolated freshwater populations of threespine stickleback after colonisation from the marine environment speaks in favour of natural selection [42]. Our study provides additional evidence supporting the hypothesis of repeated phenotypic divergence when facing similar ecological challenges, as we find the same significantly lower sociability tendency in freshwater fish compared to marine fish as previously found for entirely different marine (Pacific Ocean in Japan) and freshwater populations (benthic population from Paxton Lake in Canada) of this species [21]. This repeated strong behavioural difference associated with colonisation of the freshwater environment supports the hypothesis that this divergence in behaviour enhances fitness in the new ecological conditions. Of course, other stickleback marine-freshwater population pairs must be characterised for their behavioural phenotypes in both habitat types in order to confirm that behavioural differences measured in this study between a marine and freshwater population were due to divergent natural selection rather than being merely difference between any two populations, regardless of ecological pressures.

Predation pressure is one of the main ecological factors that act as agents of selection, which can cause phenotypic divergence among prey populations [1]. In threespine stickleback, contrasting predation pressures between marine and freshwater populations

are correlated with divergence in lateral plate armour development [31]. The marine and freshwater habitats from which breeding adults were sampled to create the two juvenile laboratory lines studied differed in predation regime (based on armour morphology, see methods) thus suggesting that predation could be a selection agent acting differentially on behaviour in the marine and the freshwater population, similar to what is found for predator defence morphology [31]. Indeed, studies comparing freshwater populations showed an association between predatory regime differences and divergence in behaviour in sticklebacks in sociability [25], activity [26,27], aggressiveness [27,28], and boldness [26,28,29]. The strength of our results, although based on only two populations, is that they are in accordance with the predictions of behavioural divergence in sociability, aggressiveness, and activity between marine and freshwater fish we made *a priori* based on these studies of freshwater populations. For example, in a high-predation habitat such as the marine environment, schooling provides fish with several benefits such as a better foraging efficiency and a greater antipredator defence [43]. It has been hypothesised to lead to a reduction of aggression between individuals because high levels of fighting might offset the benefits of schooling [44]. However, we keep in mind that marine and freshwater habitats differ in several other ecological aspects, such as habitat structure and stability [26,45], food resource availability [46], intraspecific competition [47] and temperature [30], which could also explain the behavioural divergence observed. The fact that boldness did not differ between populations contrary to our prediction, and that both populations reacted more to the bird predator, suggests that other factors must also be considered. The behaviour of freshwater juveniles could also represent heritable behavioural characteristics of the marine individuals that preferentially invaded freshwater after the glacial retreat to form the new lake population. A greater swimming activity and a higher boldness could be associated with a higher propensity to use novel habitats [48,49]. The freshwater individuals were significantly more active than marine fish, although they did not differ in boldness. This is in line with the hypothesis that the freshwater behaviour stems from standing genetic variation in the marine population, as very bold individuals could also be found in the marine population, as observed in previous studies for morphological traits in threespine stickleback [50]. In this case, natural selection acting on these individuals once in the new freshwater environment is also possible.

Our study suggests that mean behaviours in a population are associated with their own costs and benefits modulated according to the habitat. Taken together, our results provide an integrated view of the divergence in mean behaviour of individuals colonising a novel environment, potentially resulting from the new ecological constraints encountered, in contrast with the behaviour of the ancestral population, information which are of crucial importance considering the role of behavioural phenotype-environment associations in adaptive divergence.

## Supporting Information

**Table S1　Dataset used for the study by Di Poi et al.** (TXT)

**Table S2　Statistically significant correlations found between behaviours in each population.** Correlation coefficients and p-values calculated before and after corrections for multiple testing (Benjamini-Horchberg) are presented. (TXT)

## Acknowledgments

The authors gratefully acknowledge the personnel of the Laboratoire Régional des Sciences Aquatiques (LARSA) at Université Laval for their help in fish rearing, Matthew Morris for help with fish and Nancy Diplo for chromosomal sex determination of sticklebacks.

## Author Contributions

Conceived and designed the experiments: NAH CDP JL. Performed the experiments: CDP JL. Analyzed the data: CDP NAH. Contributed reagents/materials/analysis tools: SMR. Wrote the paper: CDP SMR NAH.

## References

1. Schluter D (2000) The ecology of adaptive radiation. Oxford, UK: Oxford Series in Ecology and Evolution, 296.
2. Losos JB (2009) Lizards in an evolutionary tree: ecology and adaptive radiation of Anoles. Berkeley and Los Angeles, CA, USA: University of California Press, 528.
3. Cooper WJ, Parsons K, McIntyre A, Kern B, McGee-Moore A, et al. (2010) Bentho-pelagic divergence of cichlid feeding architecture was prodigious and consistent during multiple adaptive radiations within African rift-lakes. PloS ONE 5(3): e9551.
4. Rogers SM, Gagnon V, Bernatchez L (2002) Genetically based phenotype-environment association for swimming behavior in lake whitefish ecotypes (Coregonus clupeaformis Mitchill). Evolution 56: 2322–2329.
5. Grant PR, Grant BR (2011) How and why species multiply: the radiation of Darwin's finches. Princeton, NJ, USA: Princeton University Press, 218.
6. Budaev SV (1997) "Personality" in the guppy (Poecilia reticulata): a correlational study of exploratory behavior and social tendency. J Comp Psychol 111: 399–411.
7. Herczeg G, Välimäki K (2011) Intraspecific variation in behaviour: effects of evolutionary history, ontogenetic experience and sex. J Evol Biol 24: 2434–2444.
8. Herczeg G, Gonda A, Kuparinen A, Merilä J (2012) Contrasting growth strategies of pond versus marine populations of nine-spined stickleback (Pungitius pungitius): a combined effect of predation and competition? Evol Ecol 26: 109–122.
9. Brown C, Jones F, Braithwaite V (2005) In situ examination of boldness-shyness traits in the tropical poeciliid, Brachyraphis episcopi. Anim Behav 70: 1003–1009.
10. Magurran AE, Seghers BH, Carvalho GR, Shaw PW (1992) Behavioural consequences of an artificial introduction of guppies (Poecilia reticulata) in N. Trinidad: evidence for the evolution of antipredator behaviour in the wild. Proc R Soc B 248: 117–122.
11. Huizinga M, Ghalambor CK, Reznick DN (2009) The genetic and environmental basis of adaptive differences in shoaling behaviour among populations of Trinidadian guppies, Poecilia reticulata. J Evol Biol 22: 1860–1866.
12. Chamberlain NL, Hill RI, Kapan DD, Gilbert LE, Kronforst MR (2009) Polymorphic butterfly reveals the missing link in ecological speciation. Science 326: 847–850.
13. Rogers SM, Tamkee P, Summers B, Balabahadra S, Marks M, et al. (2012) Genetic signature of adaptive peak shift in threespine stickleback. Evolution 66: 2439–2450.
14. Bell MA, Foster SA (1994) The evolutionary biology of the threespine stickleback. Oxford, UK: Oxford University Press, 571.
15. Cresko WA, McGuigan KL, Phillips PC, Postlethwait JH (2007) Studies of threespine stickleback developmental evolution: progress and promise. Genetica 129: 105–126.
16. Greenwood AK, Jones FC, Chan YF, Brady SD, Absher DM, et al. (2011) The genetic basis of divergent pigment patterns in juvenile threespine sticklebacks. Heredity 107: 155–166.
17. Snyder RJ, Dingle H (1989) Adaptive, genetically based differences in life history between estuary and freshwater threespine sticklebacks (Gasterosteus aculeatus L.). Can J Zool 67: 2448–2454.
18. Kitano J, Lema SC, Luckenbach JA, Mori S, Kawagishi Y, et al. (2010) Adaptive divergence in the thyroid hormone signaling pathway in the stickleback radiation. Curr Biol 20: 2124–2130.
19. Dalziel AC, Vines TH, Schulte PM (2012) Reductions in prolonged swimming capacity following freshwater colonization in multiple threespine stickleback populations. Evolution 66: 1226–1239.
20. Bakker TCM, Feuth-de Bruijn E (1988) Juvenile territoriality in stickleback Gasterosteus aculeatus L. Anim Behav 36: 1556–1558.
21. Wark AR, Greenwood AK, Taylor EM, Yoshida K, Peichel CL (2011) Heritable differences in schooling behavior among threespine stickleback populations revealed by a novel assay. PloS ONE 6: e18316.
22. Saimoto RK (1993) Life history of marine threespine stickleback in Oyster Lagoon, British Columbia. M. Sc. thesis, University of British Columbia, Vancouver, 97.
23. Godin JGJ, Smith SA (1988) A fitness cost of foraging in the guppy. Nature 333: 69–71.
24. Réale D, Reader SM, Sol D, McDougall PT, Dingemanse NJ (2007) Integrating animal temperament within ecology and evolution. Biol Rev 82: 291–318.
25. Nomakuchi S, Park PJ, Bell MA (2009) Correlation between exploration activity and use of social information in three-spined sticklebacks. Behav Ecol 20: 340–345.
26. Brydges NM, Colegrave N, Robert JP, Braithwaite VA (2008) Habitat stability and predation pressure affect temperament behaviours in populations of three-spined sticklebacks. J Anim Ecol 77: 229–235.
27. Lacasse J, Aubin-Horth N (2012) A test of the coupling of predator defense morphology and behavior variation in two threespine stickleback populations. Curr Zool 58: 53–65.
28. Bell AM (2005) Behavioural differences between individuals and two populations of stickleback (Gasterosteus aculeatus). J Evol Biol 18: 464–473.
29. Dingemanse NJ, Wright J, Kazem AJN, Thomas DK, Hickling R, et al. (2007) Behavioural syndromes differ predictably between 12 populations of three-spined stickleback. J Anim Ecol 76: 1128–1138.
30. Barrett RDH, Paccard A, Healy TM, Bergek S, Schulte PM, et al. (2011) Rapid evolution of cold tolerance in stickleback. Proc R Soc B 278: 233–238.
31. Marchinko KB (2009) Predation's role in repeated phenotypic and genetic divergence of armor in threespine stickleback. Evolution 63: 127–138.
32. Ministry of Environment – Government of British Columbia website. Available: http://a100.gov.bc.ca/pub/fidq/fishDistributionsQuery.do. Accessed 2014 May 14.
33. Greenwood AK, Wark AR, Yoshida K, Peichel CL (2013) Genetic and neural modularity underlie the evolution of schooling behavior in threespine sticklebacks. Curr Biol 23: 1884–1888.

34. Peichel CL, Ross JA, Matson CK, Dickson M, Grimwood J, et al. (2004) The master sex-determination locus in threespine sticklebacks is on a nascent Y chromosome. Curr Biol 14: 1416–1424.

35. Dzieweczynski TL, Crovo JA (2011) Shyness and boldness differences across contexts in juvenile three-spined stickleback *Gasterosteus aculeatus* from an anadromous population. J Fish Biol 79: 776–788.

36. Aubin-Horth N, Deschênes M, Cloutier S (2012) Natural variation in the molecular stress network correlates with a behavioural syndrome. Horm Behav 61: 140–146.

37. Huntingford FA (1976) The relationship between anti-predator behaviour and aggression among conspecifics in the three-spined stickleback. Anim Behav 24: 245–260.

38. Benjamini Y, Hochberg Y (1995) Controlling the false discovery rate: a practical and powerful approach to multiple testing. J R Stat Soc B 57: 289–300.

39. Giles N, Huntingford FA (1984) Predation risk and inter-population variation in antipredator behaviour in the three-spined stickleback, *Gasterosteus aculeatus* L. Anim Behav 32: 264–275.

40. Endler JA (1986) Natural selection in the wild. Monographs in population biology. Princeton, New Jersey, USA: Princeton University Press, 354.

41. Knowles LL, Richards CL (2005) Importance of genetic drift during Pleistocene divergence as revealed by analyses of genomic variation. Mol Ecol 14: 4023–4032.

42. Schluter D, Marchinko KB, Barrett RDH, Rogers SM (2010) Natural selection and the genetics of adaptation in threespine stickleback. Phil Trans R Soc B 365: 2479–2486.

43. Pitcher TJ, Parrish JK (1993) Functions of shoaling behaviour in teleosts. In: Behaviour of Teleost Fishes (ed. Pitcher, T.J.). 2nd edn. Chapman & Hall, London, UK, 363–439.

44. Magurran AE, Seghers BH (1991) Variation in schooling and aggression amongst guppy (*Poecilia reticulata*) populations in Trinidad. Behaviour 118: 214–234.

45. Braithwaite VA, Girvan JR (2003) Use of water flow direction to provide spatial information in a small-scale orientation task. J Fish Biol 63: 74–83.

46. Laskowski KL, Bell AM (2013). Competition avoidance drives individual differences in response to a changing food resource in sticklebacks. Ecol Lett 16: 746–753.

47. Svanbäck R, Bolnick DI (2007) Intraspecific competition drives increased resource use diversity within a natural population. Proc R Soc B 274: 839–844.

48. Fraser DF, Gilliam JF, Daley MJ, Le AN, Skalski GT (2001) Explaining leptokurtic movement distributions: intrapopulation variation in boldness and exploration. Amer Nat 158: 124–135.

49. Barrett RDH, Vines TH, Bystriansky JS, Schulte PM (2009) Should I stay or should I go? The Ectodysplasin locus is associated with behavioural differences in threespine stickleback. Biol Lett 5: 788–791.

50. Colosimo PF, Peichel CL, Nereng K, Blackman BK, Shapiro MD, et al. (2004) The genetic architecture of parallel armor plate reduction in threespine sticklebacks. PLoS Biology 2: E109.

# 8

# Diversity and Distribution of Freshwater Amphipod Species in Switzerland (Crustacea: Amphipoda)

Florian Altermatt[1,2,3]*, Roman Alther[1], Cene Fišer[4], Jukka Jokela[1,2], Marjeta Konec[4], Daniel Küry[5], Elvira Mächler[1], Pascal Stucki[6], Anja Marie Westram[1,7]

1 Department of Aquatic Ecology, Eawag: Swiss Federal Institute of Aquatic Science and Technology, Dübendorf, Switzerland, 2 Department of Environmental Systems Science, ETH Zentrum, Zürich, Switzerland, 3 Institute of Evolutionary Biology and Environmental Studies, University of Zurich, Zürich, Switzerland, 4 Department of Biology, Biotechnical Faculty, University of Ljubljana, Ljubljana, Slovenia, 5 Life Science AG, Basel, Switzerland, 6 Aquabug, Neuchâtel, Switzerland, 7 Animal and Plant Sciences, University of Sheffield, Western Bank, Sheffield, United Kingdom

## Abstract

Amphipods are key organisms in many freshwater systems and contribute substantially to the diversity and functioning of macroinvertebrate communities. Furthermore, they are commonly used as bioindicators and for ecotoxicological tests. For many areas, however, diversity and distribution of amphipods is inadequately known, which limits their use in ecological and ecotoxicological studies and handicaps conservation initiatives. We studied the diversity and distribution of amphipods in Switzerland (Central Europe), covering four major drainage basins, an altitudinal gradient of >2,500 m, and various habitats (rivers, streams, lakes and groundwater). We provide the first provisional checklist and detailed information on the distribution and diversity of all amphipod species from Switzerland. In total, we found 29 amphipod species. This includes 16 native and 13 non-native species, one of the latter (*Orchestia cavimana*) reported here for the first time for Switzerland. The diversity is compared to neighboring countries. We specifically discuss species of the genus *Niphargus*, which are often receiving less attention. We also found evidence of an even higher level of hidden diversity, and the potential occurrence of further cryptic species. This diversity reflects the biogeographic past of Switzerland, and suggests that amphipods are ideally suited to address questions on endemism and adaptive radiations, post-glaciation re-colonization and invasion dynamics as well as biodiversity-ecosystem functioning relationships in aquatic systems.

**Editor:** Diego Fontaneto, Consiglio Nazionale delle Ricerche (CNR), Italy

**Funding:** The study was funded by the Swiss National Science Foundation Grants 31003A_135622 and PP00P3_150698 (to FA), the Swiss Federal Office for the Environment (BAFU, to FA), Eawag Matching funds (to FA and RA), ETH-CCES project BioChange (to JJ) and EAWAG project AquaDivers (to JJ). Life Science AG and Aquabug provided support in the form of salaries for authors DK & PS respectively, but did not have any additional role in the study design, data collection and analysis, decision to publish, or preparation of the manuscript. The specific roles of these authors are articulated in the "author contributions" section.

**Competing Interests:** DK is an employee of Life Science AG. PS is an employee of Aquabug. There are no patents, products in development or marketed products to declare.

* Email: florian.altermatt@eawag.ch

## Introduction

Understanding the diversity and distribution of organisms is a fundamental goal of ecology, and a prerequisite for using species in monitoring programs or as bioindicators. This is especially relevant for freshwater systems, which are highly diverse, but also highly threatened [1,2], and for which the occurrence of characteristic diversity patterns is postulated [3]. While the diversity and distribution of freshwater vertebrates, such as birds, fish or mammals, is generally well-known, knowledge on invertebrates is often more limited.

Amphipods (class Crustacea, order Amphipoda; Fig. 1) are an important and diverse group of macroinvertebrates [4,5], many of which inhabit freshwater environments including epibenthic, benthic and subterranean habitats. Worldwide, about 2,000 species of freshwater amphipods are known, with 70% of these species found in the Palaearctic [4]. Even though they can contribute substantially to the diversity and biomass of aquatic communities, detailed knowledge on the distribution and community composition of freshwater amphipods is lacking for many regions. While endemic species of lake Baikal or karst regions of south-eastern Europe (e.g., Italy or Slovenia) have been studied intensively (e.g., [4,6–8]), conclusive information on the distribution and diversity of amphipods is lacking for some alpine regions, especially for Switzerland (Table 1). This is unfortunate, as the European Alps represent a diversity hotspot for many groups of aquatic species. Multiple cycles of glaciation and re-colonizations from refugia and a complex geology have resulted in a mosaic of species' distributions (e.g., [9,10]). The Swiss Alps form major continental drainage systems (origin or tributaries to the rivers Rhine, Rhone, Danube, and Po), and thus have been and are open for colonization from biogeographically different regions. This has led to a different faunal composition north and south of the Swiss Alps for many groups of organisms, including frequent adaptive radiations and high degrees of endemism, for example in whitefish (*Coregonus* sp. [10]), or in may- and stoneflies [11,12].

However, recent anthropogenic changes in the connectivity of river systems and loss of dispersal limitation also resulted in a higher inflow of non-native invertebrate species [13]. Amphipods are not only among the most successful but also among the most common invasive invertebrate species [14,15], capable of shifting whole communities of aquatic macroinvertebrates. Invasive

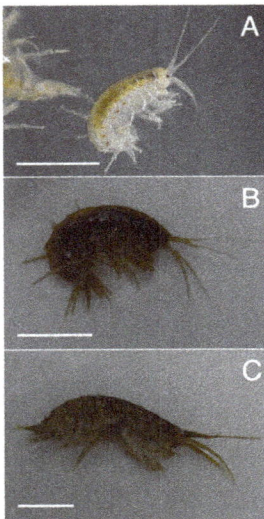

**Figure 1. Morphological diversity within the order of Amphipoda.** Three (of the in total 29) different species/species complexes known from Switzerland are shown: A) *Gammarus fossarum* complex, B) *Gammarus roeseli* and C) *Dikerogammarus villosus*. *G. fossarum* is native to Switzerland, *G. roeseli* is a non-native species that arrived in Switzerland around 1850, and *D. villosus* is a non-native species that arrived in Switzerland in the late 1990s. The scale bar is equivalent to 1 cm and gives approximate size differences between the species. The diverse color patterns visible in these pictures of living animals are completely lost in specimens preserved in alcohol. All pictures by Florian Altermatt.

species are currently changing the diversity and composition of amphipod communities in many countries, including Switzerland.

In parallel of a high ecological significance, amphipods are receiving an increasing interest in eco-toxicological and environmental biomonitoring (e.g., [16–18]). However, this work has been made difficult by major gaps in the basic distribution data and fundamental difficulties in morphological identification of amphipods: the relevant morphology-based taxonomic keys on amphipods are challenged by a very high intra- and inter-population variation in morphology (e.g., [15,19–21]). As a result, detailed information on amphipods is lacking from Switzerland (Table 1). For other European areas/countries, presence-absence checklists or large-scale distribution data are available, while more detailed

distribution data are usually only available for a subset of amphipod species (e.g., [7,15,22–24]). Subsequently, in many applied studies, correct species-level identification of amphipods is not done. This is a serious problem because different species may be inadvertently compared in ecotoxicological tests [25], or the presence and potential decline of species at a site is unrecognized, as only presence/absence of amphipods as a whole group is recorded [17].

Here, we provide the first provisional checklist and detailed information on the distribution and diversity of all amphipod species found in Switzerland to date. We include data from standardized federal and cantonal monitoring programs, literature, as well as from our own extensive fieldwork. Our database consists of >150,000 individuals collected at about 2,500 sites. Individuals were identified based on morphological and molecular methods, and include species from lakes, rivers, streams, and groundwater. We provide distribution maps and information on the altitudinal distribution of all native and non-native amphipod species known from Switzerland and compare the diversity to neighboring countries. Furthermore, we analyze community composition and co-occurrence of species, and identify diversity hotspots and invasion pathways.

## Material and Methods

### Study area

Our study area is Switzerland, covering an area of 41,285 km². Switzerland contains the origin or important tributaries of four major alpine drainage systems (Rhine, Rhone, Inn/Danube, Ticino/Po, covering 71%, 20%, 5%, and 3.5% respectively of the country), which drain into the North Sea, the Mediterranean Sea, the Black Sea and the Adriatic Sea, respectively. Thereby, Switzerland reflects the diversity and biogeography of European headwaters. The country exhibits a large altitudinal range from 193 to 4634 m a.s.l. and covers a diversity of geological substrates, including karst, granite and alluvial sediments. A temperate climate and medium to high level of precipitation result in a large number of freshwater habitats.

### Data sources and sampling methods

We compiled a database containing amphipod records from literature references, museum collections, governmental monitoring programs, as well as records from our own extensive fieldwork. First, we screened all available literature on reliable amphipod records from Switzerland. This not only included published studies

**Table 1.** Overview of the hitherto published diversity of Amphipoda in Switzerland, neighboring countries of Switzerland (Austria, Germany, Italy, France) as well as Slovenia.

| country | Nr of families | Nr of genera | Nr of species |
|---|---|---|---|
| Austria | 3/– | 6/– | 16/– |
| France | 8/– | 16/– | 67/– |
| Germany | 5/8 | 12/17 | 36/48 |
| Italy | 8/11 | 16/18 | 68/119 |
| Slovenia | 4/9 | 8/11 | 38/55 |
| Switzerland | 2/– | 4/– | 12/– |

The latter is especially well-studied and and therefore given for comparison. For each country, diversity at the family, genus and species level is given. We first give the number of taxa at each level from Fauna Europaea [77] and after the diagonal slash from other overview publications screened (when available, a list of these publications is given in the Method section). In case of missing or incomplete data at the country level (e.g., no publication considering all species within the order Amphipoda), a dash "–" is given.

but also many unpublished reports conducted by federal or cantonal agencies ("grey literature"). Literature was acquired by a Web of Science search with "amphipod" and "Switzerland" as key words, complemented by a survey targeting aquatic ecologists, consultancy companies and governmental agencies in Switzerland. In total, over 30 references were evaluated and data thereof included [23,26–55]. We only used literature records on amphipods when the identification and data source was traceable. Second, we screened museum collections for species for which we had only few records (especially Niphargidae). We screened the collections of the National History Museum in Basel, and the private collections of Aquabug (Neuchâtel) and LifeScience AG (Basel). Third, we identified all amphipod samples collected in the Biodiversity Monitoring Program of Switzerland (BDM, [12]). In this program, all macroinvertebrates are sampled based on highly standardized methods at over 500 randomly selected sites across the Swiss river and stream network since 2009. Finally, we conducted our own extensive fieldwork at >200 sites across Switzerland, targeting areas that were underrepresented by the other data-sources (e.g., Southern Switzerland/Ticino, Alpine Rhine valley, tributaries of Lake Constance, Jura mountains, and alpine valleys and alpine lakes). To access these sites and to do the sampling, no specific permission was required, as none of the sites were in protected areas and did not involve endangered or protected species. Field sampling was predominantly conducted by standardized kicknet sampling [17], following the protocols used in the BDM. Besides these standardized samplings, we also collected individuals by specifically targeting known microhabitats of amphipods, such as wells and groundwater systems, lakeshores and streams. All collected individuals were preserved in 70% ethanol. All data sources (except [23,26]) looked at all amphipod species at the sampling sites, thereby not creating biases with respect to species groups identified. Sampling efforts were not completely evenly distributed across Switzerland (except for the BDM data), and some habitat types (deep lakes, natural springs/groundwater) are underrepresented. We give information on sampling intensity (Fig. 2A), thereby also identifying "white spots" with respect to sampling efforts.

We built a database containing information on the precise geographic location, elevation, habitat type, and identification method of all amphipod individuals considered. Individuals that could not be identified to the species level with neither morphological (e.g., juveniles or damaged specimens) nor molecular methods were excluded from all analyses. The database will be integrated into the Swiss Biological Records Center (www.CSCF.ch) to become publicly available.

## Morphological identification

We aimed at identifying all individuals to the most precise and commonly accepted taxonomic level. Using standard literature [15,19–21,56], we in a first step identified individuals to the species level based on morphological characters, using a stereomicroscope at 20- to 100-fold magnification. For all individuals of the genus Niphargus, morphological analyses were made using the original description of the species. Morphologically delimited species may still contain cryptic species, and we in a second step included genetic data for species identification for the Gammarus fossarum species complex and the genus Niphargus.

## Molecular identification of Gammarus fossarum species complex

Gammarus fossarum is known to be a species complex, containing at least three species (type A, B and C) in Switzerland, which cannot be told apart based on morphological characters

only [22,23,26,57-59]. We identified G. fossarum from as many sites as possible using previously established microsatellite and SNP markers for species identification. In total, we extracted DNA from about 4,500 individuals of the G. fossarum-complex, either extracting DNA from whole individuals or from pereopods, and analyzed ten microsatellite markers using the identical method as described in Westram et al. [59]. The occurrence of specific allelic combinations in these microsatellite markers is diagnostic for each of the three cryptic species, and corresponds to both species-specific SNP as well as COI sequences (for details see [23,26,59]). The microsatellite markers diagnostic for type A is gf27 polymorphic with alleles >200 bp (but ≠205), for type B the marker is monomorphic at 205 bp. All records from type C are based on previous analyses (for details see [23,26,59]).

## Molecular identification of Niphargus

Taxonomy and systematic of the genus Niphargus is still highly disputed and not yet resolved. The genus is known for a high level of cryptic diversity, and we thus grounded our identification based on molecular methods and a phylogenetic analysis. Samples with more than one individual per site were sequenced for two nuclear markers (partial 28S rRNA gene (28S) and histone 3 gene (H3)) that were already used in previous studies of Niphargidae [60–63]. Except in one case, samples containing only a single individual were not sequenced.

Genomic DNA was extracted using the GenElute Mammalian Genomic DNA Miniprep Kit (Sigma-Aldrich), following the Mammalian tissue preparation protocol. A fragment of 28S gene was amplified using primers from Verovnik et al. [64] (primer 5'-CAAGTACCGTGAGGGAAAGTT-3') and Zakšek et al. [65] (primer 5'-AGGGAAACTTCGGAGGGAACC-3'). The H3 gene was amplified using primers H3NF and H3NR from Colgan et al. [66]. PCR cycler settings are described in Fišer et al. [61]. PCR products were purified using the enzymes Exonuclease 1 and Alkaline phosphatase (both Fermentas). Incubation consisted of two steps: 37°C for 45 min and 80°C for 15 min. PCR amplification primers were also used for sequencing. Contings were assembled and edited in Geneious 5.5.6. (Biomatters). Accession numbers for all sequences uploaded to GenBank are [will be provided upon acceptance of the manuscript].

## Analysis

For all taxa except the genus Niphargus we used accepted taxonomic and phylogenetic classifications [15,19–21,56]. For Niphargus, we first compared similarity of sequences with available comparative sequences from GenBank and unpublished sequences in our database (see http://niphargus.info/references/). In order to establish the taxonomic position of the Swiss Niphargus sequences, we performed a Bayesian analysis using concatenated dataset of two genes together with available sequences from GenBank [60,62,63,67,68].

All H3 gene sequences were of equal length (331 bp) and were unambiguously aligned using a simple algorithm (Geneious Alignment). The 28S rDNA sequences were highly variable in their length (761–904 bp) and were aligned in MAFFT ver. 6 [69] using the E-INS-i option for sequences with multiple conserved domains and long gaps. The optimal substitution model for each alignment was selected according to the Akaike information criterion in JMODELTEST 0.1.1. A GTR model of nucleotide substitution was selected for both genes, with gamma distributed rate heterogeneity for 28S and gamma distributed rate heterogeneity with a significant proportion of invariable sites for H3. Both alignments were concatenated and then analyzed in MRBAYES 3.2 [70] as two partitions. Two simultaneous runs with four chains

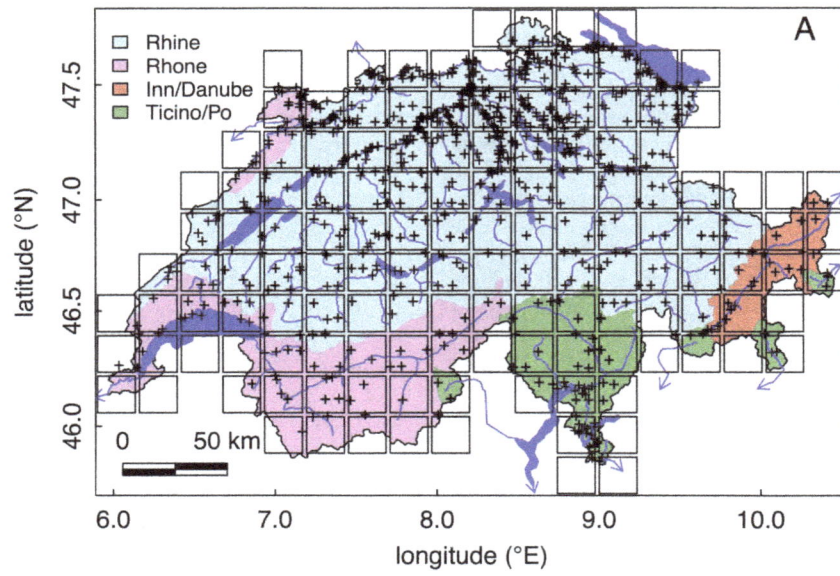

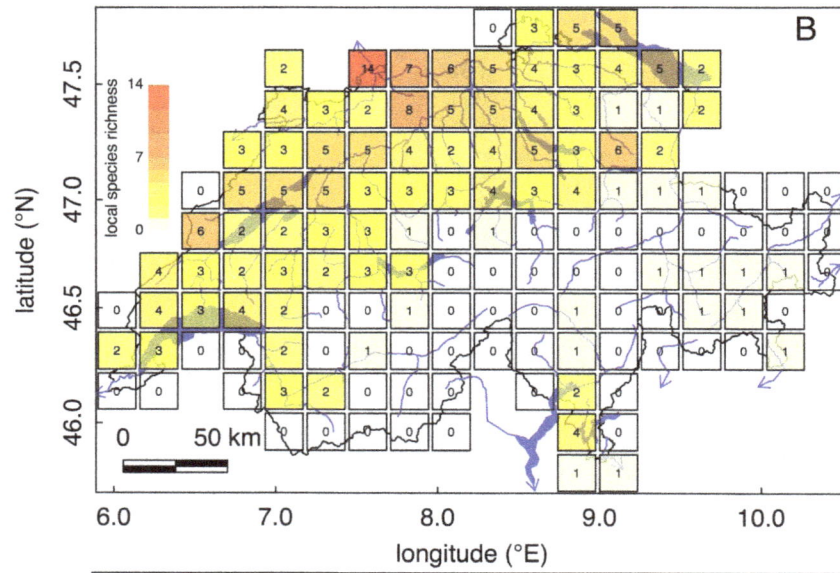

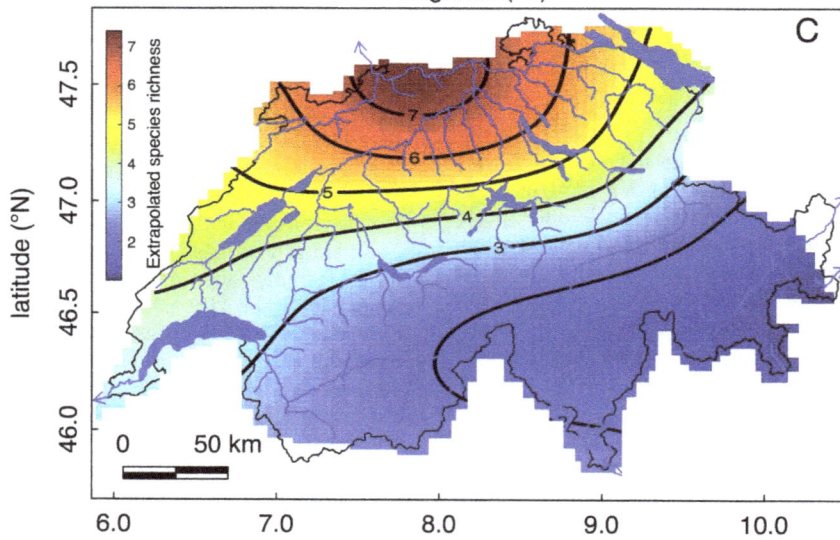

**Figure 2. Sampling locations and diversity pattern of amphipods in Switzerland.** A) Map of Switzerland showing all sampling sites included in our study (crosses). The four drainage basins (Rhine, Rhone, Inn/Danube and Ticino/Po) are given in different colors, and the major river and lakes are given in blue. The grid of the 20×20 km squares was used to calculate diversity patterns in panel B. B) Diversity of amphipods in 20×20 km squares covering all of Switzerland. Local species richness in each square is given as color gradient and a number. C) Interpolated fits of local amphipod species richness using a thin plate spline surface to irregularly spaced data.

each were run for five million generations, sampled every 100th generation. After discarding the first 25% of the sampled trees, the final topologies were constructed according to the 50% majority rule. Species identity was assigned on a basis of monophyly. We acknowledge that for identifying all potential cryptic species, a combination of further genetic markers may be recommended and

**Table 2.** Checklist of all amphipods (class Crustacea, order Amphipoda) hitherto known from Switzerland, as well as tentative year of arrival for the non-native species.

| Suprafamily | Family | Genus | Species | first record | comment |
|---|---|---|---|---|---|
| Talitroidea | Talitridae | *Orchestia* Leach, 1814 | *Orchestia cavimana* Heller, 1865 | 2013 | 1 |
| Crangonyctoidea | Crangonyctidae | *Crangonyx* Bate, 1859 | *Crangonyx pseudogracilis* Bousfield, 1958 | 2007 | |
| | | *Synurella* Wrzesniowski, 1877 | *Synurella ambulans* (F. Müller, 1846) | 2001 | |
| | Niphargidae | *Niphargus* Schiödte, 1849 | *Niphargus auerbachi* Schellenberg, 1934 | native | |
| | | | *Niphargus caspary* Pratz, 1866 | native | 2 |
| | | | *Niphargus forelii* Humbert, 1877 | native | 3 |
| | | | *Niphargus puteanus* Koch, 1836 | native | |
| | | | *Niphargus rhenorhodanensis* Schellenberg, 1937 | native | 4 |
| | | | *Niphargus setiferus* Schellenberg, 1937 | native | |
| | | | *Niphargus thienemanni* Schellenberg, 1934 | native | 5 |
| | | | *Niphargus thuringius* Schellenberg, 1934 | native | |
| | | | *Niphargus virei* Chevreux, 1896 | native | 6 |
| Gammaroidea | Gammaridae | *Gammarus* Fabricius, 1775 | *Gammarus fossarum* Koch, 1835; Type A | native | 7 |
| | | | *Gammarus fossarum* Koch, 1835; Type B | native | 7 |
| | | | *Gammarus fossarum* Koch, 1835; Type C | native | 7 |
| | | | *Gammarus wautieri* A. L. Roux, 1967 | native | 8 |
| | | | *Gammarus lacustris* Sars, 1863 | native | |
| | | | *Gammarus pulex* (Linnaeus, 1758) | native | |
| | | | *Gammarus roeseli* Gervais, 1835 | ~1850 | |
| | | | *Gammarus tigrinus* Sexton, 1939 | 1990 | |
| | | *Echinogammarus* Stebbing, 1899 | *Echinogammarus stammeri* S. Karaman, 1931 | native | |
| | | | *Echinogammarus berilloni* (Catta, 1878) | ~1900 | |
| | | | *Echinogammarus ischnus* Stebbing, 1899 | mid-1990s | |
| | | | *Echinogammarus trichiatus* (Martynov, 1932) | 2004 | |
| | | *Dikerogammarus* Stebbing, 1899 | *Dikerogammarus haemobaphes* (Eichwald, 1841) | ~1990 | 9 |
| | | | *Dikerogammarus villosus* (Sovinskij, 1894) | late 1990s | |
| Corophioidea | Corophiidae | *Chelicorophium* Bousfield & Hoover, 1997 | *Chelicorophium curvispinum* (G. O. Sars, 1895) | ~1980 | |
| | | | *Chelicorophium robustum* (G. O. Sars, 1895) | 2011 | |
| | | | *Chelicorophium sowinskyi* (Martinov, 1924) | 2011 | |

1 Ketmaier & De Matthaeis 2010 show that the continental European population is an undescribed but different species from the nominal species described from Cyprus, and will likely be given a different name. Ruffo et al. 2014 described it under the name "*Cryptorchestia garbinii*" as a new species based on specimens collected near lake Garda. For reasons of consistency and continuity, and with taxonomic work still ongoing, we use the name *Orchestia cavimana* ( = *Cryptorchestia cavimana* after Ruffo et al. 2014), but point out that the specimen reported might fall under what is now described as *Cryptorchestia garbinii*.
2 Probably comprises more than one species. Molecular analyses are needed to clarify taxonomic structure of the complex.
3 Populations in the type locality (Lake Geneva) possibly extinct.
4 A complex of at least six species (Lefébure et al. 2007), of which three are found in Switzerland.
5 A species closely related to *N. fontanus*, molecular analyses needed to clarify taxonomic structure of the complex.
6 Species complex with three morphologically similar species (Lefébure et al. 2006), of which one is found in Switzerland.
7 *G. fossarum* is a species complex with at least three species in Switzerland, called until formal description type A, B and C (Westram et al. 2011, 2013, Müller 1998, Weiss et al. 2013).
8 Karaman & Pinkster 1977 report it from the Jura mountains, and show a range-map extending into the Swiss Jura, but no specimens could be retrieved. Based on the species' distribution it is likely to occur in Switzerland (if it is not part of the *G. fossarum* complex) and its locality has been estimated from the map.
9 This species has likely been replaced by *D. villosus* and transient populations were found in Switzerland only for a few years.

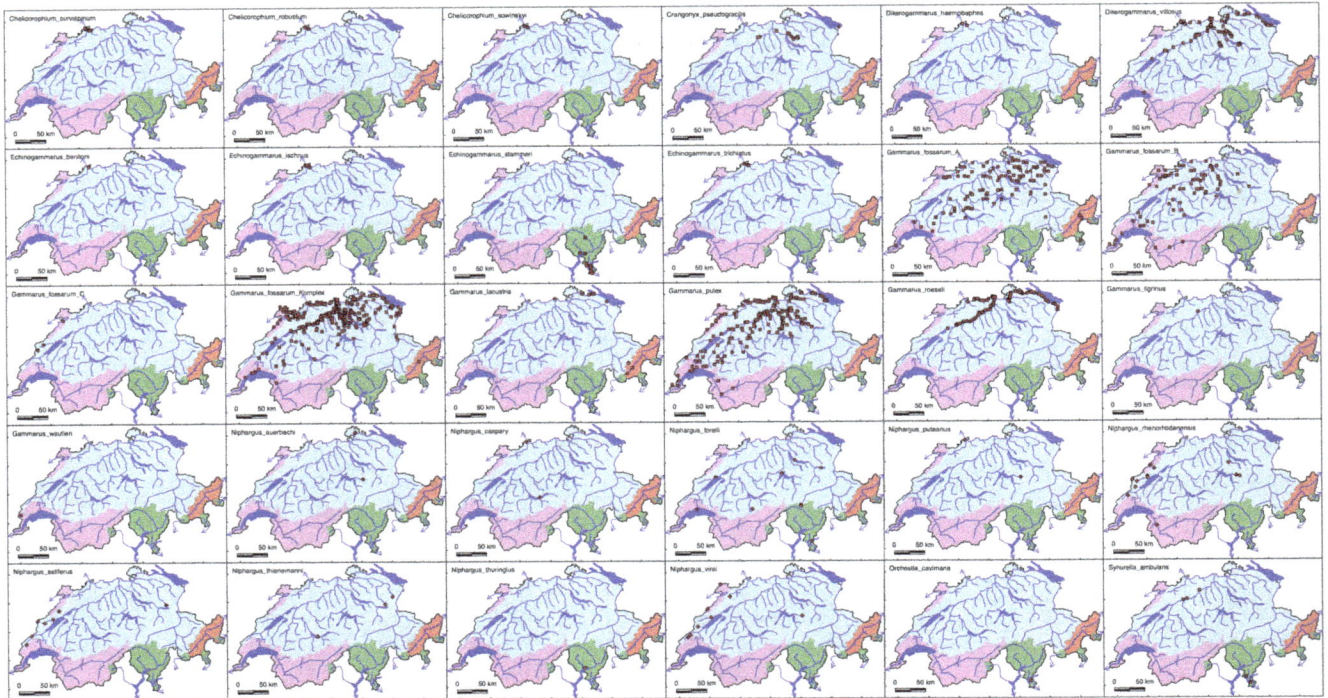

**Figure 3. Distribution maps of all 29 amphipod species of Switzerland.** Each panel gives the distribution of a species within Switzerland, in alphabetic order (see also Table 2). For *G. fossarum*, a map is given for the complex and the individual cryptic species respectively. Symbols show where the individuals were sampled: in lakes (circle), rivers and streams (square), or in the groundwater (diamond).

that our approach is rather conservative and may not resolve all possible cryptic species within *Niphargus*. However, a complete phylogeny based on several genes is beyond the focus of this work and may also require more samples.

We compared the number of amphipod taxa (family, genus and species level) with the diversity found in neighboring countries as well as Slovenia, from which the amphipod fauna is well-known. We compiled information on amphipod diversity from Fauna Europaea (http://www.faunaeur.org/) as well as from the relevant literature (including [7,15,24,56,71–74]. We used a thin plate spline surface to irregularly spaced data in order to predict species diversity patterns across Switzerland (function *Tps* in the R-package *fields* [75]), whereby the smoothing parameter is chosen by generalized cross-validation using default settings given by Nychka et al. [75]. When not mentioned differently, all statistical analyses were done with R version 3.0.1 [76].

## Results and Discussion

### Species diversity

We found a total of 29 different amphipod species in Switzerland, representing eight different genera (Fig. 1; table 2). 16 of these species (comprising three genera) are native to Switzerland, while 13 species are non-native, including five non-native genera. The herewith reported diversity of amphipods is much higher than what was previously published from Switzerland. For example, we find 100–140% higher diversity at the family, genera and species level compared to what is reported for Switzerland in Fauna Europaea [77] (Tables 1 and 2). However, we also note that several amphipod families (e.g., Ingolfiellidae, Bogiellidae, Hadziidae) found in neighboring countries (Table 1, [7,8,24,71]) are not present in Switzerland, and thus the diversity

in neighboring countries is generally higher. The lack of some major amphipod lineages may be due to the almost complete glaciation of Switzerland during the ice ages, as well as due to the lack of brackish water bodies, from which some species can invade freshwater systems.

Of the non-native species, one has been recorded in Switzerland around 1850 (*Gammarus roeseli*), while most others have been recorded in Switzerland for the first time over the last 30 years, which we interpret as a recent arrival. *Orchestia cavimana* is here reported for the first time for Switzerland (for a recent discussion of the taxonomic status of this species in northern Italy, see [78] and the footnote in Table 2). We found one individual in Lago di Lugano near Melide (Ticino, 45° 57' 10.5" N, 8° 57' 10.9" E) on July 2 2013. This species has its native range in south-eastern Europe [79], and has been found previously in the Po-region in Northern Italy [56]. Individuals from Lake Garda have recently been described as a distinct species, called *Cryptorchestia garbinii* [78]. For consistency, and lack of morphological differentiations that allow a clear assignment of the individual collected to either of these two taxa, we refer to it as *O. cavimana*. Furthermore, *O. cavimana* has also been found in the river Rhine in Southern Germany [34], but to our knowledge has not yet been confirmed from the Swiss part of the river Rhine. The most diverse amphipod genera within Switzerland are *Niphargus* (nine species) and *Gammarus* (eight species). Based on literature data (e.g., [22]), both of these genera very likely include further, overlooked cryptic species (Table 2). Such cryptic species are especially expected within the complexes of *G. fossarum*, *N. caspary*, *N. rhenorhodanensis*, *N. thienemanni* and *N. virei*.

Besides these 29 species, three further species have been previously reported for Switzerland (*N. aquilex*, *N. stygius* and *N. tatrensis*, [40–42,80]). However, these records are very likely

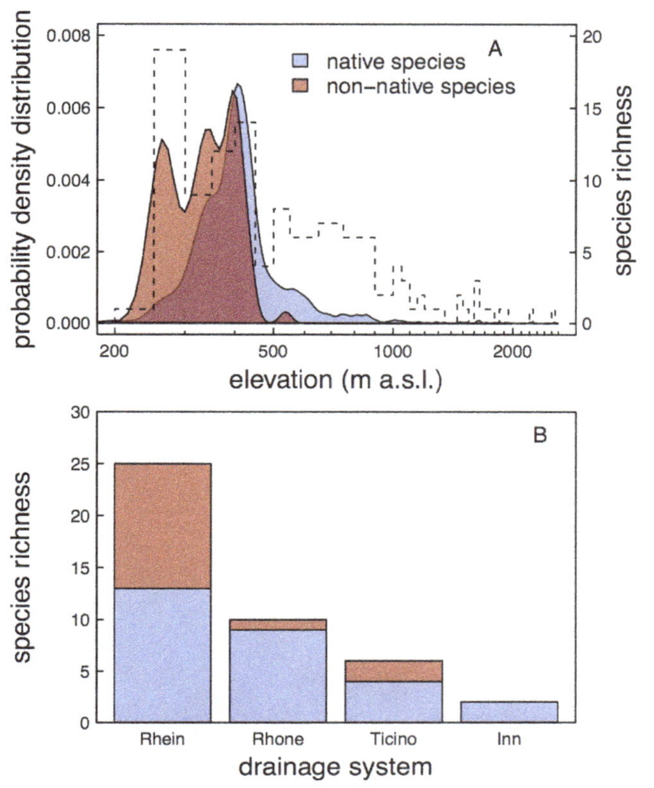

**Figure 4. Occurrence of native and non-native amphipods relative to elevation and drainage basin.** A) Occurrence of native and non-native amphipods relative to elevation. Probability density distributions are given for these two groups separately. The peaks of non-native amphipod occurrence at three elevations is linked to high sampling intensity at lakes in Ticino and River Rhine in Basel (elevation around 250 m), river Aare (elevation around 350 m) and Lake Constance (elevation 395 m). The dashed line gives the species richness at 50 m altitudinal bins. Note that the x-axis is on a $\log_{10}$-scale. B) Occurrence of native and non-native amphipod species across the four drainage basins in Switzerland.

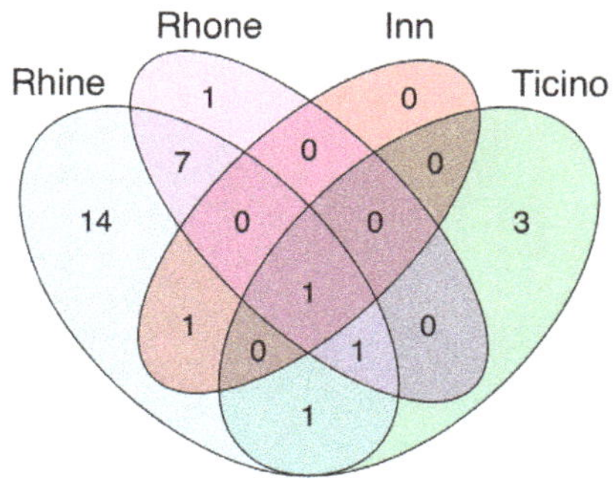

**Figure 5. Venn diagram of amphipod co-occurrences.** The Venn diagram is showing the number of co-occurring amphipod species across the four different drainage basins in Switzerland. The colors of the drainage basins follow Fig. 2.

misidentifications and relate to species that do not occur in Switzerland. *Niphargus aquilex* was described from Great Britain [31]. A recent analysis suggests it comprises of a set of unrelated species not known from Central Europe [63]. The name *N. stygius* was broadly used for many species between Italy and Romania [82]. All *N. stygius* subspecies analyzed so far proved to be good species, which are completely unrelated to the nominal species [60]. The nominal species has a rather restricted distributional range along Italian-Slovenian border [83]. Finally, *Niphargus tatrensis* is restricted to the Carpathians [71], and the identification of specimens from Switzerland is not plausible. We suggest not including these three species in the amphipod fauna of Switzerland.

## Distribution patterns

The diversity of amphipods in Switzerland is highly uneven across the country as well as across different elevations (Figs. 2–4). The highest diversity is found in the High Rhine around Basel (47° 33′ 27″ N, 7° 35′ 33″ E), in lake Constance (47° 38′ 0″ N, 9° 22′ 0″ E) and in the river Aare (47° 36′ 22″ N, 8° 13′ 26″ E) before it drains into the river Rhine. In the 20×20 km square around the city of Basel, 14 species of amphipods were found (Fig. 2). This

high diversity is due to a large number of non-native species (Fig. 3) found in the Upper Rhine [13], which are subsequently invading the High Rhine. The high diversity in the Upper Rhine and directly adjacent catchments is also supported by interpolated fits (Fig. 2C).

By far the highest number of records, but also the highest diversity of amphipods is found between 200 and 500 m a.s.l. (Fig. 4). However, amphipods could be found up to 2,540 m a.s.l. The non-native species are mostly found at lower elevations, and the altitudinal distribution of native and non-native amphipod species is significantly different (Kolmogorov-Smirnov test, D = 0.428, p<0.0001). This suggests that the non-native species originate from low-land areas (e.g., Ponto-Caspian area), and are actively invading the river system in the reverse flow direction. The distribution of records of non-native species shows three pronounced altitudinal peaks (Fig. 4A), reflecting both high sampling intensity but also high occurrence of non-native amphipods in the river Rhine around Basel (250 m a.s.l.), in the river Aare (around 350 m a.s.l.) and in Lake Constance (395 m a.s.l.). The three species with the highest elevation populations (and also with the largest altitudinal range in their distribution) were *N. forelii* (up to 2,540 m a.s.l.), *G. lacustris* (up to 1,918 m a.s.l.), and *G. fossarum A* (up to 1,850 m a.s.l.). Of the non-native species, *C. pseudogracilis* reached the highest altitudinal distribution at 538 m a.s.l.

The highest diversity in both native and non-native species was found in the river Rhine drainage (Fig. 4B). The other three drainage basins (river Rhone, river Ticino and river Inn) had lower numbers of native as well as non-native species, with the river Inn drainage basin being completely free of non-native amphipod species. We cannot exclude that part of the effect is due to some difference in sampling intensity among the drainage basins. However, the number of sites sampled (proportion of total sampling sites: 75% in Rhine, 17.5% in Rhone, 5% in Ticino, and 25% in Inn) is highly similar to the area these drainage basins cover (see Methods). Thus, we are confident that the sampling intensity between drainage basins is relatively similar, while there is some heterogeneity in the spatial location within drainage basins (see Fig. 2A). In the future, new methods such as the use of environmental DNA (eDNA) may allow to get an even better monitoring coverage [84,85]. There are multiple mutually non-exclusive explanations for the difference in species composition between drainage basins. First, the river Rhine drainage basin is by far the largest drainage basin, and therefore a higher number of species is expected [86]. Second, the altitudinal range of the drainage basins differ, such that the river Rhine and river Ticino

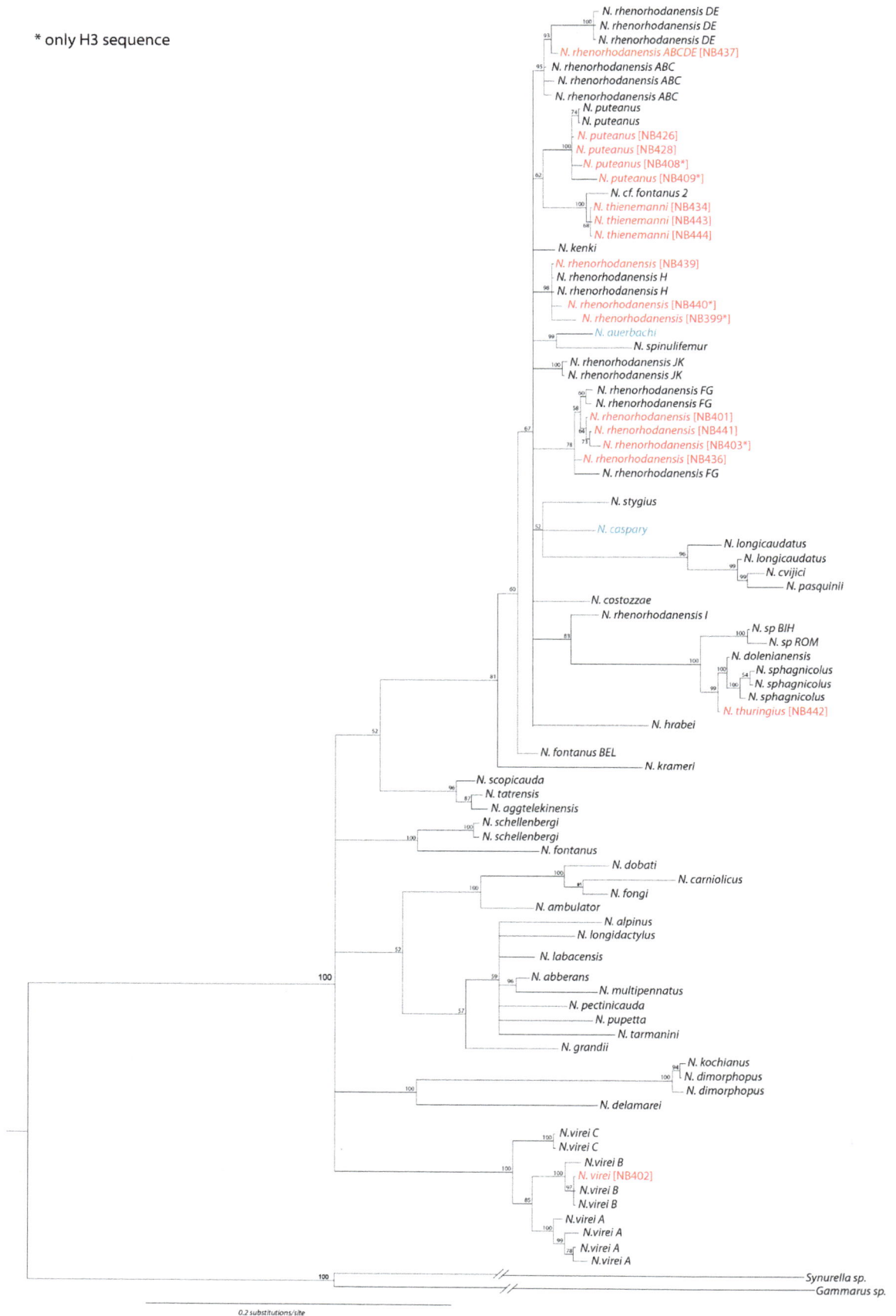

* only H3 sequence

**Figure 6. Bayesian phylogenetic tree of** *Niphargus.* Samples from this study are in red, species occurring in Switzerland, but not sequenced within this project, are in blue. Numbers above nodes indicate posterior probabilities. Asterisk denotes specimens in which sequencing of 28S gene failed.

basin reach to the lowest altitudinal levels (246 and 238 m a.s.l respectively), allowing the potential invasion of typical lowland species. A third explanation would be the existence of northern refugia (see also [9]), and finally a predominant invasion from northern drainage basins (almost exclusively through the connection to the Ponto-Caspian area through channels between the Rhine-Danube system [13]). In analogy, the set of amphipod species found in the Ticino basin is almost completely different from all other drainage basins (Fig. 5), suggesting that all of them invaded Switzerland from Northern Italy in historic or recent times [56]. It is noteworthy that no non-native amphipod species has reached the Rhone river drainage basin from Southern France. The only non-native species in the Rhone drainage basin (*D. villosus*) is most likely the result of a secondary translocation from river/lake systems in the Rhine drainage basin [32,87].

Only one amphipod species (*Gammarus fossarum* A) was found in all four drainage basins, and most of the others were restricted to one or two of the four drainage basins (Fig. 5). This indicates that processes that create and maintain species (e.g., local speciation, or persistence in refugia) as well as invasion processes unfold differently and separately in individual drainage basins. Most of the co-occurrences were found between the Rhine and the Rhone drainage basin. However, these, as well as all other co-occurrences across drainage basins, may need further investigation. Recent work on the *Gammarus fossarum* complex and on *Niphargus* shows that cryptic diversity is high across Europe, with many overlooked species [22,60]. As the populations in Switzerland strongly differ in neutral population genetic markers [23,26], we postulate that several of the existing "species" found across drainage basins may split up into different species. This has important consequences, as *Gammarus* is a model organism in ecotoxicology and commonly used as a bioindicator [16,25,88]: The largely overlapping distribution of at least three to eight *Gammarus* species in Switzerland (Fig. 3) highlights on the one hand the need for proper identification of these species and on the other hand calls for caution when individuals collected from natural populations are used for ecotoxicological tests. Translocations of individuals should be avoided due to risk of potential loss of endemic lineages and species, as observed in other freshwater organisms in (sub)alpine systems [10].

## Discussion of individual species other than the genus *Niphargus*

Of all amphipod records from Switzerland, species of the genus *Gammarus* were most common ones, both with respect to numbers of populations as well as local abundances (number of individuals, data not shown). Especially *G. fossarum* A, *G. fossarum* B, and *G. pulex* are widely distributed in the Rhine drainage basin. Interestingly, they are much less common in the other drainage basins, even though they are reported from all of them.

Of the non-native species, only *D. villosus*, *G. roeseli* and to some degree *C. pseudogracilis* have reached a wider distribution in Switzerland (Fig. 3). This is in strong contrast to the high to very high dominance of non-native species at a few sites [28], especially in the Upper and High Rhine (Fig. 2). While these non-native species have gained a lot of attention [28,29,33,34,43], their actual distribution in Switzerland is rather restricted to large rivers and lakes. It is unclear how much they are still spreading and if they would also be able to colonize most of the smaller tributaries and

water bodies at higher elevations, as evidence of such dynamics is lacking. As many of the non-native species are originating from lowland habitats, and have mostly been found in larger water bodies, it is unlikely that they would colonize the majority of small headwater reaches in Switzerland. Furthermore, some of these non-native species (all members of the family Corophiidae) are filter feeders, and thus depend on sufficient amounts of suspended particles. It is possible that their successful establishment depends or indicates changes in water quality or in the amount of suspended organic particles.

## Discussion of the genus *Niphargus*

In total, we report the occurrence of at least nine species of *Niphargus* in Switzerland. The phylogenetic and geographic position of the nine *Niphargus* species collected in Switzerland shows that they are only distantly related to each other and have clear geographic affiliations to species outside Switzerland (Fig. 6). Members of four phyletic lineages (*Niphargus virei* and *N. rhenorhodanensis* complex) belong to lineages distributed mainly west of Switzerland. Three lineages (*N. thienemanni*, *N. puteanus*, *N. caspary*) are distributed predominantly north and north-east of the Alps, whereas *N. thuringius* belongs to a south-eastern lineage. Several of these species contain cryptic species, which have not yet been formally described. As this genus has been receiving little attention so far in Switzerland, we in the following discuss the status and possible cryptic species complexes for each species of *Niphargus* found in Switzerland.

*Niphargus auerbachi.* This species is described from Switzerland (Schaffhausen, Schellenberg 1934) and reaches Southern Germany [89].

*Niphargus caspary.* This species is broadly distributed across France, Switzerland, Germany, Austria, Hungary, Serbia and Romania. Considering the low dispersal abilities of subterranean species [24], it is unlikely that it is a single species across the whole range. Moreover, the species is morphologically variable. We reviewed a single sample, unfortunately too old to obtain a sample of intact DNA.

*Niphargus forelii* is a species described from Lake Geneva, and reported from many other locations in Switzerland and deep alpine lakes from Italy and Germany [90]. An unsolved question is whether population from Lake Constance and other places in Switzerland and populations from Lake Geneva and surroundings belong to the same species.

*Niphargus puteanus.* This species was originally described from Regensburg (Germany) and molecular data are available from individuals collected in Regensburg (type locality, [91]) and Tübingen (unpublished data). The samples from Switzerland are morphologically and molecularly almost identical to those from Regensburg.

*Niphargus rhenorhodanensis.* Recently it has been found that this is not a single species, but a complex of at least six species [68], which do not even form a monophylum [63]. In 13 of our samples we found individuals belonging to the *N. rhenorhodanensis* complex. We successfully sequenced at least one sequence in eight out of fourteen individuals, and our data suggest that at least three species (lineage ABCDE, one sample; lineage FG, three samples; lineage H, four samples; lineage names according to Lefébure et al. [68]) of the complex live in Switzerland. No morphological revision exists and species identity can be assured by molecular markers only.

**Niphargus setiferus.** The species was described from the French Jura mountains and reported from Switzerland by Strinati [42].

**Niphargus thienemanni.** This species was identified in nine samples. It turned out to be molecularly closely similar to *N.* cf. *fontanus* 2 from Southern Germany [63], a sister species of *N. fontanus* from Great Britain [63,92]. Most samples from above 1,000 meters contained only this species, a pattern already noted by Schellenberg [93].

**Niphargus thuringius.** Described from Locarno, and primary considered as member of the *N. longicaudatus* species complex, it is distributed mainly in Northern Italy (Piemonte, Lombardia and Brescia). Contrary to expectations, in our molecular analyses this species turned out not to be related to the *N. longicaudatus* complex, but belongs to another clade of species that are distributed across Northern Italy (*N. dolenianesis*), Slovenia (*N. sphagnicolus* and another undescribed species), Bosnia and Herzegovina and Romania (both undescribed species), possibly also Slovakia (unpublished personal observations).

**Niphargus virei.** The species was found in a single sample from the Jura mountains (the Jura mountains are the type locality). Recently, it was shown that the name *N. virei* in fact covers three morphologically similar species [67].

The phyletic diversity of *Niphargus* relative to species diversity (6 major phylogenetic lineages with at least 9 species) is relatively high in Switzerland, and may reflect complex patterns of diversification and colonization within this genus (Figs. 3 and 6). For comparison, the much more intensely studied groundwater fauna of Slovenia (which is about half the size of Switzerland) harbors 42 species of *Niphargus* belonging to seven major phylogenetic lineages. The relative high ratio of lineage to species diversity found in Switzerland might be explained by historical effects: large parts of Switzerland were covered by ice during the Pleistocene, and the large scale distributional patterns of subterranean crustaceans may testify the devastating effects of glacial cover on the subterranean fauna. An important notion is that the ranges of all *Niphargus* species found in Switzerland extend to areas that were not covered by glaciers. Such a pattern might have been caused by mass extinctions during Pleistocene and subsequent recolonization of the emptied subterranean environment. The time for post-Pleistocene within-country speciation, which would increase the species to lineage ratio, has likely been too

short. Furthermore, it is reasonable to expect that species that colonized empty areas are good dispersers. Such species may maintain gene flow between the populations, which counteracts allopatric speciation.

The only species that might have survived all Pleistocene episodes under ice-sheet is *N. thienemanni* (see also [68,94]). Its current altitudinal distribution reaches from 690 to almost 1,640 meters above the sea, suggesting it might tolerate a broad range of temperatures. Molecular and physiological analyses, however, are needed to test whether it truly survived glaciation episodes in the Alps, or whether it is merely a successfully dispersing species.

## Conclusions

Amphipods are important for ecosystem processes and trophic dynamics in freshwater ecosystems and increasingly important for eco-monitoring and ecotoxicology. Still, accurate data on the occurrence and distribution of amphipods are only available for some European countries (Table 1). We provide the first conclusive overview of the amphipod fauna of Switzerland. We found not only a much higher diversity than previously known, but also a highly uneven distribution of species across spatial and altitudinal gradients. Switzerland contains potentially important refugia and boreo-alpine relict populations, and is prone to large-scale invasions of amphipods from different parts of Europe.

## Acknowledgments

The Swiss Federal Office for the Environment provided the amphipod samples from the BDM program. Furthermore, the cantons of Thurgau, Bern, Aargau and Zurich provided samples of amphipods. We thank L. Caduff, P. Ganesanandamoorthy, R. Illi, M. Koster, F. Leporio, K. Liljeroos, N. Martinez, P. Steinmann, and A. Stöckli for help during collecting and processing amphipod samples or for providing specimens. We thank U. Mürle and J. Ortlepp for data from the River Rhine and literature references. We thank Mat Seymour for help with R-functions, Valerija Zakšek for her help in the laboratory and Peter Trontelj and Fabio Stoch for discussion and helpful comments on the manuscript.

## Author Contributions

Conceived and designed the experiments: FA. Performed the experiments: FA RA CF JJ MK DK EM PS AMW. Analyzed the data: FA RA CF JJ MK DK EM PS AMW. Contributed reagents/materials/analysis tools: FA RA CF JJ MK DK EM PS AMW. Wrote the paper: FA.

## References

1. Vorosmarty CJ, McIntyre PB, Gessner MO, Dudgeon D, Prusevich A, et al. (2010) Global threats to human water security and river biodiversity. Nature 467: 555–561.

2. Altermatt F (2013) Diversity in riverine metacommunities: a network perspective. Aquatic Ecology 47: 365–377.

3. Carrara F, Altermatt F, Rodriguez-Iturbe I, Rinaldo A (2012) Dendritic connectivity controls biodiversity patterns in experimental metacommunities. Proceedings of the National Academy of Sciences 109: 5761–5766.

4. Väinölä R, Witt J, Grabowski M, Bradbury J, Jazdzewski K, et al. (2008) Global diversity of amphipods (Amphipoda; Crustacea) in freshwater. Hydrobiologia 595: 241–255.

5. Balian E, Segers H, Lévêque C, Martens K (2008) The Freshwater Animal Diversity Assessment: an overview of the results. Hydrobiologia 595: 627–637.

6. Macdonald KS, Yampolsky L, Duffy JE (2005) Molecular and morphological evolution of the amphipod radiation of Lake Baikal. Molecular Phylogenetics and Evolution 35: 323–343.

7. Zagmajster M, Eme D, Fišer C, Galassi DMP, Marmonier P, et al. (2014) Geographic variation in range size and beta diversity of groundwater crustacean: insights from habitats with low thermal seasonality. Global Ecology and Biogeography in press.

8. Ruffo S, Stoch F (2006) Checklist and distribution of the Italian fauna. 10,000 terrestrial and inland waters species. Memorie del Museo Civico di Storia Naturale di Verona, 2serie, Sezione Scienze della Vita 17: 1–307.

9. Eme D, Malard F, Konecny-Dupré L, Lefébure T, Douady CJ (2013) Bayesian phylogeographic inferences reveal contrasting colonization dynamics among European groundwater isopods. Molecular Ecology 22: 5685–5699.

10. Vonlanthen P, Bittner D, Hudson AG, Young KA, Muller R, et al. (2012) Eutrophication causes speciation reversal in whitefish adaptive radiations. Nature 482: 357–362.

11. Lubini V, Knispel S, Sartori M, Vicentini H, Wagner A (2012) Rote Listen Eintagsfliegen, Steinfliegen, Köcherfliegen. Gefährdete Arten der Schweiz, Stand 2010. [Red list of the Ephemeroptera, Plecoptera and Trichoptera of Switzerland, in German]. Bern: Feder Office for the Environment BAFU. 111 p.

12. Altermatt F, Seymour M, Martinez N (2013) River network properties shape α-diversity and community similarity patterns of aquatic insect communities across major drainage basins. Journal of Biogeography 40: 2249–2260.

13. Leuven RSEW, van der Velde G, Baijens I, Snijders J, van der Zwart C, et al. (2009) The river Rhine: a global highway for dispersal of aquatic invasive species. Biological Invasions 11: 1989–2008.

14. Grabowski M, Bacela K, Konopacka A (2007) How to be an invasive gammarid (Amphipoda: Gammaroidea): A comparison of life history traits. Hydrobiologia 590: 75–84.

15. Eggers TO, Martens A (2001) Bestimmungsschlüssel der Süßwasser-Amphipoda (Crustacea) Deutschlands [A key to the freshwater Amphipoda (Crustacea) of Germany] Lauterbornia 42: 1–68.

16. Bundschuh M, Schulz R (2011) Population response to ozone application in wastewater: an on-site microcosm study with *Gammarus fossarum* (Crustacea: Amphipoda). Ecotoxicology 20: 466–473.

17. Stucki P (2010) Methoden zur Untersuchung und Beurteilung der Fliessgewässer: Makrozoobenthos Stufe F. Bundesamt für Umwelt, Bern Umwelt-Vollzug 1026: 61.

18. Gerhardt A (2011) GamTox: A Low-Cost Multimetric Ecotoxicity Test with *Gammarus* spp. for In and Ex Situ Application. International Journal of Zoology 2011.

19. Karaman GS, Pinkster S (1977) Freshwater *Gammarus* species from Europe, North Africa and adjacent regions of Asia (Crustacea-Amphipoda): Part I. *Gammarus pulex*-group and related species. Bijdragen tot de Dierkunde 47: 1–97.

20. Karaman GS, Pinkster S (1977) Freshwater *Gammarus* species from Europe, North Africa and adjacent regions of Asia (Crustacea-Amphipoda): Part II. *Gammarus roeseli*-group and related species. Bijdragen tot de Dierkunde 47: 165–196.

21. Karaman GS, Pinkster S (1987) Freshwater *Gammarus* species from Europe, North Africa and adjacent regions of Asia (Crustacea-Amphipoda): Part III. *Gammarus balcanicus* group and related species. Bijdragen tot de Dierkunde 57: 207–260.

22. Weiss M, Macher JN, Seefeldt MA, Leese F (2014) Molecular evidence for further overlooked species within the *Gammarus fossarum* complex (Crustacea: Amphipoda). Hydrobiologia 721: 165–184.

23. Westram AM, Jokela J, Keller I (2013) Hidden biodiversity in an ecologically important freshwater amphipod: differences in genetic structure between two cryptic species. PLoS ONE 8: e69576.

24. Malard F, Boutin C, Camacho AI, Ferreira D, Michel G, et al. (2009) Diversity patterns of stygobiotic crustaceans across multiple spatial scales in Europe. Freshwater Biology 54: 756–776.

25. Feckler A, Thielsch A, Schwenk K, Schulz R, Bundschuh M (2012) Differences in the sensitivity among cryptic lineages of the Gammarus fossarum complex. Science of The Total Environment 439: 158–164.

26. Westram AM, Jokela J, Baumgartner C, Keller I (2011) Spatial Distribution of Cryptic Species Diversity in European Freshwater Amphipods (*Gammarus fossarum*) as Revealed by Pyrosequencing. PLoS ONE 6: e23879.

27. Ortlepp J, Rey P (2003) Biologische Untersuchungen an der Aare zwischen Bielersee und Rhein. Gewässerschutzfachstellen der Kantone Bern, Solothurn und Aargau. 134 p.

28. Rey P, Ortlepp J, Küry D (2004) Wirbellose Neozoen im Hochrhein. Ausbreitung und ökologische Bedeutung. Bern: Bundesamt für Umwelt, Wald und Landschaft. 88 p.

29. Steinmann P (2006) *Dikerogammarus villosus* im Zürichsee und in der Limmat. AWEL Amt für Abfall, Wasser, Energie und Luft Baudirektion Kanton Zürich. 22 p.

30. Steinmann P (2008) Makrozoobenthos und aquatische Neozoen im Greifensee und Pfäffikersee 2008. Zürich: Baudirektion des Kantons Zuürich AWEL Amt für Abfall, Wasser, Energie und Luft. 28 p.

31. Alp M, Keller I, Westram AM, Robinson CT (2012) How river structure and biological traits influence gene flow: a population genetic study of two stream invertebrates with differing dispersal abilities. Freshwater Biology 57: 969–981.

32. Bollache L (2004) *Dikerogammarus villosus* (Crustacea, Amphipoda): Another invasive species im Lake Geneva. Revue Suisse de Zoologie 111: 309–313.

33. Baden-Württemberg LfU (2005) Wirbellose Neozoen im Bodensee; Baden-Württemberg LfU, editor. Karlsruhe: JVA Mannheim.

34. Rey P, Küry D, Weber B, Ortlepp J (2000) Neozoen im Hochrhein und im südlichen Oberrhein. Mitteilungen des Badischen Landesvereins für Naturkunde und Naturschutz 18: 19–35.

35. BAFU (2002) Koordinierte biologische Untersuchungen am Hochrhein 2000; Makroinvertebraten. Schriftenreihe Umwelt 345: 101.

36. Eberstraller J, Eberstaller-Fleischanderl D, Rey P, Becker A (2007) Monitoring Alpenrhein. 70 p.

37. AquaPlus (2008) Biologische Untersuchungen in Seitengewässern der Aare zwischen Thun und Bern. GBL, Gewässer- und Bodenschutzlabor des Kantons Bern. 121 p.

38. Mürle U, Weber B, Ortlepp J (2003) On the occurence of *Synurella ambulans* (Amphipoda: Crangonyctidae) in the River Aare, catchment area of River Rhine, Switzerland. Lauterbornia 48: 61–66.

39. Hanselmann AJ, Gergs R (2008) First record of *Crangonyx pseudogracilis* Bousfield 1958 (Amphipoda, Crustacea) in Lake Constance. Lauterbornia 62: 21–25.

40. Bernasconi R (1969) Die biospellogische Forschung im Berner Mittel- und Oberland. Actes 3eme Congr. Nat. Speleol. Interlaken 1967. Stalactite (suppl) 3: 76–89.

41. Moeschler P (1983) Recherches biospeologiques au Holloch. Stalactite 39: 75–77.

42. Strinati P (1966) Faune cavernicole de la Suisse; Scientifique CNdlR, editor. Lons-le-Saunier, France: M. Declume Imprimeur.

43. Steinmann P (2006) Abklärung von *Dikerogammarus villosus* - Vorkommen im Walensee und Aufnahme des Makrozoobenthos am Ufer. Stein am Rhein: Amt für Jagd und Fischerei Finanzdepartement des Kantons St. Gallen. 12 p.

44. Steinmann P (2006) *Dikerogammarus villosus* im Zürcher-Obersee. Stein am Rhein: Finanzdepartement des Kantons St. Gallen, Amt für Jagd und Fischerei. 19 p.

45. Küry D, Mertens M (2010) Erfolgskontrolle BirsVital. Untersuchung 2010 Benthosfauna. Basel: Kanton Basel-Landschaft. 34 p.

46. Lods-Crozet B, Reymond O (2006) Bathymetric expansion of an invasive gammarid (*Dikerogammarus villosus*, Crustacea, Amphipoda) in Lake Léman. Journal of Limnology 65: 141–144.

47. Lubini V, Vicentini H (2011) Biologische Gewässergütebeurteilung Kanton Aargau. Periodische Bestandesaufnahmen an grösseren Bächen 2001–2009. Teil Makrozoobenthos. Zürich: Kanton Aargau, Departement Bau, Verkehr und Umwelt, Abteilung für Umwelt. 51 p.

48. Moritz C, Pfister P (2001) Trübung und Schwall Alpenrhein. Einfluss auf Substrat, Benthos, Fische. Fachbericht Makrozoobenthos, Phytobenthos. Internationalen Regierungskommission Alpenrhein. 109 p.

49. Nocentini AM (1967) Presenza di *Synurella ambulans* (F. Müller) (Crustacea Amphipoda) nel Lago Maggiore. Memorie dell'Institute Italiano di Idrobiologia "Dott Marco De Marchi" 21: 213–224.

50. Stucki P (2011) Suivi de la qualité biologique des cours d'eau par la méthode IBCH (MZB–Niveau R). Canton du Jura, Département de l'environnement et de l'équipement. 4 p.

51. Lubini V (2010) Untersuchung Benthos Wynental 2010. Abteilung für Umwelt Kanton Aargau & Umwelt und Energie Kanton Luzern. 9 p.

52. Mürle U, Becker A, Rey P (2003) Ein neuer Flohkrebs im Bodensee: *Dikerogammarus villosus* (Grosser Höckerflohkrebs). In: http//http://www.bodensee-ufer.de, editor. Konstanz: Hydra. pp. 3.

53. Pinkster S, Stock JH (1970) On three new species of *Echinogammarus*, related to *E. veneris* (Heller, 1865), from Italy and Switzerland (Crustacea, Amphipoda) Beaufortia 17: 85–104.

54. Taramelli E (1956) Ricerche sul *Niphargus* (Amphipoda Gammaridae) del Lago Maggiore. Mem Ist it Idrobiol 9: 61–82.

55. Vornatscher J (1969) *Gammarus* (*Rivulogammarus*) *lacustris* G.O. Sars (Amphipoda) in Österreich. Mitteilungen der naturwissenschaftlichen Vereins Steiermark 99: 123–129.

56. Karaman GS (1993) Crustacea: Amphipoda di acqua dolce. Bologna: Edizioni Calderini Bologna. 337 p.

57. Müller J (1998) Genetic population structure of two cryptic *Gammarus fossarum* types across a contact zone. Journal of Evolutionary Biology 11: 79–101.

58. Müller J (2000) Mitochondrial DNA variation and the evolutionary history of cryptic *Gammarus fossarum* types. Molecular Phylogenetics and Evolution 15: 260–268.

59. Westram AM, Jokela J, Keller I (2010) Isolation and characterization of ten polymorphic microsatellite markers for three cryptic *Gammarus fossarum* (Amphipoda) species Conservation Genetics Resources 2: 401–404.

60. Fišer C, Sket B, Trontelj P (2008) A phylogenetic perspective on 160 years of troubled taxonomy of *Niphargus* (Crustacea: Amphipoda). Zoologica Scripta 37: 665–680.

61. Fišer C, Zagmajster M, Zakšek V (2013) Coevolution of life history traits and morphology in female subterranean amphipods. Oikos 122: 770–778.

62. Trontelj P, Blejec A, Fišer C (2012) Ecomorphological convergence of cave communities. Evolution 66: 3852–3865.

63. Trontelj P, Douady CJ, Fišer C, Gibert J, Goričiki Š, et al. (2009) A molecular test for cryptic diversity in ground water: how large are the ranges of macrostygobionts? Freshwater Biology 54: 727–744.

64. Verovnik R, Sket B, Trontelj P (2005) The colonization of Europe by the freshwater crustacean *Asellus aquaticus* (Crustacea: Isopoda) proceeded from ancient refugia and was directed by habitat connectivity. Molecular ecology 14: 4355–4369.

65. Zaksek V, Sket B, Trontelj P (2007) Phylogeny of the cave shrimp *Troglocaris*: Evidence of a young connection between Balkans and Caucasus. Molecular phylogenetics and evolution 42: 223–235.

66. Colgan DJ, Ponder WF, Eggler PE (2000) Gastropod evolutionary rates and phylogenetic relationships assessed using partial 28S rDNA and histone H3 sequences. Zoologica Scripta 29: 29–63.

67. Lefébure T, Douady CJ, Gouy M, Trontelj P, Briolay J, et al. (2006) Phylogeography of a subterranean amphipod reveals cryptic diversity and dynamic evolution in extreme environments. Molecular ecology 15: 1797–1806.

68. Lefébure T, Douady CJ, Malard F, Gibert J (2007) Testing dispersal and cryptic diversity in a widely distributed groundwater amphipod (*Niphargus rhenorhodanensis*). Molecular phylogenetics and evolution 42: 676–686.

69. Katoh K, Toh H (2008) Improved accuracy of multiple ncRNA alignment by incorporating structural information into a MAFFT-based framework. BMC bioinformatics 9: 212–212.

70. Ronquist F, Teslenko M, van der Mark P, Ayres DL, Darling A, et al. (2012) MrBayes 3.2: efficient Bayesian phylogenetic inference and model choice across a large model space. Systematic biology 61: 539–542.

71. Fišer C, Coleman C, Zagmajster M, Zwittnig B, Gerecke R, et al. (2010) Old museum samples and recent taxonomy: A taxonomic, biogeographic and conservation perspective of the *Niphargus tatrensis* species complex (Crustacea: Amphipoda). Organisms Diversity & Evolution 10: 5–22.

72. Sket B (1996) The fauna of Malacostraca (ex. Astacidae, Oniscida) - its composition and endangerment. In: Gregori J, editor. Narava Slovenije, stanje in perspektive: zbornik prispevkov o naravni dediščini Slovenije. Ljubljana: Društvo ekologov Slovenije. pp. 222–227.

73. Fišer C, Zagmajster M (2009) Cryptic species from cryptic space: the case of *Niphargus fongi* sp. n. (Amphipoda, Niphargidae). Crustaceana 82: 593–614.

74. Ruffo S, Vonk R (2001) *Ingolfiella beatricis*, new species (Amphipoda: Ingolfiellidae) from subterranean waters of Slovenia. Journal of Crustacean Biology 21: 484–491.

75. Nychka D, Furrer R, Sain S (2014) fields: Tools for Spatial data. R package version 7.1 ed.

76. R Development Core Team (2013) R: A language and environment for statistical computing. Version 3.0.1. Vienna, Austria: R Foundation for Statistical Computing.

77. de Jong YSDM (2014) Fauna Europaea 2.6 http://www.faunaeur.org. In: de Jong YSDM, editor.

78. Ruffo S, Tarocco M, Latella L (2014) *Cryptorchestia garbinii* n. sp. (Amphipoda: Talitridae) from Lake Garda (Northern Italy), previously referred to as *Orchestia cavimana* Heller, 1865, and notes on the distribution of the two species. Italian Journal of Zoology 81: 92–99.

79. Ketmaier V, De Matthaeis E (2010) Allozymes and mtDNA reveal two divergent lineages in *Orchestia cavimana* (Amphipoda: Talitridae). Journal of Crustacean Biology 30: 307–311.

80. Ward JV, Uehlinger U (2003) Ecology of a glacial flood plain. London: Kluwer Academic Publishers.

81. Karaman GS (1980) Contribution to the knowledge of the Amphipoda. 113. Redescription of *Niphargus aquilex* Schiödte and its distribution in Great Britain. Biosistematika 6: 175–185.

82. Karaman S (1952) Podrod *Stygoniphargus* u Sloveniji i Hrvatskoj, Poseban otisak iz 25 knjige Prirodoslovnih istraživanja. Academia Scientiarum et Artium Slavorum Meridionalium 1: 5–38.

83. Sket B (1974) *Niphargus stygius* (Schiodte) (Amphipoda, Gammaridae): die Neubeschreibung des Generotypus, Variabilitat, Verbreitung und Biologie der Art, I. Biološki vestnik 22: 91–103.

84. Deiner K, Altermatt F (2014) Transport distance of invertebrate environmental DNA in a natural river. PLoS One 9: e88786.

85. Mächler E, Deiner K, Steinmann P, Altermatt F (2014) Utility of environmental DNA for monitoring rare and indicator macroinvertebrate species. Freshwater Science. in press. doi:10.1086/678128

86. MacArthur RH, Wilson EO (1967) The theory of island biogeography. Princeton: Princeton University Press.

87. Bacela-Spychalska K, Grabowski M, Rewicz T, Konopacka A, Wattier R (2013) The 'killer shrimp' *Dikerogammarus villosus* (Crustacea, Amphipoda) invading Alpine lakes: overland transport by recreational boats and scuba-diving gear as potential entry vectors? Aquatic Conservation: Marine and Freshwater Ecosystems 23: 606–618.

88. Bundschuh M, Gessner MO, Fink G, Ternes TA, Södging C, et al. (2011) Ecotoxicologial evaluation of wastewater ozonation based on detritus-detritivore interactions. Chemosphere 82: 355–361.

89. Fuchs A (2007) Erhebung und Beschreibung der Grundwasserfauna in Baden-Württemberg. Koblenz: Universität Koblenz-Landau. 109 p.

90. Ruffo S, Karaman GS (1993) *Niphargus forelii* Humbert, 1876 and its taxonomic position (Crustacea Amphipoda, Niphargidae). Bollettino del Museo civico di Storia Naturale, Verona 17: 57–68.

91. Stock JH (1974) Redescription de l'amphipode hypogé *Niphargus puteanus* (Koch in Panzer, 1836), basée sur du matériel topotypique. Bijdragen tot de Dierkunde 44: 73–82.

92. Hartke TR, Fišer C, Hohagen J, Kleber S, Hartmann R, et al. (2011) Morphological and Molecular Analyses of Closely Related Species In the Stygobiontic Genus (Amphipoda). Journal of Crustacean Biology 31: 701–709.

93. Schellenberg A (1942) Die Tierwelt Deutschlands und der angrenzenden Meeresteile nach ihren Merkmalen und nach ihrer Lebensweise: 40. Teil: Krebstiere oder Crustacea: Flohkrebse oder Amphipoda: Jena Verlag von Gustav Fischer. 81–81 p.

94. Holsinger JR, Shaw DP (1987) *Stygobromus quatsinensis*, a new amphipod crustacean (Crangonyctidae) from caves on Vancouver Island, British Columbia, with remarks on zoogeographic relationships. Canadian Journal of Zoology 65: 2202–2209.

# 9

# Prevalence of Veterinary Antibiotics and Antibiotic-Resistant *Escherichia coli* in the Surface Water of a Livestock Production Region in Northern China

**Xuelian Zhang, Yanxia Li\*, Bei Liu, Jing Wang, Chenghong Feng, Min Gao, Lina Wang**

State Key Laboratory of Water Environment Simulation, School of Environment, Beijing Normal University, Beijing, China

## Abstract

This study investigated the occurrence of 12 veterinary antibiotics (VAs) and the susceptibility of *Escherichia coli* (*E. coli*) in a rural water system that was affected by livestock production in northern China. Each of the surveyed sites was determined with at least eight antibiotics with maximum concentration of up to 450 ng $L^{-1}$. The use of VAs in livestock farming probably was a primary source of antibiotics in the rivers. Increasing total antibiotics were measured from up- to mid- and downstream in the two tributaries. Eighty-eight percent of the 218 *E. coli* isolates that were derived from the study area exhibited, in total, 48 resistance profiles against the eight examined drugs. Significant correlations were found among the resistance rates of sulfamethoxazole-trimethoprim, chloromycetin and ampicillin as well as between tetracycline and chlortetracycline, suggesting a possible cross-selection for resistance among these drugs. The *E. coli* resistance frequency also increased from up- to midstream in the three rivers. *E. coli* isolates from different water systems showed varying drug numbers of resistance. No clear relationship was observed in the antibiotic resistance frequency with corresponding antibiotic concentration, indicating that the antibiotic resistance for *E. coli* in the aquatic environment might be affected by factors besides antibiotics. High numbers of resistant *E. coli* were also isolated from the conserved reservoir. These results suggest that rural surface water may become a large pool of VAs and resistant bacteria. This study contributes to current information on VAs and resistant bacteria contamination in aquatic environments particularly in areas under intensive agriculture. Moreover, this study indicates an urgent need to monitor the use of VAs in animal production, and to control the release of animal-originated antibiotics into the environment.

**Editor:** Tara C. Smith, Kent State University, United States of America

**Funding:** The research was funded by the Natural Science Foundation of China (No. 21277013, 20977010) and the Special Fund for Environmental Protection Research in the Public Interest (200909042). The funders had no role in study design, data collection and analysis, decision to publish, or preparation of the manuscript.

**Competing Interests:** The authors have declared that no competing interests exist.

\* Email: liyxbnu@bnu.edu.cn

## Introduction

Antibiotic and antibiotic-resistant microbial contamination is an issue of growing global concern, both in the public and in the research community. The rampant use of antibiotics in medicine and agriculture has resulted in the extensive detection of antibiotics, antibiotic-resistant bacteria and antibiotic resistance genes (ARGs) in the environment worldwide, including in 139 streams in the US, the Osaka area of Japan, the Haihe River and the Yangtze estuary in China [1–4].

Animals in concentrated feeding operations are the chief consumers of antimicrobial agents, which are administered for growth improvement and disease control. The amount of veterinary antibiotics (VAs) annually administered in the world reaches $10^5$–$10^6$ tonnes [5–7]. Guidelines have existed that ban the use of certain antibiotics as growth promoters in both the European Union and the US since 1999, but this prohibition has led to a corresponding increased use of VAs for disease control and improving feed efficiency [8]. Considering Denmark as an example, the total veterinary use remained as high as 107 t in 2011 [9,10]. Antibiotics are poorly absorbed in the animal gut; as a result, approximately 40–90% of these antibiotics will be excreted as parent compounds or metabolites via urine or feces [11,12], and the residue of VAs in animal wastes has been widely reported [7,13–16].

Animal wastes are usually stored using a lagoon system and/or are composted instead of intensive treatment before being discharged from animal farms [10]. However, both of these methods are limited in their ability to completely remove antibiotics. For instance, compost can reduce antibiotics by 20%–99% [17,18], and a lagoon system can decrease tylosin by no more than 75% [19]. Thus, high amounts of antibiotics might remain in animal waste and are a potential pollution source of environmental antibiotics. Residual antibiotics in post-treated animal wastes can be disseminated into the surrounding aquatic environment via runoff when utilized as fertilizer on farmlands [20,21] or in some cases, when directly released to receiving watersheds through sewage discharge or occasional leaching from animal farms. Such scattered pollution sources from operations as well as from fertilized fields will lead to more severe and complex antibiotic contamination in these rural areas compared to point-source-affected urban rivers; therefore, the antibiotic contamination in these regions requires additional attention and study.

The particular concern over VAs exposure is the increased presence of antibiotic resistant bacteria in the environment. Antibiotic-resistant bacteria might develop resistance within animal bodies from exposure to administered VAs [22], which can be excreted in feces and be subsequently released into the environment along with antibiotics [23,24]. Furthermore, long-term exposure to sub-therapeutic levels of antimicrobial agents in aquatic environments imposes selection stress on environmental microorganisms [25]. Once antibiotic-resistant microorganisms appear in the environment, they enlarge the resistance community through the transfer of corresponding ARGs among microbial populations [3,26]. In addition, antibiotic-resistant bacteria are likely to transport vertically and horizontally through physical or biological media [27]. Since livestock breeding and related agricultural activity is a potential source of antibiotics and resistant bacteria in the environmental, there is an urgent need to elucidate how antibiotics, and particularly antibiotic-resistant fecal indicator bacteria (*E. coli*) contamination in rural aquatic environments, might be impacted by intensive concentrated animal feeding operations (CAFOs).

China is the largest animal feeding country in the world and is experiencing the expansion and intensification of animal feeding operations in many areas. Due to the absence until now of relevant regulation, the residues of a variety of VAs have been consistently reported at high levels in animal waste [7,13,15]. Nevertheless, limited information is available providing baseline data about surface water pollution with antibiotics and antibiotic-resistant bacteria in typical livestock production regions. Therefore, this study investigated the Jiyun River, Beijing, which flows through Pinggu County, which is one of the primary meat-providing counties in Beijing, to (1) determine the occurrence and spatial distribution of antibiotics in the water system, (2) examine the antibiotic susceptibility of *E. coli* in this basin and (3) explore the correlation between *E. coli* resistance and antibiotic concentrations. We seek to provide useful information of the influence that is exerted on the aquatic ecosystem by VAs administration in the animal industry.

## Materials and Methods

### Study area, sampling sites and sample collection

The study area of Pinggu County is located east of Beijing, China. This county has the highest animal feeding density among the suburbs of Beijing and provided $3.4 \times 10^5$ and $7.11 \times 10^6$ swine and poultry products, respectively, to capital market in 2011 [28]. The main stream of the Jiyun River is Ju River, with a stretch of 54.4 km, which flows across Pinggu County over an area of 1352 km$^2$ and eventually enters Bohai Bay.

The Jiyun River has two tributaries: Cuo River and Jinji River. The sampling sites were selected along the upper, middle and lower reaches of the main stream and two tributaries. Eight of the 9 sites were directly adjacent to animal farms. These animal farms are small-scale operations, and none of the 8 farms was equipped with a professional lagoon system or fecal treatment procedure. The effluents of these farms are directly discharged into adjacent water bodies. The fecal wastes were simply piled up with maize straw in the air, but usually without any bulking agents, and then applied to nearby farmlands during fertilizer season. Six fresh fecal samples were collected from six of the eight animal operations with the permission of farmers, but samples from the other two farms were not available. From each farm, multiple points of fresh feces were obtained from the pigpen and were mixed to obtain a representative sample. Two intersection sites (JC and JCJJ) in the main stream Ju River separately receiving flow from Cuo River

and Jinji River, respectively, and a site from the Haizi reservoir were also selected. The GPS coordinates of each location in this study are listed in Table S1. The sampling locations and sampling activity in this study did not require specific permissions and the sampling did not involve endangered or protected species. The study area and sample sites distributions are presented in Figure 1. The sampling was conducted on May 16, 2013, and a total of 12 water samples and 6 fresh fecal samples were obtained.

The samples were separated into two parts: one part received sodium azide immediately to inhibit microbial activity and prevent antibiotic biodegradation for further antibiotic analysis [29,30], and the other part was free of sodium azide for microbiological property determination and pure *E. coli* clone isolation. The samples were preserved in cold boxes, transported to the laboratory within one day and maintained at 4°C until use. The microbial samples were incubated and the pH values were measured immediately upon return to the laboratory. The antibiotics were detected beginning on the second day.

### Sample preparation and analysis

Twelve target antibiotics in three different groups (tetracyclines (TCs): tetraycline (TC), oxytetracycline (OTC), chlortetracycline (CTC) and doxycycline (DOC); quinolones (QLs): ciprofloxacin (CFC), enrofloxacin (EFC) and ofloxacin (OFC); and sulfonamides (SAs): sulfadiazine (SDZ), sulfamethoxazole (SMX), sulfamono-methoxine (SMM), sulfameter (SM), sulfachinoxaline (SCX)) were analyzed in this study.

The samples were pretreated according to the method of [12,15] with slight modification. Briefly, a 2-L water sample was filtered through 0.45-µm pore glass fiber membrane filter. The filtered water sample was acidified to a pH of 3.0 with 30% H$_2$SO$_4$ (V/V), followed by the addition of Na$_2$-EDTA (0.5 g L$^{-1}$) as a chelating agent. The acidified water samples were concentrated by solid-phase extraction (SPE) over an Oasis HLB cartridge (6cc, 500 mg, Waters, USA) at a flow rate of 5 ml min$^{-1}$. The antibiotics that were retained on the HLB cartridges were eluted with 10 ml of methanol (0.1% formic acid) after being dried with nitrogen gas at a flow rate of 5 ml min$^{-1}$. The eluent was concentrated to near dryness under a gentle nitrogen stream in a 37°C water bath. A mixture of internal standard containing tetracycline-D6, enrofloxacin-D5 and sulfamethazine-D4 was added to compensate for the variations of TCs, FQs and SAs, respectively, during detection. Methanol was supplemented to a final volume of 1 ml. The final solution was filtered through a PTFE filter (Millipore) for liquid chromatography - tandem mass spectrometry (LC-MS-MS) analysis. For the solid samples, 3 g of fresh fecal samples were extracted with a 20-ml mixture of methanol and EDTA-McIlvaine buffer (V:V = 1:1) with 30 minutes of rotation and ultrasonication, respectively, three times. The extracts were centrifuged and the supernatant was collected and diluted with deionized water to maintain a methanol proportion below 10%. The substantial purification and concentration followed the steps of the water samples.

The extracted samples were analyzed by LC-MS-MS. The separation of the target compounds was performed on a C18 column (250×4.6 mm, 5 mm; Akzo Nobel, Sweden) in a HPLC system (DIONEX UltiMate 3000). A binary elution gradient consisting of A (0.1% formic acid water solution) and B (methanol) was used in the following program: 0–2 min: 15–30% B, 2–5 min: 30–40% B, 5–15 min: 40–70% B, 15–17 min: 70–100% B, and 17–21 min 100% B. A triple quadrupole mass spectrometer (API 3200, AB-SCIEX, Framingham, MA) that was equipped with an electrospray ionization (ESI) source in positive ion mode was used. The multiple reaction monitoring (MRM) mode was used for the

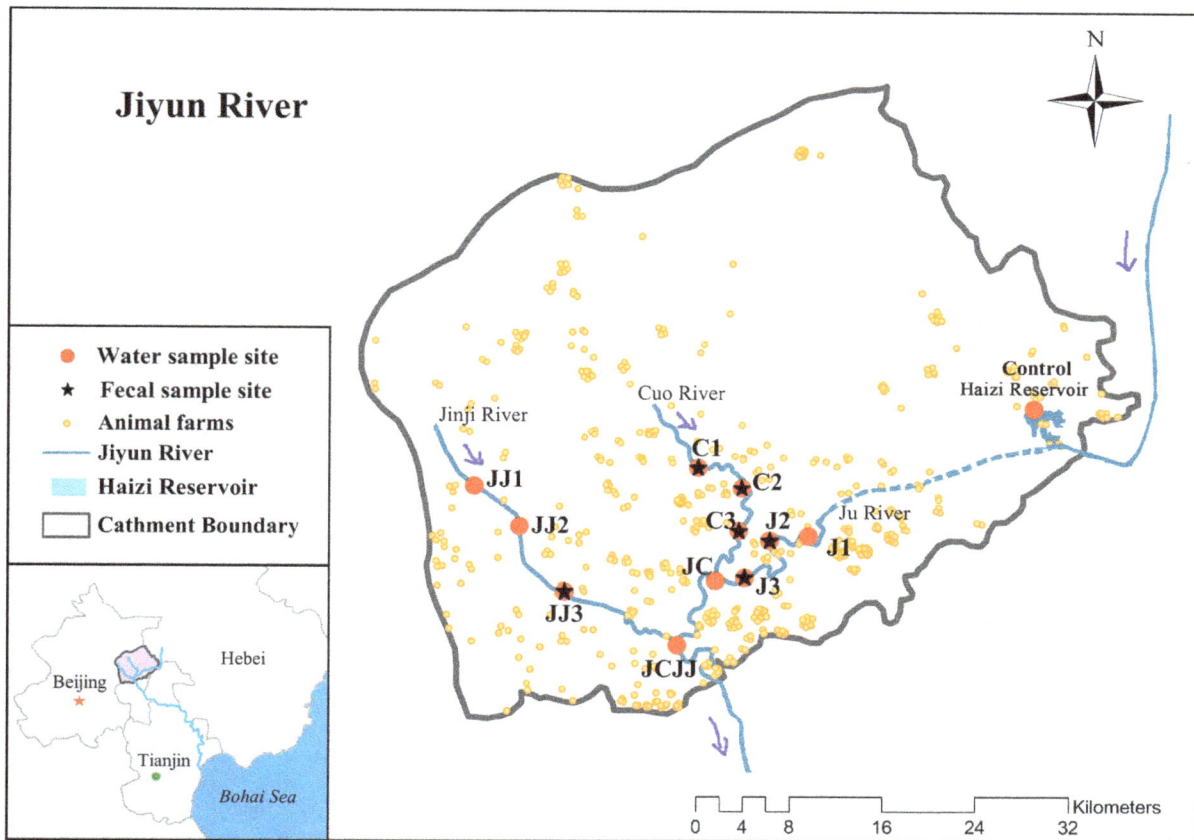

**Figure 1. Map of sampling sites in the Jiyun River.**

quantitation. The most intensive ion pairs together with retention time were used to identify the targeted antibiotics. Multipoint internal calibration curves were used to quantify the antibiotics. The recoveries of the 12 monitored compounds were in the range of 71 to 100%.

## E. coli isolation and confirmation

*E. coli* in the water was detected using the enzyme substrate method [31]. Briefly, Colilert substrates (IDEXX Laboratories, Inc., Westbrook, ME) were added to 100 ml water samples and were mixed thoroughly until all of the substrate dissolved per the manufacturer's instructions. The dissolved water was sealed in an IDEXX Quanti-Tray/2000 (IDEXX Laboratories, Inc., Westbrook, ME) and incubated at 37°C for 24 h. The wells that changed to yellow and produced additional fluorescence were identified as harboring *E. coli*.

The *E. coli* isolates were obtained and confirmed according to previous descriptions [26,31] with some minor revision. Briefly, 10 μl of a $10^{-4}$–$10^{-5}$ dilution of liquid from 10–15 wells that were positive for *E. coli* growth on the plate for each site was streaked onto a Mueller-Hinton (MH) agar plate for incubation of 18 h at 37°C to obtain 10–50 clones on each plate. Approximately 4 isolates were randomly selected from well-separated colonies on each plate and purified on Mueller-Hinton (MH) agar plates using the crossed dilution method. To confirm these isolates, they were separately inoculated in *Escherichia coli* broth (EC broth) containing 4-methylumbelliferyl-, BD-glucuronide (EC-MUG) and tryptophan separately. The isolates that generated a blue fluorescence in EC-MUG and produced indole (identified by a

rose-colored product after the addition of Kovacs agent) after 24 h of incubation at 37°C were confirmed as *E. coli*. In total, 218 isolates were collected from the study area. The confirmed isolates were re-streaked onto MH agar plates for subsequent susceptibility tests.

## Antibiotics susceptibility test

In total, eight drugs were tested for *E. coli* susceptibility primarily based on their detection in this study and their clinical importance. CTC and SMM tests were performed via growth measurement in Mueller-Hinton Broth in the presence and absence of antibiotics. The other six antibiotics were analyzed via the disk diffusion method [32]. The CTC and SMM test concentration was set to 25 μg and 500 μg per ml MH broth, respectively, according to a combination of prescribed doses in CLSI guidelines and the levels that have been used in other studies [33–35]. The CTC stock was directly prepared in MH broth. The SMM was dissolved in MH broth with a few drops of NaOH solution. A pure *E. coli* culture was suspended in sterile sodium chloride solution to 0.5 McFarland turbidity level after the incubation for 18 hours at 37°C. Fifty μl of a $10^{-2}$ suspension was used to separately inoculate wells containing 50 μl of MH broth with or without target antibiotic. The preliminary test indicated that the amount of NaOH remaining in SMM wells had no effect on *E. coli* growth. Therefore, the first and second wells in each group of the 4 wells were always the negative control (MH broth) and the positive control (MH broth with *E. coli* inoculum), respectively. After incubation for 22 hours at 37°C, the 96-well plate was read spectrophotometrically at 600 nm. The isolates were recorded as resistant to a particular antibiotic if the growth, as measured through OD values, was inhibited by less than 15% compared to that of the positive control or were recorded as susceptible when the growth was reduced by at least half [27,33,36].

The six antibiotic disks (Oxoid, UK) included ampicillin (AMP, 10 μg), chloromycetin (C, 30 μg), levofloxacin (LEV, 5 μg), tetracycline (TE, 30 μg), gentamycin (CN, 10 μg) and sulfamethoxazole-trimethoprim (SXT, 25 μg/1.25 μg). A 0.5 McFarland turbidity level of the *E. coli* isolate suspension was obtained using sterile sodium chloride solution after the isolates were incubated on MH agar plates at 37°C. The suspension was evenly inoculated onto an MH agar plate using autoclaved gauze. Three different disks were placed onto each inoculated plate. The disks were far away from each other and from the plate rim to avoid overlapping inhibition zones. The inhibition zone diameter was precisely measured using a ruler after 16–18 h of incubation at 37°C. The diameter was compared to the diameter of the susceptible, moderate and resistant standards as listed in CLSI 2009. The antibiotic resistance frequency was calculated as the ratio of *E. coli* isolates that were resistant to antibiotics to the total number that were isolated from each site. The multi-antibiotic resistance (MAR) index was estimated by the equation $a/(b \times c)$, where a is the total antibiotic resistance score of all of the isolates from the sample, b is the number of tested drugs, and c is the number of isolates from the sample [26]. The significance of difference in average *E. coli* resistance frequencies and MAR among the three rivers was examined through a non-parametric k independent samples Kruskal-Wallis test using SPSS 17.0. The correlations between *E. coli* counts and resistance frequencies or MAR as well as among the percent resistances to the eight drugs in the seven sites were examined via a Spearman correlation analysis using SPSS 17.0. The two sites, including the control and C1, were removed during the correlation analysis because of low antibiotic levels and few *E. coli* isolates, respectively.

## Results and Discussion

### Occurrences and levels of antibiotics in the Jiyun River

The detection frequencies and concentration profiles of the 12 monitored antibiotics are summarized in Table 1. The selected VAs were extensively detected in the investigated water system at percentages ranging from 58% to 100%. The concentrations varied largely among these sites from less than 1 ng L$^{-1}$ to 450 ng L$^{-1}$. Based on our determination, the SAs (SM, SMX and SDZ) and QLs (OFC, EFC and CFC) might be the predominant antibiotics in the water, as they were present in all of these investigated sites at an average concentration range of 3.79 to 92.97 ng L$^{-1}$. In comparison, the TCs were generally one or two orders of magnitude lower (mean 2.17–16.12 ng L$^{-1}$) than were the SAs (except for SCX) in concentration.

Figure 2 depicts the antibiotic distributions in the Jiyun River. The three rivers displayed varying antibiotic abundances with approximately 4- and 2-fold greater average antibiotic levels at the 3 sites in the Cuo River and Jinji River tributaries than that in the main stream Ju River. The total antibiotic contents at mid- and downstream (JJ2 and JJ3, and C2 and C3) in the two tributaries increased compared to the upstream site (JJ1 and C1). Our results contrast with those of other studies investigating antibiotic variation in point-source-affected water bodies, in which decreased antibiotic concentrations were detected in the water over distances from the site receiving a point-source effluent due to natural attenuation by adsorption, dilution, photolysis, hydrolysis and biodegradation [37,38].

The monitored sites, with the exception of Haizi reservoir, J1, JC and JCJJ, were immediately close to swine production farms, from which the sewage was directly discharged into receiving rivers, and the animal feces were used as a fertilizer in the nearby farmlands. Residual antibiotics in manure can concentrate, migrate in agricultural soil and finally end up in the aquatic environment through runoff when the feces are applied as fertilizer [20,21]. Therefore, the potential continuous input of antibiotics from the discharge of animal farms as well as from the runoff of fertilized farmland was inferred to account for the accumulated antibiotic contamination along the river. In addition, high amounts of antibiotics were also recorded at the two sites where the Ju River intersected with the two tributaries: JC (Ju River/Cuo River) and JCJJ (Ju River/Cuo River/Jinji River) (Figure 2). The antibiotic concentrations at the three sites J1, J2 and J3 were almost the lowest of the surveyed sites in the three rivers, except for JJ1. One possible reason is due to the dilution and rapid transfer of antibiotics in large water flow of the main stream. However, the specific reason requires further study. High concentrations of antibiotics at the two intersection sites likely indicate potential compound transfer with flow from the tributaries into the main stream.

As seen in Figure 2, most of the surveyed sites in the Jiyun River were generally determined to have higher levels of SAs (SM, SMM, SMX and SDZ) and QLs (OFC and CFC) than TCs. These findings agree with the results of many other studies [4,39,40] and may be chiefly explained by the discrepant partition characteristics among different antibiotic classes. The TCs have strong combination abilities with soil/sediment and their mineral or organic components via cation bridging and/or cation exchange [41,42], which causes their retention in the soil or dispersion to the sediment after being discharged into rivers. This result is supported by detection of TCs in the sediment of this river (Table S3). In contrast, most SA compounds predominantly exist in anionic species with negative charges at environmental pH values of >7 [43,44] and are consequently less frequently adsorbed onto solid-phase material. The pH values of these water samples were in the range of 7.29 to 9.24. This range facilitates the dissolution and migration of SAs. Thus, it is reasonable that the SAs were primarily observed in the river water.

A parallel survey of the antibiotic residues in six fecal samples of the animal operations that were directly adjacent to the water sample sites was conducted to explore the association of surface water antibiotic contamination with livestock production. The 12 target antibiotics were widely detected in these feces at levels of 1.03–56200 ng g$^{-1}$. The water and fecal samples from the study area were generally coincident in the antibiotic composition of the two classes of VAs: SAs and QLs (Figure 3). Both of the matrixes were dominated by SM and SMM, and SCX were seldom detected. In addition, these matrixes presented the same order of QLs: OFC>CFC>EFC. In contrast, although large amounts of TCs were detected in these animal feces, they were found at relatively low levels in the water. This finding may be explained by the strong adsorption of TCs in the environment as discussed above and is also supported by the detection of TCs in the river sediment (Table S3).

The detection concentrations of VAs in the current study are lower than or comparable to the levels in other surface waters that are proximate to animal operations in Jiangsu, China (560–2420 ng L$^{-1}$) [45] and in the US (1000–1500 ng L$^{-1}$) [46]. These reported results and the concentrations in this study were significantly higher than in rivers receiving effluent from WWTPs, such as the Victoria River, HongKong, China (<20 ng L$^{-1}$); the Elbe River, Germany (30–70 ng L$^{-1}$); and the Ebro River, Spain

**Table 1.** Frequencies and concentrations of the 12 target antibiotics in the Jiyun River (n = 12).

| Class | Compound | Frequency(%) (%) | Range (Mean)(ng L$^{-1}$) (ng/L) | MDLs((ng L$^{-1}$)) |
|---|---|---|---|---|
| Tetracyclines (TCs) | Tetrayclinc (TC) | 83.33 | n.d-11.00 (2.17) | 2.14 |
| | Oxytetracycline (OTC) | 91.67 | n.d-100.00 (16.12) | 2.35 |
| | Chlortetracycline (CTC) | 83.33 | n.d-40.60 (12.92) | 2.87 |
| | Doxycycline (DOC) | 58.33 | n.d-11.75 (2.84) | 2.45 |
| Quinolones (QLs) | Ciprofloxacin (CFC) | 100.00 | 3.56–24.80 (11.61) | 2.15 |
| | Enrofloxacin (EFC) | 100.00 | 0.55–13.41 (3.79) | 0.25 |
| | Ofloxacin (OFC) | 100.00 | 1.34–102.00 (27.89) | 1.10 |
| Sulfonamides (SAs) | Sulfadiazine (SDZ) | 100.00 | 0.03–385.70 (62.45) | 0.01 |
| | Sulfamethoxazole (SMX) | 100.00 | 4.29–230.00 (54.65) | 1.15 |
| | Sulfamonomethoxin (SMM) | 75.00 | n.d-450.00 (147.64) | 1.10 |
| | Sulfameter (SM) | 100.00 | 0.51–387.00 (92.97) | 0.16 |
| | Sulfachinoxalin (SCX) | 91.67 | n.d-13.95 (2.90) | 0.57 |

n.d: non-detected.
MDLs: method detection limitations for the 12 compounds.

(0.2–35.6 ng L$^{-1}$) [47–49]. These results indicate that the influences of the rural aquatic ecosystem caused by livestock production are much greater than are those in urban areas caused by human populations.

## Antibiotic-resistance for *E. coli* clones that were isolated from the Jiyun River

The detected *E. coli* counts in the Jiyun River were shown in Table S2. A total of 218 *E. coli* clones were isolated from the Jiyun River that not only exhibited a high percentage of resistance to the drugs that are frequently detected in these rivers but also to those that were not monitored but are of clinical concern. In total, 88% of the 218 isolates exhibited resistance to one or more of the tested drugs. The most frequent resistance appeared for CTC (61.01%), followed by AMP and TE (approximately 50%), CN (45.91%), SXT and SMM (39.09% and 36.24%), LEV (30.45%) and C (16.82%) (Figure 4). High levels of *E. coli* resistance to CTC, TE and AMP have been found in other aquatic environments [27,33] and in WWTP [50]. The great resistance frequency against AMP in the environment was possibly because of its comparatively older utilization history (over years) [51]. This resistance also reflects the common use of AMP in agricultural activities [52]. The high resistance rate to CTC maybe is due to the potential long-term exposure of *E. coli* to CTC in these aquatic environments. The frequent TE resistance despite its low concentration in water column may be caused by the possibility that the majority of TE-resistant *E. coli* have developed resistance against TE before they enter rivers because high levels of TCs were identified in the surveyed animal feces. In addition, the tet genes (e.g., tet(A), tet(C) and tet(G)) readily spreading among gram-negative bacteria such as *E. coli* through transposons and smaller plasmids might also facilitate the abundance of TE resistant *E. coli* [39]. The relatively lower resistance frequency against C agrees with the observations of an earlier previous study [53].

Approximately 72% of these bacteria exhibited multiple antibiotic resistances, and, in total, 48 resistance profiles were determined (Figure 5). The five primary profiles were AMP, CN, CTC, TE/CTC and TE/SMM/CTC. The total rate of resistance to the five profiles for *E. coli* that were isolated from the Jiyun River was approximately 24%. It is worth noting that 3.67% of the

isolates from the 9 sites exhibited eight-drug resistance. Ongoing multiple antibiotic exposure must have occurred in the study area as it has been repeatedly demonstrated that bacterial antibiotic resistance diversity is closely related to antibiotic contamination frequency [50].

Table 2 lists the relationships of the percent resistances among the eight drugs. Interestingly, significant correlations were obtained among the three drugs SXT, C, and AMP (p<0.01) and between TE and CTC (p<0.01). Positive but statistically non-significant relationships were also identified for CN and LEV with SXT, LEV and C, LEV and AMP, CN and AMP, and SMM and SXT (p>0.05, r >0.5). Strongly positive correlations were also reported in other drugs, such as nalidixic acid MICs with cefotaxime and ciprofloxacin MICs [27] and vancomycin with daptomycin resistance in *Staphylococcus aureus* (*S. aureus*) [51]. These correlations indicate that the *E. coli* clones that were tested in this study probably developed co-selection by containing one drug resistance gene on the plasmids that were resistant to other antimicrobials [54]. These correlations also indicate possible similar resistance mechanisms for the different drugs, as reported for *S. aureus*, showing strongly positive correlation between vancomycin and daptomycin susceptibility likely due to the common physical barrier of a thickened cell [51].

Table 3 presents the antibiotic resistance rate and profiles of *E. coli* from the sites of the Jiyun River. Although comparatively less *E. coli* were detected in the two sites of Jinji River (Table S2), the isolated cultures exhibited high resistance rates and MAR (Table 3). Spearman correlation analysis revealed no statistical correlations between *E. coli* MPN and resistance rate (r = 0.07, p = 0.85) and between *E. coli* MPN and MAR (r = −0.42, p = 0.27). It suggests the prevalence of antibiotic resistance for *E. coli* in the Jiyun River regardless of the *E. coli* contamination extent. The *E. coli* isolates from each of the three rivers displayed generally consistent increases in both the resistance rates and MAR indexes from up- to midstream. Statistically no-significant difference was found among the three rivers in resistance rates and resistance diversity (MAR indexes) based on a non-parametric k independent sample Kruskal-Wallis test (p = 0.50 for resistance rate and p = 0.10 for MAR). The *E. coli* isolates from different water systems showed varying drug numbers of resistance. The Ju

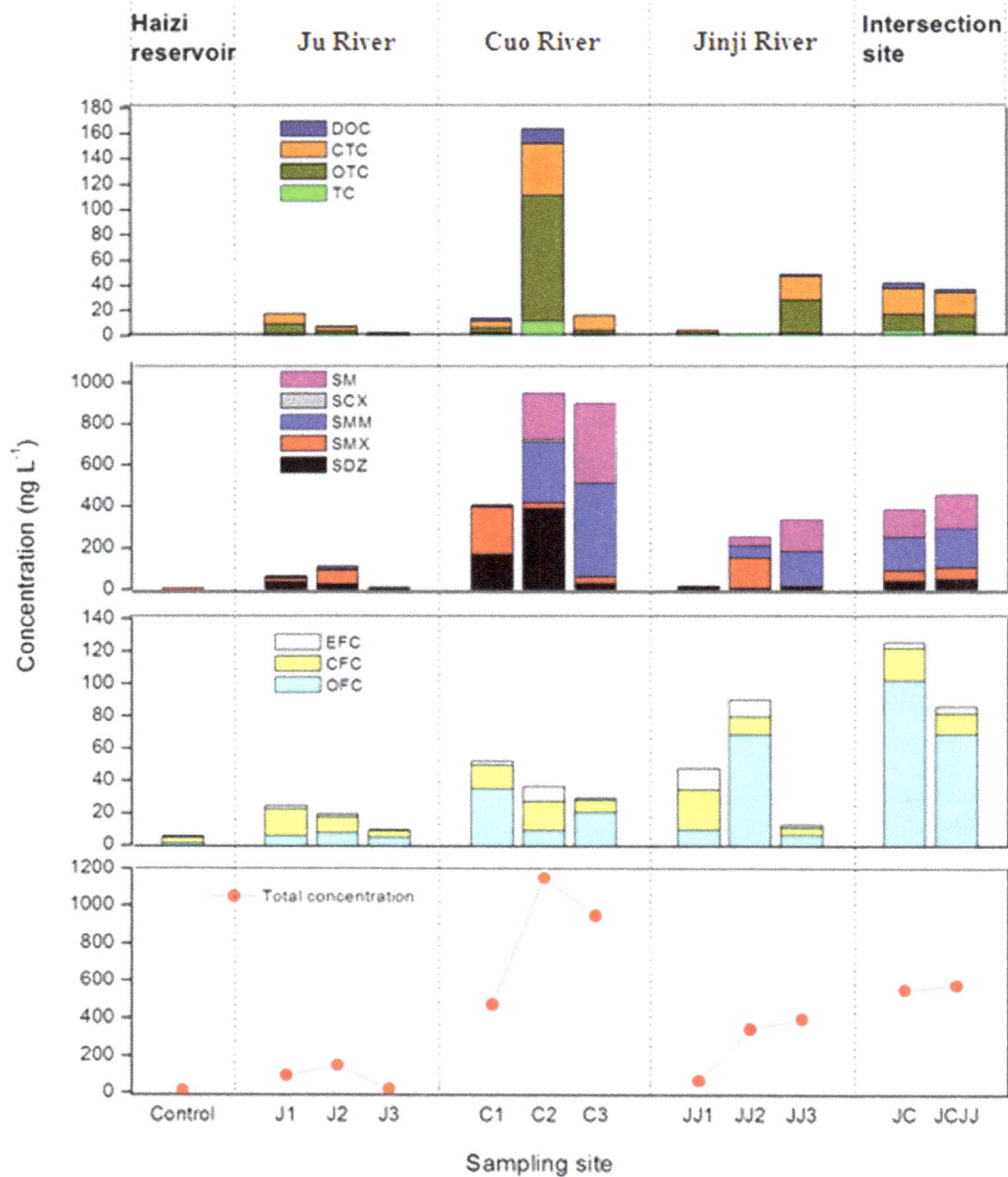

**Figure 2. Distributions of the three classes of antibiotics in the 12 sampling sites.** Control, Haizi reservoir; J1, J2 and J3, up-, mid- and downstream of Ju River; C1 C2, and C3, up-, mid- and downstream of Cuo River; JJ1, JJ2, and JJ3, up-, mid- and downstream of Jinji River; JC and JCJJ, intersection sites of Ju River with Cuo River and Jinji River, respectively. TC, tetracycline; OTC, oxytetracycline; CTC, chlortetracycline; DOC, doxycycline; CFC, ciprofloxacin; EFC, enrofloxacin; OFC, ofloxacin; SDZ, sulfadiazine; SMX, sulfamethoxazole; SMM, sulfamonomethoxine; SM, sulfameter; SCX, sulfachinoxaline.

River presented all of the resistance profiles from single- to eight-drug, dominated by one- to five-drug resistance. The Cuo River primarily displayed one-, two- and three-drug resistance. In contrast, the *E. coli* from Jinji River was more chiefly resistant to more than five drugs. In particular, 15% and 22% of *E. coli* isolates in JJ1 and JJ2, respectively, exhibited eight-drug resistance. Similar to the sites prior to intersection, the resistance profiles of *E. coli* at the two intersection sites also covered nearly all of the patterns from one- to seven-drug resistance.

The extent of the antibiotic-resistance of *E. coli* in the Jiyun River maybe is the highest currently reported and was significantly higher than the reports for other agricultural watersheds of British

Columbia (Resistance rate: 20–50%) [27], the Beijing Wenyu River (Average rate: 48%, MAR: 0.11–0.14) [26], and the Dongjiang in Guangzhou (MAR: 0.17–0.50) [55]. Furthermore, the extent of resistance in all of these agriculture areas or both agriculture and human-affected areas was greater than the urban surface water in Japan (resistance rate: 37%) and the US (resistance rate: 5–38%) [31,56]. The generally greater average resistance rate and MAR in surface water of the study agricultural region relative to reported results is probably related to generally higher levels of antibiotic contamination in such areas. In addition to the cause of frequent and long-term exposure to various different levels of antibiotics, special water conditions, such as low

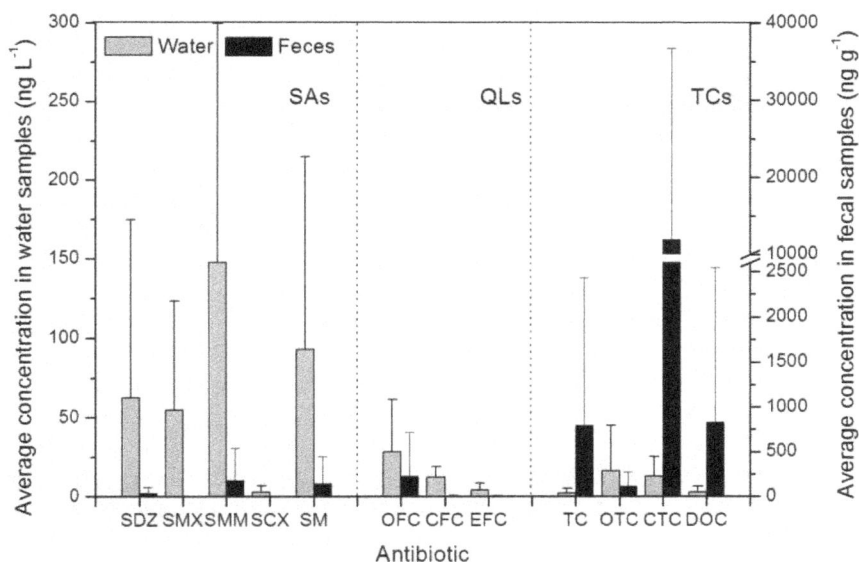

**Figure 3. The means and standard deviations (SD) for antibiotics in water samples (n = 12) and in animal fecal samples (n = 6).** TCs, tetracyclines; QLs, quinolones; SAs, sulfonamides; TC, tetracycline; OTC, oxytetracycline; CTC, chlortetracycline; DOC, doxycycline; CFC, ciprofloxacin; EFC, enrofloxacin; OFC, ofloxacin; SDZ, sulfadiazine; SMX, sulfamethoxazole; SMM, sulfamonomethoxine; SM, sulfameter; SCX, sulfachinoxaline.

flow and high nutrient concentrations, might also contribute to the high antibiotic resistance in the Jiyun Rivers [39,57,58].

## Relationships between the resistance frequencies of E. coli and the corresponding antibiotics

The correlations between the resistance rate and the detection concentration for the five drugs SMM, CTC, TE, LEV and SXT

were separately examined. In general, no clear relationship was observed for any of these drugs (Figure S1). The relationship between the extent of antibiotic resistance or ARG abundance with the antibiotic contents has also been examined in other studies, and contradictory observations have been reported. The abundances of resistance genes are positively related to the antibiotic concentrations in the Haihe and Huangpu Rivers in China [2,39]. However, no clear correlation was noted between

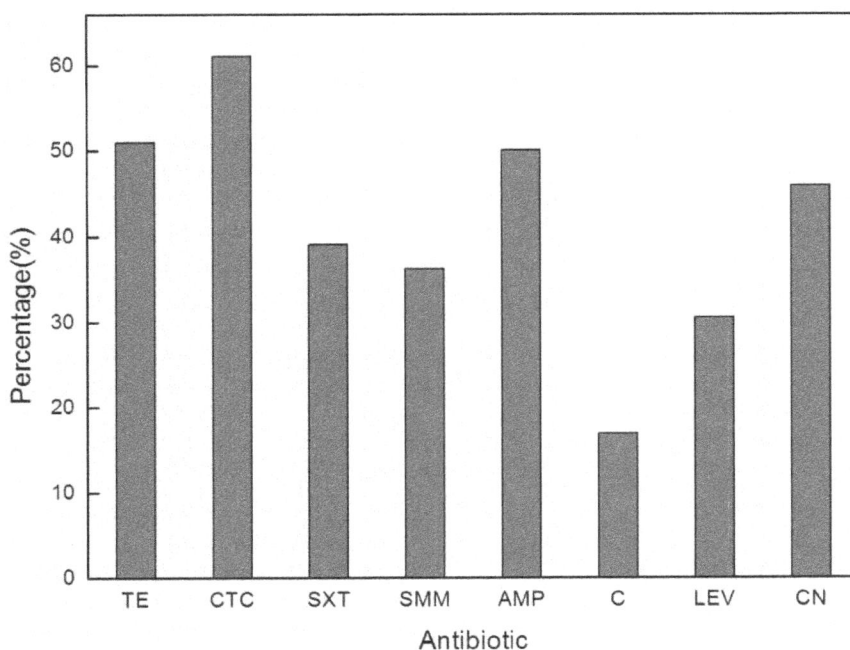

**Figure 4. Resistance percentage to the eight tested drugs for all the E. coli isolates that were isolated from the Jiyun River (n = 218).**
TE, tetracycline; CTC, chlortetracycline; SXT, sulfamethoxazole-trimethoprim; SMM, sulfamonomethoxine; AMP, ampicillin; C, chloromycetin; LEV, levofloxacin; CN, gentamycin.

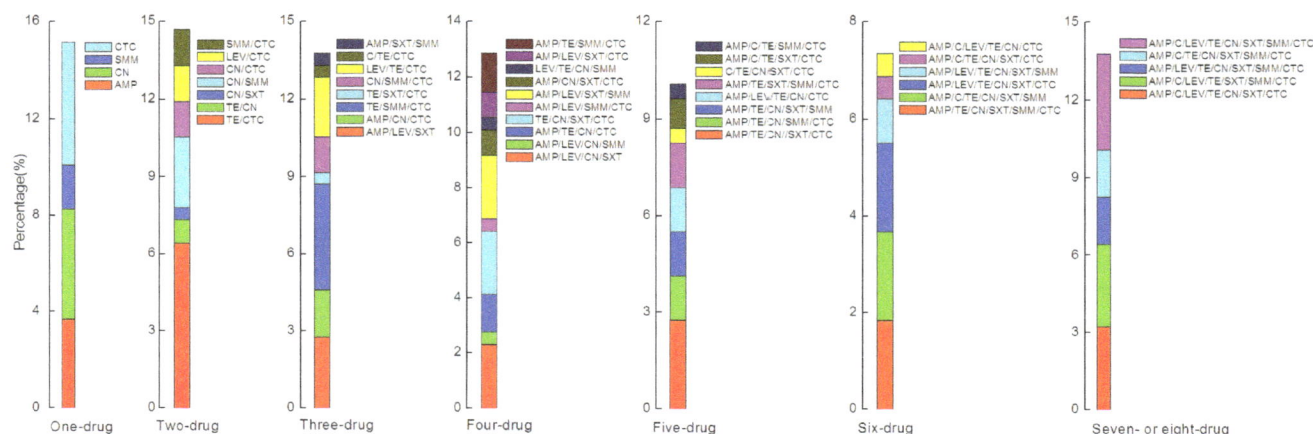

**Figure 5. Percentage of *E. coli* number in each antibiotic-resistance pattern accounting for the total *E. coli* isolates from the Jiyun River.** TE, tetracycline; CTC, chlortetracycline; SXT, sulfamethoxazole-trimethoprim; SMM, sulfamonomethoxine; AMP, ampicillin; C, chloromycetin; LEV, levofloxacin; CN, gentamycin.

fluoroquinolones and the corresponding resistant *E. coli* in the rivers of Osaka, Japan [3], in agreement with the findings in the present study. The lack of correlation between the antibiotic concentrations and the extent of resistance may be caused by several confounding effects. Antibiotic-resistant bacteria are linked not only to the exposure to corresponding antibiotics but also to other factors, such as the co-selection of toxic heavy metals [59] and the cross-resistance among different types of antibiotics [54], as well as certain water quality variables, such as nutrients, temperature, DO and salinity [27]. In addition to the direct formation of resistance for *E. coli* when exposed to the selective pressure of antibiotics and other factors, *E. coli* can also indirectly obtain resistance through the horizontal ARG transfer among microbial populations [3,26]. Antibiotics and resistant *E. coli* have different fates and persistence in the environment. For example, antibiotic resistance will remain for at least one month after the disappearance of antibiotics [60]. Considering the complexity of the antibiotic-resistance-formation mechanisms and the influences of various water quality parameters in aquatic environments, more in-depth and systematic studies should be conducted in the future. Regardless, it is clear that a large amount of VAs have been abused in animal production farms and reside in the surrounding aquatic environment, which is likely a primary driver for the emergence of resistant strains in the environment.

The observations in the present study are a cause for concern in terms of the following aspects. First, the Jiyun River is upstream of the Haihe River, which is the largest water system in north China and the main river flowing into Bohai Bay. Thus, high levels of antibiotics and resistant *E. coli* in the Jiyun River serve as potential contamination sources of the Bohai Bay. Second, in most instances, the surface water in rural areas is directly used for agricultural irrigation, which will likely result in the uptake and translocation of antibiotics by plants. After being irrigated with carbamazepine-containing wastewater for 60 and 110 days, carbamazepine was detected in soybean roots and beans [61]. Finally, given that more than 90% of the antibiotic-resistant *E. coli* in aquatic environments is conveyed by corresponding ARGs [26], this resistant *E. coli* might be the potential reservoir of various ARGs, which is particularly worth noting.

## Antibiotics and antibiotic-resistant *E. coli* in the potential drinking water source

Only 12.15 ng L$^{-1}$ antibiotics were detected at the site of the Haizi reservoir, which is the potential drinking water source of the Beijing population. However, 38 *E. coli* isolates were also retrieved from this site that displayed multiple antibiotic resistances (Table 3). The presence of *E. coli* and resistance in the well-protected Haizi reservoir may be caused by input from the upper stretch in Hebei province. Wildlife is also a suspected source because wildlife has been reported as the primary *E. coli* contributor in headwaters and can acquire antibiotic resistance through crossover with livestock animals [62–65]. In addition, stationary water flow in such reservoirs is also favorable for resistant bacterial growth and ARG transfer [57,58]. However, the specific reason is still unclear and requires further examination. The occurrence of antibiotic-resistant microorganisms and ARGs in drinking water sources or headwaters has been reported in other regions, such as the Huangpu River, China [66] and the Elk Creek Watershed, British Columbia [27]. The risk from *E. coli* could be completely eliminated if used for human consumption because the water is required to be treated by the Beijing Water Group Company to reach drinking standards. These results underscore the urgent need to explore the source and risks of resistant bacteria and ARG contamination in drinking water sources.

## Conclusions

This study demonstrated the extensive use and partial abuse of VAs in livestock farming in the study area, which probably is responsible for the prevalence of VAs, particularly hydrophilic VA compounds, in the surrounding surface water. A total of 88% of *E. coli* isolates from these rivers were resistant to one or more antimicrobial agents. Significant correlations were found among the resistance rate of SXT, C and AMP as well as between TE and CTC, suggesting a possible cross-selection for resistance among these drugs. An increasing tendency was observed for total antibiotic contents along the two tributaries, which was most likely attributed to the potential continuous input from animal operations discharge and field runoff. Also, the antibiotics resistance frequency for *E. coli* from midstream is greater than those from upstream for the three rivers. *E. coli* isolates from

**Table 2.** Spearman correlation matrix of the percent resistances to the eight drugs in seven of the nine sampling sites.

| | AMP | C | LEV | TE | CN | SXT | SMM | CTC |
|---|---|---|---|---|---|---|---|---|
| AMP | 1.000 | | | | | | | |
| C | 0.898** | 1.000 | | | | | | |
| LEV | 0.503 | 0.524 | 1.000 | | | | | |
| TE | 0.048 | 0.286 | 0.286 | 1.000 | | | | |
| CN | 0.587 | 0.452 | −0.048 | 0.333 | 1.000 | | | |
| SXT | 0.934** | 0.905** | 0.548 | 0.214 | 0.595 | 1.000 | | |
| SMM | 0.311 | 0.405 | 0.000 | 0.071 | 0.262 | 0.524 | 1.000 | |
| CTC | 0.036 | 0.286 | 0.143 | 0.952** | 0.357 | 0.119 | −0.071 | 1.000 |

The sites of Haizi reservoir and Cuo River upstream have been removed based on the low antibiotic levels and few *E. coli* isolates, respectively. AMP = ampicillin, C = chloromycetin, LEV = levofloxacin, TE = tetracycline, CN = gentamycin, SXT = sulfamethoxazole-trimethoprim, SMM = sulfamonomethoxine, CTC = chlortetracycline). ** indicate the significance level of $p < 0.01$.

**Table 3.** Overall antibiotic-resistance frequency (%), MAR index and one-, two-, three-, four-, five-, six-drug, seven-drug and eight-drug resistance frequency (%) for *E. coli* isolated from each of the nine sampling sites.

| Parameter | Control (n = 38) | J1 (n = 25) | J2 (n = 40) | C1 (n = 2) | C2 (n = 24) | JJ1 (n = 26) | JJ2 (n = 9) | JC (n = 29) | JCJ (n = 25) |
|---|---|---|---|---|---|---|---|---|---|
| Frequency (%) | 82 | 88 | 98 | 50 | 88 | 77 | 100 | 93 | 84 |
| MAR index | 0.47 | 0.37 | 0.40 | 0.13 | 0.31 | 0.46 | 0.79 | 0.45 | 0.29 |
| One-drug (%) | 3 | 23 | 26 | 0 | 10 | 20 | 0 | 7 | 43 |
| Two-drug (%) | 3 | 27 | 15 | 100 | 29 | 15 | 0 | 22 | 14 |
| Three-drug (%) | 23 | 5 | 13 | 0 | 52 | 0 | 0 | 19 | 5 |
| Four-drug (%) | 29 | 18 | 15 | 0 | 5 | 0 | 11 | 11 | 19 |
| Five-drug (%) | 10 | 9 | 18 | 0 | 0 | 5 | 22 | 19 | 10 |
| Six-drug (%) | 6 | 5 | 5 | 0 | 0 | 30 | 11 | 15 | 0 |
| Seven-drug (%) | 23 | 9 | 8 | 0 | 0 | 15 | 33 | 7 | 10 |
| Eight-drug (%) | 3 | 5 | 0 | 0 | 5 | 15 | 22 | 0 | 0 |

Control, Haizi reservoir; J1 and J2, up- and midstream of Ju River; C1 and C2, up- and midstream of Cuo River; JJ1 and JJ2, up- and midstream of Jinji River; JC and JCJJ, intersection sites of Ju River with Cuo River and Jinji River, respectively.

different water systems showed varying drug numbers of resistance. No obvious correlation was found between the antibiotic resistant rate of *E. coli* and the corresponding antibiotic concentrations, indicating that the resistance formation process must be affected by aquatic factors besides antibiotics. These results provide baseline data on the antibiotics and antibiotic-resistant bacteria contamination that are associated with widespread livestock production. For improved contamination control and environmental protection in rural areas, there is an urgent need to develop management protocols in the animal industry. In addition, the occurrence of resistance genes and potential threats to agricultural ecosystem safety should also be noted, as water is the main interface of antibiotics, resistant bacteria and ARG transfer in the ecosystem.

## Supporting Information

**Figure S1** Relationships between CTC resistance rate versus CTC concentration (a), TE resistance rate versus TC concentration (b), SMM resistance rate versus SMM concentration (c), SXT resistance rate versus SMX concentration (d) and LEV resistance rate versus OFC concentration (e) in seven of the nine sampling sites. The sites of Haizi reservoir and Cuo River upstream have been removed based on the low antibiotic levels and few *E. coli* isolates, respectively. TE, tetracycline; CTC, chlortetracycline; SXT, sulfamethoxazole-trimethoprim; SMM, sulfamonomethoxine; LEV, levofloxacin; TC, tetracycline; CTC, chlortetracycline; OFC, ofloxacin; SMX, sulfamethoxazole; SMM, sulfamonomethoxine. (TIF)

**Table S1** GPS coordinates (deg./min./sec.) of the 12 water sample sites in study region (Control, Haizi reservoir; J1, J2 and J3, up-, mid- and downstream of Ju River; C1 C2, and C3, up-, mid- and downstream of Cuo River; JJ1, JJ2, and JJ3, up-, mid- and downstream of Jinji River; JC and JCJJ, intersection sites of Ju River with Cuo River and Jinji River, respectively). (DOCX)

**Table S2** Most possible number (MPN) of the *E. coli* in the Jiyun River (MPN/100 ml) (Control, Haizi reservoir; J1, and J2, up- and midstream of Ju River; C1 and C2, up- and midstream of Cuo River; JJ1 and JJ2, up- and midstream of Jinji River; JC and JCJJ, intersection sites of Ju River with Cuo River and Jinji River, respectively.). (DOCX)

**Table S3** Detection frequencies, ranges and means of the 12 target antibiotics in sediment of the Jiyun River. (DOCX)

## Author Contributions

Conceived and designed the experiments: XZ. Analyzed the data: XZ. Wrote the paper: XZ. Collected and analyzed samples: XZ BL JW MG LW. Directed the experiment design, implementation, and manuscript writing: YL CF.

## References

1. Kolpin DW, Furlong ET, Meyer MT, Thurman EM, Zaugg SD, et al. (2002) Pharmaceuticals, hormones, and other organic wastewater contaminants in US streams, 1999–2000: A national reconnaissance. Environ Sci Technol 36: 1202–1211.
2. Luo Y, Mao DQ, Rysz M, Zhou QX, Zhang HJ, et al. (2010) Trends in antibiotic resistance genes occurrence in the Haihe River, China. Environ Sci Technol 44: 7220–7225.
3. Adachi F, Yamamoto A, Takakura KI, Kawahara R. (2013) Occurrence of fluoroquinolones and fluoroquinolone-resistance genes in the aquatic environment. Sci Total Environ 444: 508–514.
4. Yan CX, Yang Y, Zhou JL, Liu M, Nie MH, et al. (2013) Antibiotics in the surface water of the Yangtze Estuary: Occurrence, distribution and risk assessment. Environ Pollut 175: 22–29.
5. Levy SB. (1998) The challenge of antibiotic resistance. Sci. Am 278: 46–53.
6. Sarmah AK, Meyer MT, Boxall ABA. (2006) A global perspective on the use, sales, exposure pathways, ocurrence, fate and effects of veterinary antibiotics (VAs) in the environment. Chemosphere 65: 725–759.
7. Li YX, Zhang XL, Li W, Lu XF, Liu B, et al. (2013) The residues and environmental risks of multiple veterinary antibiotics in animal faeces. Environ Monit Assess 185: 2211–2220.
8. Cogliani C, Goossens H, Greko C. (2011) Restricting antimicrobial use in food animals: lessons from Europe. Microbe 6: 274.
9. DANMAP. Danmap 2011-use of antimicrobial agents and occurrence of antimicrobial resistance in bacteria from food animals, food and humans in Denmark. Available: http://danmap.org/Downloads/~/media/Projekt%20sites/Danmap/DANMAP%20reports/Danmap _2011.ashx/. Accessed 2013 July 26.
10. Hong PY, Al-Jassim N, Ansari MI, Mackie RI. (2013) Environmental and public health implications of water reuse: Antibiotics, antibiotic resistant bacteria, and antibiotic resistance genes. Antibiotics 2: 367–399.
11. Kumar KC, Gupta S, Chander Y, Singh AK. (2005) Antibiotic use in agriculture and its impact on the terrestrial environment. Adv Agron 87: 1–54.
12. Gutiérrez IR, Watanabe N, Harter T. (2010) Effect of sulfonamide antibiotics on microbial diversity and activity in a Californian Mollic Haploxeralf. J Soil Sediment 10: 537–544.
13. Martínez-Carballo E, González-Barreiro C, Scharf S, Gans O. (2007) Environmental monitoring study of selected veterinary antibiotics in animal manure and soils in Austria. Environ Pollut 148: 570–579.
14. Zhao L, Dong YH, Wang H. (2010) Residues of veterinary antibiotics in manure from feedlot livestocks in eight provinces of China. Sci Total Environ 408: 1069–1075.
15. Motoyama M, Nakagawa S, Tanoue R, Sato Y, Nomiyama K, et al. (2011) Residues of pharmaceutical products in recycled organic manure produced from sewage sludge and solid waste from livestock and relationship to their fermentation level. Chemosphere 84: 432–438.
16. Pan X, Qiang ZM, Ben WW, Chen MX. (2011) Residual veterinary antibiotics in swine manure from concentrated animal feeding operations in Shandong Province, China. Chemosphere 84: 695–700.
17. Dolliver H, Gupta S, Noll S. (2008) Antibiotic degradation during manure composting. J Environ Qual 37: 1245–1253.
18. Bao Y, Zhou Q, Guan L, Wang Y. (2009) Depletion of chlortetracycline during composting of aged and spiked manures. Waste Manage 29: 1416–1423.
19. Ali M, Wang JJ, DeLaune RD, Seo DC, Dodla SK, et al. (2013) Effect of redox potential and pH status on degradation and adsorption behavior of tylosin in dairy lagoon sediment suspension. Chemosphere 91: 1583–1589.
20. Kay P, Blackwell PA, Boxall A. (2005) Transport of veterinary antibiotics in overland flow following the application of slurry to arable land. Chemosphere 59: 951–959.
21. Sun P, Barmaz D, Cabrera ML, Pavlostathis SG, Huang CH. (2013) Detection and quantification of ionophore antibiotics in runoff, soil and poultry litter. J Chromatogr A 1312: 10–17.
22. Pezzotti G, Serafin A, Luzzi I, Mioni R, Milan M, et al. (2003) Occurrence and resistance to antibiotics of *Campylobacter jejuni* and *Campylobacter coli* in animals and meat in northeastern Italy. Int J Food Microbiol 82: 281–287.
23. Heuer H, Schmitt H, Smalla K. (2011) Antibiotic resistance gene spread due to manure application on agricultural fields. Curr opin microbiol 14: 236–243.
24. Marti R, Scott A, Tien YC, Murray R, Sabourin L, et al. (2013) Impact of manure fertilization on the abundance of antibiotic-resistant bacteria and frequency of detection of antibiotic resistance genes in soil and on vegetables at Harvest. Appl Environ Microbiol 79: 5701–5709.
25. Gullberg E, Cao S, Berg OG, Ilbäck C, Sandegren L, et al. (2011) Selection of resistant bacteria at very low antibiotic concentrations. PLOS pathogens 7: e1002158.
26. Hu JY, Shi JC, Chang H, Li D, Yang M, et al. (2008) Phenotyping and genotyping of antibiotic-resistant *Escherichia coli* isolated from a natural river basin. Environ Sci Technol 42: 3415–3420.
27. Maal-Bared R, Bartlett KH, Bowie WR, Hall ER. (2013) Phenotypic antibiotic resistance of Escherichia coli and *E. coli* O157 isolated from water, sediment and biofilms in an agricultural watershed in British Columbia. Sci Total Environ 443: 315–323.
28. China Agricultural Statistic Year Book. 2012.
29. Batt AL, Kim S, Aga DS. (2007) Comparison of the occurrence of antibiotics in four full-scale wastewater treatment plants with varying designs and operations. Chemosphere 68: 428–435.
30. Zhou LJ, Ying GG, Zhao JL, Yang JF, Wang L et al. (2011) Trends in the occurrence of human and veterinary antibiotics in the sediments of the Yellow River, Hai River and Liao River in northern China. Environ Pollut 159: 1877–1885.

31. Akiyama T, Savin MC. (2010) Populations of antibiotic-resistant coliform bacteria change rapidly in a wastewater effluent dominated stream. Sci Total Environ 408: 6192–6201.

32. CLSI. (2009) Performance Standards for Antimicrobial Susceptibility Testing; Sixteenth Informational Supplement, M100-S16. Clinical and Laboratory Standards Institute (National Committee for Clinical Laboratory Standards), Wayne, PA.

33. Parveen S, Murphree RL, Edmiston L, Kaspar CW, Portier KM. (1997) Association of multiple-antibiotic-resistance profiles with point and nonpoint sources of Escherichia coli in Apalachicola Bay. Appl. Environ. Microbiol 7: 2607–2612.

34. Whitlock JE, Jonesb DT, Harwooda VJ. (2002) Identification of the sources of fecal coliforms in an urban watershed using antibiotic resistance analysis. Water Res 36: 4273–4282.

35. CLSI. (2012) Performance Standards for Antimicrobial Susceptibility Testing; Sixteenth Informational Supplement, M100- S16. Clinical and Laboratory Standards Institute (National Committee for Clinical Laboratory Standards), Wayne, PA.

36. Walczak JJ, Xu SP. (2011) Manure as a Source of Antibiotic-Resistant Escherichia coli and Enterococci: a Case Study of a Wisconsin, USA Family Dairy Farm. Water Air Soil Pollut 219: 579–589.

37. Tamtam F, Mercier F, Le Bot B, Eurin J, Tuc Dinh Q, et al. (2008) Occurrence and fate of antibiotics in the Seine River in various hydrological conditions. Sci Total Environ 393: 84–95.

38. Luo Y, Xu L, Rysz M, Wang YQ, Zhang H, et al. (2011) Occurrence and Transport of Tetracycline, Sulfonamide, Quinolone, and Macrolide Antibiotics in the Haihe River Basin, China. Environ Sci Technol 45: 1827–1833.

39. Jiang L, Hu XL, Yin DQ, Zhang HC, Yu ZY. (2011) Occurrence, distribution and seasonal variation of antibiotics in the Huangpu River, Shanghai, China. Chemosphere 82: 822–828.

40. Zou SC, Xu WH, Zhang RJ, Tang JH, Chen YJ, et al. (2011) Occurrence and distribution of antibiotics in coastal water of the Bohai Bay, China: Impacts of river discharge and aquaculture activities. Environ Pollut 159: 2913–2920.

41. Pils JRV, Laird DA. (2007) Sorption of tetracycline and chlortetracycline on K- and Ca-saturated soil clays, humic substances, and clay-humic complexes. Environ Sci Technol 41: 1928–1933.

42. Xu XR, Li XY. (2010) Sorption and desorption of antibiotic tetracycline on marine sediments. Chemosphere 78: 430–436.

43. Gao J, Pedersen JA. (2005) Adsorption of sulfonamide antimicrobial agents to clay minerals. Environ Sci Technol 39: 9509–9516.

44. Kahle M, Stamm C. (2007) Time and pH-dependent sorption of the veterinary antimicrobial sulfathiazole to clay minerals and ferrihydrite. Chemosphere 68: 1224–1231.

45. Wei RC, Ge F, Huang SY, Chen M, Wang R. (2011) Occurrence of veterinary antibiotics in animal wastewater and surface water around farms in Jiangsu Province, China. Chemosphere 82: 1408–1414.

46. Campagnolo ER, Johnson KR, Karpati A, Rubin CS, Kolpin DW et al. (2002) Antimicrobial residues in animal waste and water resources proximal to large-scale swine and poultry feeding operations. Sci Total Environ 299: 89–95.

47. Wiegel S, Aulinger A, Brockmeyer R, Harms H, Löffler J, et al. (2004) Pharmaceuticals in the river Elbe and its tributaries. Chemosphere 57: 107–126.

48. Xu WH, Zhang G, Zou SC, Li XD, Liu YC. (2007) Determination of selected antibiotics in the Victoria Harbour and the Pearl River, South China using high-performance liquid chromatography-electrospray ionization tandem mass spectrometry. Environ Pollut 145: 672–679.

49. García-Galán MJ, Díaz-Cruz MS, Barceló D. (2011) Occurrence of sulfonamide residues along the Ebro river basin: removal in wastewater treatment plants and environmental impact assessment. Environ Int 37: 462–473.

50. Reinthaler FF, Posch J, Feierl G, Wüst G, Haas D, et al. (2003) Antibiotic resistance of E. coli in sewage and sludge. Water Res 37: 1685–1690.

51. Cui LZ, Tominaga EJ, Neoh HM, Hiramatsu KC. (2006) Correlation between reduced daptomycin susceptibility and vancomycin resistance in vancomycin-intermediate Staphylococcus aureus. Antimicrob Agents Ch 50: 1079–1082.

52. China MOA (Ministry of Agriculture of the People Republic of China) (2010) Chinese Veterinary Pharmacopoeia, the 4th edition. Beijing, China: Agriculture Press.

53. Rooklidge SJ. (2004) Environmental antimicrobial contamination from terraccumulation and diffuse pollution pathways. Sci Total Environ 325: 1–13.

54. Bean DC, Livermore DM, Hall MC. (2009) Plasmids imparting sulfonamide resistance in Escherichia coli: implications for persistence. Antimicrob Agents Ch 53: 1088–1093.

55. Su HC, Ying GG, Tao R, Zhang RQ, Zhao JL, et al. (2012) Class 1 and 2 integrons, sul resistance genes and antibiotic resistance in Escherichia coli isolated from Dongjiang River, South China. Environ Pollut 169: 42–49.

56. Ham YS, Kobori H, Kang JH, Matsuzaki T, Iino M, et al. (2012) Distribution of antibiotic resistance in urban watershed in Japan. Environ Pollut 162: 98–103.

57. Schlüter A, Szczepanowski R, Pühler A, Top EM. (2007) Genomics of IncP-1 antibiotic resistance plasmids isolated from wastewater treatment plants provides evidence for a widely accessible drug resistance gene pool. FEMS Microbiol Rev 31: 449–477.

58. Rahube TO, Yost CK. (2010) Antibiotic resistance plasmids in wastewater treatment plants and their possible dissemination into the environment. Afr J Biotechnol 9: 9183–9190.

59. Ji XL, Shen QH, Liu F, Ma J, Xu G, et al. (2012) Antibiotic resistance gene abundances associated with antibiotics and heavy metals in animal manures and agricultural soils adjacent to feedlots in Shanghai; China. J of Hazard Mater 235–236: 178–185.

60. Rysz M, Alvarez PJJ. (2004) Amplification and attenuation of tetracycline resistance in soil bacteria: aquifer column experiments. Water Res 38: 3705–3712.

61. Wu CX, Spongberg AL, Witter JD, Fang M, Czajkowski KP. (2010) Uptake of pharmaceuticals and personal care products by soybean plants from soils applied with biosolids and irrigated with contaminated water. Environ Sci Technol 44: 6157–6161.

62. Hagedorn C, Robinson SL, Filtz JR, Grubbs SM, Angier TA, et al. (1999) Determining sources of fecal pollution in a rural Virginia watershed with antibiotic resistance patterns in fecal streptococci. Appl Environ Microbiol 65: 5522–5531.

63. Sayah RS, Kaneene JB, Johnson Y, Miller RA. (2005) Patterns of antimicrobial resistance observed in Escherichia coli isolates obtained from domestic and wild-animal fecal samples, human septage, and surface water. Appl Environ Microbiol 71: 1394–1404.

64. Meays CL, Broersma K, Nordin R, Mazumder A, Samadpour M. (2006) Spatial and annual variability in concentrations and sources of Escherichia coli in multiple watersheds. Environ Sci Technol 40: 5289–5296.

65. Blanco G, Lemus JA, Grande J. (2009) Microbial pollution in wildlife: linking agricultural manuring and bacterial antibiotic resistance in red-billed choughs. Environ Res 109: 405–412.

66. Jiang L, Hu XL, Xu T, Zhang H, Sheng D, et al. (2013) Prevalence of antibiotic resistance genes and their relationship with antibiotics in the Huangpu River and the drinking water sources, Shanghai, China. Sci. Total. Environ. 458: 267–272.

# 10

# Distance-Decay and Taxa-Area Relationships for Bacteria, Archaea and Methanogenic Archaea in a Tropical Lake Sediment

**Davi Pedroni Barreto[1], Ralf Conrad[3], Melanie Klose[3], Peter Claus[3], Alex Enrich-Prast[2,4]***

**1** Instituto de Microbiologia Prof. Paulo de Góes, Universidade Federal do Rio de Janeiro, Rio de Janeiro, Brazil, **2** Instituto de Biologia, Universidade Federal do Rio de Janeiro, Rio de Janeiro, Brazil, **3** Max-Planck Institute for Terrestrial Microbiology, Marburg, Hessen, Germany, **4** Department of Water and Environmental Studies, Linköping University, Linköping, Sweden

## Abstract

The study of of the distribution of microorganisms through space (and time) allows evaluation of biogeographic patterns, like the species-area index (z). Due to their high dispersal ability, high reproduction rates and low rates of extinction microorganisms tend to be widely distributed, and they are thought to be virtually cosmopolitan and selected primarily by environmental factors. Recent studies have shown that, despite these characteristics, microorganisms may behave like larger organisms and exhibit geographical distribution. In this study, we searched patterns of spatial diversity distribution of bacteria and archaea in a contiguous environment. We collected 26 samples of a lake sediment, distributed in a nested grid, with distances between samples ranging from 0.01 m to 1000 m. The samples were analyzed using T-RFLP (Terminal restriction fragment length polymorphism) targeting *mcrA* (coding for a subunit of methyl-coenzyme M reductase) and the genes of Archaeal and Bacterial 16S rRNA. From the qualitative and quantitative results (relative abundance of operational taxonomic units) we calculated the similarity index for each pair to evaluate the taxa-area and distance decay relationship slopes by linear regression. All results were significant, with *mcrA* genes showing the highest slope, followed by Archaeal and Bacterial 16S rRNA genes. We showed that the microorganisms of a methanogenic community, that is active in a contiguous environment, display spatial distribution and a taxa-area relationship.

**Editor:** Jonas Waldenström, Linneaus University, Sweden

**Funding:** This study was funded by the National Council of Scientific and Technological Development (CNPq), trough the project 477260/2011-0. (Brazil - www.cnpq.br) and by the Max Planck Society (Germany - www.mpg.de). The funders had no role in study design, data collection and analysis, decision to publish, or preparation of the manuscript.

**Competing Interests:** The authors have declared that no competing interests exist.

* Email: aeprast@biologia.ufrj.br

## Introduction

The biogeography concept is defined as the study of the distribution and the range of living organisms across space and time. Most studies in this field were traditionally performed targeting macro-organisms such as plants and animals [1]. Since the development of molecular tools, the concepts of biogeography started to be also studied in microorganisms [2–5].

A long-held concept in microbial ecology is that microorganisms are ubiquitously distributed and can be found in any habitat with favorable environmental conditions. This concept was introduced by Martinus Willem Beijerinck and concisely summarized by Lourens Gerhard Marinus Baas Becking in the quote, "Everything is everywhere, the environment selects" [6]. This statement is based on some traits of the microorganisms, such as the small size of individuals and the consequent ease of their dispersal across long distances, high rates of reproduction, short generation times, and large population sizes, leading to a small chance of local extinction.

Free-living eukaryotic microorganisms are often described as occurring ubiquitously. When they are not dominant in some specific environment, it is possible to reanimate the cryptic diversity by changing the environmental conditions *in vitro* [3,7].

A study showed that it is possible to find nearly 80% of all known species of the flagellate genus *Paraphysomonas* in just a small sample of sediment [8], meaning that the global diversity of this genus is well represented by its local diversity. This observation is mostly explained by the high dispersal rate of the flagellates (due to their low size), their extremely short generation times (leading to a low rate of extinction) and also their capacity to generate resistant forms when the environmental conditions are unfavorable [8]. The authors suggested that if eukaryotic microorganisms were ubiquitous, then prokaryotic microorganisms should be ubiquitous as well, since they have an even smaller size and larger populations. Indeed, some studies on prokaryotes suggested global distribution, for example, psychrophilic polar bacteria were found at both the South and the North poles [9].

Recent studies, however, showed that the distribution of microorganisms is not random, and that biogeography patterns of distribution are established [10,11]. For example, the genetic distances between populations of microorganisms were shown to increase with geographic distance, which might represent a speciation process driven by the geographic isolation of the microorganisms [12]. Other studies were able to identify endemic microorganisms, and true geographic isolation in extreme environments like hot springs, pristine soils, salt lakes, and hot

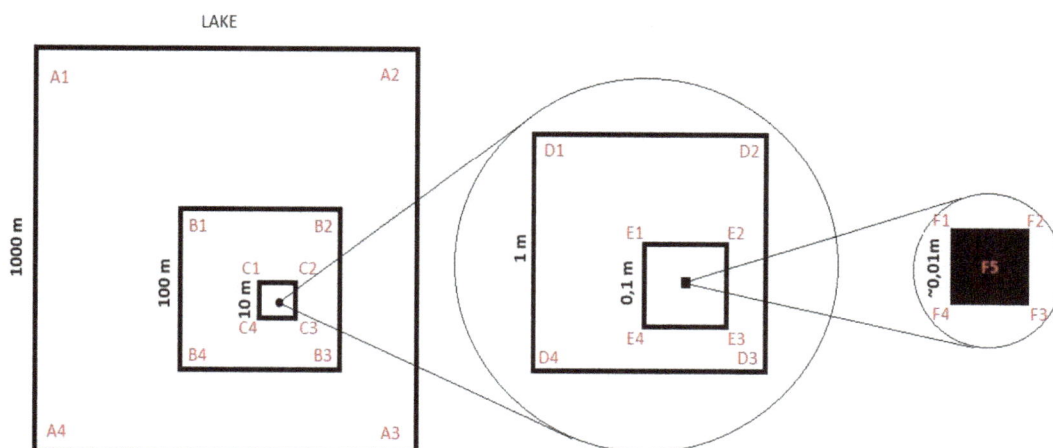

**Figure 1. Sampling grid.** Designed inside the Lago Negro area, A1-F5 represents the sampling points. (Adapted from (Horner-Devine 2004)).

and cold deserts around the world, all being strong evidence for non-cosmopolitan distribution [10,13–17]. The main difficulty is to evaluate the factors determining the geographical distribution, whether historic evolutionary events (geographic barriers for example) or contemporary ecological environmental factors. [5].

The taxa-area relationship is one of the most consistent general patterns in ecology and well described for macro-organisms [1]. It is represented by the equation $S = c\,A^z$, where $S$ is the number of species, $A$ is the area sampled, $c$ is a constant that is empirically derived from the taxon and the specific location studied, and the exponent $z$, the power law index (i.e. $z$-value), represents the rate of the increasing number of species along the increasing sampling area (graphical slope). When significant, the z-value may be strong evidence for biographical distribution of the species.

Values for the z exponent have already been described for microorganisms. Interestingly, z-values for microorganisms were often smaller than for macro-organisms. This result may be attributed: to the larger capacity of dispersion; to the lack of a clear "species" resolution; and to the use of molecular fingerprint or sequencing techniques [18–20]. Molecular fingerprint methods, such as T-RFLP (Terminal restriction fragment length polymorphism) and DGGE (Denaturing gradient gel electrophoresis), proved to be an important tool for accessing the diversity of microorganisms in different environments with relatively low costs and little time consumed [21]. However, fingerprinting methods usually are limited as they detect the most common species and thus underestimate the total diversity in a sample [22]. That is mostly because fingerprinting techniques lump different closely related "species" into a single taxonomic unit (often called Operational Taxonomical Unity – OTU), and usually ignore rare species. Nevertheless, fingerprinting techniques are still extremely valuable for rapidly comparing the microbial community composition in different environments [2].

Another parameter of species distribution through space is the distance-decay relationship, which consists of the decay of similarity between different communities as a function of the distance separating them [23], and can also be seen as evidence of a biogeographical pattern. The main difference between the distance-decay approach and the species-area relationship is based on the consideration of the relative abundance of the species in addition of their presence or absence. Bell et al. showed that bacteria in water-filled tree holes, found at the same place, displayed a significant distance-decay relationship [24].

Little is known about the geographical distribution of methanogenic archaea. So far, hot desert soil methanogenic Archaea were shown to be widely spread between different parts of the globe, and this could be found mostly by reactivation of the cryptic methanogenic process *in vitro* [25]. To our best knowledge, there is no description of distance-decay or species-area relationship for non-extremophilic archaea in a contiguous environment. Given the high ecological stability of methanogenic sediments and soils, with a continuous anaerobic environment and a regular input of organic matter, the microbial communities related to the methanogenesis processes tend to be stable through time [26].

Among the processes commonly underlying the biogeographical patterns of distribution of organisms and communities - selection, drift, dispersion and diversification - selection and dispersion are closely related with geographical distances, given the increase of habitat heterogeneity and dispersion limitations with increasing area [27]. Thus, we hypothesized that if there is a change of diversity patterns among different spots of a contiguous lake sediment, it should have a strong correlation with the geographical distance between them, and this change should also present itself differently depending on which part of the microbial community is considered. For this purpose we were targeting three different genes. The 16S rRNA genes of Bacteria and Archaea that are transcribed to generate the structural RNA of the small-subunit ribosomes are universal and strongly conserved genes and are therefore widely used as taxonomic markers [28]. The *mcrA* is a functional gene coding for the alpha subunit of the methyl-coenzyme M reductase, an enzyme being essential and characteristic for the methanogenesis biochemical pathway in Archaea [29]. Thus, we were targeting different groups of prokaryotes. By that we expected to see if there is a differential influence of geographic distance as a factor driving distribution of these three different groups of microorganisms.

We were able to show a geographical distribution pattern of the composition of Bacteria, Archaea and methanogenic Archaea in a contiguous tropical lake sediment and the presence of a significant **z-value** for all three groups. We hypothesized that intrinsic ecological differences between the communities, general diversity profiles, and technical particularities and limitations could be defining the differences in the distribution of these taxonomical groups or at least how we perceived it.

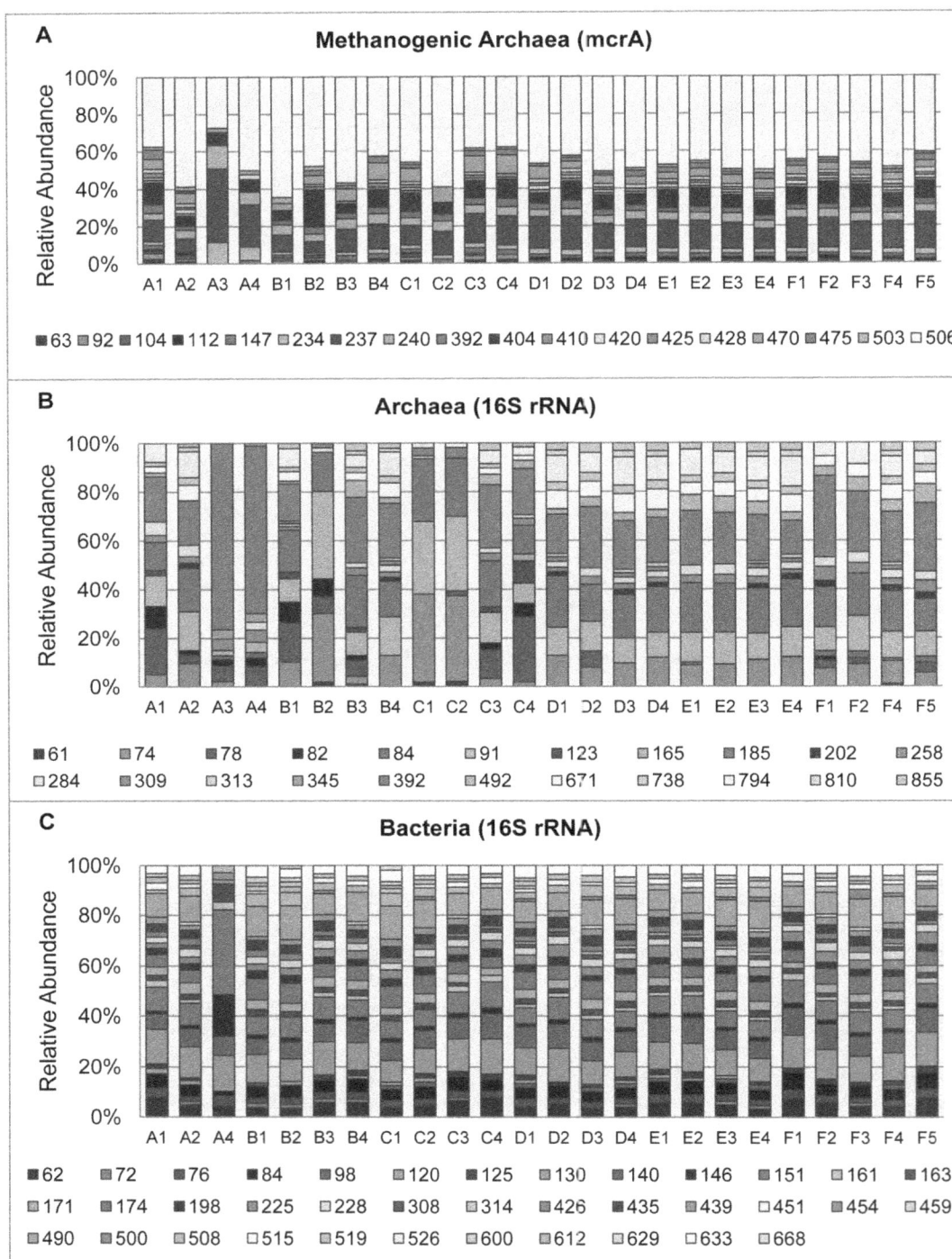

**Figure 2. T-RFLP profiles of the three different genes studied in sediment inside the Lago Negro area.** A1-F5 represent the sampling points (see Figure 1 for more details), the bar size represents the relative abundance of each OTU, which are defined by their size in base pair (bp). (A) *mcrA*, (B) Archaeal 16S rRNA gene, and (C) Bacterial 16S rRNA gene.

## Materials and Methods

### Sampling and Study Area

The Pantanal consists of the largest floodplain in South America and is periodically flooded by the Paraguay River and its tributaries. Altitude above sea level varies from 80 to 120 m and the total estimated area is around 138.123 km$^2$. The water flow continuously carries organic material, and during the flood period the sediment spreads all over the plain constituting the most important source of carbon and nutrients for the methanogenic archaea [30]. The climate is hot and wet in the summer, and cold and dry in the winter. The maximum temperature often surpasses 40°C. Between the months of May and July, the average

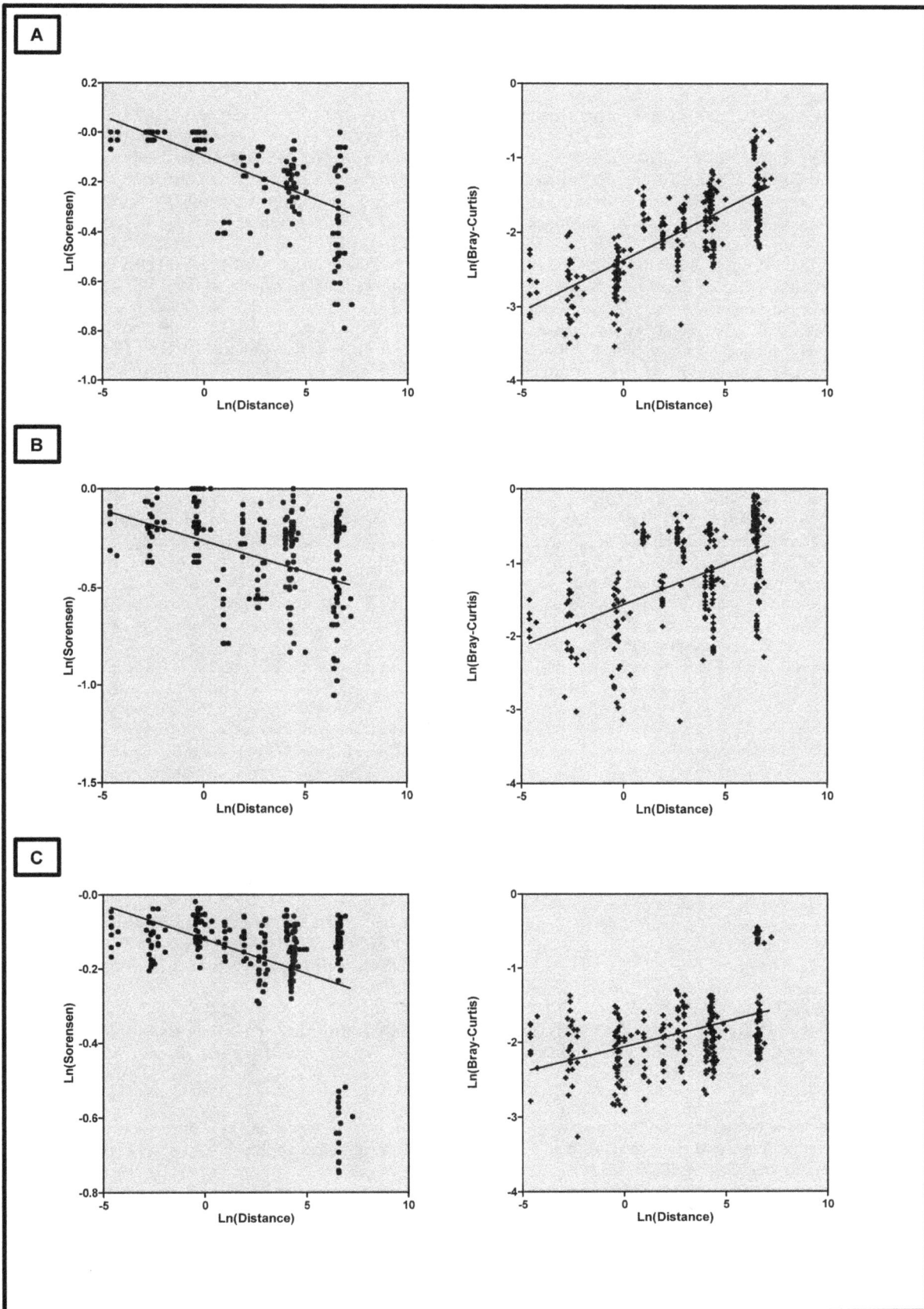

**Figure 3. Distance decay and taxa-area relationships.** Log transformed Bray-Curtis and Sørensen indices plotted against the distance separating the points, for the three different genes, lines represent a simple linear regression. (A) *mcrA*; (B) Archaeal 16S rRNA gene and (C) Bacterial 16S rRNA gene.

temperature drops below 20°C, and the minimum temperature may reach 0°C [31].

The study was conducted in a single lake (19°02.651′S 57°30.254′W) called *Lagoa Negra*. No special permission is needed for sediment sampling in non-protected areas in Brazil, so no specific permissions were required for these locations. This study did not involve endangered or protected species.

The lake is located at the west margin of the Paraguay River and a few kilometers east from the cities of Corumbá and Ladário, close to the Bolivia-Brazil border. *Lagoa Negra* is a perennial shallow freshwater lake, with water depth varying between 2 m and 3 m approximately, getting lower or higher following the flood regime. The total area of the lake is 10.8 km² (maximum length of 4.1 km and maximum width of 3.5 km, but it can also vary following the Paraguay River flood regime), water pH ranging from 6.9 to 8.3 and conductivity ranging from 140 to 240 µS cm⁻¹ at 25°C. The top layer of sediment is mostly composed of a thin clayey substrate with approximately 2% of sand (>63 µm) and total carbon and organic matter concentrations of 1.8% and 2% respectively [32]. Twenty one sediment cores (7 cm diameter) were collected in the lake in July 2011. The cores were distributed on a nested square grid with distance between points varying from 0.01 m to approximately 1,400 m (Figure 1). The 10 cm top layer of the sediment was homogenized and then 1 ml was sampled and frozen for molecular biology analyses. The central core E was sub sampled within distances of 10 cm and 1 cm between points inside the core area (also 10-cm deep with polyethylene straws). The sediment samples were immediately frozen in liquid nitrogen and then air-shipped to Germany for later molecular biology analyses at the Max-Planck Institute for Terrestrial Microbiology.

## T-RFLP

The frozen sediment samples were thawed, and the total DNA was extracted using the FastDNA SPIN Kit for Soil (MP Biomedicals) following the manufacturer's protocol and a three times treatment with 5.5 M guanidine thyocianate as an additional cleaning step during the matrix binding step. The total DNA was quantified using spectophotometry. Preparations were considered to be of good quality when resulting in more than 30 ng/µl of DNA.

Bacterial and Archaeal 16S rRNA gene fragments were PCR-amplified using the primer pairs Eub 9/27f (5′-GAG TTT GAT CMT GGC TCA G-3′) and Eub 907/926r (5′-CCG TCA ATT CMT TTR AGT TT-3′), as described by Lane and Weisburg [33] for Bacteria, and the primer pairs A109f (5′-ACK GCT

CAG TAA CAC GT-3′) and A934b (5′-GTG CTC CCC CGC CAA TTC CT-3′), as described by Großkopf et al. [34] for Archaea. For the T-RFLP analyses the forward Bacteria and the backward Archaea primers were labeled with FAM (5-carboxy-fluorescein). Each 50-µl volume of PCR reaction contained 1× GoTaq Flexi Green Buffer (Promega); 1.5 mM MgCl₂ (Promega); 200 µM of dNTP's Mix (Fermentas); 0.33 µM of each primer described (Sigma); 1 U GoTaq Flexi DNA Polymerase (Promega) and 10 µg of Bovine Serum Albumin BSA (Roche). Diluted total DNA extract was added as template (1 µl). The reaction was started with a initial denaturation step (94°C for 3 min), followed by 24–28 cycles of denaturation (94°C for 45 sec), annealing (52°C for 45 sec) and extension (72°C for 80 sec), and a final extension step (72°C for 5 min).

For amplification of *mcrA* the primer pair MCRf (5′-TAY GAY CAR ATH TGG YT-3′) and MCRb (5′-ACR TTC ATN GCR TAR TT-3′) was used as described by Springer et al. [29], the forward primer being labeled with FAM for T-RFLP analyses. Each 50-µl of PCR reaction contained 1× MasterAmp PCR PreMix B (Biozym); 0.33 µM of each primer described; 1 U GoTaq Flexi DNA Polymerase (Promega); 10 µg BSA (Roche) and 1 µl of DNA template. The reaction started with an initial denaturation step (94°C for 3 min), followed by 32 cycles of denaturation (94°C for 45 sec), annealing (50°C for 45 sec) and extension (72°C for 90 sec), finished by a final extension step (72°C for 5 min).

The quality of the PCR products was controlled using agarose gel (1.5%) electrophoresis. The DNA was then purified using the GenElute PCR Clean-Up Kit (Sigma) following the manufacturer's instructions, and stored at −20°C.

For T-RFLP, the PCR product was digested using the following restriction enzymes: MspI incubated overnight at 37°C for Bacterial 16S rRNA genes, TaqI incubated for 3 h at 65°C for Archaeal 16S rRNA genes, and Sau96 incubated for 3 h at 37°C for *mcrA*. After the incubation, the digested products were once again cleaned with the SigmaSpin Sequencing Reaction Clean-Up, Post-Reaction Purification Columns (Sigma). The samples were denatured at 94°C for 2 min and loaded into an ABI 3100 automated gene sequencer (Applied Biosystems) for separation of the TRFs. T-RFLP data were retrieved by comparison with an internal standard using GeneScan 3.71 software (Applied Biosystems).

## Taxa-area and distance-decay relationships

The T-RFLP profiles were analyzed and standardized as described in Dunbar et. al [35] resulting in T-RFs of 60 to 855

**Table 1.** Distance-decay regression coefficients based on the Bray-Curtis dissimilarity index **(Y)**; Regression coefficient based on the Sørensen Similarity index; and the exponent **z** calculated by the distance-decay approach values for each one of the target genes calculated by -(regression coefficient)/2.

| Gene | Distance-decay (Y) | Sørensen Regr. Coefficient | (z) |
|------|--------------------|-----------------------------|-----|
| *mcrA* | 0.14 | −0.032 | 0.016 |
| Archaea 16S rRNA | 0.11 | −0.031 | 0.0155 |
| Bacteria 16S rRNA | 0.067 | −0.018 | 0.009 |

**Table 2.** Z-values previously described for microorganisms, 99% or 95% represent the taxonomic resolution applied for definition of the OTUs (sequence similarity).

| Microbial Community | z-value | Diversity Access | Habitat Org. | Reference |
|---|---|---|---|---|
| Desert Soil Fungi | 0.074 | fingerprint (ARISA) | Contiguous | [19] |
| Salt-Marsh Bacteria (99% OTU) | 0.04 | sequencing | Contiguous | [18] |
| Salt-Marsh ß-proteobacteria (99% OTU) | 0.019 | sequencing | Contiguous | [18] |
| Salt-marsh Bacteria (95% OTU) | 0.019 | sequencing | Contiguous | [18] |
| Salt-Marsh ß-proteobacteria (95% OTU) | 0.008* | sequencing | Contiguous | [18] |
| Water-filled tree holes Bacteria | 0.26 | fingerprint (DGGE) | Island | [20] |
| Metal-cutting fluid sump Bacteria | 0.26-0.29 | fingerprint (DGGE) | Island | [38] |
| Soil Bacteria | 0.03 | Fingerprint (T-RFLP) | Noncontiguous | [39] |
| Tropical Forest Soil Bacteria | 0.42 and 0.47 | Fingerprint (T-RFLP) | Contiguous | [40] |
| Freshwater lake sediment Bacteria | 0.009 | fingerprint (T-RFLP) | Contiguous | this study |
| Freshwater lake sediment Archaea | 0.0155 | Fingerprint (T-RFLP) | Contiguous | this study |
| Freshwater lake sediment methanogenic Archaea | 0.016 | fingerprint (T-RFLP) | Contiguous | this study |

*not significant different from zero.

base pairs (bp) size, each representing more than 1% of the total fluorescence of that sample.

We used the 25 resulting profiles for pair wise calculation of the Bray-Curtis dissimilarity indices [36] and the Sørensen similarity indices resulting in 300 different pairs for each targeted gene. A simple linear regression of the log-transformed data of the Bray-Curtis indices plotted against the distances between the sediment samples from which the pairs originated was used to estimate the slope of the distance-decay relationship [23], in this work we are using the resulting regression coefficients (**Y**) as a parameter for discussion.

The same log-transformed linear regression was used to calculate the slope of the Sørensen indices plotted against the distance between points. The resultant similarity slope was used to calculate the **z-value** of the taxa-area relationship with the formula $\log(S_S) = \text{constant} - 2z \log(D)$, where $S_S$ is the pair-wise similarity between communities and D is the distance between two samples used in the distance-decay approach, where the **z-value** is determined by $-(\text{regression coefficient})/2$ as described by Harte et al. [37].

By using both, the taxa-area relationship based on the Sørensen index and the distance-decay based on the Bray-Curtis dissimilarity index we addressed the subject not only by "presence/absence" of a taxon (Sørensen index), but also the relative abundance of a taxon (Bray-Curtis index) provided by the T-RFLP.

In order to avoid randomization patterns that could influence the results, the same calculations were performed utilizing distances smaller than 200 m and 20 m and they showed similar slopes as the complete data set (data not shown).

## Results

The analysis of the *mcrA* T-RFLP profile resulted in a total of 18 different OTUs of methanogens throughout the lake area (Figure 2A) of which 5 OTUs were found at all sampling points (237 bp, 240 bp, 404 bp, 470 bp and 506 bp) and one OTU (63 bp), which was the least frequently retrieved OTU, was found at only 6 sampling points. The OTU with 506 bp showed the highest relative abundance in all the samples. The Archaeal 16S rRNA T-RFLP profile showed 22 different OTUs (Figure 2B)

with none of them showing dominance over the others, only 2 OTUs were found at all sampling points (91 bp and 392 bp) and the OTUs with 165 bp and 345 bp were found at only 1 and 2 sampling points, respectively. The Bacterial 16S rRNA gene T-RFLP profiles showed the largest number of OTUs among the different genes targeted with a total of 37 different OTUs found at all the sampling points across the lake area, showing a high diversity of dominant groups (Figure 2C). Eight OTUs were recovered from all the sampling points (62 bp, 76 bp, 84 bp, 130 bp, 140 bp, 146 bp, 151 bp and 439 bp); in contrast, 3 other OTUs (459 bp, 526 bp and 612 bp) were recovered from only 3 sampling points. In a literature based affiliation it was possible to relate some of the observed OTUs with previously described ones, keeping in mind that such affiliation can only be tentative (Table S1).

Linear regressions were performed to evaluate the slope coefficient of the correlations between the geographical distance and the communities' similarities, and Mantel tests (with 9999 permutations) were performed to test the significance of these correlations.

All the three different T-RFLP profiles showed a significant distance-decay relationship based on the Bray-Curtis dissimilarity (**Y**) (Figure 3). The *mcrA* gene showed the largest slope of 0.14 (Mantel test $P < 0.001$, $R = 0.69$), followed by the archaeal 16S rRNA gene with a slope of 0.11 (Mantel test $P = 0.005$, $R = 0.30$), while the slope for the bacterial 16S rRNA gene was only 0.067 (Mantel test $P = 0.006$, $R = 0.34$).

The regression coefficient based on the Sørensen similarity indices calculated for all the three genes also showed significant values (Figure 3) of $-0.032$ ($P < 0.001$, $R = -0.60$) for *mcrA*, $-0.031$ (Mantel test $P = 0.03$, $R = -0.45$) for archaeal and $-0.018$ (Mantel test $P = 0.006$, $R = -0.34$) for bacterial 16S rRNA genes. However the calculated z-values were small, thus indicating relatively flat taxa-area relationships (Table 1).

## Discussion

Our study indicates a small but significant biogeographical distribution of the microbial community composition in the lake sediment for all the three different genes targeted, which represent the Bacteria, the Archaea, and the methanogens. The conclusion is based on the distance decay relationship of operational taxonomic units of three different microbial genes. All dominant operational taxa (more than 10% relative abundance) from the three genes analyzed in this study were observed at all the sampling points within the lake. Nevertheless, a significant taxa-area relationship could be observed. These values were small if compared to others described for macro-organisms, which was expected given the high capacity of dispersion of the targeted microorganisms. However, the similarity distance-decay, which takes into consideration also the relative abundance of the groups, shows that they were not homogeneously distributed within the entire lake area. We therefore conclude that the most distant samples were the ones, which were most different from each other.

The **z-values** and the distance decay regression coefficients (**Y**) observed in this study were lower than others previously observed for microorganisms in the literature [17–19][36–38], but are the first ones described for Bacteria, Archaea and methanogenic archaea in a single contiguous environment. The existence of significant **z-values** shows that all these microbial groups have a biogeographical distribution. The low values that we found were expected given that we used a fingerprint technique for accessing the microbial diversity of our target environment, and not a high resolution ribotyping technique. At an OTU resolution of 95% sequence similarity of the 16S rRNA gene, Horner-Devine et al. described similar z-values for Bacteria ($z = 0.019$) and Beta-proteobacteria ($z = 0.008$) in salt-marshes [18]. At higher resolutions the z presented higher values (0.04 for Bacteria and 0.019 for Beta-proteobacteria at 99% sequences similarity). It is interesting to note that the lower values described by the mentioned study were presented by the lower taxonomical levels reached, while in this study we observed higher values for *mcrA* and the Archaea than for the Bacteria. Other **z-values** found for other groups of microorganisms are displayed in the Table 2.

The differences between **z-values** and distance decay values (**Y**) described for the three different genes, representing three different groups of microorganisms, may be explained by two hypotheses. The first one is that Bacteria (16S rRNA) have a larger capacity for dispersal than the Archaea (16S rRNA) in general and the methanogenic archaea (*mcrA*) in particular. By sampling the first 10 centimeters of the sediment we were able to access different communities living at different depths at the same time. We assume that the methanogens (all Archaea) were preferable located in the deeper sediment layers, since their activity is inhibited by the presence of other electron acceptors like oxygen, nitrate, iron and sulfate, potentially present in the surface layers of the sediment. On the other hand, Bacteria were not restricted to deeper sediment layers [41–43]. The Bacteria domain showed weaker species-area and distance-decay relationships than the methanogens (*mcrA*) and the Archaea (16S rRNA). The profile of Archaea and Bacteria OTUs was well distributed throughout the lake area, showing the large diversity of micro-habitats that can be exploited by these groups, and that the ability of dispersion seems not to be compromised at smaller scales.

The second hypothesis is based on the taxonomic sensitivity of the T-RFLP method. OTUs can represent a large variety of different taxonomic groups. The universal primers targeting 16S rRNA genes do not represent the entire set of conceivable species, which are many more than the OTUs derived from T-RFLP. Hence, the total diversity of the community is underestimated by T-RFLP [44]. A study targeting 16S rRNA genes of the bacterioplankton of temperate lakes showed that the sequence similarity within a single OTU varied from 73 to 100%, however, sequence homology of 97% is generally used to define bacterial species [5,45]. When utilizing a functional gene such as the *mcrA*,

the probability of reaching lower taxonomic levels is higher, because translated genes show diminished conservation and can present a higher variability of codons, and thus a higher chance for being differentiated by terminal fragment size methods [46].

This may explain why the **z-values** were higher for T-RFLP of *mcrA* than of 16S rRNA genes, but it does not explain why the Archaea distance decay and **z-values** were higher than those of the Bacterial since both of them targeted 16S rRNA genes. Several authors described that Bacteria diversity tends to be higher than Archaea diversity at different environments [47–49], thus the T-RFLP fingerprinting technique could be more efficient in accessing a larger part of the total diversity of Archaea 16S rRNA and consequentially being closer to the actual taxa distribution of the environment, thus showing higher values of **z** and **Y**.

Our work shows some important results that contribute to a better understanding of similar ones developed within different environments. Studies of bacterioplankton distribution in lakes showed similar patterns between different and not connected lakes in distant geographic regions, but with variations in their relative abundance [50]. On the other hand bacterioplankton communities distance dissimilarity inside one single lake were described as being weaker than between different lakes in North America, and mostly influenced by different water regimes and partial geographic isolation [51]. Some studies in saline lakes in China, Mongolia and Argentina showed that Bacterial biogeography in these environments was based on contemporary environmental factors ($Na^+$, $CO_3^{2-}$, and $HCO_3^-$ ion concentrations, pH and temperature) and geographic distance, while Archaeal biogeography was influenced only by environmental factors [17].

Geographical distances have been previously described as an important factor related with microbial spatial distribution corroborating our findings [52–54]. Rosselló-Mora et al. used metabolic compounds as a comparison parameter between different populations of the extremophilic bacterium *Salinibacter rubium*. They discovered that the divergence among different phenotypes was related to different geographical locations, and geographical distance between the sites[55]. Environmental and geographical factors also influenced magnetotactic bacteria biogeographical distribution [56].

Our distance-decay and taxa-area relationships can possibly be a result of some degree of dispersal limitation coupled with ecological drift, thus maintaining the taxa-area relationship and distance-decay pattern significant. It is believed that drift plays a important role for the distance-decay patterns found among microorganisms living in some degree of dispersal limitation (i.e. in subsurface habitats)[5]. A high dispersal rate is expected inside this contiguous environment with no significant geographical barriers imposing any kind of mobility limitation to microorganisms, and that could be a factor flattening the curves. But as we already stated, a group preferentially found in deeper layers of sediment (as the methanogenic archaea) could be more easily restricted.

This study did not focus on scanning environmental factors that could be driving the observed biogeographical pattern, so selection as a possible driving factor cannot be excluded, and selection processes are also related with geographical distances, as the diversity of habitats tends to increase with an increasing area [5,27]. However, given the small scale of the study and the probable absence of a clear environmental gradient within the Lagoa Negra we don't think it as a major factor. We believe that the correlation that we described between geographical distance and communities' structures are representative of the biogeographical pattern of distribution of the lake microbial community.

In conclusion, we showed that there was a significant geographic distribution of methanogenic archaea, Archaea and even Bacteria, related with geographic distance in a contiguous environment, i.e. the sediment of a tropical lake.

## Supporting Information

**Table S1 Tentative genetic affiliation of the OTUs.** For the three studied genes there are in the literature some genetic affiliation with different phylogenetic groups, we show some of these in this table. * The affiliation with the phylogenetic groups was based on literature data that used the same primers and restriction enzymes for TRFLP. All the affiliations are only tentative, and can only be interpreted as the probable main groups of the lake sediment community.

## References

1. Begon M, Townsend CR, Harper JL (2006) Ecology: from individuals to ecosystems. 4th ed. Blackwell Publishing Ltd.
2. Martiny JBH, Bohannan BJM, Brown JH, Colwell RK, Fuhrman JA, et al. (2006) Microbial biogeography: putting microorganisms on the map. Nat Rev 4: 102–112.
3. Fenchel T, Esteban GF, Finlay BJ (1997) Local versus Global Diversity of Microorganisms: Cryptic Diversity of Ciliated Protozoa. Oikos 80: 220–225.
4. Fenchel T (2003) Biogeography for bacteria. Science 301: 925.
5. Hanson Ca, Fuhrman Ja, Horner-Devine MC, Martiny JBH (2012) Beyond biogeographic patterns: processes shaping the microbial landscape. Nat Rev Microbiol 10: 497–506.
6. Baas-Becking L, Becking LGMB (1934) Geobiologie of inleiding tot de milieukunde. 18.
7. Finlay BJ (2002) Global dispersal of free-living microbial eukaryote species. Science 296: 1061.
8. Finlay BJ, Clarke KJ (1999) Ubiquitous dispersal of microbial species. Nature 400: 1999.
9. Staley JT, Gosink JJ (1999) Poles apart: biodiversity and biogeography of sea ice bacteria. Annu Rev Microbiol 53: 189–215.
10. Papke RT, Ramsing NB, Bateson MM, Ward DM (2003) Geographical isolation in hot spring cyanobacteria. Environ Microbiol 5: 650–659.
11. Papke RT, Ward DM (2004) The importance of physical isolation to microbial diversification. FEMS Microbiol Ecol 48: 293–303.
12. Diniz-Filho JAF, Telles MPDC (2000) Spatial pattern and genetic diversity estimates are linked in stochastic models of population differentiation. Genet Mol Biol 23: 541–544.
13. Cho J-CC, Tiedje JM (2000) Biogeography and Degree of Endemicity of Fluorescent Pseudomonas Strains in Soil. Appl Environ Microbiol 66: 5448–5456.
14. Oda Y, Star B, Huisman LA (2003) Biogeography of the purple nonsulfur bacterium Rhodopseudomonas palustris. Appl Environ Microbiol 69: 5186–5191.
15. Takacs-Vesbach C, Mitchell K, Jackson-Weaver O, Reysenbach A-L (2008) Volcanic calderas delineate biogeographic provinces among Yellowstone thermophiles. Environ Microbiol 10: 1681–1689.
16. Bahl J, Lau MCY, Smith GJD, Vijaykrishna D, Cary SC, et al. (2011) Ancient origins determine global biogeography of hot and cold desert cyanobacteria. Nat Commun 2: 163.
17. Pagaling E, Wang H, Venables M, Wallace A, Grant WD, et al. (2009) Microbial biogeography of six salt lakes in Inner Mongolia, China, and a salt lake in Argentina. Appl Environ Microbiol 75: 5750–5760.
18. Horner-Devine MC, Lage M, Hughes JB, Bohannan BJM (2004) A taxa-area relationship for bacteria. Nature 432: 750–753.
19. Green JL, Holmes AJ, Westoby M, Oliver I, Briscoe D, et al. (2004) Spatial scaling of microbial eukaryote diversity. Nature 432: 747–750.
20. Bell T, Ager D, Song J-I, Newman Ja, Thompson IP, et al. (2005) Larger islands house more bacterial taxa. Science 308: 1884.
21. Head IM, Saunders JR, Pickup RW (1998) Microbial evolution, diversity, and ecology: a decade of ribosomal RNA analysis of uncultivated microorganisms. Microb Ecol 35: 1–21.
22. Woodcock S, Curtis TP, Head IM, Lunn M (2006) Taxa-area relationships for microbes: the unsampled and the unseen. Ecology 9: 805–812.
23. Nekola JC, White PS (2004) The distance decay of similarity in biogeography and ecology. J Biogeogr 26: 867–878.
24. Bell T (2010) Experimental tests of the bacterial distance-decay relationship. ISME J 4: 1357–1365.
25. Angel R, Claus P, Conrad R (2012) Methanogenic archaea are globally ubiquitous in aerated soils and become active under wet anoxic conditions. ISME J 6: 847–862.
26. Conrad R (2007) Microbial ecology of methanogens and methanotrophs. Adv Agron 96: 1–63.
27. Nemergut DR, Schmidt SK, Fukami T, O'Neill SP, Bilinski TM, et al. (2013) Patterns and processes of microbial community assembly. Microbiol Mol Biol Rev 77: 342–356.
28. Stackebrandt E, Goebel BM (1994) Taxonomic Note: A Place for DNA-DNA Reassociation and 16S rRNA Sequence Analysis in the Present Species Definition in Bacteriology. Int J Syst Bacteriol 44: 846–849.
29. Springer E, Sachs MS, Woese CR, Boone DR (1995) Partial gene sequences for the A subunit of methyl-coenzyme M reductase (mcrI) as a phylogenetic tool for the family Methanosarcinaceae. Int J Syst Bacteriol 45: 554–559.
30. Marani L, Alvalá PCC (2007) Methane emissions from lakes and floodplains in Pantanal, Brazil. Atmos Environ 41: 1627–1633.
31. Guerrini V (1978) Bacia do alto rio Paraguai: estudo climatológico. Brasília: EDIBAP/SAS.
32. Bezerra M de O, Mozeto A (2008) Deposição de carbono orgânico na planície de inundação do rio Paraguai durante o Holoceno Médio. Oecol Bras 12: 155–171.
33. Weisburg WG, Barns SM, Pelletier Da, Lane DJ (1991) 16S ribosomal DNA amplification for phylogenetic study. J Bacteriol 173: 697–703.
34. Großkopf R, Janssen PH, Liesack W (1998) Diversity and structure of the methanogenic community in anoxic rice paddy soil microcosms as examined by cultivation and direct 16S rRNA gene sequence retrieval.
35. Dunbar J, Ticknor LO, Kuske CR (2001) Phylogenetic specificity and reproducibility and new method for analysis of terminal restriction fragment profiles of 16S rRNA genes from bacterial communities. Appl Environ Microbiol 67: 190–197.
36. Roger Bray J, Curtis JTT, Bray JRR (1957) An ordination of the upland forest communities of southern Wisconsin. Ecol Monogr 27: 325–349.
37. Harte J, McCarthy S, Taylor K, Kinzig A, Fischer ML (1999) Estimating Species-Area Relationships from Plot to Landscape Scale Using Species Spatial-Turnover Data. Oikos 86: 45.
38. Van der Gast CJ, Lilley AK, Ager D, Thompson IP (2005) Island size and bacterial diversity in an archipelago of engineering machines. Environ Microbiol 7: 1220–1226.
39. Fierer N, Jackson RB (2006) The diversity and biogeography of soil bacterial communities. Proc Natl Acad Sci U S A 103: 626.
40. Noguez AMM, Arita HTT, Escalante AEE, Forney LJJ, Garcia-Oliva F, et al. (2005) Microbial macroecology: highly structured prokaryotic soil assemblages in a tropical deciduous forest. Glob Ecol Biogeogr 14: 241–248.
41. Falz KZ, Holliger C, Großkopf R, Liesack W, Nozhevnikova AN, et al. (1999) Vertical distribution of methanogens in the anoxic sediment of Rotsee (Switzerland). Appl Environ Microbiol 65: 2402–2408.
42. Lovley DR, Klug MJ (1986) Model for the distribution of sulfate reduction and methanogenesis in freshwater sediments. Geochim Cosmochim Acta 50: 11–18.
43. Chan OC, Claus P, Casper P, Ulrich A, Lueders T, et al. (2005) Vertical distribution of structure and function of the methanogenic archaeal community in Lake Dagow sediment. Environ Microbiol 7: 1139–1149.
44. Liu WT, Marsh TL, Cheng H, Forney LJ (1997) Characterization of microbial diversity by determining terminal restriction fragment length polymorphisms of genes encoding 16S rRNA. Appl Environ Microbiol 63: 4516–4522.
45. Eiler A, Bertilsson S (2004) Composition of freshwater bacterial communities associated with cyanobacterial blooms in four Swedish lakes. Environ Microbiol 6: 1228–1243.
46. Marsh TL (1999) Terminal restriction fragment length polymorphism (T-RFLP): an emerging method for characterizing diversity among homologous populations of amplification products. Curr Opin Microbiol 2: 323–327.
47. Inagaki F, Nunoura T, Nakagawa S, Teske A, Lever M, et al. (2006) Biogeographical distribution and diversity of microbes in methane hydrate-bearing deep marine sediments on the Pacific Ocean Margin. Proc Natl Acad Sci U S A 103: 2815–2820.
48. Ochsenreiter T, Selezi D, Quaiser A, Bonch-Osmolovskaya L, Schleper C (2003) Diversity and abundance of Crenarchaeota in terrestrial habitats studied by 16S RNA surveys and real time PCR. Environ Microbiol 5: 787–797.
49. Bowman JP, McCuaig RD (2003) Biodiversity, community structural shifts, and biogeography of prokaryotes within Antarctic continental shelf sediment. Appl Environ Microbiol 69: 2463.

(DOCX)

## Acknowledgments

We thanks, the colleagues Roberta Peixoto, Juliana Valle, Ruan Andrade and the Federal University of Mato Grosso do Sul - Brazil (UFMS) for the support during the sampling.

## Author Contributions

Conceived and designed the experiments: DPB RC AEP. Performed the experiments: DPB PC MK. Analyzed the data: DPB PC MK. Contributed reagents/materials/analysis tools: RC PC MK AEP. Wrote the paper: DPB RC AEP.

50. Lindström ES, Leskinen E (2002) Do neighboring lakes share common taxa of bacterioplankton? Comparison of 16S rDNA fingerprints and sequences from three geographic regions. Microb Ecol 44: 1–9.

51. Yannarell AC, Triplett EW (2004) Within and between Lake Variability in the Composition of Bacterioplankton Communities: Investigations Using Multiple Spatial Scales. Appl Environ Microbiol 70: 214.

52. Schauer R, Bienhold C, Ramette A, Harder J (2010) Bacterial diversity and biogeography in deep-sea surface sediments of the South Atlantic Ocean. ISME J 4: 159–170.

53. Galand PE, Potvin M, Casamayor EO, Lovejoy C (2010) Hydrography shapes bacterial biogeography of the deep Arctic Ocean. ISME J 4: 564–576.

54. Xiong J, Liu Y, Lin X, Zhang H, Zeng J, et al. (2012) Geographic distance and pH drive bacterial distribution in alkaline lake sediments across Tibetan Plateau. Environ Microbiol 14: 2457–2466.

55. Rosselló-Mora R, Lucio M (2008) Metabolic evidence for biogeographic isolation of the extremophilic bacterium Salinibacter ruber. ISME J 2: 242–253.

56. Lin W, Wang Y, Gorby Y, Nealson K, Pan Y (2013) Integrating niche-based process and spatial process in biogeography of magnetotactic bacteria. Sci Rep 3: 1643.

# 11

# Correlates of Zooplankton Beta Diversity in Tropical Lake Systems

**Paloma M. Lopes[1]\*, Luis M. Bini[2], Steven A. J. Declerck[3], Vinicius F. Farjalla[1], Ludgero C. G. Vieira[4], Claudia C. Bonecker[5], Fabio A. Lansac-Toha[5], Francisco A. Esteves[1,6], Reinaldo L. Bozelli[1]**

**1** Laboratório de Limnologia, Departamento de Ecologia, Instituto de Biologia, Universidade Federal do Rio de Janeiro, CCS, Cidade Universitária, Rio de Janeiro, RJ, Brazil, **2** Departamento de Ecologia, Instituto de Ciências Biológicas, Universidade Federal de Goiás, Goiânia, GO, Brazil, **3** Netherlands Institute of Ecology (NIOO-KNAW), Department of Aquatic Ecology, Wageningen, The Netherlands, **4** Faculdade UnB Planaltina, Universidade de Brasília, Área Universitária n. 1 - Vila Nossa Senhora de Fátima, Planaltina, Distrito Federal, Brazil, **5** Núcleo de Pesquisas em Limnologia, Ictiologia e Aqüicultura (NUPELIA), Universidade Estadual de Maringá, Jd. Universitário, Maringá, Paraná, Brazil, **6** Núcleo de Pesquisas em Ecologia e Desenvolvimento Sócio Ambiental de Macaé, Rodovia Amaral Peixoto, Macaé, RJ, Brazil

## Abstract

The changes in species composition between habitat patches (beta diversity) are likely related to a number of factors, including environmental heterogeneity, connectivity, disturbance and productivity. Here, we used data from aquatic environments in five Brazilian regions over two years and two seasons (rainy and dry seasons or high and low water level periods in floodplain lakes) in each year to test hypotheses underlying zooplankton beta diversity variation. The regions present different levels of hydrological connectivity, where three regions present lakes that are permanent and connected with the main river, while the water bodies of the other two regions consist of permanent lakes and temporary ponds, with no hydrological connections between them. We tested for relationships between zooplankton beta diversity and environmental heterogeneity, spatial extent, hydrological connectivity, seasonality, disturbance and productivity. Negative relationships were detected between zooplankton beta diversity and both hydrological connectivity and disturbance (periodic dry-outs). Hydrological connectivity is likely to affect beta diversity by facilitating dispersal between habitats. In addition, the harsh environmental filter imposed by disturbance selected for only a small portion of the species from the regional pool that were able to cope with periodic dry-outs (e.g., those with a high production of resting eggs). In summary, this study suggests that faunal exchange and disturbance play important roles in structuring local zooplankton communities.

**Editor:** Frédéric Guichard, McGill University, Canada

**Funding:** Our research was financed by the Vale, Mineração Rio do Norte and PELD site 6/CNPq. P.M.L is grateful to CAPES, CNPq and FAPERJ institutions for post-graduate scholarships. L.M.B., C.C.B., V.F.F., F.A.E. and R.L.B. are partially supported by CNPq productivity grants. The funders had no role in study design, data collection and analysis, decision to publish, or preparation of the manuscript.

**Competing Interests:** The authors have declared that no competing interests exist.

\* Email: paloma.marinho@gmail.com

## Introduction

Beta diversity, that is, the change in species composition between habitat patches, is directly related to local or alpha-diversity (i.e., number of species within a particular habitat) and to regional or gamma-diversity (i.e., diversity in the different habitats within a region) [1,2]. Understanding the mechanisms underlying beta diversity is one of the main goals in community ecology and interest in this area has increased substantially in the last decade [3,4,5].

Both deterministic and stochastic processes may affect beta diversity patterns. Deterministic processes are based on niche theory and assume that environmental filtering and biotic interactions play major roles in shaping local community composition. According to this theory, species have different ecological requirements, determining different responses to environmental gradients [6]. Stochastic processes are more related to the importance of colonization rates, random extinction and

disturbance [7]. It is increasingly recognised that these processes shape the community structure simultaneously [8].

A number of factors have been shown to be important in predicting beta diversity. High spatial variation in environmental conditions allows species with different ecological requirements to occur in different sites, increasing beta diversity [9,10,11,12]. Another important driver of beta diversity is the degree of spatial connectivity [10,13,14,15]. Communities that are highly connected (e.g., by hydrological connections and smaller distance between habitats) may have lower beta diversity due to the higher exchange of individuals between these communities via active and passive dispersal. In addition, when connectivity is high, beta diversity can also decrease due to environmental homogenization [16,17,18,19]. The degree of connectivity within a region can also vary over time. Lakes in river-floodplain systems, for example, tend to be highly connected to each other and with the main river during flood periods, leading to more similar communities and environmental conditions [17]. In addition, disturbances are, in general, responsible for a decrease in beta diversity [14,20,21,22], as they impose environmental filters that select only portions of the species from the regional pools that are disturbance-tolerant [14,21,22]. On the other hand, disturbance can increase beta diversity in more isolated ponds as a result of stochastic recolonization and priority effects [23]. Temporary ponds undergo drastic disturbances, as they periodically dry up completely [24]. Finally, a positive relationship between beta diversity and primary productivity is expected due to a greater contribution of stochastic processes relative to deterministic ones in high productivity-environments [25,26]. In regions with high productivity, a greater number of species can coexist and the composition of communities is likely to depend on the colonization history, such as priority effects [25,27].

In this study, we tested the effects of environmental heterogeneity, spatial extent, hydrological connectivity (isolated vs. floodplain lakes), seasonality (wet season or high water period vs. dry season or low water period), disturbance (periodic dry-outs) and productivity (mean chlorophyll-a concentration) on beta diversity of zooplankton communities in five geographic regions of Brazil. We predicted that (a) lake systems that are more heterogeneous in their environmental conditions would have higher beta diversity, (b) regions where the spatial extent is larger (i.e., larger distances between studied local communities) would have higher values of beta diversity, (c) connected environments would have lower beta diversity than isolated environments due to their greater similarity in environmental conditions and/or greater dispersal rates, (d) beta diversity would be higher during the low water period or dry season, when there is less connectivity between environments, (e) temporary environments would have lower beta diversity as they undergo disturbances (periodic dry-outs), selecting species tolerant to these extreme conditions and (f) regions with higher primary productivity would have a higher beta diversity.

## Materials and Methods

### Ethics Statement

Collecting permits were provided by the Instituto Chico Mendes de Conservação da Biodiversidade – ICMBio (Brazilian Ministry of Environment). None of the species collected are considered threatened.

### Study Area

This study was based on the analysis of data collected in aquatic environments of five regions in Brazil: 24 lakes in Trombetas River floodplain (Amazonian region, Northern Brazil), 20 lakes in the Upper Paraná River floodplain (Southern Brazil), 32 lakes in the Middle Araguaia River floodplain (Central Brazil), 21 coastal lakes and ponds located in the Restinga de Jurubatiba National Park (Macaé, Southeast Brazil) and 23 lakes and ponds in the Amazonian upland region (Carajás, mean altitude 710 m, Northern Brazil) (Figure S1). The Trombetas, Paraná and Araguaia floodplain lakes are permanent and connected with the main river (at least during the high water periods in the case of the Paraná River), while the water bodies of Macaé and Carajás regions consist of permanent lakes and temporary ponds, respectively, with no hydrological connections between them, except for small and sporadic connections during the rainy season. Ponds were considered temporary if they completely dried up at least once during our study period.

### Field Sampling

Each region was sampled in two wet and two dry seasons (or high and low water periods), except for Araguaia, which was sampled in one high and one low water period (see Table S1 for more details about the sampling schedule). For each lake, water samples were analyzed for total nitrogen ($\mu$mol L$^{-1}$), total phosphorus ($\mu$mol L$^{-1}$) and chlorophyll-a ($\mu$g L$^{-1}$). In the field, we also measured pH, dissolved oxygen (mg L$^{-1}$) and conductivity ($\mu$S cm$^{-1}$). Details of the methods employed for the determination of these environmental variables are described in [28,29,30].

In Trombetas, Macaé and Carajás, samples of zooplankton were collected either by filtering 100 L of water (collected with a bucket in the case of shallow lakes; i.e. <1 m depth) through a 50 $\mu$m mesh plankton net or by directly taking vertical hauls with a 50 $\mu$m plankton net (for deep lakes; i.e. >1 m depth). In Paraná and Araguaia regions, zooplankton samples were obtained by pumping 600 L and 1000 L of water, respectively, over a 50 $\mu$m mesh plankton net. Samples were immediately fixed with 4% formaldehyde. In the laboratory, zooplankton individuals were identified to the lowest possible taxonomic unit. Triplicate aliquots (1 ml) of zooplankton samples were counted in either a Sedgewick-Rafter cell under a microscope (for rotifers and nauplii) or in open chambers under a stereomicroscope (for cladocerans and copepods). At least 100 individuals per group (rotifers, nauplii, cladocerans and copepods) were counted in each aliquot, but in most of samples, these numbers were exceeded. Entire samples (rather than aliquots) were analysed to identify rare species. We verified that most species were sampled within each region, as the cumulative richness curves tended to reach an asymptote (Figure S2).

### Data Analysis

We made an *a priori* distinction between three categories of lakes based on their degree of hydrological isolation and permanency. These categories consisted of permanent and connected (all lakes in Trombetas, Araguaia and Paraná), permanent isolated and temporary isolated lakes (the latter two groups being present only in Macaé and Carajás). For each category of lakes in each region, we separately calculated beta diversity for each of the sampling periods. Beta diversity was calculated as the mean distance of individual observations (e.g., lakes) to the group centroid (combination of lake category and sampling period), using a permutational analysis of multivariate dispersions (PERMDISP [31]). PERMDISP can be used for calculating beta diversity with the use of any dissimilarity measure [12,31]. Because dissimilarity measures have different properties and can generate different beta diversity patterns, we applied PERMDISP to multiple distance measures, such as Bray-Curtis (DistC$_{BC}$), 1-Jaccard (DistC$_{Jac}$) and Simpson (DistC$_{Sim}$) indices. Bray-Curtis is an abundance-based index, while Jaccard is based on presence/absence data. The pairwise Simpson dissimilarity coefficient [32] is also based on presence/absence data, but unlike Jaccard, it is independent of differences in local richness. Species abundance data were log (x+1) transformed prior to analysis.

Environmental heterogeneity and spatial extent were calculated by applying PERMDISP on Euclidean distances from standardized environmental variables (total nitrogen, total phosphorus, pH, dissolved oxygen and electrical conductivity) and geographic coordinates (decimal degrees), respectively. In all analyses, environmental heterogeneity and spatial extent were continuous variables, while connectivity, seasonality and disturbance were dummy variables with two levels (connected vs. isolated, high water vs. low water period, with (temporary) vs. without drought (permanent), respectively).

Linear mixed-effects models (LMMs) with restricted maximum likelihood estimation were used to examine the effects of connectivity (connected vs. isolated), seasonality (high water vs. low water period), environmental heterogeneity and spatial extent on each of the beta diversity measures (DistC$_{BC}$, DistC$_{Jac}$ and DistC$_{Sim}$). In these analyses, region was specified as a random factor to account for non-independence of data. The effect of connectivity was evaluated only for permanent lakes because there were no connected temporary lakes in our dataset. Using data from the regions with both temporary and permanent lakes (Macaé and Carajás; $n = 16$, 8 temporary and 8 permanent), we applied LMM to test for the effect of disturbance (periodic dry out) on beta diversity also incorporating seasonality, environmental heterogeneity and spatial extent as explanatory variables in the model. Connectivity, disturbance, season, spatial extent and region may affect beta diversity directly as well as indirectly through their effect on environmental heterogeneity. To obtain a better understanding of the potential importance of such indirect effects, we also performed LMMs to analyze the dependence of environmental heterogeneity on the other explanatory variables. Significance ($P<0.05$) was tested with using Type II tests (Wald $F$ tests with Kenward-Roger degrees of freedom). Linear mixed-effects models were used to test the relationship between beta diversity measures and mean chlorophyll-$a$ concentration (a proxy for productivity) and were performed separately because this variable was not available for all sampling times.

We also examined the effect of seasonality (high water vs. low water period) and disturbance (permanent vs. temporary lakes) on zooplankton communities by calculating the local contribution to beta diversity (LCBD) as a measure of ecological uniqueness of each lake in terms of species composition [33]. LCBS values "indicate the sites that contribute more (or less) than the mean to beta diversity" [33]. To test the effect of seasonality on zooplankton beta diversity of floodplain lakes (i.e., Trombetas, Paraná and Araguaia regions) we calculated LCBD for each region and for each year of sampling (i.e., data from one sampling during low and high water periods). According to our hypothesis, one would expect higher values of LCBD during the low water periods. Similarly, using data obtained in Carajás and Macaé regions from each sampling period, we calculated LCBD in order to test the effect of disturbance on beta diversity (especially, whether permanent lakes contribute more to the overall beta diversity than temporary lakes). All LCBD values were tested by permutation (999 runs) according to the procedures described in Legendre and De Cáceres [33].

All analyses were carried out in R 2.15.1 [34]. PERMDISP was performed using the *vegan* package [35], the pairwise Simpson dissimilarity coefficient was calculated using the *betapart* package [36], LMMs were fitted with the *lmer* function from the package *lme4* [37], Wald tests were carried out using the *car* package [38] and marginal R$^2$ values (variance explained by fixed factors) [39] were calculated using the *MuMIn* package [40]. LCBD values were calculated using the functions provided by Legendre and De Cáceres [33].

## Results

We detected a total of 156 zooplankton species in Trombetas River floodplain, 208 in Paraná River floodplain, 128 in Araguaia River floodplain, and 168 and 120 species in the isolated lakes of Macaé and Carajás regions, respectively (for species lists and information on dominance, see Table S2). In general, species richness (according to a specific number of samples as, for instance, 40 samples, or to the mean values; see Figures S2 and S3, respectively) tended to be higher in lakes from the Paraná River floodplain and the Macaé region than in lakes from the other floodplains (Trombetas and Araguaia) and the Carajás region. Rotifers, when compared to other broad taxonomic groups,

dominated the communities both in terms of species richness and abundance (except in the lakes from the Araguaia River floodplain, where cladocerans were more abundant; Figure S4).

Patterns of zooplankton beta diversity were similar independently of the community dissimilarity metric used (see Figures 1A and S5). For this reason, we only present results based on Jaccard distances. Beta diversity in regions with permanent lakes responded significantly to connectivity (Table 1; for models based on other dissimilarity coefficients, see Table S3), where regions with connected lakes showed lower values than regions with isolated lakes (Figure 1A). The lowest beta diversities were found in the lakes associated to the Trombetas River floodplain (Figure 1A), which has the highest connectivity level of all regions. Environmental heterogeneity, seasonality and spatial extent had no significant effect on beta diversity (Table 1).

Beta diversity in the isolated lakes of Carajás and Macaé responded significantly to disturbance, but not to seasonality, environmental heterogeneity or spatial extent (Table 2; for models based on other dissimilarity coefficients, see Table S4). Within these regions, beta diversity was consistently higher in permanent than in temporary lakes (Figure 1A). The global models were similar to the reduced models in showing that only connectivity and disturbance were significant correlates of beta diversity ($t = 3.72$, $P = 0.037$ and $t = -6.40$, $P<0.0001$, respectively). The relationship between zooplankton beta diversity and the mean chlorophyll-$a$ concentration was not statistically significant ($t = -1.28$, $P = 0.22$).

In agreement with the results of the LMMs, the LCBD values for lakes sampled during low and high water periods were similar (Figure S6), suggesting no effect of seasonality on zooplankton beta diversity in these regions. Moreover, the contribution of permanent lakes to zooplankton beta diversity in Macaé and Carajás regions was higher than the contribution of temporary ponds, with the latter showing lower and mostly non-significant values (Figure S7).

Environmental heterogeneity tended to be lower in connected than in isolated lakes (Figure 1B), but the difference was not statistically significant ($t = 1.69$, $P = 0.08$). Environmental heterogeneity was also unrelated to seasonality, spatial extent and disturbance (Table S5).

## Discussion

The hypothesis that beta diversity would be lower in hydrologically connected lakes was corroborated, independently of the measure of beta diversity utilized (i.e., Bray-Curtis, Jaccard or Simpson dissimilarity coefficients). Connectivity is known to increase the similarity in species composition in aquatic systems in two different ways. First, higher connectivity increases the similarity in environmental conditions among lakes [16,17,19]. Second, hydrological connectivity can also facilitate the exchange of organisms, via passive dispersal, among connected lakes, increasing the similarity in species composition [13,14,16,41,42,43], even when environmental conditions are heterogeneous among lakes. The effects of environmental heterogeneity could be ruled out, as these effects were not significant, nor was there a significant relationship between environmental heterogeneity and connectivity. Hydrological connectivity therefore likely influences beta diversity primarily by increasing dispersal rates among the lakes, homogenizing zooplankton composition.

Our results also indicated that beta diversity was lower among temporary environments than among permanent ones within the same region. Environments that undergo disturbances, such as

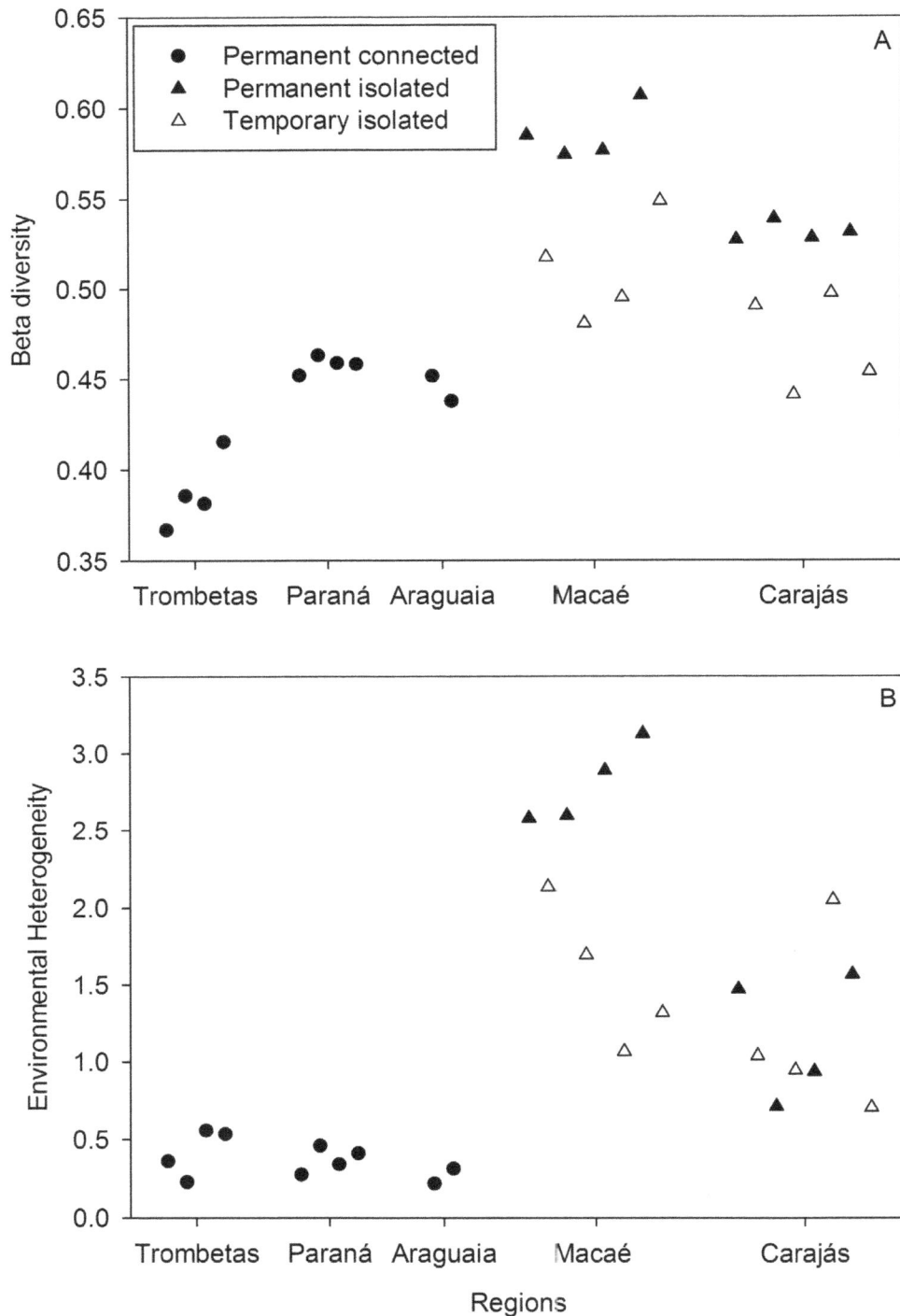

**Figure 1. Zooplankton beta diversity for each studied region.** (A) Zooplankton beta diversity (as the mean Jaccard distance to group centroid) for each region, sampling time (for each region, the different data points in the X-axis represent the different sampling times) and lake categories (permanent connected, permanent isolated and temporary isolated). (B) Environmental heterogeneity for each region, sampling time and lake categories.

periodic dry-outs, are considered extreme environments. Thus, a particular group of species can persist under these conditions, representing a strong environmental filter [14,21,22,24,44,45], but see [46]. The species that are able to produce more resting eggs and can hatch more rapidly after refilling are more likely to be positively selected to occur in temporary ponds [24]. Indeed, some studies have shown that rotifers of temporary ponds are characterized by a higher production of resting eggs than rotifers from permanent lakes [47,48,49]. Moreover, an enclosure experiment (Lopes et al., in preparation) performed in the region

**Table 1.** Summary of the linear mixed-effects model of zooplankton beta diversity measured as the mean1-Jaccard distance to group (DistC$_{Jac}$) centroid for connectivity data (connected and isolated permanent lakes from all study regions).

**Sample size: n = 18 observations**

**Group: regions = 5**

**Marginal R$^2$ = 0.68**

| DistC$_{Jac}$ | Random effect | Variance component | | | | |
|---|---|---|---|---|---|---|
| | Region | 0.00191 | | | | |
| | Residual | 0.00018 | | | | |
| | **Fixed effects** | **Estimate** | **SE** | **df** | **t** | **P** |
| | Intercept | 0.421 | 0.048 | | 8.701 | |
| | Connectivity (isolated) | 0.119 | 0.051 | 1 | 2.344 | 0.047 |
| | Environmental heterogeneity | 0.007 | 0.012 | 1 | 0.638 | 0.571 |
| | Spatial extent | 0.010 | 0.110 | 1 | 0.096 | 0.932 |
| | Seasonality (wet) | 0.004 | 0.006 | 1 | 0.615 | 0.551 |

The intercept corresponds to expected beta diversity in connected lakes during the dry season, when environmental heterogeneity and spatial extent are zero. Marginal R$^2$ represents the variance explained by fixed factors.

of Macaé showed that most common and dominant species present in the studied temporary environments of this region are able to produce resting eggs that rapidly hatch after the ponds are refilled (e.g., rotifers *Cephalodella gibba*, *Lecane bulla*, *L. leontina*, *Lepadella patella*, cladocerans *Coronatella monacantha*, *Ephemeroporus barroisi*, *Diaphanosoma birgei*, *Ilyocryptus spinifer* and the calanoid copepod *Diaptomus azureus*). Some of them, in particular, rotifers, were also able to colonize the ponds rapidly by aerial dispersal (less than 20 days after refilling). Priority effects may also play a key role in structuring these communities. The sequence in which species are added to an environment can facilitate or inhibit the establishment of other species, thus affecting the composition of communities [23,50,51]. Species that first established in an environment are more likely to be

competitively superior to those arriving later [52,53]. This effect is even stronger in communities that have dense resting egg banks [23]. In short, although the majority of zooplankton produce resting eggs, some species produce more than others. Species also differ in the speed with which they respond to hatching cues and in the viability of the eggs. Thus, we infer that the communities of temporary ponds would be assembled mainly by species that simultaneously have higher production of resting eggs and respond more quickly to hatching cues. Finally, although we did not test the effect of hydroperiod on zooplankton beta diversity, it is important to point out that the degree of water permanency is likely to influence beta diversity. According to the intermediate disturbance hypothesis, species diversity is higher at intermediate levels of disturbance, assuming a unimodal relationship between

**Table 2.** Summary of the linear mixed-effects model of zooplankton beta diversity measured as the mean1-Jaccard distance to group (DistC$_{Jac}$) centroid for disturbance data (permanent and temporary aquatic systems from Macaé and Carajás).

**Sample size: n = 16 observations**

**Group: regions = 2 (Macaé and Carajás)**

**Marginal R$^2$ = 0.78**

| DistC$_{Jac}$ | Random effect | Variance component | | | | |
|---|---|---|---|---|---|---|
| | Region | 0.00007 | | | | |
| | Residual | 0.00046 | | | | |
| | **Fixed effects** | **Estimate** | **SE** | **df** | **t** | **P** |
| | Intercept | 0.626 | 0.071 | | 8.846 | |
| | Disturbance (temporary) | −0.060 | 0.012 | 1 | −4.864 | <0.0001 |
| | Environmental heterogeneity | 0.014 | 0.010 | 1 | 1.286 | 0.293 |
| | Spatial extent | −0.720 | 0.419 | 1 | −1.718 | 0.808 |
| | Seasonality (wet) | 0.003 | 0.011 | 1 | 0.258 | 0.719 |

The intercept corresponds to expected beta diversity in permanent lakes during the dry season, when environmental heterogeneity and spatial extent are zero. Marginal R$^2$ represents the variance explained by fixed factors.

diversity and disturbance [54]. On the other hand, the relationship between disturbance level and beta diversity is more likely to be negative, where temporary ponds have the most similar species composition (low beta diversity). Therefore, semi-permanent ponds are expected to have higher species richness and intermediate values of beta diversity in comparison to permanent and temporary ones [14]. Further studies should test these predictions (but see [55] for a strong criticism of the intermediate disturbance hypothesis).

According to niche theory, environmental factors and biotic interactions act as strong filters that select species that can persist within a community [56,57,58]. This implies that beta diversity should increase with environmental heterogeneity. Indeed, a relationship between beta diversity and environmental heterogeneity has been demonstrated by studies conducted at different spatial scales [59,60,61,62,63]. In contrast, we did not find an effect of environmental heterogeneity on zooplankton beta diversity. We cannot exclude the possibility that we may have missed some important environmental variables for zooplankton beta diversity. However, although environmental heterogeneity should intuitively have an influence on beta diversity, there are also other studies that failed to show a significant relationship between these variables (see [64] and references therein). Given the uncertainty on the relative importance of environmental heterogeneity in predicting beta diversity, we are of the opinion that this question is still open to further research.

According to the neutral theory of biodiversity [65], beta diversity should increase with increasing distance between the habitat patches exclusively due to dispersal limitation. There is also evidence, including in the systems we studied here (see [17,43]), that floods increase environmental similarity among floodplain lakes within a region, decreasing beta diversity. However, neither spatial extent nor floods were significant correlates of beta diversity. The distance between aquatic environments may not be large enough to create effective barriers to dispersal and zooplankton dispersal can be effective enough to occur even during events of low hydrological connectivity (i.e., low water period or dry season) and/or species establishment in new habitats may be a consequence of their dispersal during periods of high connectivity. The degree of isolation of an aquatic environment is perceived differently by different groups of organisms and will depend, for example, on their dispersal abilities, which may be related to characteristics such as body size. Zooplankters are believed to be efficient passive (overland) dispersers, especially over small scales (see [66,67]). Besides being small-bodied, most species have the ability to produce resting eggs, which increases the likelihood of dispersal by wind and animal vectors [66,68]. However, dispersal abilities may vary between the zooplankton groups. For instance, while copepods reproduce only sexually, rotifers and cladocerans are parthenogenetic organisms, thus probably increasing their chances of establishment, especially in more isolated and temporary environments [69]. Furthermore, we showed that the presence of hydrological connections can highly increase dispersal rates between environments in relation to overland dispersal. The latter may still be effective but dispersal rates are higher between connected aquatic ecosystems (but see [70]).

We expected beta diversity to be higher in regions with higher productivity (i.e., higher mean chlorophyll-a concentration) due to the greater contribution of stochastic factors in relation to deterministic ones in regions with higher productivity [25,27,71]. However, this relationship was not observed in this study. The lack of relationship between beta diversity and mean chlorophyll-a (a proxy for productivity) cannot be explained by the lack of sufficient variation in chlorophyll-a concentrations, which was wide between the regions. Thus, further studies are needed to understand the role of productivity in determining zooplankton beta diversity.

In conclusion, we showed that zooplankton beta diversity in the five regions studied here is mainly associated with hydrological connectivity and disturbances caused by droughts. The hydrological connectivity in the studied areas may act by facilitating the exchange of species among habitats, whereas droughts impose a strong environmental filter that selects for species that can cope with this disturbance. A fruitful avenue for further research would be to extend the approach used here to test specific correlates of zooplankton beta diversity patterns in reservoirs (which are environments highly different from those analysed in our study). For instance, the datasets obtained by [72,73] could be used to test the role of reservoir trophic status and hydrological variation in predicting zooplankton beta diversity.

## Supporting Information

**Figure S1   Map showing the location and number of lakes (n) sampled in each study region in Brazil.**
(TIF)

**Figure S2   Cumulative species richness curves for each study region.**
(TIF)

**Figure S3   Gamma and alpha diversities for each region.** (A) Zooplankton gamma diversity for each region, sampling time (for each region, the different data points in the X-axis represent the different sampling times) and lake categories (permanent connected, permanent isolated and temporary isolated). (B) Mean zooplankton alpha diversity (as the mean Simpson distance to group centroid) for each region, sampling time and lake categories. The error bars represent the standard errors of the mean over aquatic environments.
(TIF)

**Figure S4   Mean species richness and abundance for each zooplankton group.** (A) Mean number of species for each zooplankton group (Rotifera, Cladocera, Copepoda), region (Trom, Trombetas; Par, Paraná; Arag, Araguaia; Mac, Macaé; Car, Carajás) and lake category (P, permanent; T, temporary). (B) Mean abundance (ind/mL) for each group, region and lake category. The error bars represent the standard errors of the mean over aquatic environments.
(TIF)

**Figure S5   Zooplankton beta diversities (as the mean Bray-Curtis and Simpson distances to group centroid) for each region.** (A) Zooplankton beta diversity (as the mean Bray-Curtis distance to group centroid on biological data based on $\log_{10}$ transformation) for each region, sampling time (for each region, the different data points in the X-axis represent the different sampling times) and lake categories (permanent connected, permanent isolated and temporary isolated). (B) Zooplankton beta diversity (as the mean Simpson distance to group centroid) for each region, sampling time and lake categories.
(TIF)

**Figure S6   Local contribution to beta diversity (LCBD) for Trombetas, Paraná and Araguaia regions.** Maps of Trombetas, Paraná and Araguaia regions during high and low water periods showing the local contributions to beta diversity (LCBD) of the zooplankton community at the study lakes. Size of the circles is proportional to the LCBD. Lakes in red have

significant LCDB indices ($P < 0.05$). 1 = first sampling year, 2 = second sampling year.
(TIF)

**Figure S7  Local contribution to beta diversity (LCBD) for Macaé and Carajás regions.** Maps of Macaé and Carajás regions during dry and wet seasons showing the local contributions to beta diversity (LCBD) of the zooplankton community at the study lakes. Size of the circles is proportional to the LCBD. Lakes in red have significant LCBD indices ($P < 0.05$). All lakes in red are permanent lakes, except for the lakes in Macaé 3 and Macaé 4 labelled with T (temporary). 1 = first sampling time, 2 = second sampling time, 3 = third sampling time, 4 = fourth sampling time. (TIF)

**Table S1  Sampling schedule for each study region.** Type of environment, date of collection and number of sampled aquatic environments ($n$) in each region, season and sampling time. Con, connected; Isol, isolated; Perm, permanent; Temp, temporary. (DOCX)

**Table S2  Species list for each region.** List of species per region and lake category. Dominant species are highlighted in yellow. Tr, Pr, Ar, Ma and Ca = Trombetas, Paraná, Araguaia, Macaé and Carajás, respectively. PC, TI, PI = permanent connected, temporary isolated, permanent isolated. (DOCX)

**Table S3  Linear mixed-effects models of zooplankton beta diversity measured as the mean Bray-Curtis and Simpson distances to group centroid for connectivity data.** Summary of the linear mixed-effects models of zooplankton beta diversity measured as the mean Bray-Curtis ($DistC_{BC}$) and Simpson ($DistC_{Sim}$) distance to group centroid for connectivity data (connected and isolated permanent lakes from all study regions). Marginal $R^2$ represents the variance explained by fixed factors.

(DOCX)

**Table S4  Linear mixed-effects models of zooplankton beta diversity measured as the mean Bray-Curtis and Simpson distances to group centroid for disturbance data.** Summary of the linear mixed-effects models of zooplankton beta diversity measured as the mean Bray Curtis ($DistC_{BC}$) and Simpson ($DistC_{Sim}$) distance to group centroid for disturbance data (permanent and temporary aquatic systems from Macaé and Carajás). Marginal $R^2$ represents the variance explained by fixed factors. (DOCX)

**Table S5  Linear mixed-effects models of environmental heterogeneity for connectivity and disturbance datasets.** Summary of the linear mixed-effects models of environmental heterogeneity (Env) measured as the mean Euclidean distance to group centroid for connectivity (connected and isolated permanent lakes from all studied regions) and disturbance datasets (permanent and temporary aquatic systems from Macaé and Carajás). Marginal $R^2$ represents the variance explained by fixed factors. (DOCX)

## Acknowledgments

We are indebted to Instituto Chico Mendes for permission for sampling. We thank Laboratório de Limnologia Básica Nupélia – UEM for providing environmental data from the Paraná region and Tom Reed (University College Cork) for revising the English in the manuscript.

## Author Contributions

Conceived and designed the experiments: PML LMB RLB FAE. Performed the experiments: PML RLB VFF LCGV CCB FALT. Analyzed the data: PML LMB SAJD. Contributed reagents/materials/analysis tools: RLB FAE CCB FALT VFF LMB. Contributed to the writing of the manuscript: PML LMB.

## References

1. Whittaker RH (1972) Evolution and Measurement of Species Diversity. Taxon 21: 213–251.
2. Whittaker RH (1960) Vegetation of the Siskiyou Mountains, Oregon and California. Ecological Monographs 30: 279–338.
3. Anderson MJ, Crist TO, Chase JM, Vellend M, Inouye BD, et al. (2011) Navigating the multiple meanings of b diversity: a roadmap for the practicing ecologist. Ecology Letters 14: 19–28.
4. Jurasinski G, Retzer V, Beierkuhnlein C (2009) Inventory, differentiation, and proportional diversity: a consistent terminology for quantifying species diversity. Oecologia 159: 15–26.
5. Tuomisto H (2010) A diversity of beta diversities: straightening up a concept gone awry. Part 1. Defining beta diversity as a function of alpha and gamma diversity Ecography 33: 2–22.
6. Chase JM, Leibold MA (2003) Ecological niches: linking classical and contemporary approaches. Chicago: University of Chicago Press.
7. Chase JM, Myers JA (2011) Disentangling the importance of ecological niches from stochastic processes across scales. Philosophical Transactions of the Royal Society of London Series B-Biological Sciences 366: 2351–2363.
8. Adler PB, HilleRisLambers J, Levine JM (2007) A niche for neutrality. Ecology Letters 10: 95–104.
9. Chase JM (2003) Experimental evidence for alternative stable equilibria in a benthic fond food web. Ecology Letters 6: 733–741.
10. Verleyen E, Vyverman W, Sterken M, Hodgson DA, De Wever A, et al. (2009) The importance of dispersal related and local factors in shaping the taxonomic structure of diatom metacommunities. Oikos 118: 1239–1249.
11. Veech JA, Crist TO (2007) Habitat and climate heterogeneity maintain beta-diversity of birds among landscapes within ecoregions. Global Ecology and Biogeography 16: 650–656.
12. Anderson MJ, Ellingsen KE, McArdle BH (2006) Multivariate dispersion as a measure of beta diversity. Ecology Letters 9: 683–693.
13. Akasaka M, Takamura N (2012) Hydrologic connection between ponds positively affects macrophyte a and g diversity but negatively affects b diversity. Ecology 93: 967–973.
14. Chase JM (2003) Community assembly: when should history matter? Oecologia 136: 489–498.
15. Declerck SAJ, Coronel JS, Legendre P, Brendonck L (2011) Scale dependency of processes structuring metacommunities of cladocerans in temporary pools of High-Andes wetlands. Ecography 34: 296–305.
16. Cottenie K, Michels E, Nuytten N, De Meester L (2003) Zooplankton metacommunity structure: Regional vs. local processes in highly interconnected ponds. Ecology 84: 991–1000.
17. Thomaz SM, Bini LM, Bozelli RL (2007) Floods increase similarity among aquatic habitats in river-floodplain systems. Hydrobiologia 579: 1–13.
18. Mouquet N, Loreau M (2003) Community patterns in source-sink metacommunities. American Naturalist 162: 544–557.
19. Gonzalez A (2009) Metacommunities: Spatial Community Ecology. Encyclopedia of Life Sciences (ELS). Chichester: John Wiley & Sons.
20. Vanschoenwinkel B, Waterkeyn A, Jocqué M, Boven L, Seaman M, et al. (2010) Species sorting in space and time-the impact of disturbance regime on community assembly in a temporary pool metacommunity. Journal of the North American Benthological Society 29: 1267–1278.
21. Chase JM (2007) Drought mediates the importance of stochastic community assembly. Proceedings of the National Academy of Sciences of the United States of America 104: 17430–17434.
22. Lepori F, Malmqvist B (2009) Deterministic control on community assembly peaks at intermediate levels of disturbance. Oikos 471–479.
23. De Meester L, Gomez A, Okamura B, Schwenk K (2002) The Monopolization Hypothesis and the dispersal-gene flow paradox in aquatic organisms. Acta Oecologica-International Journal of Ecology 23: 121–135.
24. Wellborn GA, Skelly DK, Werner EE (1996) Mechanisms creating community structure across a freshwater habitat gradient. Annual Review of Ecology and Systematics 27: 337–363.
25. Chase JM (2010) Stochastic community assembly causes higher biodiversity in more productive environments. Science 328: 1388–1391.
26. Chase JM, Ryberg WA (2004) Connectivity, scale-dependence, and the productivity-diversity relationship. Ecology Letters 7: 676–683.
27. Chase JM, Leibold MA (2002) Spatial scale dictates the productivity–biodiversity relationship. Nature 416: 427–430.
28. Lopes PM, Caliman A, Carneiro LS, Bini LM, Esteves FA, et al. (2011) Concordance among assemblages of upland Amazonian lakes and the

structuring role of spatial and environmental factors. Ecological Indicators 11: 1171–1176.

29. Roberto MC, Santana NF, Thomaz SM (2009) Limnology in the Upper Paraná River floodplain: large-scale spatial and temporal patterns, and the influence of reservoirs. Brazilian Journal of Biology 69: 717–725.

30. Nabout JC, Siqueira T, Bini LM, Nogueira ID (2009) No evidence for environmental and spatial processes in structuring phytoplankton communities. Acta Oecologica-International Journal of Ecology 35: 720–726.

31. Anderson MJ (2006) Distance-Based Tests for Homogeneity of Multivariate Dispersions. Biometrics 62: 245–253.

32. Baselga A (2010) Partitioning the turnover and nestedness components of beta diversity. Global Ecology and Biogeography 19: 134–143.

33. Legendre P, De Cáceres M (2013) Beta diversity as the variance of community data: dissimilarity coefficients and partitioning. Ecology Letters 16: 951–963.

34. R Core Team (2012) R: A language and environment for statistical computing. R Foundation for Statistical Computing, Vienna, Austria. ISBN 3-900051-07-0, Available: http://www.R-project.org/.

35. Oksanen J, Blanchet FG, Kindt R, Legendre P, Minchin PR, et al. (2012) Vegan: Community Ecology Package. R package version 20-4 Available: http://CRANR-projectorg/package=vegan.

36. Baselga A, Orme D, Villeger S (2013) betapart: Partitioning beta diversity into turnover and nestedness components. R package version 12 Available: http://CRANR-projectorg/package=betapart.

37. Bates D, Maechler M, Bolker B, Walker S (2012) lme4: Linear mixed-effects models using Eigen and S4 R package version 10-4 Available: http://CRANR-projectorg/package=lme4.

38. Fox J, Weisberg S (2011) An {R} Companion to Applied Regression, Second Edition. Thousand Oaks CA: Sage. Available: http://socserv.socsci.mcmaster.ca/jfox/Books/Companion.

39. Nakagawa S, Schielzeth H (2012) A general and simple method for obtaining $R^2$ from Generalized Linear Mixed-effects Models. Methods in Ecology and Evolution 4: 133–142.

40. Barton K (2014) MuMIn: Multi-model inference. R package version 1100 Available: http://CRANR-projectorg/package=MuMIn.

41. Pedruski MT, Arnott SE (2011) The effects of habitat connectivity and regional heterogeneity on artificial pond metacommunities. Oecologia 166: 221–228.

42. Bozelli RL (1992) Composition of the zooplankton community of Batata and Mussurá lakes and the Trombetas River, State of Pará, Brazil. Amazoniana 12: 239–261.

43. Simões NR, Dias JD, Leal CM, Braghin LSM, Lansac-Tôha FA, et al. (2013) Floods control the influence of environmental gradients on the diversity of zooplankton communities in a neotropical floodplain. Aquatic Sciences 75: 607–617.

44. Silver CA, Vamosi SM, Bayley SE (2012) Temporary and permanent wetland macroinvertebrate communities: Phylogenetic structure through time. Acta Oecologica 39: 1–10.

45. Lindo Z, Winchester NN, Didham RK (2008) Nested patterns of community assembly in the colonisation of artificial canopy habitats by oribatid mites. Oikos 117: 1856–1864.

46. Vanschoenwinkel B, Buschke F, Brendonck L (2013) Disturbance regime alters the impact of dispersal on alpha and beta diversity in a natural metacommunity. Ecology 94: 2547–2557.

47. Schroder T, Howard S, Arroyo ML, Walsh EJ (2007) Sexual reproduction and diapause of Hexarthra sp. (Rotifera) in short-lived ponds in the Chihuahuan Desert. Freshwater Biology 52: 1033–1042.

48. Gilbert JJ, Dieguez MC (2010) Low crowding threshold for induction of sexual reproduction and diapause in a Patagonian rotifer. Freshwater Biology 55: 1705–1718.

49. Smith HA, Snell TW (2012) Rapid evolution of sex frequency and dormancy as hydroperiod adaptations. Journal of Evolutionary Biology 25: 2501–2510.

50. Connell JH, Slatyer RO (1977) Mechanisms of succession in natural communities and their role in community stability and organization. American Naturalist 111: 1119–1144.

51. Lawler SP, Morin PJ (1993) Temporal overlap, competition, and priority effects in larval anurans. Ecology 74: 174–182.

52. Beaver RA (1977) Nonequilibrium island communities – Diptera breeding in dead snails. Journal of Animal Ecology 46: 783–798.

53. Hodge S, Arthur W, Mitchell P (1996) Effects of temporal priority on interspecific interactions and community development. Oikos 76: 350–358.

54. Connell JH (1978) Diversity in tropical rain forests and coral reefs. Science 199: 1302–1310.

55. Fox JW (2013) The intermediate disturbance hypothesis should be abandoned. Trends in Ecology & Evolution 28: 86–92.

56. Leibold MA, Holyoak M, Mouquet N, Amarasekare P, Chase JM, et al. (2004) The metacommunity concept: a framework for multi-scale community ecology. Ecology Letters 7: 601–613.

57. Van der Gucht K, Cottenie K, Muylaert K, Vloemans N, Cousin S, et al. (2007) The power of species sorting: Local factors drive bacterial community composition over a wide range of spatial scales. Proceedings of the National Academy of Sciences of the United States of America 104: 20404–20409.

58. Farjalla VF, Srivastava DS, Marino NAC, Azevedo FD, Dib V, et al. (2012) Ecological determinism increases with organism size. Ecology 93: 1752–1759.

59. Melo AS, Rangel TFLVB, Diniz-Filho JAF (2009) Environmental drivers of beta-diversity patterns in New-World birds and mammals. Ecography 32: 226–236.

60. Mysák J, Horsák M (2011) Floodplain corridor and slope effects on land mollusc distribution patterns in a riverine valley. Acta Oecologica 37: 146–154.

61. Condit R, Pitman N, Leigh EG Jr, Chave J, Terborgh J, et al. (2002) Beta-Diversity in Tropical Forest Trees. Science 295: 666–669.

62. Karp DS, Rominger AJ, Zook J, Ranganathan J, Ehrlich PR, et al. (2012) Intensive agriculture erodes b-diversity at large scales. Ecology Letters 15: 963–970.

63. McKnight MW, White PS, McDonald RI, Lamoreux JF, Sechrest W, et al. (2007) Putting beta-diversity on the map: Broad-scale congruence and coincidence in the extremes. Plos Biology 5: 2424–2432.

64. Bini LM, Landeiro VL, Padial AA, Siqueira T, Heino J (2014) Nutrient enrichment is related to two facets of beta diversity of stream invertebrates across the continental US. Ecology 95: 1569–1578.

65. Hubbell SP (2001) The Unified Neutral Theory of Biodiversity and Biogeography. Princeton, NJ: Princeton University Press.

66. Havel JE, Shurin JB (2004) Mechanisms, effects, and scales of dispersal in freshwater zooplankton. Limnology and Oceanography 49: 1229–1238.

67. Soininen J, Kokocinski M, Estlander S, Kotanen J, Heino J (2007) Neutrality, niches, and determinants of plankton metacommunity structure across boreal wetland ponds. Ecoscience 14: 146–154.

68. Shurin JB, Cottenie K, Hillebrand H (2009) Spatial autocorrelation and dispersal limitation in freshwater organisms. Oecologia 159: 151–159.

69. Gray DK, Arnott SE (2012) The role of dispersal levels, Allee effects and community resistance as zooplankton communities respond to environmental change. Journal of Applied Ecology 49: 1216–1224.

70. Beisner BE, Peres PR, Lindstrom ES, Barnett A, Longhi ML (2006) The role of environmental and spatial processes in structuring lake communities from bacteria to fish. Ecology 87: 2985–2991.

71. Steiner CF. Stochastic sequential dispersal and nutrient enrichment drive beta diversity in space and time. Ecology: in press.

72. Pinto-Coelho RMP, Pinel-Alloul B, Methot G, Havens K (2005) Crustacean zooplankton in lakes and reservoirs of temperate and tropical regions: variations with trophic status. Canadian Journal of Fisheries and Aquatic Sciences 62: 348–361.

73. Sousa W, Attayde JL, Rocha EDS, Eskinazi-Sant'Anna EM (2008) The response of zooplankton assemblages to variations in the water quality of four man-made lakes in semi-arid northeastern Brazil. Journal of Plankton Research 30: 699–708.

# 12

# New Insights into Phosphorus Mobilisation from Sulphur-Rich Sediments: Time-Dependent Effects of Salinisation

Josepha M. H. van Diggelen[1,2]*, Leon P. M. Lamers[2], Gijs van Dijk[1,2], Maarten J. Schaafsma[1¤], Jan G. M. Roelofs[2], Alfons J. P. Smolders[1,2]

1 B-WARE Research Centre, Radboud University Nijmegen, Mercator 3, Nijmegen, The Netherlands, 2 Institute for Water and Wetland Research, Department of Aquatic Ecology and Environmental Biology, Radboud University Nijmegen, Nijmegen, The Netherlands

## Abstract

Internal phosphorus (P) mobilisation from aquatic sediments is an important process adding to eutrophication problems in wetlands. Salinisation, a fast growing global problem, is thought to affect P behaviour. Although several studies have addressed the effects of salinisation, interactions between salinity changes and nutrient cycling in freshwater systems are not fully understood. To tackle eutrophication, a clear understanding of the interacting effects of sediment characteristics and surface water quality is vital. In the present study, P release from two eutrophic sediments, both characterized by high pore water P and very low pore water iron ($Fe^{2+}$) concentrations, was studied in a long-term aquarium experiment, using three salinity levels. Sediment P release was expected to be mainly driven by diffusion, due to the eutrophic conditions and low iron availability. Unexpectedly, this only seemed to be the driving mechanism in the short term (0–10 weeks). In the long term (>80 weeks), P mobilisation was absent in most treatments. This can most likely be explained by the oxidation of the sediment-water interface where $Fe^{2+}$ immobilises P, even though it is commonly assumed that free $Fe^{2+}$ concentrations need to be higher for this. Therefore, a controlling mechanism is suggested in which the partial oxidation of iron-sulphides in the sediment plays a key role, releasing extra $Fe^{2+}$ at the sediment-water interface. Although salinisation was shown to lower short-term P mobilisation as a result of increased calcium concentrations, it may increase long-term P mobilisation by the interactions between sulphate reduction and oxygen availability. Our study showed time-dependent responses of sediment P mobilisation in relation to salinity, suggesting that sulphur plays an important role in the release of P from $FeS_x$-rich sediments, its biogeochemical effect depending on the availability of $Fe^{2+}$ and $O_2$.

**Editor:** Todd Miller, University of Wisconsin Milwaukee, United States of America

**Funding:** This study was part of the National Research Programme "Wormer- en Jisperwater," funded by the Dutch Ministry of Agriculture, Nature and Food Quality (LNV), within the framework of "Nota Ruimte." The water management authority "Hoogheemraadschap Hollands Noorderkwartier" facilitated this programme. The authors did not require further external funding source for this study. The funders had no role in study design, data collection and analysis, decision to publish, or preparation of the manuscript.

**Competing Interests:** The authors have declared that no competing interests exist.

* Email: J.vanDiggelen@b-ware.eu

¤ Current address: Royal Haskoning DHV, Nijmegen, The Netherlands

## Introduction

The eutrophication of surface waters is an urgent problem worldwide [1]. Increased P concentrations have led to a strong decline of the biodiversity in freshwater wetlands, due to the resulting dominance of highly competitive macrophytes, and of algae and cyanobacteria, monopolising light [1–3]. Salinisation of freshwater systems has received increasing attention, especially in relation to climate change and sea level rise [4]. With increasing salinity, higher P concentrations are often found in surface waters (e.g. [5–8]), which may affect P cycling in freshwater systems. Therefore, salinisation is expected to enhance eutrophication in coastal, freshwater wetlands, leading to water quality deterioration and loss of biodiversity.

Internal mobilisation of P from eutrophic aquatic sediments is an important process adding to eutrophication problems in wetlands [9–12]. The classic theoretical framework suggests that sufficiently high oxygen ($O_2$) concentrations in the surface water can prevent P release from the sediment [13–14]. According to this, the oxidation of dissolved iron ($Fe^{2+}$) in the sediment will result in the formation of iron oxides and hydroxides ($Fe(OH)_x$) at the sediment surface, effectively binding P and thereby preventing its release to the surface water. Under anaerobic conditions, these ferric compounds will be mobilised by Fe-reducing bacteria, and part of the P is released to the surface water.

Besides anaerobic conditions, increased sulphate ($SO_4^{2-}$) reduction rates are also known to be able to increase P mobilisation by decoupling Fe - P interactions at the sediment-water interface [12,15–19]. Sulphide ($S^{2-}$) binds efficiently to dissolved $Fe^{2+}$ in sediment pore water, and most $Fe^{2+}$ can become bound as iron sulphides ($FeS_x$) in the sediment, strongly decreasing $Fe^{2+}$ sediment pore water concentrations [18,19]. Geurts et al. [19] found that, in aerobic surface waters, P mobilisation from sediments with low pore water Fe:P ratios ($<1$ mol $mol^{-1}$) was a linear function of sediment pore water P concentrations. As a result, one would expect a release of P irrespective of the $O_2$ concentration in the surface water of $SO_4^{2-}$ enriched wetlands [10,12,18,20]. In addition, dissolved P concentrations might further increase due to the enhanced anaerobic breakdown of organic matter linked to $SO_4^{2-}$ reduction and concomitant mineralisation of P [2,12,19].

Salinisation of freshwater systems can enhance $SO_4^{2-}$ reduction rates due to a higher $SO_4^{2-}$ availability [21], which may strongly affect P mobilisation as described above. Moreover, increasing $Cl^-$ and $SO_4^{2-}$ concentrations might enhance P release from sediments by competition for anion binding sites [10,22]. At the same time, an increase in salinity also leads to increased $Ca^{2+}$ concentrations [21], which may result in the immobilisation of P by co-precipitation with $Ca^{2+}$ and calcium carbonate ($CaCO_3$) [4,9,23]. Salinity changes affect a suite of biogeochemical processes in freshwater systems, where the net effect on P mobilisation is the combined result of these processes. Moreover, a time-dependent shift in dominance of each process on P release can be expected [18]. Most studies regarding P release focus on relative short-term effects ranging from one day to 90 days [6,19,20,24], while long-term experiments are mostly lacking. In this paper we explore the time-dependent release of P from eutrophic sediments under different salinities, which is highly relevant regarding the worldwide interest in salinisation effects on freshwater wetland functioning.

To test time-dependent interactions between salinisation and P mobilisation, a controlled aquarium experiment was set up that lasted two years. Two $FeS_x$-rich sediments from a coastal freshwater wetland were subjected to three naturally occurring water types characterised by different salinities. Pore waters of the peat sediments were typically rich in P and S, and very poor in Fe, and the low total Fe:S ratios in the sediment suggested that most Fe was bound to reduced S [2]. In such sediments, a very high release of P from the sediment to the surface water can be expected, predominantly depending on pore water P concentrations [10,12,18-20]. By monitoring biogeochemical changes in porewater and surface water under controlled conditions, we try to reveal how salinity affects short-term and long-term P release, in these type of sediments common for coastal wetlands.

## Materials and Methods

### 2.1 Sampling area

In this study, peat sediments were used from the coastal lowland fen area Wormer- and Jisperveld ($52° 30' 42.7644''$; $4° 52' 27.3756''$) in the Netherlands. Due to historic intrusion of brackish water, peat rich in minerals such as S, Ca and Fe has accumulated in this area. After more than 50 years of desalinisation resulting from altered hydrological conditions, it gradually became a freshwater system. The peatland comprises ca. 500 ha of open water and ca. 1660 ha of peat meadows, predominantly used for agricultural purposes and partly for nature conservation. Drainage is a standard procedure in this area, leading to peat decomposition and land subsidence. As a result, risks of flooding events and salinisation are increasing in this freshwater peatland.

### 2.2 Experimental design

On 18 March 2008, two types of submerged peat sediment were collected from a ditch at a depth of 0–20 cm (ca. 25 L in total), using a sediment multi sampler (Eijkelkamp Agrisearch Equipment). Although both sediments were relatively rich in organic S and P, they differed in P availability (sediment characteristics are given in Table 1). To minimise $O_2$ intrusion, the sediments were

stored anaerobically at $4°C$ in large, closed containers. The next day, 12 glass cylinders (diameter 15 cm, height 60 cm) were filled with 15 cm of sediment A and another 12 cylinders with 15 cm of sediment B. Next 40 cm of water was carefully poured on top of the sediments, avoiding re-suspension of sediment particles. Artificially composed surface water, based on site conditions (control treatment, Table 2), was used for all sediments during an acclimatisation period of 4 weeks. The experiment was carried out in the dark at a constant and environmentally relevant temperature of $15°C$. To allow oxygen diffusion to the surface water, an open cylinder system was used.

After this acclimatisation period, three different surface water types were applied as salinity treatments: rainwater (low salinity; 100 μmol $Cl$ $L^{-1}$), brackish water (high salinity; 85 mmol $Cl$ $L^{-1}$) and freshwater (control; 7 mmol $Cl$ $L^{-1}$). All treatments were artificially composed, based on field measurements (Table 2). Control water simulated water quality in the current conditions that exist in the wetland, brackish water composition was based on the historic conditions reported by Reigersman in 1946 [25]. Rainwater quality equalled the chemical composition of atmospheric deposition as measured in the Netherlands [26]. No P was added in order to be able to estimate the release of P from the sediment. For each treatment and sediment type, 4 replicates were used (24 cylinders in total).

Treatment solutions were stored in polyethylene containers (10 L), from which they were pumped into the cylinders using Masterflex L/S multichannel pumps (model 7535-08). The treatments were started by replacing the control water with the appropriate treatment water during 4 weeks, to ensure that all treatment solutions were added properly. Directly after treatment addition (week 10), stagnant conditions were created in order to measure short-term effects of P and S release from the sediment. Short-term mobilisation rates were calculated from the linear increase of the surface water P and S concentrations (0–10 weeks of the stagnant period). After a stagnant period of 26 weeks, pumps were running with a hydraulic retention time of 25 weeks for the treatment solutions during 48 weeks in order to maintain the appropriate treatment conditions. To measure long-term effects of salinity changes on the release of P and S from the sediment, pumps were stopped again (week 81) to create another stagnant period for 32 weeks. Long-term mobilisation rates were again calculated from the linear increase of the S and P concentrations in the surface water during this stagnant period.

Intact peat cores from the same location as the main experiment were collected separately, to test the effects of aerobic versus anaerobic conditions of the surface water on P release. Water and sediment oxygen ($O_2$) profiles were measured, using a fixed fiber optical oxygen microsensor (optode) in combination with a Microx TX3 transmitter (PreSens Precision Sensing GmbH). The peat sediment cores were monitored during 18 weeks of either aerobic conditions similar to those of the main experiment, or anaerobic conditions by gently supplying $N_2$ to the surface water. During both aerobic and anaerobic conditions, P mobilisation rates were calculated from the linear increase in surface water P concentrations.

### 2.3 Chemical analyses

To monitor water quality, samples of surface water and pore water were collected every 2 months and analysed during the experiment. Pore water was collected anaerobically, using 30 mL vacuum bottles connected to Rhizon SMS-10 cm samplers that were fixed in the upper 10 cm of the sediment (Eijkelkamp Agrisearch Equipment). Disturbance of the sediment and water was minimised by the low frequency of sampling and small sample

**Table 1.** Characteristics of the two sediments used.

| Sediment | Organic content | Bulk Density | Total amounts bound to sediment | | | | | | | |
| | % | kg DW L⁻¹ FW | Total - P mmol L⁻¹ FW | Org - P mmol L⁻¹ FW | Inorg - P mmol L⁻¹ FW | Total - Al mmol L⁻¹ FW | Total - Ca mmol L⁻¹ FW | Total - S mmol L⁻¹ FW | Total - Fe mmol L⁻¹ FW | Fe:S ratio mol mol⁻¹ |
|---|---|---|---|---|---|---|---|---|---|---|
| A Mean | 59.7 | 0.17 | 2.9[a] | 1.8 | 1.2[a] | 44.5 | 54.9 | 99.3 | 33.1 | 0.33[a] |
| SEM | 4.8 | 0.03 | 0.4 | 0.3 | 0.1 | 12.6 | 5.4 | 8.1 | 3.9 | 0.024 |
| B Mean | 59.6 | 0.13 | 4.7[b] | 2.4 | 2.3[b] | 43.4 | 44.0 | 83.6 | 38.3 | 0.46[b] |
| SEM | 7.4 | 0.02 | 0.3 | 0.1 | 0.1 | 4.4 | 0.8 | 4.7 | 2.3 | 0.001 |

Significant differences between the sediment types are indicated by different letters.

sizes (max. 25 mL). Sulphide concentrations were determined directly after the collection by fixing 10.5 mL pore water with 10.5 mL Sulphide Anti Oxidant Buffer (SAOB), and using an Orion sulphide-electrode and a Consort Ion meter (type C830) [27]. The pH and alkalinity of all samples were measured within 24 hours after sampling, using a combined pH electrode (Radiometer) in combination with a TIM840 pH meter and a Titration Manager Titralab Autoburette. Dissolved total inorganic carbon (TIC) was measured within 24 hours after sampling by injecting 0.2 mL pore water or surface water in a closed chamber containing 0.2 M $H_3PO_4$ solution, converting all dissolved TIC into $CO_2$. A continues gas flow ($N_2$) directly transports the $CO_2$ to an ABB Advance optima Infrared Gas Analyzer (IRGA) to measure total inorganic C concentrations. A calibration curve was made by injecting different volumes (0.1–1.0 mL) of 1.25 mM $HCO_3^-$ solution. Prior to storage at 4°C until elemental analysis, 0.1 mL $HNO_3^-$ (65%) was added to 10 mL of each sample to prevent metal precipitation. Concentrations of dissolved Ca, Fe, P, S, and Al in these stored samples were measured using an Inductively Coupled Plasma Spectrophotometer (ICP IRIS Intrepid II XDL; Thermo Electron Corporation). Due to the anaerobic sampling of pore water, measured Fe predominantly consisted of dissolved $Fe^{2+}$ rather than far less mobile $Fe^{3+}$. The remaining samples were stored at −20°C in order to determine the following ion concentrations colourimetrically on Auto Analyzer 3 systems (Bran and Luebbe): $NO_3^-$ [28], $NH_4^+$ [29], ortho-$PO_4^{3-}$ [30] and $Cl^-$ [31]. $Na^+$ and $K^+$ were determined with a Technicon Flame Photometer IV Control (Technicon Corporation).

For both sediments gravimetric water contents were determined by drying for 48 h at 70°C. Organic matter contents were estimated by loss on ignition for 4 h at 550°C. A homogenized portion of 200 mg dry sediment was digested in 5 mL $HNO_3$ (65%) and 2 mL $H_2O_2$ (30%), using an Ethos 1 Advanced microwave digestion system (Milestone Inc.). Digestates were diluted and analysed by ICP as described above. In order to distinguish between the organic and inorganic P fraction, a P-fractionation procedure was carried out adapted after Golterman [32].

## 2.4 Statistical analyses

For statistical analysis, SPSS Statistics for Windows (Version 21.0. IBM Corp. Armonk, NY; 2012) was used. To test for differences among treatments in sediment analyses (single measurements) or differences in calculated mobilisation rates, the General Linear Model (GLM) univariate procedure combined with Tukey's-b post-hoc test was used.

To test for significant differences among treatments in repeated measurements, a GLM mixed model procedure was used. When significant differences between the two sediments were found, using a 2-way GLM mixed model with treatment as fixed factor, sediment as random factor and time as repeated measures, both sediments were analysed separately. In this separate model for sediments, time was used as repeated measures and treatment as fixed factor, with AR(1) heterogeneous as the covariance type. A Bonferroni post-hoc test was used to test for differences between treatments.

## 2.5 Ethics statement

This study was part of the National Research Programme 'Wormer- en Jisperwater', funded by the Dutch Ministry of Agriculture, Nature and Food Quality (LNV), within the framework of 'Nota Ruimte'. The water management authority 'Hoogheemraadschap Hollands Noorderkwartier' facilitated this

**Table 2.** Chemical composition of the surface water used for the different treatments (low, normal or high salinity).

| Element | Low salinity | Normal salinity | High salinity |
| | Rain water | Fresh water | Brackish water |
| | $\mu$mol L$^{-1}$ | $\mu$mol L$^{-1}$ | $\mu$mol L$^{-1}$ |
|---|---|---|---|
| Na$^+$ | 100 | 7000 | 85000 |
| Cl$^-$ | 100 | 7000 | 85000 |
| SO$_4^{2-}$ | 5 | 1500 | 5500 |
| K$^+$ | 30 | 500 | 1000 |
| Ca$^{2+}$ | 10 | 2000 | 2500 |
| Mg$^{2+}$ | 10 | 1250 | 3750 |
| HCO$_3^-$ | 0 | 4000 | 4000 |
| NO$_3^-$ | 50 | 50 | 50 |
| NH$_4^+$ | 50 | 50 | 50 |

programme and the nature management authority 'Natuurmonumenten' gave permission to take samples in their reserve.

## Results

### 3.1 Pore water chemistry

As expected, pore water chemistry was strongly affected by changes in surface water salinity (Fig. 1). Under brackish conditions, Na$^+$ showed a highly significant ($p<0.005$) gradual increase in the pore water over time. For the low salinity and control treatment, no significant changes in pore water Na$^+$ concentrations occurred in sediment A, while Na$^+$ concentrations showed a significant decrease ($p<0.05$) over time at a low salinity in sediment B. An interaction between treatment and sediment type was found for both Na$^+$ and S concentrations, which means that the treatments had a significant, but different, effect on the two sediments. Pore water S concentrations also showed a highly significant ($p<0.005$) gradual increase (Fig. 1). At a low salinity, S concentrations remained at a steady level while the control treatment showed a small, but not significant, increase. Moreover, no clear differences in sulphide concentrations were found between sediments or treatments (average values ranged between 0–50 $\mu$mol L$^{-1}$ for sediment A, and between 0–500 $\mu$mol L$^{-1}$ for sediment B; data not shown).

As a result of a higher salinity, Ca$^{2+}$ was mobilised in the sediment, as shown by significantly ($p<0.005$) increased pore water Ca$^{2+}$ concentrations (Fig. 1). This increase in the pore water, well above the added concentration of 2500 $\mu$mol L$^{-1}$ to the surface water, started directly after the onset of the high salinity treatment and Ca$^{2+}$ concentrations remained at a steady high level during the course of the experiment.

Dissolved Fe$^{2+}$ concentrations were low and showed a gradual decrease for all treatments over time in both sediment types (Fig. 2). A significantly higher ($p<0.005$) Fe$^{2+}$ concentration was found for the low salinity treatment at sediment A when compared to the control and higher salinity treatment. In contrast, no differences in pore water Fe$^{2+}$ concentrations between treatments were found for sediment B. Pore water HCO$_3^-$ concentrations were significantly higher ($p<0.05$) under brackish conditions at sediment B, while no differences were found between treatments at sediment A (data not shown).

Pore water P concentrations showed a gradual decrease at sediment A for all treatments. Moreover, a significantly ($p<0.05$) stronger decrease of P in pore water was found for the control treatment, compared to the high and low salinity treatment at sediment A. In strong contrast, P concentrations showed a gradual increase in the pore water of sediment B for all salinity treatments (Fig. 2), with the significantly ($p<0.05$) lowest P concentrations in the high salinity treatment.

### 3.2 Surface water chemistry

A higher salinity led to gradually increased Na$^+$ and S concentrations in the surface water, and showed significant ($p<0.005$) differences among all treatments, which eventually equalled the concentrations added (S: Fig. 3; Na$^+$: data not shown). For the low salinity treatment, however, S concentrations in the surface water reached much higher concentrations than the concentrations of the treatment water, which suggests S mobilisation from the sediment. These S mobilisation rates were calculated (Table 3) for both a short term, showing significantly ($p<0.005$) negative rates at a high salinity (high S consumption) for sediment A, and for a long term, still showing significantly ($p<0.005$) negative S mobilisation rates at a high salinity in both sediments. In the surface water, Ca$^{2+}$ concentrations also increased and differed significantly ($p<0.005$) among all treatments for both sediments (data not shown). However, both the low and high salinity treatment led to much higher concentrations than the added concentrations.

In the low salinity and control treatment, P concentrations in the surface water increased directly after onset of the treatments (after 10 weeks; t = 0). For the high salinity treatments, P concentrations of the surface water showed a strong and significant ($P<0.05$) decrease immediately after the onset of the treatments (after 10 weeks; t = 0). After this temporary decrease, P concentrations started to increase gradually. As a result, significantly lower P concentrations ($p<0.05$) were found for the high salinity treatment compared to the low salinity treatment in both sediments at a short term (after 20 weeks; t = 10), and a trend was found when compared to the control treatment ($p<0.1$) at sediment A. When P mobilisation rates were calculated for the short term, however, no differences among salinity treatments were found (Table 3).

More than 80 weeks after the start of the experiment, P concentrations in the surface water above sediment A were

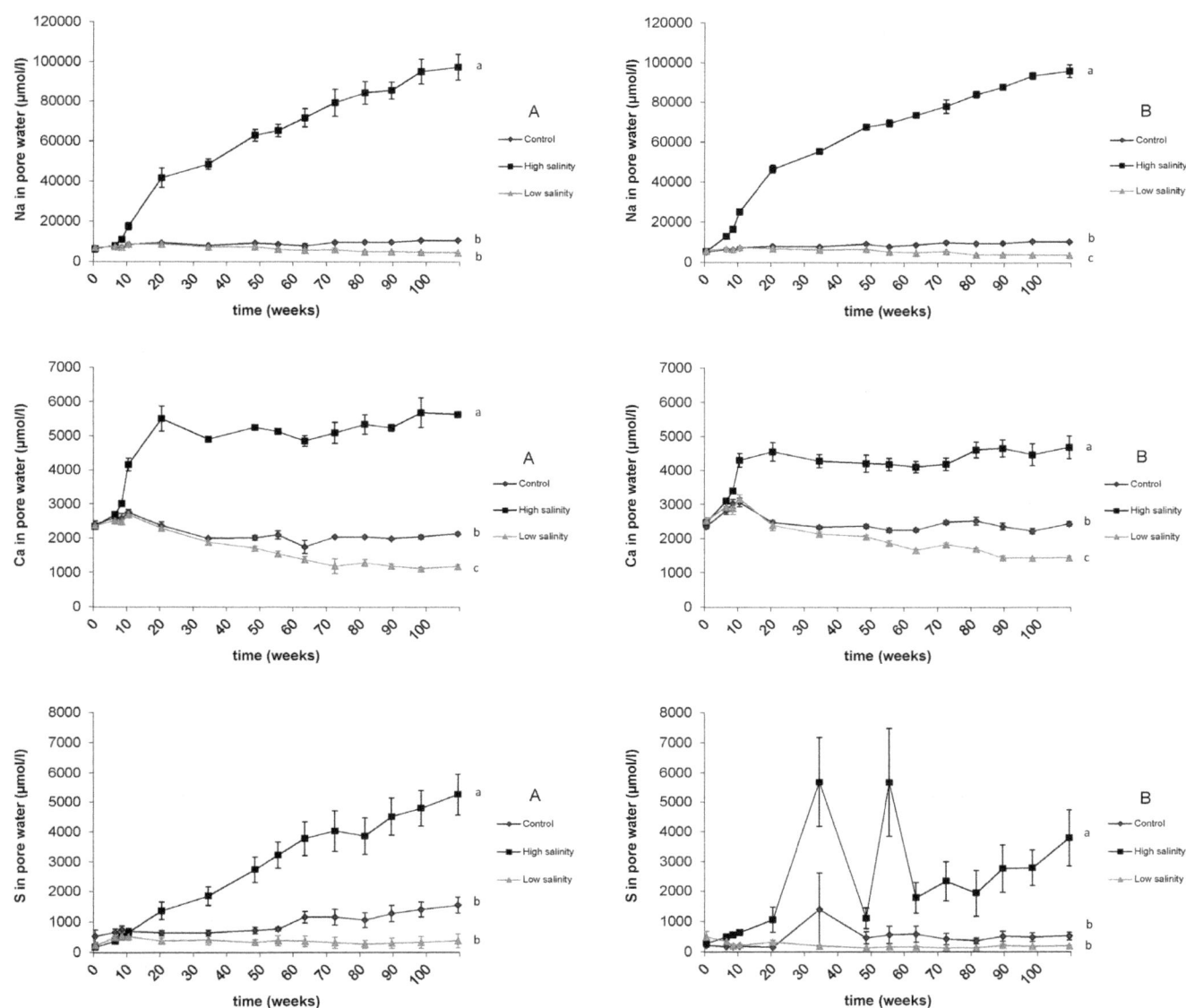

**Figure 1. Sodium (Na⁺), calcium (Ca²⁺) and sulphur (S) pore water concentrations ($\mu$mol L$^{-1}$) in both sediments (A: left, B: right).** Significant differences between treatments are indicated with different letters.

significantly higher ($p<0.01$) at a high salinity (Fig. 3), which was totally opposite to the short-term effect. Calculated P mobilisation rates were also significantly higher ($p<0.05$) with a high salinity compared to a low salinity at sediment A. For the control and low salinity treatment, P concentrations in surface water remained low, or even showed a decrease in the long term. At sediment B, however, no change of P in the surface water was found for any of the salinity treatments. The long-term P mobilisation rates with a high salinity were similar to the short-term rates at sediment A, while no long-term P mobilisation was observed for the low salinity and control treatments.

### 3.3 Aerobic versus anaerobic surface water

The O₂ concentration profile (Fig. 4) shows that under aerobic conditions, O₂ is still available in the sediment to an average depth of 7 mm (sediment A) and 3 mm (sediment B). The cores of both sediment A and B showed a significant ($p<0.001$) higher

mobilisation rate of P during anaerobic conditions (Fig. 5). At sediment A, P mobilisation was on average 3 times higher during anaerobic conditions compared to aerobic conditions, while this was almost 4 times higher at sediment B. These aerobic mobilisation rates were well within range of the short-term mobilisation rates found in the main experiment (control treatment; Table 3).

### Discussion

#### 4.1 Short-term effects (0–10 weeks)

**4.1.1 P mobilisation.** In the short term, no differences in net mobilisation rates of P were found among the different treatments. During this first stagnant period, moderate P mobilisation rates of 7–103 $\mu$mol m$^{-2}$ d$^{-1}$ were found that fitted within the range of Geurts et al. [19], who found mobilisation rates of 10–150 $\mu$mol m$^{-2}$ d$^{-1}$ for sediments of which pore water Fe:P and total sediment Fe:S ratios were <1. Diffusion was most likely the main

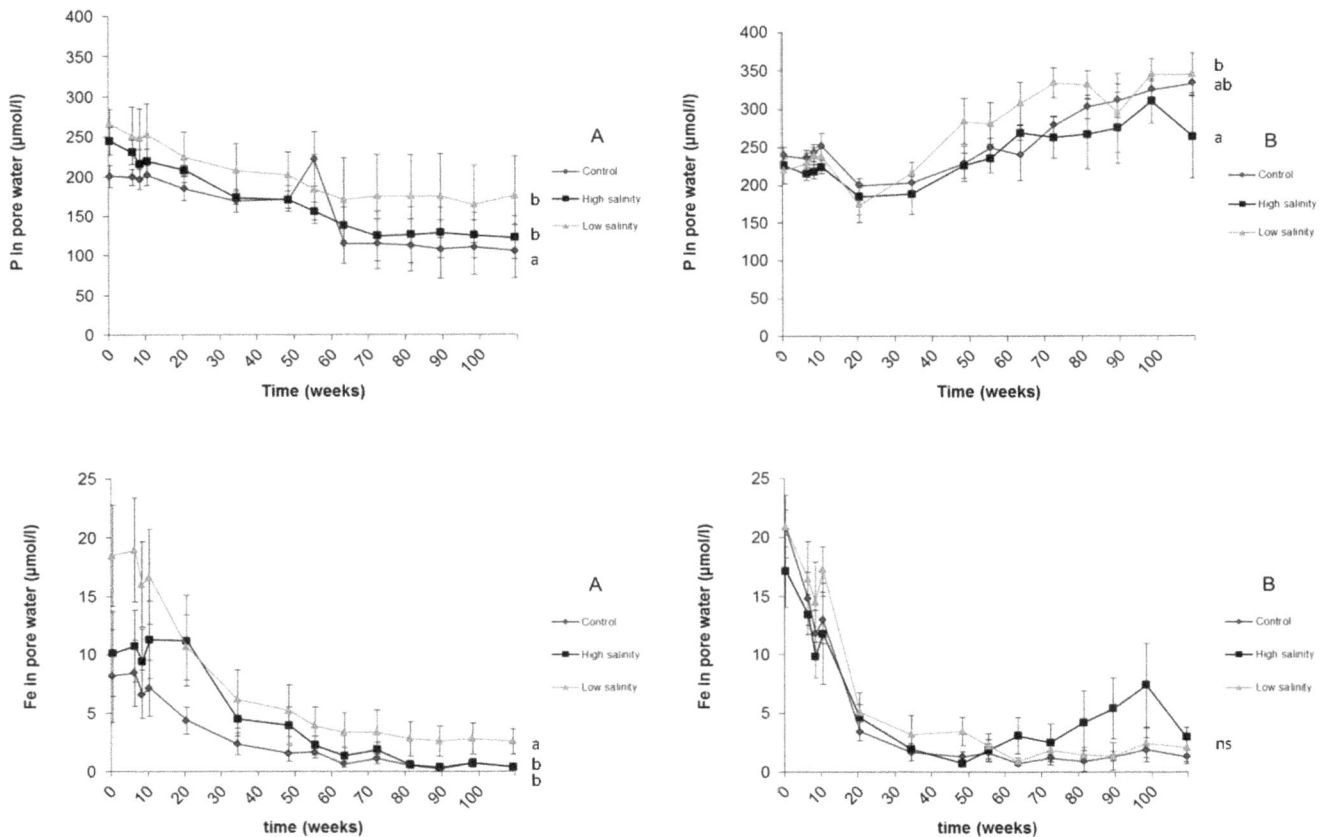

**Figure 2. Phosphorus (P) and iron (Fe$^{2+}$) concentrations (µmol L$^{-1}$) in pore water of both sediments (A: left, B: right).** Significant differences between treatments are indicated with different letters.

mechanism driving P release [19,33], since the sediments used in this experiment were not subjected to bioturbation or resuspension [7], nor to a changed pH or temperature [24]. Moreover, both sediments were characterised by total Fe:S ratios below 0.5 (Table 1), which indicates that most Fe was bound to reduced S [20]. Indeed, dissolved pore water Fe$^{2+}$ concentrations were low in this study (and ranged between 0–20 µmol L$^{-1}$ for both sediments), and showed an even further decrease over time, resulting in very low pore water Fe:P ratios (<0.1) during the entire experimental period.

**4.1.2 Salinity effects.** Although increased salinity may lead to increased desorption of P from anion exchange sites [10], or by increased S$^{2-}$ production and enhanced mineralisation rates [2,12], we did not find higher pore water P concentrations in the high salinity treatment. Instead, during the addition of the salinity treatments (between week 6 and 10), P concentrations in the surface water showed a short, strong drop for both sediments. This immediate drop of P observed upon a change of the surface water chemistry strongly points at a chemical, rather than a microbiological, explanation. It can most likely be explained by the co-precipitation of P with Ca$^{2+}$ or CaCO$_3$ at the sediment-water interface [4,9], as Ca$^{2+}$ concentrations directly and strongly increased in both surface and pore water upon the high salinity treatment (0–10 weeks). Accordingly, Suzumura *et al.* [7] found a fast chemical P (im)mobilisation response within minutes, due to adsorption-desorption processes after a changed salinity. Van Dijk *et al.* [34] found a similar immobilisation of P with increased salinity, explained by co-precipitation with Ca$^{2+}$ in the sediment.

Degassing of carbon dioxide (CO$_2$) and possibly also the presence of microbial mats [35] may well have contributed to the precipitation of CaCO$_3$ at the sediment surface, as HCO$_3^-$ concentrations were up to three times higher in pore water than in the surface water. After the initial drop of P, concentrations started to gradually increase, which shows that the short-term overall net P mobilisation to the surface water was higher than its immobilisation due to co-precipitation with Ca$^{2+}$.

## 4.2 Long-term effects (1.5–2 years)

**4.2.1 P mobilisation.** In contrast to the short-term results, and rather unexpectedly for eutrophic sediments, P mobilisation to the surface water was absent in 5 out of 6 treatments in the longer term (after 80 weeks). This is remarkable, as a strong net diffusive P release in both sediments was expected given the very low pore water Fe$^{2+}$ concentrations and the still very high pore water P concentrations [19]. Although a gradual decrease of P in the pore water of sediment A was observed, concentrations still remained sufficiently high for diffusive P release (>100 µmol L$^{-1}$) [19,20]. Sediment B even showed a gradual increase of pore water P concentrations during the experiment, without any increase of the P mobilisation to the surface water. Such results can only be explained by assuming that processes preventing net P release at the sediment-water interface become active in the long term, at least under the conditions that were created during our experiment. Possible explanations for this phenomenon are: (1) precipitation of P with Fe$^{3+}$ or Fe(OH)$_x$ by the oxidation of the sediment surface [13,14], (2) storage of P by the microbial

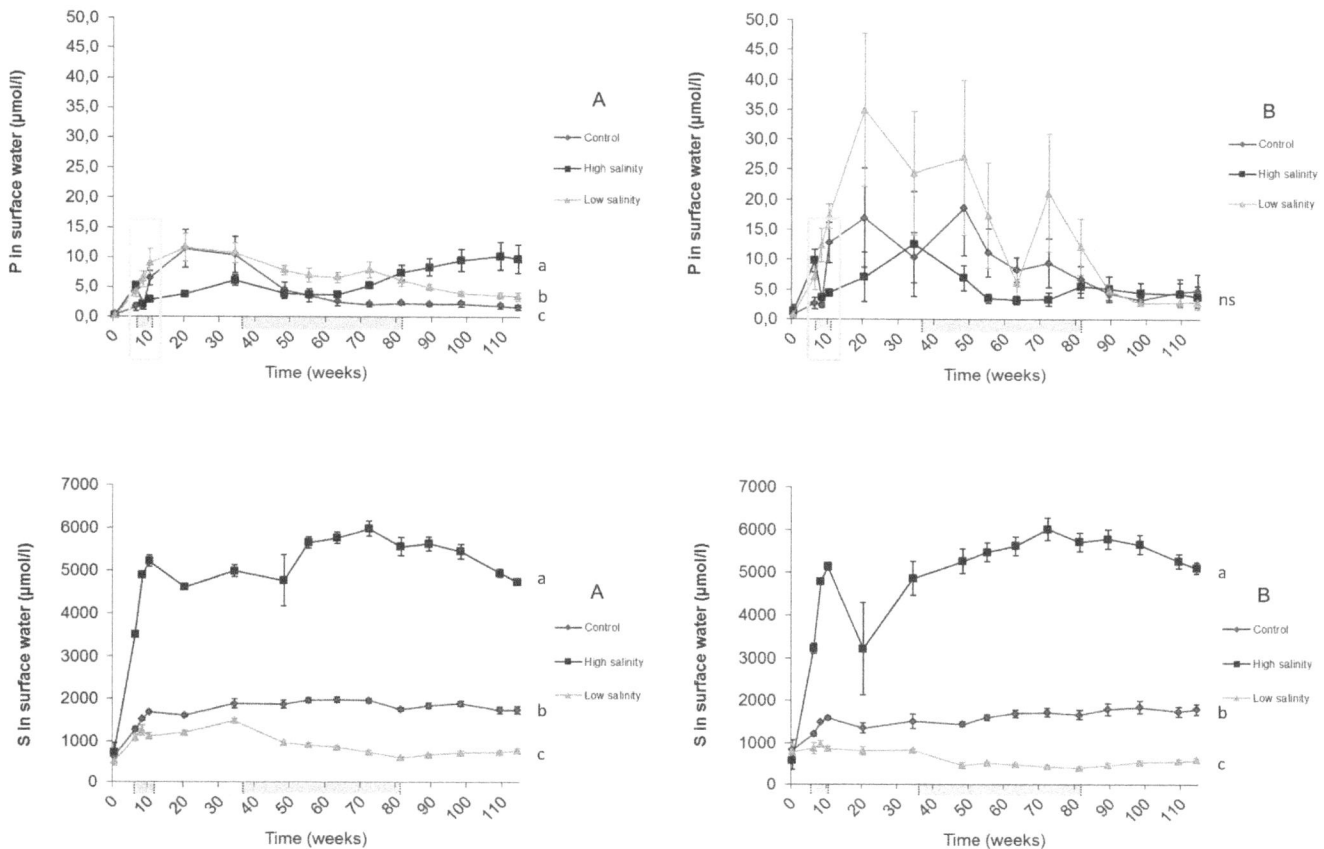

**Figure 3. Phosphorus (P) and sulphur (S) concentrations (µmol L$^{-1}$) in the surface water above both sediments (A: left, B: right).** Significant differences between treatments are indicated with different letters. Grey shadings under the x-axis indicate periods with through-flow (see Materials and Methods).

**Table 3.** P and S mobilisation rates (µmol m$^{-2}$ day$^{-1}$) during stagnant conditions in the short term (0–10 weeks) and in the long term (80–110 weeks).

| | | | P mobilisation (µmol m$^{-2}$ day$^{-1}$) | | S mobilisation (µmol m$^{-2}$ day$^{-1}$) | |
|---|---|---|---|---|---|---|
| | | | short term | long term | short term | long term |
| A | High salinity | Mean | 6.8 | 4.1 [a] | −1916.4 [a] | −1447.5 [a] |
| | | SEM | 1.6 | 2.4 | 86.7 | 243.0 |
| | Control | Mean | 37.7 | −1.1 [ab] | 140.9 [b] | −87.0 [b] |
| | | SEM | 14.3 | 0.6 | 216.0 | 86.4 |
| | Low salinity | Mean | 19.8 | −4.0 [b] | −6.5 [b] | 244.2 [b] |
| | | SEM | 12.9 | 2.3 | 425.0 | 69.2 |
| B | High salinity | Mean | 16.3 | −2.9 | −8536.5 | −1121.7 [a] |
| | | SEM | 24.2 | 1.0 | 5474.0 | 141.3 |
| | Control | Mean | 53.3 | −2.2 | −840.4 | 123.6 [b] |
| | | SEM | 36.6 | 2.5 | 517.8 | 91.1 |
| | Low salinity | Mean | 102.6 | −13.1 | −597.4 | 294.8 [b] |
| | | SEM | 74.3 | 6.5 | 236.3 | 83.3 |

Significant differences between treatments are indicated by different letters.

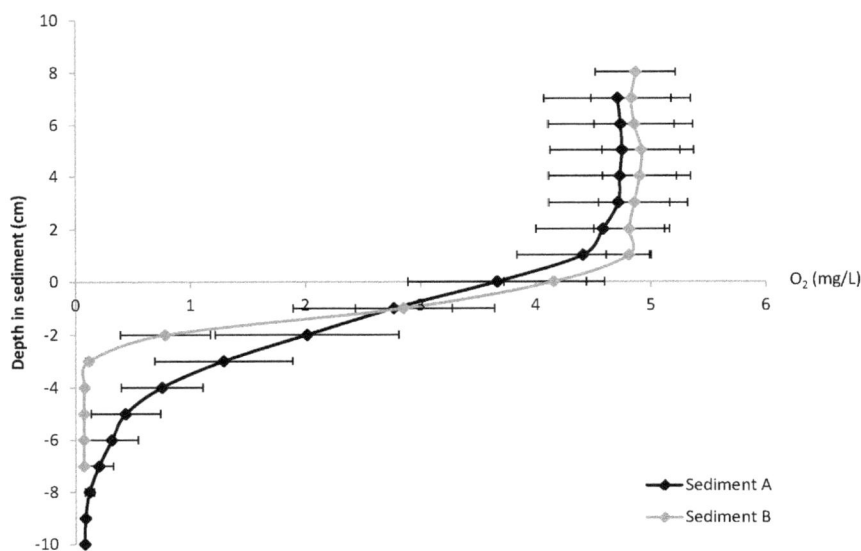

**Figure 4. Oxygen (O₂) concentration (mg L⁻¹) profile per mm of both sediments (A and B), at the sediment-water interface (indicated by vertical dotted line) during aerobic and anaerobic conditions.**

community at the sediment surface during aerobic conditions [36,37], (3) precipitation of P with calcium-minerals [9,23], although the latter would mainly be expected in the high salinity treatment.

An explanation for the lack of P release in the long term might be the uptake of P by microbial mats growing on top of the sediment [35–37]. These mats can develop over time and might also benefit from stable sediment conditions that developed in the experimental set-up. However, our experiment was carried out in the dark, excluding photosynthetically active organisms, and no visible signs of such mats were observed. Nevertheless, the potential role of microbial sequestration of P on the long term cannot be ruled out.

Most likely Fe redox cycling played a dominant role in the absence of P mobilisation, as was also indicated by the strongly increased P release under anaerobic conditions compared to aerobic conditions (Fig. 5). It has been demonstrated that diffusive

P release should be prevented under aerobic conditions if pore water Fe:P ratios are relatively high (at least >1) [19,38,39]. In our sediments, however, pore water Fe:P ratios were very unfavourable. Nevertheless, oxidation processes might be able to mobilise $Fe^{2+}$ from $FeS_x$ at a spatial micro-scale in the sediment surface at relatively low $O_2$ levels [40], catalysed by S oxidising microbes [41]. Our $O_2$ profiles showed that $O_2$ was available in the surface water and in the top millimetres of the sediment. The observed high S mobilisation rates in the low salinity treatment, where no S was added, indeed showed that $SO_4^{2-}$ is being mobilised from the sediment by the oxidation of $FeS_x$. Simultaneously, $Fe^{2+}$ thus becomes available to be oxidised [40], and is able to sequester dissolved P. So the intrusion of $O_2$ in reduced sediments may mobilise S bound Fe at a millimetre spatial scale, providing dissolved $Fe^{2+}$ for the formation of ferric $Fe(OH)_x$ at the sediment surface (Fig. 6). This mechanism may very well explain the

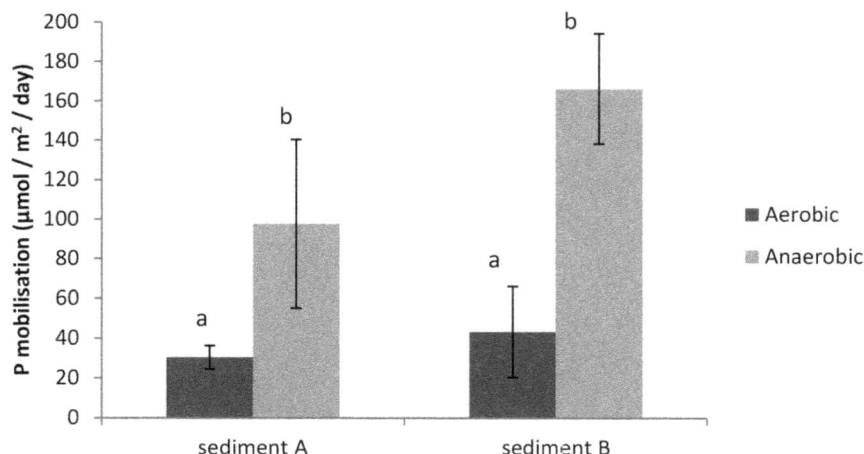

**Figure 5. P mobilisation rates (μmol m⁻² day⁻¹) during aerobic and anaerobic conditions for both sediment cores (A and B).** Significant differences between treatments are indicated with different letters.

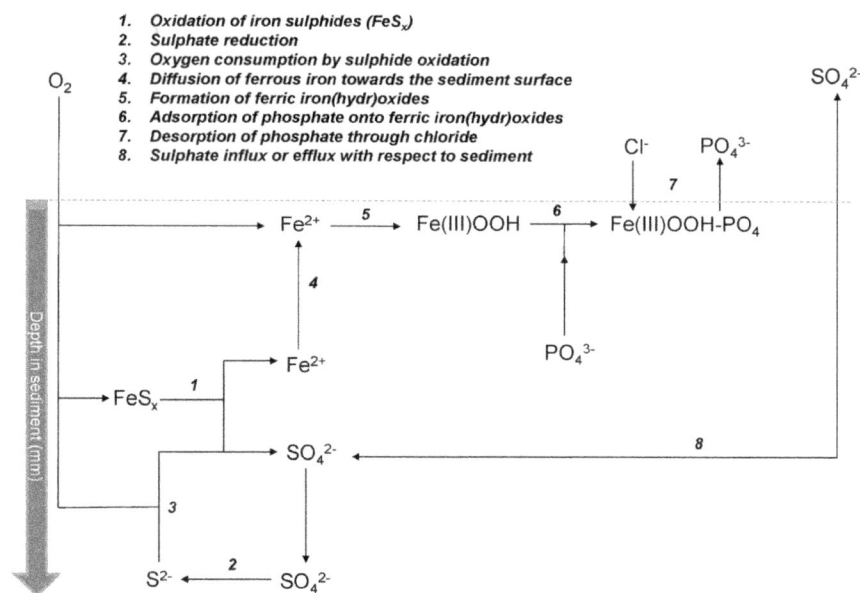

**Figure 6. Schematic overview of the proposed mechanism, showing key processes in the upper millimetres of the S-rich, peat sediments involved in P mobilisation.** Salinisation leads to an increased $SO_4^{2-}$ influx, affecting Fe diffusion to the sediment surface, enabling increased P mobilisation in the longer term.

unexpected lack of P release from the sediments in the long term under aerobic conditions.

Our experimental set-up, without sediment disturbance and with relatively low biochemical $O_2$ demand (BOD) due to the absence of fresh organic matter input, will certainly have contributed to the long-term outcome of the experiment. Nevertheless, it seems plausible that it took a relatively long time before the sediment surface became sufficiently oxidised, or before the microbial population was sufficiently developed, to completely prevent P mobilisation in the experiment. These results in the longer term may represent field situations with stable non-bioturbated $FeS_x$-rich sediments or sediments with stagnant, hypolimnetic water. During anaerobic conditions, P mobilisation was strongly enhanced (Fig. 5), which clearly highlights the importance of $O_2$ availability to prevent P release. Field experiments are, therefore, necessary to validate our experimental results and suggested mechanism for the lack of P release from S-rich aquatic sediments.

**4.2.2 Salinity effects.** For the high salinity treatment, one of the sediments showed an increase of the surface water P concentration also in the long term (80 weeks). In saline or estuarine systems, P is often found to be easily released from soil particles [5,7], and dissolved P concentrations are usually higher with increasing salinity [6,8,39]. At a high salinity, $SO_4^{2-}$ concentrations increased in both surface water and pore water and a considerable part may be reduced deeper in the sediment, since it was not released to the surface water (Fig. 6). Produced $S^{2-}$ will react with $O_2$ and interfere with the oxidation of $FeS_x$, or again immobilise $Fe^{2+}$. As expected, the net mobilisation of $Fe^{2+}$ will be less, leading to insufficient formation of ferric $Fe(OH)_x$ to prevent the release of P to the surface water [18,19]. This decoupling of the Fe and P cycle [10] at a micro-scale diminishes the P-binding capacity at the water-sediment interface. In sediment A, $O_2$ penetrated deeper into the sediment, suggesting that less $O_2$ was consumed, less $FeS_x$ was oxidised, and less $Fe^{2+}$ was mobilised. This may partly explain the long-term release of P

from sediment A in the high salinity treatment. Desorption of P from ferric $Fe(OH)_x$ due to the high Cl concentrations [10,22] might have increased this effect.

## 4.3 Implications for water management

Although the mobilisation of P from the S-rich and relatively Fe-poor sediments (typical for coastal wetlands) was mainly driven by diffusion, the build-up of a stable oxidised sediment surface may have prevented the release of P under the experimental conditions. We hypothesise that the oxidation of $FeS_x$ in the sediment surface delivers the $Fe^{2+}$ necessary for the precipitation of P at the sediment-water interface (Fig. 6). Disturbance of the sediment-water interface due to wind, ebullition of gases from the sediment, and bioturbation can, however, prevent this build-up of a protective Fe-rich sediment surface and potentially increase the release of P [9,33]. Although such processes might also mix the sediment surface with $O_2$ and have an opposite effect. Moreover, our results indicate that an increased salinity may lead to a long-term P release, probably by interfering with the $Fe^{2+}$ mobilisation due to increased $SO_4^{2-}$ reduction rates in the anaerobic sediment. They also point out that sediments may react differently upon increased salinity. Therefore, $O_2$ and BOD, but also the actual concentration of $SO_4^{2-}$ play a key role in the mobilisation of P from $FeS_x$-rich sediments. This might have important implications for water management and nature management of eutrophic peatlands in relation to salinisation.

More research, especially field measurements, is necessary to further confirm the experimental results we found for these $FeS_x$-rich sediments. Our experiment was carried out at 15°C and without the continuous input of reactive organic material. Warmer conditions, e.g. during warm episodes in summer will lead to increased mineralisation rates, and also to higher $O_2$ consumption rates and lower solubility of $O_2$. Especially when there is a high input of reactive organic matter, this will lead to strongly decreased $O_2$ concentrations in the surface water, which may prevent adequate oxidation of the sediment surface. Under such conditions this biogeochemical mechanism is expected to fail, leading to strong P mobilisation from the sediment as was shown in this study and also found by Smolders *et al.* [12]. As a result, floating-leaved species, or floating beds of algae or cyanobacteria may develop, which will further decrease the $O_2$ concentrations in the surface water and enhance sediment P mobilisation. This explains why $FeS_x$-rich sediments that show very high dissolved P concentrations and low dissolved $Fe^{2+}$ concentrations tend to show a high P release mainly in summer, which has important implications for water management.

## Conclusions

- Low pore water Fe:P ratios indicated a decoupling of the Fe and P cycle. Although these $FeS_x$-rich sediments were expected to release significant amounts of P by diffusion, this only seemed to be the case in the short term under aerobic conditions.

- Increased salinity led to co-precipitation of P with $Ca^{2+}$ in the short term, lowering actual P concentrations. However, short-term P mobilisation rates were found to be similar for all treatments, regardless of salinity.

- Our experimental results suggest that the classic theoretical framework of oxidative conditions in the surface water that prevent P release from the sediment, may also hold in sediments showing unfavourable total Fe:S ratios but high $FeS_x$ concentrations. In our $FeS_x$-rich, eutrophic sediments, typical for coastal wetlands, $O_2$ availability still seemed to be the most important determinant of sediment P release, at least under stable sediment conditions.

- We suggest a controlling mechanism in which the partial oxidation of $FeS_x$ mobilises sufficient $Fe^{2+}$ at micro-scale for the precipitation of P at the sediment-water interface.

- Next to $O_2$, $SO_4^{2-}$ plays a key role in P mobilisation, as high concentrations may counteract the oxidising effect by immobilising $Fe^{2+}$. In the longer term, an increased salinity may, as a result, led to P mobilisation despite oxidation of the sediment surface.

## Acknowledgments

We would like to thank Jeroen Graafland and Rick Kuiperij for their practical assistance in the field and chemical analyses, Jelle Eygensteyn, Paul van der Ven and Sebastian Krosse for their help with chemical analyses, and Leon van den Berg for his help with statistical analyses. We are grateful to the water management authority 'Hoogheemraadschap Hollands Noorderkwartier' for the facilitation of this programme and to the nature management authority 'Natuurmonumenten' for giving their kind permission to take samples in their reserve.

## Author Contributions

Conceived and designed the experiments: JMHVD AJPS LPML. Performed the experiments: JMHVD MJS. Analyzed the data: JMHVD AJPS LPML MJS. Contributed reagents/materials/analysis tools: JMHVD AJPS LPML JGMR. Wrote the paper: JMHVD AJPS LPML JGMR GVD.

## References

1. Smith VH (2003) Eutrophication of Freshwater and Coastal Marine Ecosystems. A Global Problem. Environ Science & Pollution Research 10 (2): 126–139.
2. Lamers LPM, Falla S-J, Samborska EM, van Dulken IAR, van Hengstum G, et al. (2002) Factors Controlling the Extent of Eutrophication and Toxicity in Sulfate-Polluted Freshwater Wetlands. Limnology and Oceanography 47 (2): 585–593.
3. Geurts JJM, Sarneel JM, Willers BJC, Roelofs JGM, Verhoeven JTA, et al. (2009) Interacting effects of sulphate pollution, sulphide toxicity and eutrophication on vegetation development in fens: A mesocosm experiment. Environmental Pollution 157 (7): 2072–2081.
4. Nielsen DL, Brock MA, Rees GN, Baldwin DS (2003) Effects of increasing salinity on freshwater ecosystems in Australia. Australian Journal of Botany 51: 655–665.
5. Carpenter PD, Smith JD (1984) Effect of pH, iron and humic acid on the estuarine behaviour of phosphate. Environmental Technology Letters 6: 65–72.
6. Gunnars A, Blomqvist S (1997) Phosphate exchange across the sediment-water interface when shifting from anoxic to oxic conditions – an experimental comparison of freshwater and brackish-marine systems. Biogeochemistry 37: 203–226.
7. Suzumura M, Udea S, Sumi E (2000) Control of phosphate concentration through adsorption and desorption processes in groundwater and seawater mixing at sandy beaches in Tokyo Bay, Japan. Journal of Oceanography 56: 667–673.
8. Jordan TE, Cornwell JC, Boynton WR, Anderson JT (2008) Changes in phosphorus biogeochemistry along an estuarine salinity gradient: The iron conveyer belt. Limnology and Oceanography 53 (1): 172–184.
9. Boström B, Andersen JM, Fleisher S, Jansson M (1988) Exchange of phosphorus across the sediment-water interface. Hydrobiologia 170: 229–244.
10. Caraco NF, Cole JJ, Likens GE (1989) Evidence for sulphate-controlled phosphorus release from sediments of aquatic systems. Nature 341: 316–317.
11. Lamers LPM, Tomassen HBM, Roelofs JGM (1998) Sulfate-Induced Eutrophication and Phytotoxicity in Freshwater Wetlands. Environmental Science and Technology 32: 199–205.
12. Smolders AJP, Lamers LPM, Lucassen ECHET, van der Velde G, Roelofs JGM (2006) Internal eutrophication: How it works and what to do about it – a review. Chemistry and Ecology 22 (2): 93–111.
13. Einsele W (1936) Über die Beziehungen des Eisenkreislaufs zum Phosphatkreislauf im eutrophen See. Archiv für Hydrobiologie 29: 664–686.
14. Mortimer CH (1941, 1942) The exchange of dissolved substances between mud and water in lakes. J. Ecology 29: 280–329, Journal of Ecology 30: 147–201.
15. Roelofs JGM (1991) Inlet of alkaline river water into peaty lowlands: effects on water quality and Stratiotes aloides L. stands. Aquatic Botany 39: 267–293.
16. Caraco NF, Cole JJ, Likens GE (1993) Sulfate control of phosphorus availability in lakes. Hydrobiologia 253: 275–280.
17. Smolders AJP, Roelofs JGM (1995) Internal eutrophication, iron limitation and sulphide accumulation due to the inlet of river Rhine water in peaty shallow waters in The Netherlands. Archiv für Hydrobiologie 133: 349–365.
18. Hupfer M, Lewandowski J (2008) Oxygen Controls the Phosphorus Release from Lake Sediments –a Long-Lasting Paradigm in Limnology. International Review of Hydrobiology 93: 414–432.
19. Geurts JJM, Smolders AJP, Banach AM, van de Graaf JPM, Roelofs JGM, et al. (2010) The interaction between decomposition, N and P mineralization and their mobilisation to the surface water in fens. Water Research 44: 3487–3495.
20. Smolders AJP, Lamers LPM, Moonen M, Zwaga K, Roelofs JGM (2001) Controlling phosphate release from phosphate-enriched sediments by adding various iron compounds. Biogeochemistry 54: 219–228.
21. Wetzel RG (2001) Limnology: Lake and River Ecosystems. Academic Press 3, An Imprint of Elsevier, USA: 1006 pag.
22. Beltman B, Rouwenhorst TG, Van Kerkhoven MB, Van Der Krift T, Verhoeven JTA (2000) Internal eutrophication in peat soils through competition between chloride and sulphate with phosphate for binding sites. Biogeochemistry 50: 183–194.
23. House WA (1999) The physio-chemical conditions for the precipitation of phosphate with calcium. Environmental Technology 20: 727–733.
24. Wu Y, Wen Y, Zhou J, Wu Y (2014) Phosphorus release from lake sediments: Effects of pH, temperature and dissolved oxygen. KSCE Journal of Civil Engineering 18 (1): 323–329.
25. Reigersman CJA (1946) Ontzilting van Noord-Holland. (Desalinisation of North-Holland.) Rapport van de Commissie inzake het zoutgehalte der boezemen polderwateren van Noord-Holland, ingesteld bij Besluit van den Minister van Waterstaat van 24 april 1939. Rijksuitgeverij, 's-Gravenhage: 191 pp.
26. Boxman AW, Peters RCJH, Roelofs JGM (2008) Long term changes in atmospheric N and S throughfall deposition and effects on soil solution chemistry in a Scots pine forest in the Netherlands. Environmental Pollution 156: 1252–1259.
27. Van Gemerden H (1984) The sulphide affinity of phototrophic bacteria in relation to the location of elemental sulphur. Archiv für Mikrobiologie 139: 289–294.
28. Kamphake LJ, Hannah SA, Cohen JM (1967) Automated analysis for nitrate by hydrazine reduction. Water Research 1: 205–206.
29. Grasshof K, Johannsen H (1972) A new sensitive and direct method for the automatic determination of ammonia in sea water. Journal du Conseil Permanent International pour l'Exploration de la Mer 34: 516–521.
30. Henriksen A (1965) An automated method for determining low-level concentrations of phosphate in fresh and saline waters. Analyst 90: 29–34.

31. O'Brien JE (1962) Automation in sanitary chemistry part 4: automatic analysis of chloride in sewage. Wastes Engineering 33: 670–682.

32. Golterman HL (1996) Fractionation of sediment phosphate with chelating compounds: a simplification, and comparison with other methods. Hydrobiologia 335: 87–95.

33. Boström B, Pettersson K (1982) Different patterns of phosphorus release from lake sediments in laboratory experiments. Hydrobiologia 92: 415–429.

34. Van Dijk G, Loeb R, Smolders AJP, Westendorp PJ (2013) Verbrakking in voormalig brak laag Nederland, bedreiging of kans? (Salinisation in former brackish Dutch lowlands, threat or opportunity?) H2O 46 (3): 1–5.

35. Dupraz C, Reid RP, Braissant O, Decho AW, Normanc RS, et al. (2009) Processes of carbonate precipitation in modern microbial mats. Earth-Science Reviews 96: 141–162.

36. Deinema MH, Habets LHA, Scholten J, Turkstra E, Webers HAAM (1980) The accumulation of polyphosphate in Acinetobacter spp. FEMS Microbiology Letters 9: 275–279.

37. Hupfer M, Uhlmann D (1991) Microbially mediated phosphorus exchange across the mud-water interface. Verhandlungen des Internationalen Verein Limnologie 24: 2999–3003.

38. Gunnars A, Blomqvist S, Johansson P, Andersson C (2002) Formation of Fe(III) oxyhydroxide colloids in freshwater and brackish seawater, with incorporation of phosphate and calcium. Geochimica et Cosmochimica Acta 66 (5): 745–758.

39. Blomqvist S, Gunnars A, Elmgren R (2004) Why the limiting nutrient differs between temperate coastal seas and freshwater lakes: A matter of salt. Limnology and Oceanography 49 (6): 2236–2241.

40. Roden EE (2012) Microbial iron-redox cycling in subsurface environments. Biochemical Society Transactions 40: 1249–1256.

41. Imhoff A, Schneider A, Podgorsek L (1995) Correlation of viable cell counts, metabolic activity of sulphur-oxidizing bacteria and chemical parameters of marine sediments. Helgoländer Meeresunters 49: 223–236.

# 13

# Modeling the Evolution of Riparian Woodlands Facing Climate Change in Three European Rivers with Contrasting Flow Regimes

Rui P. Rivaes[1]*, Patricia M. Rodríguez-González[1], Maria Teresa Ferreira[1], António N. Pinheiro[2], Emilio Politti[3], Gregory Egger[3], Alicia García-Arias[4], Felix Francés[4]

1 Forest Research Center, Instituto Superior de Agronomia, Universidade de Lisboa, Lisbon, Portugal, 2 CEHIDRO, Instituto Superior Técnico, Universidade de Lisboa, Lisbon, Portugal, 3 Environmental Consulting Klagenfurt, Klagenfurt, Austria, 4 Research Institute of Water and Environmental Engineering, Universitat Politècnica de València, Valencia, Spain

## Abstract

Global circulation models forecasts indicate a future temperature and rainfall pattern modification worldwide. Such phenomena will become particularly evident in Europe where climate modifications could be more severe than the average change at the global level. As such, river flow regimes are expected to change, with resultant impacts on aquatic and riparian ecosystems. Riparian woodlands are among the most endangered ecosystems on earth and provide vital services to interconnected ecosystems and human societies. However, they have not been the object of many studies designed to spatially and temporally quantify how these ecosystems will react to climate change-induced flow regimes. Our goal was to assess the effects of climate-changed flow regimes on the existing riparian vegetation of three different European flow regimes. Cases studies were selected in the light of the most common watershed alimentation modes occurring across European regions, with the objective of appraising expected alterations in the riparian elements of fluvial systems due to climate change. Riparian vegetation modeling was performed using the *CASiMiR-vegetation* model, which bases its computation on the fluvial disturbance of the riparian patch mosaic. Modeling results show that riparian woodlands may undergo not only at least moderate changes for all flow regimes, but also some dramatic adjustments in specific areas of particular vegetation development stages. There are circumstances in which complete annihilation is feasible. Pluvial flow regimes, like the ones in southern European rivers, are those likely to experience more pronounced changes. Furthermore, regardless of the flow regime, younger and more water-dependent individuals are expected to be the most affected by climate change.

**Editor:** Robert Guralnick, University of Colorado, United States of America

**Funding:** This work was supported by the IWRM Era-Net Funding Initiative through the RIPFLOW project (references ERAC-CT-2005-026025, ERA-IWRM/0001/2008, CGL2008-03076-E/BTE), http://www.old.iwrm-net.eu/spip.php. Rui Rivaes benefited from a PhD grant sponsored by UTL - Universidade Técnica de Lisboa (www.utl.pt) and Patricia María Rodriguez-González benefited from a post-doctoral grant sponsored by FCT - Fundação para a Ciência e Tecnologia (www.fct.pt) (SFRH/BPD/47140/2008). The Spanish team would like to thank the Spanish Ministry of the Economy and Competitiveness the support provided through the SCARCE project (Consolider-Ingenio 2010 CSD2009-00065). The funders had no role in study design, data collection and analysis, decision to publish, or preparation of the manuscript. Environmental Consulting Klagenfurt provided support in the form of salaries for authors EP and GE, but did not have any additional role in the study design, data collection and analysis, decision to publish, or preparation of the manuscript. The specific roles of these authors are articulated in the 'author contributions' section.

**Competing Interests:** Emilio Politti and Gregory Egger are employed by Environmental Consulting Klagenfurt. There are no patents, products in development or marketed products to declare.

\* Email: ruirivaes@isa.ulisboa.pt

## Introduction

For decades scientists have been raising awareness about ongoing global climate change brought about by anthropic greenhouse gas (GHG) emissions into the atmosphere (e.g. [1,2,3,4]). While at first it was possible to raise doubts in relation to the alleged global climate change process, the development and continued improvement of global circulation models (GCM) has allowed the scientific community to project with a high level of confidence that global mean surface temperature will increase over the course of the 21st century [5]. What is more, this trend will be followed by an increase in global averaged mean water vapor, evaporation and precipitation [6]. In Europe, regional circulation models (RCM) forecast climate warming above the projected global mean temperature rise, with precipitation pursuing contrasting tendencies according to region and season [7]. In Northern Europe, annual rainfall is expected to increase, while the opposite trend is expected for southern Mediterranean areas [8]. Nevertheless, seasonal precipitation estimates in these regions are not straightforward. If winter precipitation in northern and central Europe is very likely to rise, in southern Europe there are some uncertainties, with different rainfall projections depending on the emissions scenario. On the other hand, it is consensual that summer rainfall will decrease all over Europe, and the same is true for snow, which is predicted to decrease throughout this continent [9].

Such meteorological changes will significantly affect European river flow regimes, essentially through more pronounced low flow magnitudes in the Mediterranean climate zone and major modifications in high flow magnitudes in snow climates [10]. In summer, higher temperatures and evaporation rates, combined in a number of cases with less precipitation, will reduce runoff in many European regions [11,12,13]. Even in nival or glacier-affected basins, runoff is expected to decrease due to a decline in melt water [14], leading to important reductions in floodplain inundations in the summer season. In contrast, higher runoff values in the wet season can enhance the risk of flooding caused by increased heavy rain events in a Mediterranean climate, or sleet (commonly known as "rain on snow events") in snow ones [5]. This will be further aggravated by the likelihood that modifications in river flow regimes and their associated ecosystems will be amplified by future climate change interactions with anthropogenic pressures, such as increased water withdrawals to satisfy human needs [15,16].

Rivers have a natural flow regime, on the basis of which aquatic and riparian communities have evolved in reliance on the ecological integrity of their ecosystems [17]. Flow regime alterations can thus have numerous impacts – geomorphological [18], ecological [19] and biological [20] – on those communities. Depending on the severity of changes, it may be that thresholds will eventually be crossed with unforeseeable consequences for mankind [21], given that ecosystems provide ecological services that are critical to the functioning of Earth's life-support system and give a very important contribution to human welfare [22].

Riparian ecosystems are particularly vulnerable to flow regime changes [23], since they are governed mostly by that regime and its stream flow components [24,25,26]. Riparia forms a transitional boundary that connects aquatic and terrestrial communities [27,28,29], consequently presenting high biodiversity and production [28,30] while simultaneously harboring the most endangered ecosystems on earth [31,32]. Additionally, riparian areas perform important hydrologic, geomorphic and biological functions to a greater degree than upland areas, considering the proportional area they cover within a watershed [29]. Indeed, researchers have documented several benefits to freshwater environment occasioned by the presence of riparian vegetation (e.g. [33,34]), as well as evidence of the effects of its deterioration on instream species [35]. Riparian ecosystems also provide goods and services that are directly valued by human societies, such as reductions in damage from floodwaters [36,37], supplying suitable areas for bird watching, wildlife enjoyment and game hunting [38,39,40], or providing fish for food and recreation [41,42]. Thus, if decision-makers want to ensure that river restoration and administration produce successful results, they must consider riparian management to be an emerging environmental issue that plays an essential role in water and landscape planning.

Given that flood cycles are paramount in influencing riparian forest patterns [43], new tools are urgently needed to provide a long-term quantification of the predictable effects of stream hydrological re-setting on riparian dynamics [44,45]. Also, a valid assessment of spatiotemporal shifts in different functional types of vegetation might become essential to forecast feedbacks in stream flow changes and associated disturbance processes [20]. Although valuable, some of the latest approaches to riparian vegetation modeling still lack a spatial output of the functional type dynamics, which is essential for predicting and managing riparian ecosystems as a whole (e.g. [46,47,48,49,50,51]).

In the present paper we endeavor to assess riparian vegetation structural changes caused by climate-changed flow regimes in different climatic and hydrogeomorphic contexts across Europe, as

well as to consider responses to emerging topics that are yet insufficiently studied in fluvial ecosystems (see [52] for a better understanding), particularly with regard to riparian vegetation.

Preliminary results addressing such issues have been presented by the authors [53,54], but not as comprehensively and using old-fashioned scenarios in some cases. The present work goes beyond the scope of those earlier results, inasmuch as it further analyzes riparian patch amendments in accordance with climate-driven hydrologic changes. Moreover, this is the first time that a joint effort to ascertain the spatiotemporal response of riparian ecosystems to climate-changed flow regimes, considering the latest climate change scenarios with available regional hydrologic forecasts and on a European scale basis, has been made.

## Methods

### Ethics Statement

This study was conducted on hydric public domain locations at the three considered countries. No specific permits were necessary for the described field studies as the performed observational assessments do not qualify as a procedure requiring a license under the national legislation of any of the mentioned countries. Field studies didn't involve elimination or removal of any endangered or protected species.

### Study site selection

Three study sites were selected with a view to encompass the principal watershed alimentation modes occurring across Europe. Although this was the primary criterion, we also attempted to consider an existing climatic gradient, determined by variables such as latitude, altitude or air temperature. Study sites (river reaches) were thus located in different countries with diverse climates and flow regimes (by both main water alimentation mode and transient pattern of discharge), namely Austria, Portugal and Spain (Figure 1).

**Kleblach reach (Drau River, Austria).** The Austrian case study is representative of the central Europe flow regimes, where maximum flows occur in spring and are attributable to snow-melt and glacial thaw. The study site is located at an altitude of approximately 570 meters in the upper river Drau, next to the village of Kleblach. Study site length is about 700 meters, and bank protection had been removed during an earlier river restoration project. Riparian vegetation comprises several species, most importantly including purple reed grass [*Calamagrostis pseudophragmites* (Haller f.) Koeler], German tamarisk [*Myricaria germanica* (L.) Desv], several willow species (*Salix triandra* L., *Salix purpurea* L., *Salix eleagnos* Scop. and *Salix alba* L.), grey alder [*Alnus incana* (L.) Moench] and European ash (*Fraxinus excelsior* L.). The river flow regime typifies a permanent temperate river, characterized by a mixed nivo-glacial regime [55] with significant flow (mean discharge between 1951 and 2008 equal to 74 m$^3$/s) and a high degree of predictability. Although considered a mixed regime, only one real maximum occurs – in June-July, when the highest water levels occur as a result of watershed melt water flow-off. Conversely, minimum discharges occur in winter, due to solid precipitation and nival retention (Figure 2).

**Ribeira reach (Odelouca River, Portugal).** The Portuguese case study exemplifies the South-Western Europe flow regimes, with minimum flows in summer due to the seasonal lack of rain. This study site is located in the Odelouca River, near Ribeira village, with a studied length of close to 400 meters, at an altitude of about 132 meters. Riparian vegetation is typically Mediterranean, inhabited mostly by tamarisk (*Tamarix africana* Poir.), willow (*Salix salviifolia* Brot.) and narrow-leaved ash

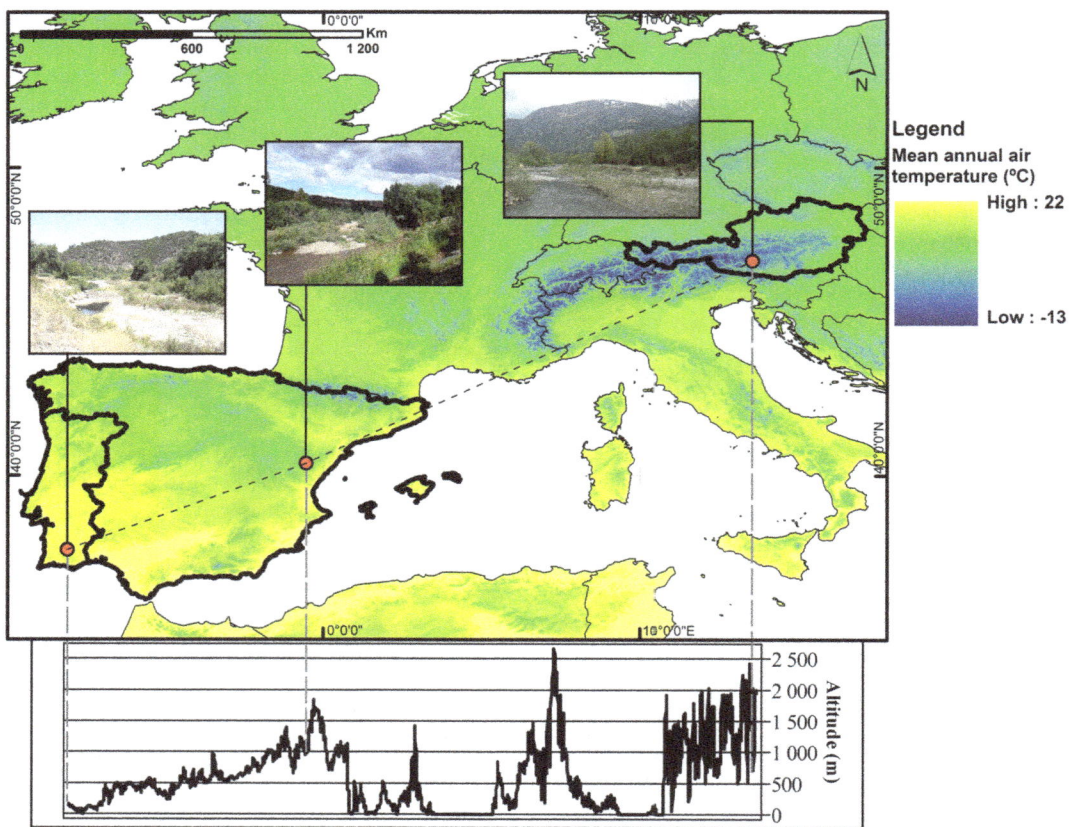

**Figure 1. Study site location.** Study site location showing the spatial variation in mean annual air temperature and an altitude profile across the three study sites (Digital Elevation Model and Mean annual air temperature data source: EDIT Geoplatform, [January, 2013], (CC BY-NC-SA 2.5 ES), http://edit.csic.es/).

(*Fraxinus angustifolia* Vahl.). In the outermost floodplain areas it is also possible to find the emergence among riparian species of terrestrial species like cork oak (*Quercus suber* L.) or holm oak

(*Quercus ilex* L. subsp. ballota). This case study features an intermittent river with a simple pluvial regime, where maximum mean monthly discharges occur in winter, while minimum

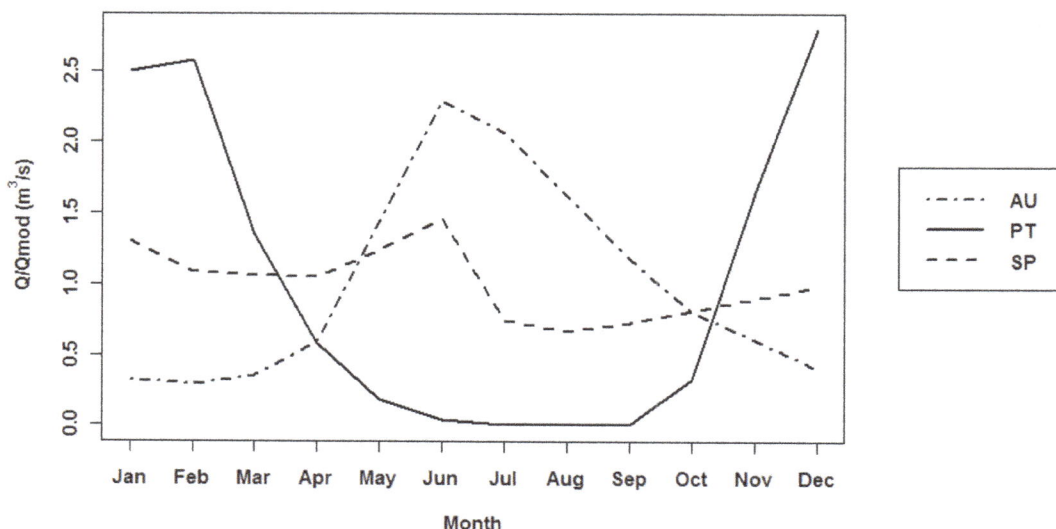

**Figure 2. River flow regimes in the three considered study sites.** River flow regimes in the three considered study sites (Austria – AU, Portugal – PT and Spain – SP). Mean monthly discharges are presented as ratio Discharge (Q)/Mean annual discharge ($Q_{av}$) for 1960–1990 year period.

discharges (commonly null) take place in summer. River flow is generally low, but discharge is highly responsive to rainfall and flash floods happen whenever there are heavy rain events (although mean discharge is 2.5 m³/s, flash floods range between 80 and 480 m³/s). This hydrological regime thus displays a great intra and inter-annual variability (Figure 2).

**Terde reach (Mijares River, Spain).** Typical river flow regimes of mountain-fed catchments are illustrated by the Spanish case study, located in the Mijares River, between the villages of Sarrión and Mora de Rubielos. This site lies at an altitude of approximately 850 meters, where it presents a permanent river, 540 meters of which were surveyed. The floodplain vegetation is generally characterized by different willow species (*Salix eleagnos* Scop., *Salix purpurea* L. and *Salix alba* L.), black poplar (*Populus nigra* L.) and common reed [*Phragmites australis* (Cav.) Trin. ex Steud.]. Terrestrial species like juniper (*Juniperus spp.*), kermes oak (*Quercus coccifera* L.) or holm oak (*Quercus ilex* L. subsp. ballota) are also found within the one hundred-year flood area. This case study is characterized by a mixed pluvio-nival river regime with a low mean monthly discharge coefficient amplitude. This river flow regime displays two mean monthly discharge maximums, one in January due to precipitation, and a more pronounced one in late spring originated by snowmelt (Figure 2).

## Climate change scenarios and expected hydrologic changes

In order to determine the deviation in riparian ecosystems caused by climate change, it is necessary to adopt a reference riparian patch mosaic from which to calculate riparian alterations linked to this stressor. To that end we considered a reference scenario, taking into account the popular and commonly used World Meteorological Organization (WMO) climate reference period of 1961–1990. This period is usually selected because it allows the comparison of future climate change regarding near present climatological conditions while having generally the best observational climate data coverage and availability from the periods considered meaningfully free from anthropogenic trends embedded [56].

The climate change scenarios adopted in this study were based on the latest IPCC emission scenarios from which hydrologic modeling have been performed. As described in its Special Report on Emission Scenarios (SRES) [57], this set of emission scenarios (A1 – *medium-high emission levels*, A2 – *high emission levels*, B1 – *low emissions* and B2 – *medium-low emissions*) attempts to reproduce the current knowledge in climate change science in order to characterize the range of probable driving forces and GHG emissions until 2100. Two of the above emission scenarios were selected for use as scenario templates in each case study, reflecting different intensities of climate change severity (Optimist and Pessimist scenarios) and spanning the existing uncertainties about future socioeconomic developments. In the light of the available data, the emissions scenario selection in each case study was determined in accordance with the Global and Regional Circulation Model scenarios whose results have been most consistent with the historical observations for each country, as regards temperature and rain forecasts in diverse climate change circumstances (see [58,59,60]). Corresponding discharge anomalies in the study site flow regimes were then obtained from national climate change assessments in which hydrology was also envisaged. The anomalies were applied to the existing reference flow regime data for each study site by multiplicative factors obtained in those studies to obtain the corresponding study site scenario data series.

As a result, for the Kleblach reach study site, SRES B1 and SRES A2 emission scenarios were selected as Optimist and Pessimist respectively. The GCM model used as a basis for these scenarios was GCM ECHAM5 [59]. The expected flow regime changes due to the projected meteorological alterations was determined by hydrological models based on information produced by the REMO-UBA regional climate model [59]. The climate change scenarios for the Ribeira reach were grounded in the RCM HadRM3 results for the Optimist SRES B2 scenario and the Pessimist SRES A2 scenario, as presented for Portugal by Santos *et al.* (2002, 2006) [58,61]. The impact of climate change on freshwater assets was assessed using the Temez model – a simplification of the Stanford Watershed Model [62,63]. Finally, for the Terde study site, the selected emission scenarios were also SRES B2 as the Optimist, and SRES A2 as the Pessimist. These were obtained from the Spanish modeling with the Hadley Centre Global Climate Model (HadCM3) as boundary conditions and regionalized with the PROMES regional climate model [60]. Hydrological scenarios were obtained from PATRICAL precipitation-runoff model results [64]. A summary of the hydrological changes considered for the aforementioned climate change scenarios for each study site is presented in Table 1.

## Riparian vegetation modeling

For this task we used the state-of-the-art *Computer Aided Simulation Model for In-stream Flow and Riparian vegetation model*, commonly known as the *CASiMiR-vegetation* model [65]. This tool is a dynamic rule-based spatially distributed model that supports its computation on fluvial disturbance in riparian vegetation – a concept that has been increasingly recognized since the late 1980's [66] and whose influence is known to be a key cause of spatiotemporal variability in streams [27,67,68,69,70,71]. More precisely, this tool relates ecologically relevant hydrological elements [17] with riparian vegetation features that directly respond to chronic hydrologic alteration [25], thus being able to reproduce local fluvial disturbance on an annual time step basis and determine the expected succession/retrogression phenomena in vegetation patches, depending on the fluvial physical driving forces to which they are subjected. The structure of *CASiMiR-vegetation* [65] consists of grid-based modules (*Recruitment*, *Morphodynamic disturbance* and *Flood duration*) functioning with a Boolean logic framed by hard thresholds derived from expert judgment. Together, those modules mimic the succession/retrogression episodes experienced by patches when subjected to a particular fluvial disturbance stress.

A huge asset of this model is that modeling is performed by succession phase instead of site-specific features. This permits worldwide application [53,54,72,73,74] and eliminates divergences (e.g. species composition, ecoregion differences) that make generalized application unfeasible in many other models (see [25]). Using this approach it is possible to obtain a homogeneous vegetation classification for the three case studies and thus permit a common appraisal of the modeling results. The adopted classification was first presented by García-Arias *et al.* (2013) [75] (see this reference for a more detailed explanation of the vegetation types/succession phases transformation process), in which thirteen succession phases embedded on four succession stages and three succession series were acknowledged (Figure 3).

With this classification the model presented substantial positive results at the calibration/validation stage and also proved that a study site comparison analysis using standardized succession phases is possible. In addition, model uncertainty due to estimation errors in estimated parameter thresholds was determined not to be significant [76]. *CASiMiR-vegetation* model calibration/validation

**Table 1.** Hydrological regime modifications accounted for the riparian vegetation modeling in the considered climate changes scenarios.

| | Austria | Portugal | | | Spain | |
| --- | --- | --- | --- | --- | --- | --- |
| | Change SRES A2 (Pessimist) | Change SRES B1 (Optimist) | Change SRES A2 (Pessimist) | Change SRES B2 (Optimist) | Change SRES A2 (Pessimist) | Change SRES B2 (Optimist) |
| **Mean monthly discharge (%)** Winter (DJF) | 38 | 26 | -60 | 30 | -30 | -27 |
| Spring (MAM) | 9 | 12 | -80 | -25 | -24 | -25 |
| Summer (JJA) | -17 | -12 | -80 | -50 | -32 | -27 |
| Autumn (SON) | -11 | -3 | -80 | -60 | -33 | -30 |
| **Minimum watertable elevation (m)** | NE | NE | -4 | -1 | -0.27 | -0.25 |
| **Flood duration** | NE | NE | NE | NE | NE | NE |

Values stand for deviation from Reference period (1960-1990). NE stands for non-expected changes.

for these cases is not presented here, as it is already thoroughly explained in previous studies [75,77].

The input data needed to run this tool are grid-based topography, maximum annual discharge shear stress, flood duration and mean/base water table elevation files. Our topography inputs were obtained by topographic surveys and were considered to be fixed during the modeling runs, so that riparian change evaluation could be endorsed solely to the hydrologic regime changes. Shear stresses and water table elevations in each study site were obtained by 2D hydraulic modeling, while flood duration was retrieved from daily recorded discharge data [77]. Among the input data, shear stress stood out in terms of intra-scenario variability and was therefore analyzed for significant differences between scenarios. On the other hand, because minimum annual water table elevation and flood duration were considered unchanged within scenarios, we did not examine them by these means.

A simple method for appraising significant differences related to shear stress disturbance is to build confidence intervals for shear stress sample means in each scenario. We did this using two sample t-tests from the *R Stats package* in R environment [78].

Riparian vegetation modeling considered three modeling runs for each study site – namely the Reference, Optimist and Pessimist scenarios – starting from the same initial condition provided by the model. The expected 1990 riparian vegetation map was considered as the Reference scenario and was intended for use as a benchmark for assessing riparian deviations in the climate-change scenarios. The climate change scenarios (both Optimist and Pessimist) were characterized by the expected riparian vegetation maps at year 2100, under the corresponding climate-changed flow regimes. Once again, expected climate-changed riparian vegetation maps were obtained by modeling riparian vegetation under the likely river flow regimes in the 2071–2100 period. Riparian vegetation changes were analyzed by proportional change in total study site area and within each succession phase area, further denominated "specific area cover anomaly", and referring to the difference between specific areas of succession phases in the Reference and climate-changed scenarios.

## Results

For all study sites, the expected flow regime in each climate change scenario follows a pattern similar to that of its reference regimes (Figure 4). Having said this, some changes are noticeable and can lead to structural modifications in riparian woodlands. In the Austrian case, both scenarios forecast similar changes in the hydrological regimes. Winter and early spring mean discharges are likely to be higher than those in the reference period, whereas in the remaining months mean monthly discharge is expected to be lower. Nonetheless, water table elevations and flood durations are not expected to change significantly (Table 1). In the Portuguese case study, changes in the flow regime differ depending on the climate change scenario. This discharge variability is found in winter, when river flows are expected to be higher in the Optimist scenario, but lower in the Pessimist one. In the remaining seasons, both scenarios predict a reduced discharge compared to the corresponding Reference scenario, which in turn will contribute to a water table drop of about 1 and 4 meters in the Optimist and Pessimist scenarios, respectively. No flood duration changes are expected in this flow regime, as floods occur on a very short period of time (Table 1). Finally, in the Spanish case study both scenarios show a decreased discharge throughout the hydrological year, with very similar changes. Fluvial disturbance is attenuated, and reduced water availability will be experienced in the floodplains

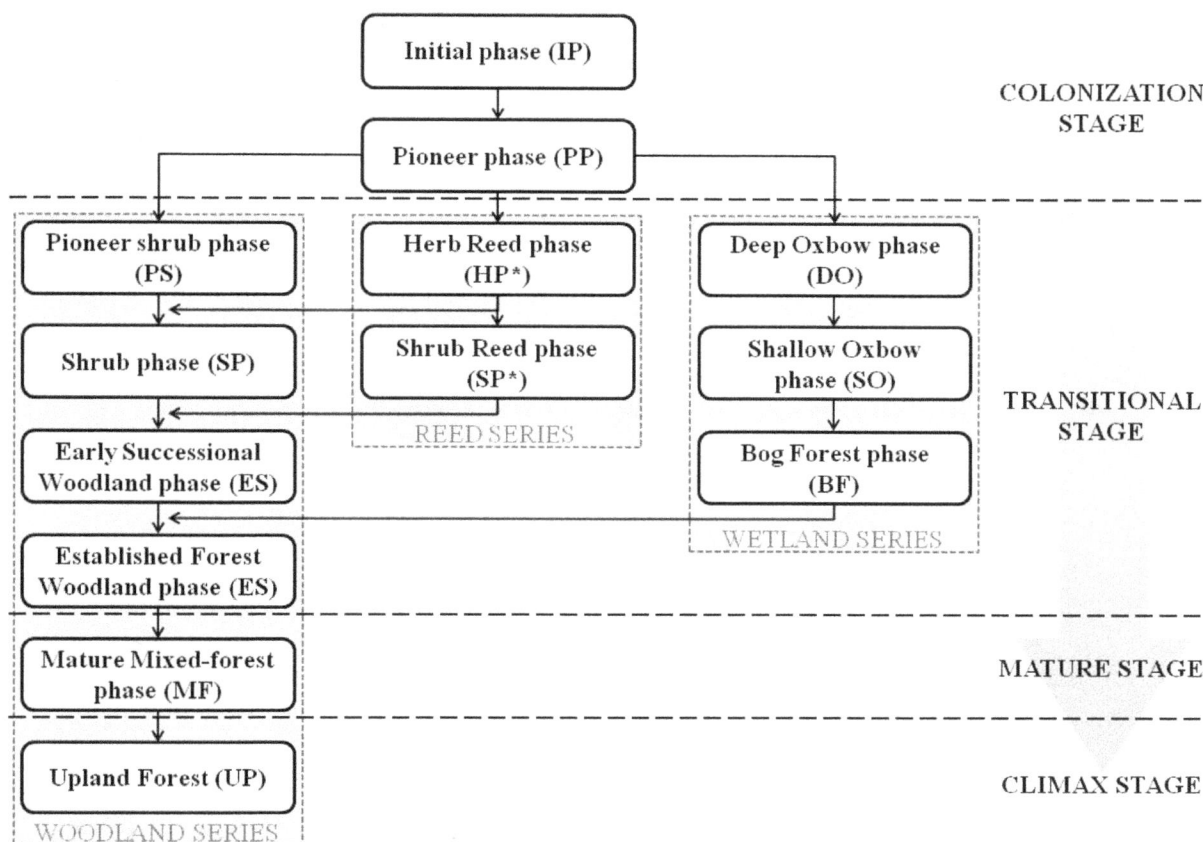

**Figure 3. Common vegetation classification adopted for the three case studies.** Common vegetation classification (by succession phase and stage) adopted for the three case studies, according to the existing vegetation series in each case study. Adapted from [74].

all year long. Water table elevations are expected to decline about 0.25 m in the Optimist scenario and 0.27 m in the Pessimist, while no changes were predicted concerning flood duration (Table 1).

Consistent with the expected climate change-induced flow regimes in each case study, maximum annual shear stress modifications in the study sites are also predicted. In fact, shear stress differences between scenarios proved significant with a 99% confidence level and corroborated earlier affirmations (Figure 5 and Table S1).

Riparian vegetation modeling results show that, under the influence of climate-changed flow regimes, all the studied riparian ecosystems will experience structural changes in their riparian patch mosaics. Despite the fact that for the same modeling area (100-year flooded area), the three case studies achieved different stages in terms of vegetation development, the same tendency is perceptible in all of them. Novel succession phases are replaced by older and more hydric stress-tolerant ones in most cases; and wherever that replacement is not possible, riparian vegetation fades away, giving way to a complete retrogression to the Initial phase (Figure 6).

Table 2 illustrates the proportional area covered by succession phases in each study site scenario. Austrian Reference scenario is characterized by the existence of three different vegetation series, mostly in a Transitional Stage (approximately 95%) and with little Colonization stage (near 5%). Riparian corridor is composed mainly of Woodland series (almost 87% of total area), the most common phase being Early Successional Woodland (ES) with about 82% of total area. Wetland series cover around 8% of total

study site area, with Deep Oxbow phase (DO) with 1.5%, Shallow Oxbow phase (SO) with nearly 6%, and Bog Forest phase (BF) with 0.5%. The Initial phase (IP) represents almost 5% of total study site area. In opposition to the Reference scenario, slight changes are predicted in succession phases. As an example, in both Optimist and Pessimist scenarios the Woodland series Shrub Woodland Phase (SP) converts into Early Successional Woodland Phase (ES) with a consequent decline of approximately 4% in total area. In the case of the Wetland series, despite maintaining its cover area in all modeled scenarios, its succession phases adjust towards improved hydric stress adaptation. In fact, in both climate change scenarios the Deep Oxbow Phase (DO) decreases by 0.7% of total area, in favor of the Shallow Oxbow Phase (SO), which increases by the same amount in both scenarios. Reed series appear in the form of the Herb Reed phase (HP*), taking over areas once occupied by the Initial phase (IP) and where fluvial disturbance previously precluded vegetation establishment. In a climate change scenario, this succession phase achieves a habitat settlement ranging from 0.2% (in the Pessimist scenario) to 0.4% (in the Optimist scenario) of the total study site area.

The Portuguese case study presents a Reference scenario composed of Colonization and Transitional stages, each occupying approximately half the total area. Succession phases are present in different proportions, with Initial phase (IP) and Established forest phase (EF) occupying the majority of the study area (nearly 40% of total area each). In the considered climate change scenarios, the increase in the Colonization stage is proportional to climate change severity, due to the retrogression of younger phases, which

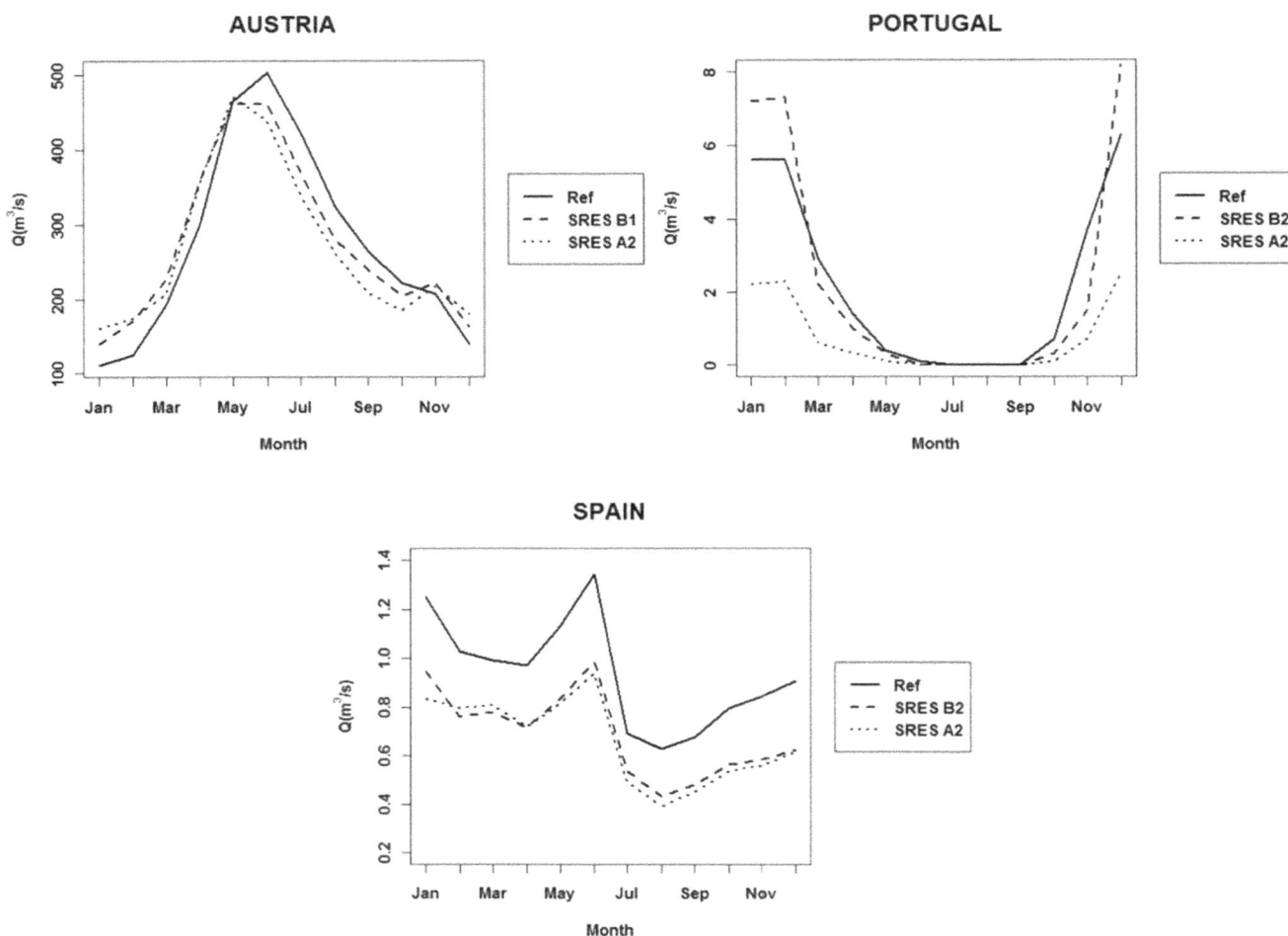

**Figure 4. Reference and expected climate-changed hydrologic regimes in the considered study sites.** Reference and expected climate-changed hydrologic regimes in the considered study sites (Discharge values stand for mean monthly discharges).

attain growth of more than 60% in the Pessimist scenario. On the other hand, the Transitional stage deviation takes an inverse route, with a reduction to 33% in the Optimist scenario, and to less than 26% in the Pessimist one. Considering the specificity of the succession phase, it is noticeable that all succession phases are expected to experience moderate changes, ranging from around 3 to 24% of total area. Initial (IP) and Mature Mixed-forest (MF) phases swell in both scenarios, with the former responsible for the biggest change in the riparian patch mosaic, specifically in the Pessimist scenario, where just this phase is responsible for a change in almost a quarter of the studied landscape. This increase occurs at the expense of the remaining phases, and even entails the total disappearance of the Pioneer (PP) and Early Succession Woodland (ES) phases in the worst scenario. The Established forest phase (EF) also reduces its cover area in the study site (roughly 13 and 12% in Optimist and Pessimist scenarios), but this time due to aging towards the Mature Mixed-forest phase (MF).

The Reference scenario in the Spanish study site is character-ized by the existence of all the successional stages mentioned earlier and two succession series. Here, the Colonization, Transitional, Mature and Climax stages respectively comprise around 26, 3, 19 and 52% of the total area. With particular reference to the succession phases of the Woodland series, this case study is mainly represented by the Upland Terrestrial Forest (UF)

and Mature Mixed-forest phases (together occupying nearly 70% of total area), while the remaining phases cover areas ranging 11% to 19% of the total area. Reed series cover almost 2% of the total area, namely in the form of a Shrub Reed Phase (SP*). In an Optimist climate change scenario, the Colonization stage increases by 5%, with a corresponding decrease in the Transitional and Mature Stages (0.4% drop-off in the former and nearly 5% in the latter). The Climax stage remains unaltered in this scenario. On the other hand, in the Pessimist scenario, Colonization and Transitional stages decline by approximately 7 and 1% of total area respectively, but the Mature and Climax Stages enlarge by approximately the same proportion of total area – namely 4% for the former and 4.5% for the latter. Where succession phases are concerned, minor changes are expected for the riparian patch mosaic in all the considered climate change scenarios, as none of the adjustments attain 5% of the total area. Major changes occur with the Pioneer (PP) and Mature Mixed-forest (MF) phases, but with no consistent trend. In fact, whereas in the Optimist scenario PP is expected to rise by nearly 3% and MF to decrease by about 5% of total area, in the Pessimist scenario PP faces a drop of almost 5%, but MF is reduced by more than 4% of total area. Changes in the Reed series represent a minute proportion of the total study area in both climate change scenarios.

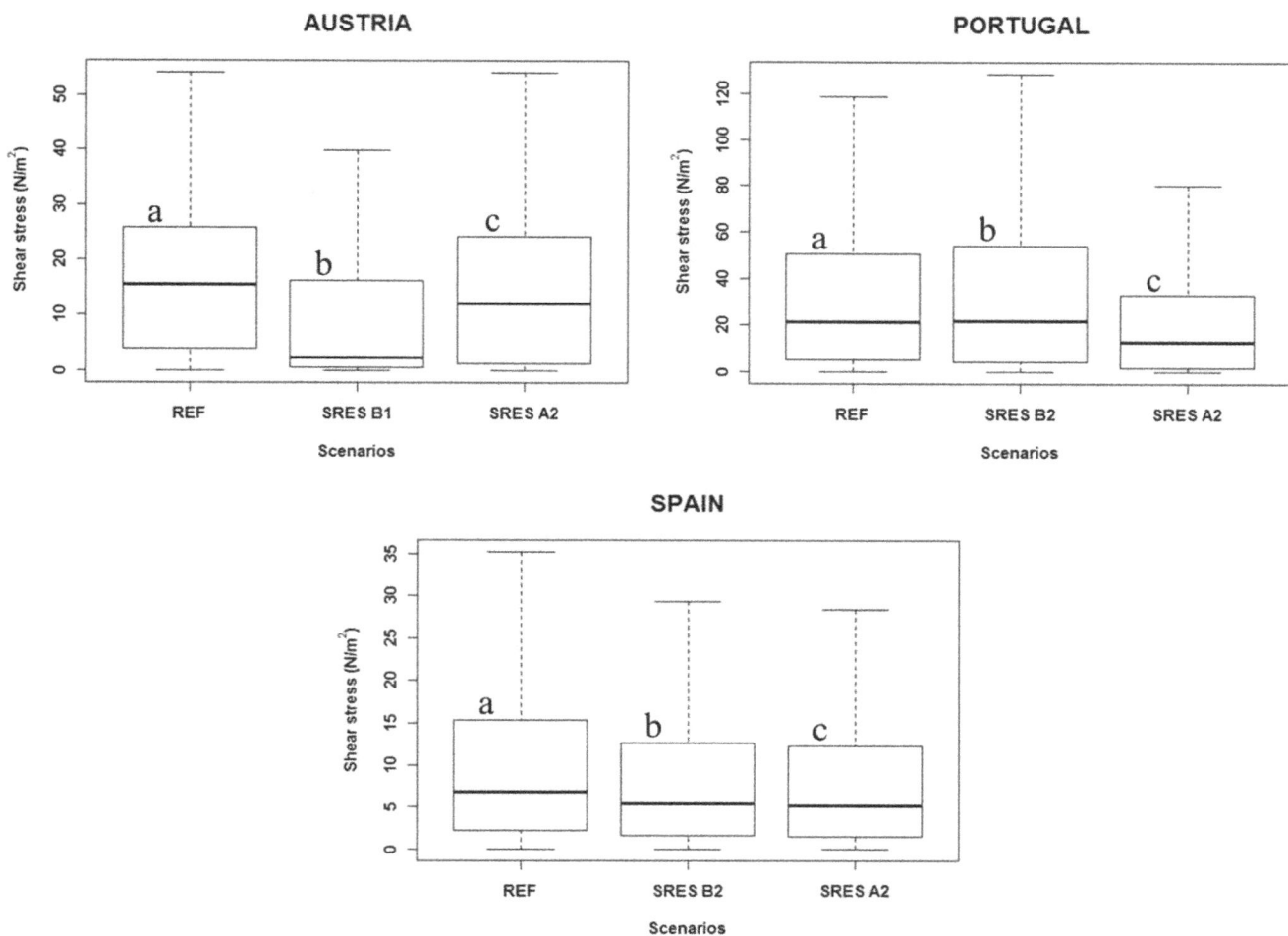

**Figure 5. Scenarios of maximum annual discharge shear stress in each study site.** Expected microhabitat shear stress of the maximum annual discharges in each study site according to the Reference, Optimist and Pessimist scenarios (whiskers stand for non-outlier extremes, box for 1st and 3rd quartiles, thick line for mean, and letters for significantly different groups).

However, although succession phase adjustments may not greatly change the riparian patch mosaic, standalone analysis does reveal profound alterations in the specific habitat area of each succession phase (Figure 7). This means that in the Austrian case some succession phases suffer extensive losses – e.g. the Shrub Woodland Phase (SP) experiences a specific decline in area of almost 90% in the Optimist scenario and faces near extinction (97.9% decline) in the Pessimist one. For the Wetland series, the Deep Oxbow Phase (DO) also faces a decrease in area of nearly 48% in both scenarios. In the Reed series there are noteworthy variations as well, but this time the Herb Reed phase (HP*) is expected to see a tenfold increase in area in the Optimist scenario and a fivefold one in the Pessimist scenario. In Portugal, at least a quarter of the areas of all the existing succession phases in the Reference scenario are expected to be modified in a climate change situation. The Pioneer (PP) and Early Successional Woodland (ES) phases are the most susceptible in this ecosystem, respectively suffering a specific area deprivation of approximately 43% and 28% in the Optimist scenario, while in the Pessimist scenario total retrogression may even occur. These decreases also lead to more than double the expansion of the Initial phase (IP) in the latter scenario. The Spanish case study is no exception to the other two, experiencing considerable succession phase changes.

The area of Pioneer (PP) and Established Forest Woodland (EF) phases clearly increase, by almost 27 and 63% respectively in the Optimist scenario, while the Herb Woodland (HP) and Mature Mixed-forest (MS) phases are expected to suffer shrinkages in area of around 86 and 26% respectively. Succession phases in the Reed series are also prone to extensive reduction, with the Herb Reed (HP*) and Shrub Reed (SP*) phases losing roughly 89 and 41% of their specific areas. What is more, in the Pessimist scenario the Pioneer (PP) and Early Successional Woodland (ES) phases undergo a notable contraction in area of nearly 39 and 75% respectively. In this scenario the area of the Shrub Reed phase (SP*) is also likely to fall by approximately 58%, but it is estimated that the Herb Reed phase (HP*) will increase by almost 78%.

## Discussion

In all the considered cases there are expected changes in river flow regimes that can lead to significant effects on the hydraulic and hydrological conditions of riparian vegetation habitats, namely flood disturbance and hydric stress, which are effectively two of the most important conditioning factors in riparian dynamics [20,79,80,81]. River regimes powered mainly by snow melt or glacial thaw will experience minor increases in discharge.

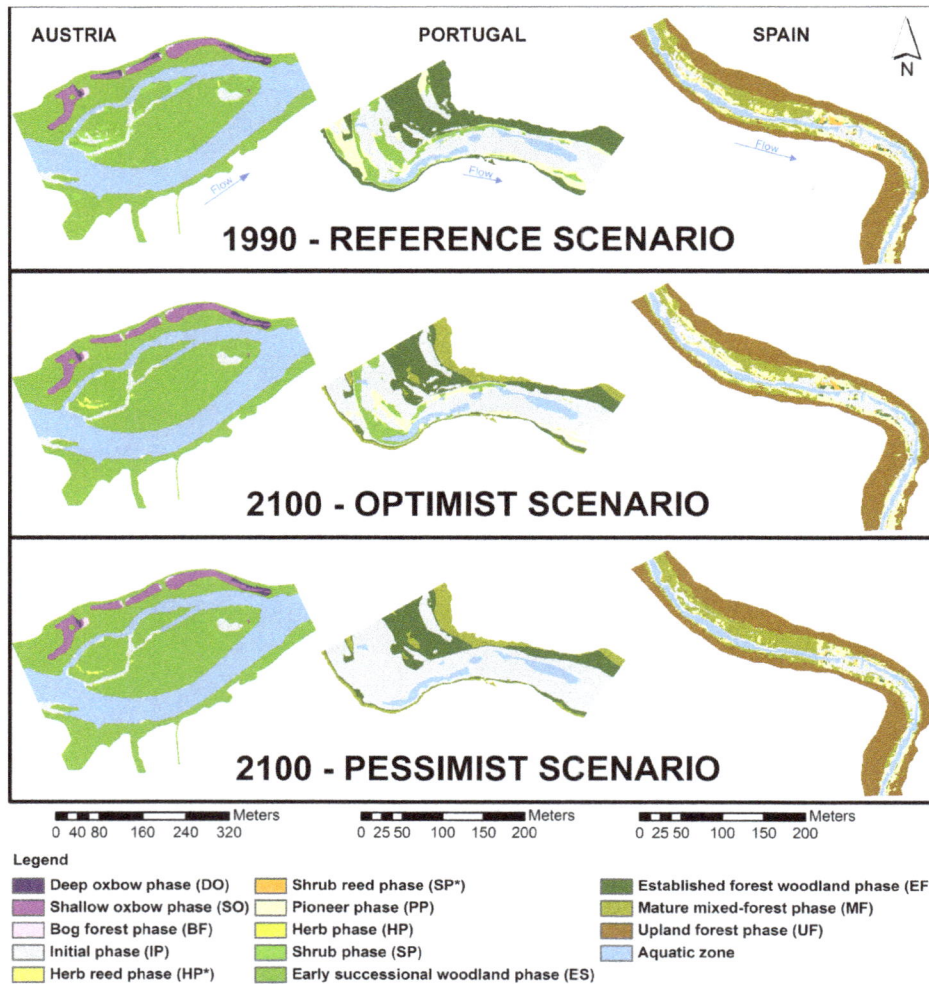

**Figure 6. Riparian vegetation modeling results in each study site for the considered scenarios.** Riparian vegetation modeling results in each study site for the Reference, Optimist and Pessimist scenarios.

Winter discharges will be higher due to less nival retention, whereas summer discharges will fall due to the depletion of snow storage and the resulting decrease in melt water. In river regimes where rainwater is the main form of water alimentation, there is some uncertainty with regard to winter months, as not all rain forecasts agree [7] and different flood disturbances are thus expected for this season, depending on the scenario. Nonetheless, both climate change scenarios expect riparian vegetation to be subjected to lower discharges and accentuated hydric stress in the remaining months of the year.

Accordingly, analyses of the microhabitat shear stresses of maximum discharges in each case study revealed significant differences between scenarios, proving that there will be a meaningful variation in flood morphodynamic disturbance in a climate change scenario.

The riparian vegetation modeling was performed using three different case studies contrasting in flow regime. Such flow regimes encompass the three main water alimentation forms of European rivers, according to Pardé's [82] and L'vovich's [83] typologies, recently upheld by Wrzesiński [84]. However, these case studies are representative of specific flow regime sub-types, which are not sufficient to make assumptions for the general trend of riparian

vegetation changes driven by climate-changed flow regimes in Europe. Nevertheless, this study represents a first approach to portray that evolution.

To analyze the outcomes of the riparian vegetation model, one must regard a number of assumptions that first must be acknowledged. For this study, results should be understood within the context of vegetation patch dynamics, facing a certain scenario created by specific *CASiMiR-vegetation* model settings. The obtained forecasts need to be interpreted more as an indicative trend rather than an exact prevision, due to the shortcomings of modeling such a high dynamic and complex system. The model was calibrated for each basin, considering that the vegetation patches evolution is essentially conditioned by the maximum discharge and by the minimum water table elevation registered in each year. To forecast that for different climate change scenarios the maximum annual discharge series in each basin were multiplied by a factor and the water table elevations were changed, according to the literature considered for the climate change scenarios [58,59,60,61,64].

Despite the inherent stochasticity of fluvial systems, we opted by a deterministic modeling approach. Although, the non-consideration of the discharge sequence stochasticity of a flood event being

**Table 2.** Changes in succession phase cover area according to the considered scenarios.

| Succession series | Succession stage | Succession phase | Austria | | | | | | Portugal | | | | | | Spain | | | | | |
|---|---|---|---|---|---|---|---|---|---|---|---|---|---|---|---|---|---|---|---|---|
| | | | Reference scenario | Optimist scenario | | Pessimist scenario | | | Reference scenario | Optimist scenario | | Pessimist scenario | | | Reference scenario | Optimist scenario | | Pessimist scenario | | |
| | | | % | % | Δ | % | Δ | | % | % | Δ | % | Δ | | % | % | Δ | % | Δ | |
| Any | Colonization stage | IP | 4.9 | 4.1 | −0.8 | 4.1 | −0.8 | | 37.3 | 46.6 | 9.3 | 61.3 | 24.0 | | 14.0 | 16.0 | 2.0 | 11.3 | −2.7 | |
| Any | Colonization stage | PP | - | - | - | - | - | | 13.1 | 7.5 | −5.6 | 0.0 | −13.1 | | 11.7 | 14.8 | 3.1 | 7.1 | −4.6 | |
| Woodland series | Transitional stage | HP | - | - | - | - | - | | - | - | - | - | - | | 0.1 | 0.0 | −0.1 | 0.1 | 0.0 | |
| Woodland series | Transitional stage | SP | 4.3 | 0.5 | −3.8 | 0.1 | −4.2 | | - | - | - | - | - | | - | - | - | - | - | |
| Woodland series | Transitional stage | ES | 82.4 | 86.6 | 4.2 | 87.2 | 4.8 | | 11.6 | 8.3 | −3.3 | 0.0 | −11.6 | | 0.3 | 0.2 | −0.1 | 0.1 | −0.2 | |
| Woodland series | Transitional stage | EF | - | - | - | - | - | | 38.0 | 24.8 | −13.2 | 25.8 | −12.2 | | 0.9 | 1.4 | 0.5 | 0.7 | −0.2 | |
| Woodland series | Mature stage | MP | - | - | - | - | - | | 0.0 | 12.8 | 12.8 | 12.9 | 12.9 | | 19.0 | 14.2 | −4.8 | 23.1 | 4.1 | |
| Woodland series | Climax stage | UF | - | - | - | - | - | | - | - | - | - | - | | 52.3 | 52.3 | 0.0 | 56.8 | 4.5 | |
| Reed series | Transitional stage | HP* | 0.0 | 0.5 | 0.5 | 0.2 | 0.2 | | - | - | - | - | - | | 0.0 | 0.0 | 0.0 | 0.1 | 0.1 | |
| Reed series | Transitional stage | SP* | - | - | - | - | - | | - | - | - | - | - | | 1.7 | 1.0 | −0.7 | 0.7 | −1.0 | |
| Wetland series | Transitional stage | DO | 1.5 | 0.8 | −0.7 | 0.8 | −0.7 | | - | - | - | - | - | | - | - | - | - | - | |
| Wetland series | Transitional stage | SO | 6.4 | 7.1 | 0.7 | 7.1 | 0.7 | | - | - | - | - | - | | - | - | - | - | - | |
| Wetland series | Transitional stage | BF | 0.5 | 0.5 | 0.0 | 0.5 | 0.0 | | - | - | - | - | - | | - | - | - | - | - | |

Succession phase changes (area cover) in accordance with the considered case studies and scenarios. See Figure 5 for succession phase acronyms; percentage values relative to the total modeling area in each case study; Δ stands for scenario variation when compared to the Reference scenario.

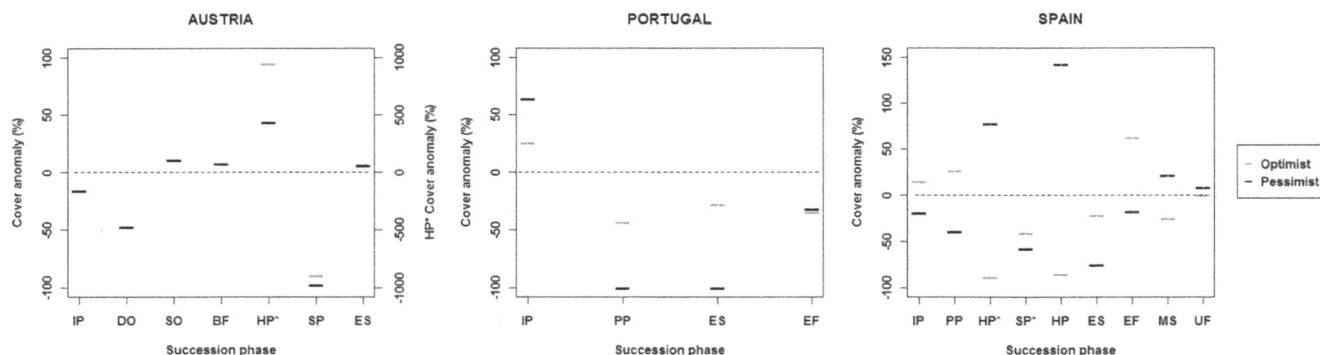

**Figure 7. Specific area cover anomaly of succession phases.** Specific area cover anomaly (%) of the succession phases in each study site and for the considered scenarios (see Figure 5 for succession phase acronyms).

a simplification, the maximum instantaneous discharge registered in each year seems to be the ultimate circumstance of morphodynamic forces driving the succession/retrogression dynamics of riparian woodlands (see [20,27,28,79,81,85,86]).

This restriction enhances the appreciation of broad features or general trends and allows the understanding on how specific components of the flow regime affect riparian vegetation (see [17] for a better understanding). Consequently, based on this deterministic approach, we were able to eliminate the response variability caused by flow regime stochasticity and thus be able to address riparian responses to the discharges that are really important to condition the riparian vegetation dynamics.

Moreover, the fact that our modeling approach considers a fixed topographic input between years obviously represents a simplification of the multifaceted complex fluvial processes occurring within the riverbed. The flow patterns occurring over the banks of a river with riparian galleries and movable bed are very complex and difficult to model with accuracy. The models that consider the movable bed are still relatively inaccurate, due to the use of different empirical formulas, and to the difficulty in obtaining the representative granulometric curve of the different sediment patches and of the different sediment layers of the river bed, not to mention the possible occurrence of layers armoring. In the same line, the hydrodynamic patterns through the riparian galleries are also very difficult to model, due to the vegetation heterogeneity and to the different bending resistance of vegetation species and of their succession phases to the flow velocity. The interaction between vegetation and sediment transport is, of course, still more complex. One example is the vegetation feedbacks, influencing the creation of fluvial landforms, trapping or stabilizing sediments, organic matter and the propagules of other plant species, i.e. acting as physical ecosystem engineers [87]. Another effect particularly relevant in these case studies is the retrogression of transitional and mature stages, which are retrogressed mainly by side erosion and bank failure rather than mechanical disturbance. This is an aspect that will be very difficult to model and that was not considered in the present research.

In this context, the authors believe these complex effects should not be considered, so that the obtained results can reflect the influence of the main succession driving factors: maximum annual discharge and minimum water table elevation.

Besides, despite the recent recognition of those issues concerning the modeling of interactions between flow regime, vegetation and morphology [88,89], such processes were not yet implemented in the CASiMiR-vegetation model and would call for a specific research effort aiming at their integration in future model developments, not only within the climate change effects modeling but more generally within the riparian vegetation modeling context [88]. But, the development of suitable models to simulate and analyze the biogeomorphologic feedbacks is still a priority in ecogeomorphology science agenda [90], as limited capacity remains to predict flow properties in vegetated channels, due to the great difficulty of linking complex dynamic vegetation structures to non-homogeneous hydrogeomorphic processes [91]. Notwithstanding, in a similar study Politti et al. [92] suggested to consider a modeling period ranging from 5 to 25 simulated years, in order to work around those issues. According to this author, within this time frame the effect of the initial riparian landscape condition fades away after the 5th year while the non-consideration of the river morphological changes is not relevant before the 25th year.

Notwithstanding the previously stated, the performed vegetation modeling demonstrates that, for the considered flow regimes, contradictory changes are expected to occur in riparian ecosystems. While in snow-powered flow regimes succession is most likely to occur right across the transversal gradient of the river, in rain-fed watersheds a more complex situation is expectable, with retrogression prevailing inside the channel and succession occurring in areas further from the river. In typical river flow regimes fed by mountain catchments, greater changes will likely occur in the older phases of the ecological succession, but, as other authors have pointed out (e.g. [93,94]), results are not linearly correlated to any of the imposed stresses. In fact, lower flood disturbance and increased hydric stress do not result in a clear tendency in riparian vegetation structural amendment terms, thus showing that in this case shear stress and hydric stress don't explain successional dynamics by themselves.

Nor is the extent of change equal across the considered flow regimes. In both nivo-glacial and mountain-fed flow regimes, moderate changes in total area do occur, but some particular smaller variations in certain succession phases may not be enough to say whether this adjustment is due to model causal effects rather than model uncertainty or input errors. In fact, such a detailed analysis should be conducted carefully as the average model area balance error of succession phases in the three case studies was about 7% [77], especially in smaller and highly disturbed patches like younger succession phases. On the contrary, in Mediterranean pluvial flow regimes, succession area changes can be substantial and rivers with flow intermittency seem to be the most affected [13], where succession phases can change per se almost a quarter of the total riparian patch mosaic.

Nonetheless, small changes in total area can mask dramatic habitat changes in succession phases within all the considered flow regimes. In fact, in the nivo-glacial regime-characteristic site, changes in succession phases can represent almost a tenth of the entire wetland areas, with large declines in some wetland succession phases, thus demonstrating that climate change will favor less water-dependent species. The same occurs in mountain-fed catchments, with succession phases experiencing specific area changes ranging from declines to near extinction, or to area boosts of about 50%. However, considering the variability of riparian responses to the climate-changed flow regimes in this case study, we are led to assume that in small river basins other factors may greatly influence riparian communities. These can include the availability of habitats provided by the river cross-section and the channel breath, or even human-related pressures [80,95,96].

All in all, climate-changed river flow regimes will most probably cause riparian vegetation amendments across rivers with similar flow regimes and even a general reduction in the areas covered by this vegetation. A common feature in all our case studies is that younger and more water-dependent phases are the most affected in a climate change scenario, whatever the forceful climate change or local environmental harshness may be. In snow-fed watersheds the main pathway for riparian vegetation appears to be succession, as minor summer floods cause less fluvial disturbance and greater hydric stress, which in turn allow vegetation to establish itself and develop to maturity, resulting in less water table-dependent phases. In pluvial flow regimes the tendency is consistently the opposite, despite some climatic uncertainties [7]. In this case, retrogression seems to be the main succession pathway for these communities, with large areas near river channels retrogressed to bare soil. Nevertheless, herein changes are not only due to the process of vegetation recycling to the Colonization stage, but also because of its aging to the Mature stage in the farthest floodplain areas. In mountain-fed catchments with mixed flow regimes the tendencies are not so clear and may reflect the existence of insufficient changes in flow regimes for there to be a clear change in their riparian communities. Meticulous analysis of the specific change in area in each succession phase showed that changes that may appear moderate when considering the total riparian patch mosaic can expose dramatic modifications when we look at the specific area changes in each succession phase. This means that many succession phases may face a serious threat in the future, when some of them will be confronted with complete annihilation. This outcome raises the question of maintaining viable populations of species that are important to conservation and are dependent on instream habitats. Additionally, more pronounced modifications – like the ones taking place in Southern European countries – are likely to occur in riparian communities that are dependent on pluvial flow regimes. These results are feasible expectations, inasmuch as similar riparian responses have been documented in vegetation assessments related to past flow regime events [20,44,48,67,97,98,99,100]. There are also existing forecasts that support our findings [50,60,101,102,103,104,105], although gen-erally more superficially and with less detail regarding inner riparian community structure diversity. Climate change can therefore endanger specific riparian species, drive shifts in which exotics become dominant [100,106], or completely disrupt ecological succession in riparian ecosystems – something that can also lead to an increased risk to instream species survival [33] and flood hazards in downstream populations [36].

These results also pave the way for improved knowledge about emerging topics that are as yet insufficiently studied in fluvial ecosystems [52]. In this sense our results help substantiate the metacommunity Patch Dynamics Concept, which can be traced to Hutchinson's [107] seminal ideas about non-equilibrium communities, and reinforces the notion that competitively inferior species are favored by patch disturbance, without which they would be replaced by competitively superior ones. It also helps understand the effects of patch dynamics across different river gradients, as well as the fact that species' life history attributes can influence community dynamics in response to disturbed flow regimes and changed habitat characteristics.

Finally, the results obtained by us through vegetation dynamics simulation can generate new questions stemming from riparian ecology concepts. The expected changes in the spatial ratio of different riparian types, with the likely suspension of succession in some cases, could lead to reflection on the interplay between the fluvial setting and vegetation (e.g. [108]) – i.e. the relative dominance of non-equilibrium versus quasi-equilibrium processes [79]. Our work also suggests new scientific questions regarding the potential feedbacks of novel habitats associated with an altered riparian vegetation mosaic, leading to changes in shear stress disturbance and hydrogeomorphic processes [109,110], or in relation to potential alterations in the global functioning of the ecosystem and thus the services it provides.

## Supporting Information

**Table S1   Confidence intervals for mean shear stress differences.** Confidence intervals for mean shear stress differences between scenarios in each case study.
(DOCX)

## Acknowledgments

The Portuguese team would like to thank António Albuquerque for his priceless support and experienced judgment in fieldwork and data treatment.

## Author Contributions

Conceived and designed the experiments: RR PR MTF AP EP GE AG FF. Performed the experiments: RR EP AG. Analyzed the data: RR PR MTF AP EP GE AG FF. Contributed reagents/materials/analysis tools: RR PR MTF AP EP GE AG FF. Wrote the paper: RR PR MTF AP EP GE AG FF.

## References

1. Bach W (1976) Global air pollution and climatic change. Reviews of Geophysics 14: 429–474.
2. Benton GS (1970) Carbon dioxide and its role in climate change. National Academy of Sciences. pp. 898–899.
3. Hansen J, Johnson D, Lacis A, Lebedeff S, Lee P, et al. (1981) Climate Impact of Increasing Atmospheric Carbon Dioxide. Science 213: 957–966.
4. Lovelock JE (1971) Air pollution and climatic change. Atmospheric Environment (1967) 5: 403–411.
5. IPCC (2008) Climate change 2007: Synthesis Report. Contribution of Working Groups I, II, and III to the Fourth Assessment Report of the Intergovernmental Panel on Climate Change; Team CW, Pachauri RK, Reisinger A, editors. Geneva, Switzerland: Intergovernmental Panel on Climate Change. 104 p.
6. Meehl GA, Stocker TF, Collins WD, Friedlingstein P, Gaye AT, et al. (2007) Global Climate Projections. In: Solomon S, Qin D, Manning M, Chen Z, Marquis M, et al., editors. Climate Change 2007: The Physical Science Basis Contribution ofWorking Group I to the Fourth Assessment Report of the Intergovernmental Panel on Climate Change. Cambridge, United Kingdom and New York, NY, USA: Cambridge University Press. pp. 747–845.
7. Alcamo J, Moreno JM, Nováky B, Bindi M, Corobov R, et al. (2007) Europe. In: Parry ML, Canziani OF, Palutikof JP, van der Linden PJ, Hanson CE, editors. Climate Change 2007: Impacts, Adaptation and Vulnerability

Contribution of Working Group II to the Fourth Assessment Report of the Intergovernmental Panel on Climate Change. Cambridge: Cambridge University Press. pp. 541–580.

8. Christensen J, Christensen OB (2007) A summary of the PRUDENCE model projections of changes in European climate by the end of this century. Climatic Change 81: 7–30.

9. Christensen JH, Hewitson B, Busuioc A, Chen A, Gao X, et al. (2007) Regional Climate Projections. In: Solomon S, Qin D, Manning M, Chen Z, Marquis M, et al., editors. Climate Change 2007: The Physical Science Basis Contribution of Working Group I to the Fourth Assessment Report of the Intergovernmental Panel on Climate Change. Cambridge, United Kingdon and New York, NY, USA: Cambridge University Press. pp. 848–940.

10. Schneider C, Laizé CLR, Acreman MC, Flörke M (2013) How will climate change modify flow regimes in Europe? Hydrology and Earth System Sciences 17: 325–339.

11. Nijssen B, O'Donnell G, Hamlet A, Lettenmaier D (2001) Hydrologic Sensitivity of Global Rivers to Climate Change. Climatic Change 50: 143–175.

12. Serrat-Capdevila A, Valdés JB, Pérez JG, Baird K, Mata LJ, et al. (2007) Modeling climate change impacts - and uncertainty - on the hydrology of a riparian system: The San Pedro Basin (Arizona/Sonora). Journal of Hydrology 347: 48–66.

13. Serrat-Capdevila A, Scott RL, James Shuttleworth W, Valdés JB (2011) Estimating evapotranspiration under warmer climates: Insights from a semi-arid riparian system. Journal of Hydrology 399: 1–11.

14. Verzano K, Menzel L (2007) Snow conditions in mountains and climate change - a global view. In: Marks D, Hock R, Lehning M, Hayashi M, Gurney R, editors; Perugia, IT. IAHS Proceedings and Reports. pp. 147–154.

15. Alcamo J, Flörke M, Märker M (2007) Future long-term changes in global water resources driven by socio-economic and climatic changes. Hydrological Sciences Journal 52: 247–275.

16. Murray SJ, Foster PN, Prentice IC (2012) Future global water resources with respect to climate change and water withdrawals as estimated by a dynamic global vegetation model. Journal of Hydrology 448–449: 14–29.

17. Poff LN, Allan JD, Bain MB, Karr JR, Prestegaard KL, et al. (1997) The natural flow regime. Bioscience 47: 769–784.

18. Lloyd NJ, Quinn G, Thoms MC, Arthington AH, Gawne B, et al. (2004) Does flow modification cause geomorphological and ecological response in rivers? A literature review from an Australian perspective. Technical report 1/2004. Canberra, Australia: CRC for Freshwater Ecology. 0975164202. 57 p. http://www.library.adelaide.edu.au/cgi-bin/director?id=V1114450.

19. Poff LN, Zimmerman JKH (2010) Ecological responses to altered flow regimes: a literature review to inform the science and management of environmental flows. Freshwater Biology 55: 11.

20. Stromberg JC, Lite SJ, Dixon MD (2010) Effects of stream flow patterns on riparian vegetation of a semiarid river: implications for a changing climate. River Research and Applications 26: 712–729.

21. Jenkins M (2003) Prospects for Biodiversity. Science 302: 1175–1177.

22. Costanza R, d'Arge R, de Groot R, Farber S, Grasso M, et al. (1997) The value of the world's ecosystem services and natural capital. Nature 387: 253–260.

23. Perry LG, Andersen DC, Reynolds LV, Nelson SM, Shafroth PB (2012) Vulnerability of riparian ecosystems to elevated CO2 and climate change in arid and semiarid western North America. Global Change Biology 18: 821–842.

24. Karrenberg S, Edwards PJ, Kollmann J (2002) The life history of Salicaceae living in the active zone of floodplains. Freshwater Biology 47: 733–748.

25. Merritt DM, Scott ML, Poff LN, Auble GT, Lytle DA (2010) Theory, methods and tools for determining environmental flows for riparian vegetation: riparian vegetation-flow response guilds. Freshwater Biology 55: 206–225.

26. Rood SB, Braatne JH, Hughes FMR (2003) Ecophysiology of riparian cottonwoods: stream flow dependency, water relations and restoration. Tree Physiology 23: 1113–1124.

27. Junk WJ, Bayley PB, Sparks RE (1989) The Flood Pulse Concept in River-Floodplain Systems. In: Dodge DP, editor. Canadian Special Publication of Fisheries and Aquatic Sciences. pp. 110–127.

28. Naiman RJ, Décamps H (1997) The ecology of interfaces: Riparian zones. Annual Review of Ecology and Systematics 28: 621–658.

29. NRC NRC (2002) Riparian Areas: Functions and Strategies for Management. Washington, D.C., USA: The National Academies Press. 444 p.

30. McClain M, Boyer E, Dent L, Gergel S, Grimm N, et al. (2003) Biogeochemical Hot Spots and Hot Moments at the Interface of Terrestrial and Aquatic Ecosystems. Ecosystems 6: 301–312.

31. Tockner K, Stanford J (2002) Riverine flood plains: present state and future trends. Environmental Conservation 29: 308–330.

32. Tockner K, Bunn SE, Gordon C, Naiman RJ, Quinn GP, et al. (2008) Flood plains: critically threatened ecosystems. In: Polunin NVC, editor. Aquatic Ecosystems: trends and global prospects. New York, USA: Cambridge University Press. pp. 482.

33. Broadmeadow S, Nisbet TR (2004) The effects of riparian forest management on the freshwater environment: a literature review of best management practice. Hydrology & Earth System Sciences 8: 286–305.

34. Naiman RJ, Decamps H, Pollock M (1993) The Role of Riparian Corridors in Maintaining Regional Biodiversity. Ecological Applications 3: 209–212.

35. Casatti L, Teresa FB, Gonçalves-Souza T, Bessa E, Manzotti AR, et al. (2012) From forests to cattail: how does the riparian zone influence stream fish? Neotropical Ichthyology 10: 205–214.

36. Blackwell MSA, Maltby E, editors (2006) How to use floodplains for flood risk reduction. Luxembourg, Belgium: European Communities. 144 p.

37. Daily GC, editor (1997) Nature's Services - Societal Dependence on Natural Ecosystems. Washington D. C., USA: Island press. 392 p.

38. Berges SA (2009) Ecosystem services of riparian areas: stream bank stability and avian habitat. Ames, Iowa, USA: Iowa State University. 106 p.

39. Flather CH, Cordell HK (1995) Outdoor Recreation: Historical and Anticipated Trends. In: Knight RL, Gutzwiller KJ, editors. Wildlife and Recreationists - Coexistence through management and research. Washington D. C., USA: Island press. pp. 372.

40. Holmes TP, Bergstrom JC, Huszar E, Kask SB, Orr Iii F (2004) Contingent valuation, net marginal benefits, and the scale of riparian ecosystem restoration. Ecological Economics 49: 19–30.

41. Naiman RJ, Bilby RE, Bisson PA (2000) Riparian Ecology and Management in the Pacific Coastal Rain Forest. Bioscience 50: 996–1011.

42. Nehlsen W, Williams JE, Lichatowich JA (1991) Pacific Salmon at the Crossroads: Stocks at Risk from California, Oregon, Idaho, and Washington. Fisheries 16: 4–21.

43. Loučková B (2012) Vegetation-landform assemblages along selected rivers in the Czech Republic, a decade after a 500-year flood event. River Research and Applications 28: 1275–1288.

44. Stromberg JC, Tluczek MGF, Hazelton AF, Ajami H (2010) A century of riparian forest expansion following extreme disturbance: Spatio-temporal change in Populus/Salix/Tamarix forests along the Upper San Pedro River, Arizona, USA. Forest Ecology and Management 259: 1181–1198.

45. Wohl E, Angermeier PL, Bledsoe B, Kondolf GM, MacDonnell L, et al. (2005) River restoration. Water Resources Research 41: W10301.

46. Auble G, Scott M, Friedman J (2005) Use of individualistic streamflow-vegetation relations along the Fremont River, Utah, USA to assess impacts of flow alteration on wetland and riparian areas. Wetlands 25: 143–154.

47. Camporeale C, Ridolfi L (2006) Riparian vegetation distribution induced by river flow variability: A stochastic approach. Water Resources Research 42: W10415.

48. Dixon MD, Turner MG (2006) Simulated recruitment of riparian trees and shrubs under natural and regulated flow regimes on the Wisconsin River, USA. River Research and Applications 22: 1057–1083.

49. Orellana F, Verma P, Loheide SP II, Daly E (2012) Monitoring and modeling water-vegetation interactions in groundwater-dependent ecosystems. Reviews of Geophysics 50: RG3003.

50. Primack AB (2000) Simulation of climate-change effects on riparian vegetation in the Pere Marquette River, Michigan. Wetlands 20: 538–547.

51. Tealdi S, Camporeale C, Ridolfi L (2013) Inter-species competition-facilitation in stochastic riparian vegetation dynamics. Journal of Theoretical Biology 318: 13–21.

52. Winemiller KO, Flecker AS, Hoeinghaus DJ (2010) Patch dynamics and environmental heterogeneity in lotic ecosystems. Journal of the North American Benthological Society 29: 84–99.

53. Politi E, Egger G, Angermann K, Blamauer B, Klösch M, et al. (2011) Evaluating climate change impacts on Alpine floodplain vegetation. In: C. . Chomette & Steiger E, editor; 15–17 June; Clermont-Ferrand, France. pp. 177–182.

54. Rivaes R, Rodríguez-González PM, Albuquerque A, Pinheiro AN, Egger G, et al. (2012) Climate change impacts on Mediterranean riparian vegetation; 5th International Perspective on Water Resources & the Environment (IPWE 2012). January 4th-7th; Marrakech, Morocco.

55. Mader H, Steidl T, Wimmer R (1996) Abflußregime österreichischer Fließgewässer. Wien, AUT: Umweltbundesamt. 192 p.

56. Means LO, Hulme M, Carter TR, Leemans R, Lal M, et al. (2001) Climate Scenario Development. In: Houghton JT, Ding Y, Griggs DJ, Noguer M, van der Linden PJ, et al., editors. Climate Change 2001: The Scientific Basis. Cambridge, UK: Cambridge University Press. pp. 739–768.

57. Nakicenovik N, Swart R, editors (2000) Emission Scenarios - Intergovernmental Panel on Climate Change (IPCC) Special Report on Emission Scenarios. Cambridge, UK: Cambridge University Press. 570 p.

58. Santos FD, Forbes K, Moita R, editors (2002) Climate Change in Portugal, Scenarios, Impacts and Adaptation Measures - SIAM project. Lisbon, Portugal: Gradiva. 454 p.

59. Stanzel P, Nachtnebel HP (2010) Mögliche Auswirkungen des Klimawandels auf den Wasserhaushalt und die Wasserkraftnutzung in Österreich. Österreichische Wasser- und Abfallwirtschaft 62: 180–187.

60. Moreno JM, Aguiló E, Alonso S, Cobelas MÁ, Anadón R, et al. (2005) A Preliminary Assessment of the Impacts in Spain due to the Effects of Climate Change. Madrid, SP: Ministerio del Medio Ambiente.

61. Santos FD, Miranda P, editors (2006) Alterações climáticas em Portugal cenários, impactos e medidas de adaptação, Projecto SIAM II. Lisbon, Portugal: Gradiva. 506 p.

62. Crawford NH, Linsley RK (1966) Digital simulation in hydrology: Stanford Watershed Model IV. Department of Civil Engineering, Stanford University. 210 p.

63. Linsley RK, Crawford NH (1960) Computation of a synthetic streamflow record on a digital computer. International Association of Scientific Hydrology 5: 526–538.

64. Hernández L (2007) Efectos del Cambio Climático en los Sistemas Complejos de Recursos Hídricos. Aplicación a la Cuenca del Jucar. Valenvia, SP: Universidad Politécnica de Valencia.

65. Benjankar R, Egger G, Jorde K, Goodwin P, Glenn NF (2011) Dynamic floodplain vegetation model development for the Kootenai River, USA. Journal of Environmental Management 92: 3058–3070.

66. Stanley EH, Powers SM, Lottig NR (2010) The evolving legacy of disturbance in stream ecology: concepts, contributions, and coming challenges. Journal of the North American Benthological Society 29: 67–83.

67. Stromberg JC (2001) Restoration of riparian vegetation in the south-western United States: importance of flow regimes and fluvial dynamism. Journal of Arid Environments 49: 17.

68. Lake PS (2000) Disturbance, patchiness, and diversity in streams. The North American Benthological Society 19: 573–592.

69. Resh VH, Brown AV, Covich AP, Gurtz ME, Li HW, et al. (1988) The Role of Disturbance in Stream Ecology. Journal of the North American Benthological Society 7: 433–455.

70. White PS (1979) Pattern, process, and natural disturbance in vegetation. The Botanical Review 45: 229–299.

71. Tockner K, Malard F, Ward JV (2000) An extension of the flood pulse concept. Hydrological processes 14: 2861–2883.

72. Benjankar R, Egger G, Jorde K (2009) Development of a dynamic floodplain vegetation model for the Kootenai river, USA: concept and methodology. 7th ISE and 8th HIC.

73. Egger G, Politti E, Woo H, Cho K, Park M, et al. (2011) A dynamic vegetation model as a tool for ecological impact assessments of dam operation. Journal of Hydro-environment Research 6: 151–161.

74. García-Arias A, Francés F, Andrés-Doménech I, Vallés F, Garófano-Gómez V, et al. (2011) Modeling the spatial distribution and temporal dynamics of Mediterranean riparian vegetation in a reach of the Mijares River (Spain). In: C. Chomette & Steiger E, editor; EUROMECH Colloquium 523. 15–17 June; Clermont-Ferrand, France. pp. 153–157.

75. García-Arias A, Francés F, Ferreira T, Egger G, Martínez-Capel F, et al. (2013) Implementing a dynamic riparian vegetation model in three European river systems. Ecohydrology 6: 635–651.

76. Rivaes R, Rodríguez-González PM, Albuquerque A, Pinheiro AN, Egger G, et al. (2013) Riparian vegetation responses to altered flow regimes driven by climate change in Mediterranean rivers. Ecohydrology 6: 413–424.

77. RIPFLOW (2011) Riparian vegetation modelling for the assessment of environmental flow regimes and climate change impacts within the WFD. 238 p. http://www.iiama.upv.es/RipFlow/publications/08_RIPFLOW%20Project%20-%20Final%20Report.pdf.

78. R Development Core Team (2011) R: A language and environment for statistical computing. Vienna, AT: R Foundation for Statistical Computing.

79. Bendix J, Hupp CR (2000) Hydrological and geomorphological impacts on riparian plant communities. Hydrological processes 14: 2977–2990.

80. Johnson WC (1999) Response of Riparian Vegetation to Streamflow Regulation and Land Use in the Great Plains. Great Plains Research Vol. 9: pp. -357–369.

81. Tabacchi E, Correll DL, Hauer R, Pinay G, Planty-Tabacchi A-M, et al. (1998) Development, maintenance and role of riparian vegetation in the river landscape. Freshwater Biology 40: 497–516.

82. Pardé M (1955) Fleuves et rivières. Paris: Armand Colin. 241 p.

83. L'vovich MI (1979) World water resources and their future. Chelsea, Michigan, USA: American Geophysical Union. 415 p.

84. Wrzesiński D (2013) Uncertainty of flow regime characteristics of rivers in europe. QUAESTIONES GEOGRAPHICAE 32: 43–53.

85. Friedman JM, Lee VJ (2002) Extreme floods, channel change, and riparian forests along ephemeral streams. Ecological Monographs 72: 16.

86. Whited DC, Lorang MS, Harner MJ, Hauner FR, Kimball JS, et al. (2007) Climate, hydrologic disturbance, and succession: drivers of floodplain pattern. Ecology 88: 940–953.

87. Gurnell A (2014) Plants as river system engineers. Earth Surface Processes and Landforms 39: 4–25.

88. Camporeale C, Perucca E, Ridolfi L, Gurnell AM (2013) MODELING THE INTERACTIONS BETWEEN RIVER MORPHODYNAMICS AND RIPARIAN VEGETATION. Reviews of Geophysics 51: 379–414.

89. Gurnell AM, Bertoldi W, Corenblit D (2012) Changing river channels: The roles of hydrological processes, plants and pioneer fluvial landforms in humid temperate, mixed load, gravel bed rivers. Earth-Science Reviews 111: 129–141.

90. Corenblit D, Baas ACW, Bornette G, Darrozes J, Delmotte S, et al. (2011) Feedbacks between geomorphology and biota controlling Earth surface processes and landforms: A review of foundation concepts and current understandings. Earth-Science Reviews 106: 307–331.

91. Corenblit D, Tabacchi E, Steiger J, Gurnell AM (2007) Reciprocal interactions and adjustments between fluvial landforms and vegetation dynamics in river corridors: A review of complementary approaches. Earth-Science Reviews 84: 56–86.

92. Politti E, Egger G, Angermann K, Rivaes R, Blamauer B, et al. (2014) Evaluating climate change impacts on Alpine floodplain vegetation. Hydrobiologia: 1–19.

93. Auble G, Scott M (1998) Fluvial disturbance patches and cottonwood recruitment along the upper Missouri River, Montana. Wetlands 18: 546–556.

94. Johnson SL, Swanson FJ, Grant GE, Wondzell SM (2000) Riparian forest disturbances by a mountain flood — the influence of floated wood. Hydrological processes 14: 3031–3050.

95. Aguiar FC, Ferreira MT (2005) Human-disturbed landscapes: effects on composition and integrity of riparian woody vegetation in the Tagus River basin, Portugal. Environmental Conservation 32: 30–41.

96. Ferreira MT, Aguiar CF, Nogueira C (2005) Changes in riparian woods over space and time: Influence of environment and land use. Forest Ecology and Management 212: 145–159.

97. Pettit NE, Froend RH, Davies PM (2001) Identifying the natural flow regime and the relationship with riparian vegetation for two contrasting western Australian rivers. Regulated Rivers: Research & Management 17: 201–215.

98. Shafroth PB, Stromberg JC, Patten DT (2002) Riparian vegetation response to altered disturbance and stress regimes. Ecological Applications 12: 107–123.

99. Stromberg JC, Tiller R, Richter B (1996) Effects of Groundwater Decline on Riparian Vegetation of Semiarid Regions: The San Pedro, Arizona. Ecological Applications 6: 113–131.

100. Stromberg JC, Lite SJ, Marler R, Paradzick C, Shafroth PB, et al. (2007) Altered stream-flow regimes and invasive plant species: the Tamarix case. Global Ecology and Biogeography 16: 381–393.

101. Frederick KD, Major DC (1997) Climate Change and Water Resources. Climatic Change 37: 7–23.

102. Hoffman MT, Rohde RF (2011) Rivers Through Time: Historical Changes in the Riparian Vegetation of the Semi-Arid, Winter Rainfall Region of South Africa in Response to Climate and Land Use. Journal of the History of Biology 44: 59–80.

103. Schneider C, Flörke M, Geerling G, Duel H, Grygoruk M, et al. (2011) The future of European floodplain wetlands under a changing climate. Journal of Water and Climate Change 2: 106–122.

104. Tague C, Seaby L, Hope A (2009) Modeling the eco-hydrologic response of a Mediterranean type ecosystem to the combined impacts of projected climate change and altered fire frequencies. Climate Change 93: 137–155.

105. Watson RT, Zinyowera MC, Moss RH, editors (1996) Climate Change 1995 - Impacts, adaptations and mitigation of climate change: scientific-technical analyses. Contribution of Working Group II to the Second Assessment Report of the Intergovernmental Panel on Climate Change. New York, USA: Cambridge University Press. 879 p.

106. Hultine KR, Bush SE (2011) Ecohydrological consequences of non-native riparian vegetation in the southwestern United States: A review from an ecophysiological perspective. Water Resources Research 47: W07542.

107. Hutchinson GE (1953) The Concept of Pattern in Ecology. Proceedings of the Academy of Natural Sciences of Philadelphia 105: 1–12.

108. Muneepeerakul R, Rinaldo A, Rodriguez-Iturbe I (2007) Effects of river flow scaling properties on riparian width and vegetation biomass. Water Resources Research 43: W12406.

109. Gran K, Paola C (2001) Riparian vegetation controls on braided stream dynamics. Water Resources Research 37: 3275–3283.

110. Johnson WC (2002) Riparian vegetation diversity along regulated rivers: contribution of novel and relict habitats. Freshwater Biology 47: 749–759.

# 14

# Rapid Response of Hydrological Loss of DOC to Water Table Drawdown and Warming in Zoige Peatland: Results from a Mesocosm Experiment

**Xue-Dong Lou[1,2], Sheng-Qiang Zhai[1], Bing Kang[2], Ya-Lin Hu[3], Li-Le Hu[1]\***

**1** Chinese Research Academy of Environmental Sciences, Beijing, China, **2** College of Life Sciences, Northwest Agriculture & Forestry University, Yangling, Shaanxi, China, **3** Institute of Applied Ecology, Chinese Academy of Sciences, Shenyang, China

## Abstract

A large portion of the global carbon pool is stored in peatlands, which are sensitive to a changing environment conditions. The hydrological loss of dissolved organic carbon (DOC) is believed to play a key role in determining the carbon balance in peatlands. Zoige peatland, the largest peat store in China, is experiencing climatic warming and drying as well as experiencing severe artificial drainage. Using a fully crossed factorial design, we experimentally manipulated temperature and controlled the water tables in large mesocosms containing intact peat monoliths. Specifically, we determined the impact of warming and water table position on the hydrological loss of DOC, the exported amounts, concentrations and qualities of DOC, and the discharge volume in Zoige peatland. Our results revealed that of the water table position had a greater impact on DOC export than the warming treatment, which showed no interactive effects with the water table treatment. Both DOC concentration and discharge volume were significantly increased when water table drawdown, while only the DOC concentration was significantly promoted by warming treatment. Annual DOC export was increased by 69% and 102% when the water table, controlled at 0 cm, was experimentally lowered by −10 cm and −20 cm. Increases in colored and aromatic constituents of DOC (measured by $Abs_{254\ nm}$, $SUVA_{254\ nm}$, $Abs_{400\ nm}$, and $SUVA_{400\ nm}$) were observed under the lower water tables and at the higher peat temperature. Our results provide an indication of the potential impacts of climatic change and anthropogenic drainage on the carbon cycle and/or water storage in a peatland and simultaneously imply the likelihood of potential damage to downstream ecosystems. Furthermore, our results highlight the need for local protection and sustainable development, as well as suggest that more research is required to better understand the impacts of climatic change and artificial disturbances on peatland degradation.

**Editor:** Shiping Wang, Institute of Tibetan Plateau Research, China

**Funding:** The authors received funding from the Nature Science Foundation of China under Grant Nos. 41103041 and 41271318 to support this study. The funders had no role in study design, data collection and analysis, decision to publish, or preparation of the manuscript.

**Competing Interests:** All authors have declared that no competing interests exist.

\* Email: hulile@craes.org.cn

## Introduction

Generally, peat-accumulating wetlands provide waterlogged conditions where carbon accumulation is encouraged [1], and therefore have huge carbon storage potential. However, there is increasing concern that carbon storage in peatlands is unstable and may be susceptible to water table drawdown and higher temperatures over the next two centuries due to projected climatic change [2–6]. Furthermore, the water table in peatlands may also be significantly lowered by drainage resulting from human activities [7,8]. As the largest highland wetland in the world [9] and the largest peat storage area in China, the Zoige alpine wetland serves as a natural barrier and prevents desertification in Northwest China, extending farther toward Southeast China, and is very sensitive to climate change [10]. The peatland in Zoige is also the major water source of the world's largest plateau reserve (i.e., Three-Rivers Source Nature Reserve), supplying water for the three most important rivers in East Asia (i.e. the Yellow, Yangtze, and Lancang rivers) [11]. The Zoige wetland is

particularly closely associated with the ecological security of the Yellow River drainage basin [12] because it provides about 40% of the total flow of the Yellow River [13]. Zoige peatland covers an estimated area of 0.5 million hectares and accounts for 47.53% of the total organic carbon reserves in Chinese peatland. Thus, it accounts for the highest organic carbon accumulation of any peatland in China [14].

Unfortunately, due to climate warming, artificial drainage for pastures, and peat exploitation since the 1970s, Zoige peatland has suffered extensive biodiversity loss and ecosystem degradation, including severe peat deterioration [9]. The Zoige wetland has decreased by 30% in the past 30 years due to water table drawdown [15], and artificial drainage has been regarded to be the most important cause of Zoige wetland (including peatland) degradation [9]. Previous studies suggested that the carbon cycle in peatland could change rapidly with climate change [16–18] and that is sensitive to water table [4,19–21]. Therefore, climate warming and a lowered of the water table could potentially create a carbon storage and ecosystem stability crisis in Zoige peatland.

Dissolved organic carbon (DOC) is the most active and sensitive indictor in the carbon cycle [22], and connects the biogeochemical cycle from terrestrial to aquatic ecosystems [23]. Hydrological losses of aquatic carbon can be of significant concern when determining carbon storage in peatlands [24] and may be increasing [25,26]. Among the aquatic constituents of peatlands, DOC is generally considered to have the largest aquatic carbon flux [27,28]. The peculiar water–peat interaction system and strong hydrological connectivity in peatlands ensures that the export of DOC from peatland to downstream plays a key role in the regional redistribution of terrestrial carbon [29] and the carbon balance [30]. Furthermore, the transfer of carbon from terrestrial peatland to fluvial downstream locations has a large influence on the water quality in aquatic ecosystems [31,32]. Previous studies have warned that larger amounts of DOC feeding into downstream locations could increase the levels of aquatic organic acids, decrease the buffering ability of the water, and attenuate the penetration of visible and ultraviolet (UV) light due to changes in the water color [33]. This is likely to cause damage to the sustainable and stable development of aquatic ecosystems, such as their net primary productivity [34] and production of bacteria [2,35]. A large body of literature has reported changes in the color or aromatic components of water in peatlands that has occurred in recent years [11,31,36,37]. $SUVA_{254 nm}$ was a useful parameter for determining the aromatic characteristic of DOC [38], and absorbance at 400 nm was used as a measure of the color composition [36] and could further indicate changes in DOC composition when combined with specific absorbance [39]. Therefore, DOC is likely to be an important part of the carbon cycle linking peatland and downstream ecosystems, although it is not the only pathway of carbon loss from an upland peatland.

The amount of DOC exported from peatlands is believed to depend on interactions between discharged water through peatland and the production and consumption of DOC within the peatland [4]. However, it has also been reported to increase with a higher discharge [25,40,41] without any effect on DOC concentration. Climate change can regulate the import and export of DOC [42,43], mainly by controlling the most important environmental factors (i.e., temperature and the water table) affecting the peatland carbon cycle. A high water table and low soil temperatures are believed to be major reasons for the low decay rates, which could restrict the production of DOC compounds [44–46]. However, previous observations have indicated that DOC concentrations in peat could be either elevated [47–49] or lowered [4,37,50,51] with a decline in the water level, which could be contributed to the complicated mechanisms and processes involved in the production, consumption, and transport of DOC in peat along with inevitable site-specific characteristics [52]. Similarly, high temperatures can not only improve DOC production through enhanced phenol oxidase activity but also increase the consumption of DOC [25,43]. Thus, it is difficult to determine DOC concentrations in specific regions without performing practical experiments. Moreover, some studies have observed significant changes in water color and aromatic content with shifting water tables and soil temperatures at a range of sites [28,36,49,52–54]. Many previous studies have produced inconsistent results regarding the effects of changes to the water table and/or warming on aquatic DOC release, with both factors able to impact DOC concentrations, the amount of discharge, or both, in a confounding way. Specifically, the response of these variables in peatlands could depend on the length of the observation period. For example, the response of DOC production to drought conditions in the year of drought may differ from that a few years after the drought [55,56], and in a Tibetan alpine meadow experiment, the response of the aboveground environment to warming treatments in the third year was found to be different from the trend of the first two years [57]. Therefore, our observations in the year immediately after a controlled experiment are helpful for understanding how DOC export might react to climate change and anthropogenic drainage.

Zoige peatland is known to be undergoing a warming and drying climate trend [46], and severe artificial drainage [58]. Several studies in Zoige recently have reported that changes of temperature and/or water table could cause effect on the emissions of CH4 and CO2 [23,26,58,59], and Luo et al. [60] has noticed that DOC could response to experimental warming and grazing. However, there is currently knowledge of the potential response of hydrological DOC loss to the variation of temperature and water table in Zoige peatland. Furthermore, most previous studies on DOC have been conducted countries other than China, particularly in Europe and North America. Therefore, investigating regarding the response of DOC export to warming and water table treatments could provide insight into the impact and mechanisms of climate warming and artificial drainage on the regional carbon budget of Zoige peatland, as well as provide guidance for the local protection and restoration of this deteriorating natural environment. Thus, we undertook a mesocosm experiment to investigate how the hydrological loss of DOC would respond to climatic warming and artificial drainage. The specific objectives were to determine whether the export quantity and concentration, as well as the qualities of DOC and the discharge of flow water, could respond significantly to water table and temperature manipulations. In terms of potential changes of DOC export, the study provides evidence of possible changes to the carbon cycle and storage under the impact of climate change or artificial disturbance and provides evidence for the need to protect and further restore the Zoige peatland.

## Methods

### Field Site

The peat columns used for mesocosms were collected from Zoige peatland in Hongyuan County, Sichuan Province, on the northeastern margin of the Qinghai–Tibet Plateau (32.76°N, 102.5°E), with a mean altitude of about 3,500 m. Peat was extracted in the area for energy production until 2003, which has left a peat layer of approximately 2 m deep and created severe long-term water shortages [61]. The vegetation community mainly consists of *Carex muliensis* (relative coverage of 41%) and *Kobresia setchwanensis* (39%), as well as a small number of scattered *Potentilla anserina* (15%) and *Plantago depressa* (11%). The topography and vegetation characteristics of the study area are shown in Figure S1. During the period 2002–2011, the site experienced a mean annual temperature and precipitation level of 2.27°C and 700 mm year$^{-1}$, respectively. During that period, the mean temperature and precipitation from May to October were 8.26°C and 596.34 mm, respectively (data obtained from the China Meteorological Data Sharing Service System at http://cdc.cma.gov.cn/home.do). This study was conducted from May to October in 2012, when the mean temperature and precipitation were 8.65°C and 808 mm, respectively. Therefore, the site experienced higher rainfall and higher temperatures than the average of the previous 10 years.

The study was carried out on the private land of Mr. Jiang in Hongyuan County. Please contact the author first if further information is required. No further permits were required for the locations/activities in the study, and our work did not involve any endangered or protected species.

## Mesocosm Experiment

All peat columns were extracted intact from the source plot in December 2011, when the peat was totally frozen and easy to move and reset. Frozen peat cores (cuboid-shaped, with intact vegetation and peat structure) with a surface area of 1 m$^2$ and a depth of 50–66 cm were carefully placed into stainless-steel barrels with only an open top. We used perforated stainless steel (diameter 9.0 cm) as a pocket sand filter (gravel particle size <4 mm), passing water through its inlet to maintain a near-natural infiltration rate. The perforated stainless-steel filter was buried into the peat column and connected by a drainage system to 5-L tanks in the closed bottom used to store the discharge [62]. The drainage system was connected to a manostat device with a similar pocket sand filter in the interface to lessen the peat outflow. Eighteen mesocosms were constructed for the manipulation of temperature and water table levels (three water table levels, two temperature, and three replicates, n = 18) in a crossed factorial experiment that commenced in May 2012. Positions of the water table level were controlled by hanger loops of the drainage system and set to 0 cm (W0), −10 cm (W1), and −20 cm (W2). They were calibrated using engraved rulers placed adjacent to the bottom of the steel barrels (i.e., the height of the water table was equal to the depth of the peat column plus the observed value). Warming treatment was achieved by using open top chambers (OTCs) during the snow-free period following Walker et al. [63], with 0.43-m-high polycarbonate solar panels placed outside of the mesocosms instead of infrared lamps. Actually, OTCs realize warming mainly through reducing both wind-speed and air convection and increasing incoming solar radiation [64,65]. It can be confirmed by results of previous studies [63,65,66]. During the first growing season, we observed an overall temperature increase of 1.35°C (on an annual basis) was observed for the peat with a −10 cm water table in the warming mesocosms (T1) compared to the ambient mesocosms (T0) (Figure S2). The details on the experiment design in the study were shown in Figure S4.

To closely monitor the output–input water budget in the mesocosms, water discharged from the mesocosms, rainfall, and recharge water were measured using a gauge at least once a week, and more frequently for the first two measurements when rainfall occurred. Water in each mesocosm was mainly supplied by natural precipitation and supplemented by water pumped from a nearby drainage ditch to maintain the preset water table level when necessary. As the drainage ditch extended from the same continuous *C. muliensis* peatland, thus this supplementary water had a similar attributes to the water at the field site. We buried four HOBOPro data-loggers to record peat temperature at −10 cm depth in the mesocosms: two in warming mesocosms with a water table level of 0 cm and two in control mesocosms with a water table level of −20 cm. Monthly weather data from the Hongyuan County weather station were collected for reference.

## Sample Analysis

During the study period, discharged water was collected every month for DOC analysis during the growing season (May–October) in 2012. Water samples collected from the manostat tanks were mixed well before sampling, stored in sterile containers (volume 100 ml), and then filtrated through a syringe microfilter (0.45 µm) as preparation for further testing. The DOC concentration was equivalent to total carbon (TC) minus dissolved inorganic carbon (DIC), and both were determined directly using a TOC/TN analyzer (Multi N/C3100TOC/TN; Analytik Jena, Germany). TC was measured by wet combustion, and DIC was measured after sample acidification by 10% $H_3PO_4$ as proposed

by Guo et al. [67]. The water budget data and the measured DOC concentrations were used to estimate DOC export by Method 3 proposed by Walling and Webb [68]. The UV absorbances of filtered water samples at wavelengths of 254 nm and 400 nm were determined using a UV-visible spectrophotometer (UV-2600; Shimadzu, Kyoto, Japan). UV absorption characteristics of DOC are generally measured to obtain information regarding changes in the composition of DOC compounds. We thus determined the characteristics of DOC composition by means of four related measurements of specific- and UV absorption (i.e., Abs$_{254 nm}$, SUVA$_{254 nm}$, Abs$_{400 nm}$, and SUVA$_{400 nm}$).

## Statistical Analysis

Statistical analysis was done using a three-way ANOVA, including the effect of interactions between the time variable (month) and the two treatments on the monthly changes of DOC. Then a sequential full model of two-way repeated-measures ANOVA and main effect analysis and a Sidak post hoc comparison of means test were successively conducted to determine the effects of two treatments. All of these analyses were conducted after testing for essential homogeneity of variance ($p>$ 0.05, meaning that variances were homogenous; see Table. 1). Further correlation and regression analyses were conducted to determine the relationships of the monthly mean DOC concentrations and DOC export values with the corresponding peat temperature and precipitation. We also conducted linear regression analysis to determine what proportion of the two treatments and discharged volumes respectively. All statistical analyses were performed using SPSS 17.0 (SPSS Inc., Chicago, IL, USA).

## Results

### Microclimate in Mesocosms and Its Correlation with DOC

**Soil Temperature in Peat and Precipitation.** The mean monthly temperatures in the mesocosms showed that peat temperature (from May to October) was significantly higher on average in warmed (11.95 ± 4.03°C) than in the controlled mesocosms (10.60 ± 4.25°C, $p = 0.003$; n = 164; Figure S2). The mean monthly precipitation measured with the rain gauge in the mesocosms was 139.14 mm (139.14 L m$^{-2}$) during the growing season, which was very similar to the value of 134.92 mm recorded by the meteorological station in Hongyuan County. In general, precipitation in all mesocosms was nearly identical regardless of the possible changes in discharge and evapotranspiration among mesocosms located at the same site. A mean number of 17.4 rainy days per month occurred from May to September. Specifically, October only had five rainy days, and June and July each had twenty-one rainy days. The average discharge in the mesocosms was 281.19 L m$^{-2}$ year$^{-1}$, which accounted for 32% of the water input (precipitation and water recharge) and 34% of the rainfall. In contrast, the annual precipitation of temperate biomes (e.g. temperate forest) in China ranges from 400 to 650 mm, i.e., less than the average precipitation (834.84 mm for 6 months) at Zoige. This suggests an abundance of precipitation in Zoige peatland, which is necessary to maintain its year-round spongy condition. It was found that the effect of warming on discharge was nonsignificant in the experiment ($p>0.05$; Table 1). Meanwhile, the interactive effect of warming and water table on discharge was also insignificant ($p>0.05$; Table 1).

**Correlation of Peat Temperature and Precipitation with DOC.** One year study provides limited perspective on the interannual patterns of DOC hydrological export. However, we obtained extra information by the correlation analysis between the two main microclimatic factors and the mean monthly concen-

**Table 1.** P-values of a two-way ANOVA and Levene's test for the effects of the water table level, temperature, and their interactions on the amount of annual DOC export, DOC concentration, absorbance and specific absorbance, and water discharge.

| Treatment | DOC | | Discharge | Absorbances and specific absorbances | | | |
|---|---|---|---|---|---|---|---|
| | Export | Concentration | Discharge | $Abs_{254\ nm}$ | $SUVA_{254}$ | $Abs_{400\ nm}$ | $SUVA_{400}$ |
| Water table | <0.001** | <0.001** | 0.037* | 0.001** | 0.005** | 0.003** | 0.008** |
| Temperature | 0.075 | 0.012** | 0.764 | 0.007** | 0.008** | 0.010** | 0.018** |
| Water table × Temperature | 0.734 | 0.735 | 0.553 | 0.689 | 0.077 | 0.439 | 0.431 |
| Levene's Test | 0.179 | 0.318 | 0.235 | 0.520 | 0.318 | 0.616 | 0.110 |

*indicates a significant difference ($p<0.05$, n = 18)
**indicates a highly significant difference ($p<0.01$, n = 18).

tration and export of DOC, see Figure S3. Monthly DOC export was positively correlated with peat temperature ($R^2 = 0.4984$, $p < 0.01$; n = 24) and precipitation ($R^2 = 0.8982$, $p<0.01$; n = 6), while the mean monthly DOC concentration was significantly correlated with peat temperature ($R^2 = 0.4025$, $p<0.01$; n = 24), but had a nonsignificant correlation with precipitation ($R^2 = 0.3046$, $p = 0.128$; n = 6).

## Annual Export of DOC

Neither the water table × warming nor the month × two controlling-factor interactions were statistically significant. Thus, we examined the single effect of water table and temperature manipulation on the measured variables in the study (i.e. DOC export, concentration, and quality, and the water budget).

We found a difference between the experimental warming treatment and water table treatment in terms of the annual amount of DOC exported. During the period of the study, the manipulation of water table depth in the mesocosms significantly influenced DOC export ($p<0.001$; Figure 1). The export of DOC displayed an upward trend with decreasing water levels. DOC export was $5.76 \pm 0.63$ g C m$^{-2}$ year$^{-1}$ when the water table was at 0 cm, significantly lower than the levels of $9.75 \pm 0.84$ g C m$^{-2}$ year$^{-1}$ ($-10$ cm water-level; $p = 0.004$) and $11.65 \pm 1.68$ g C m$^{-2}$ year$^{-1}$ ($-20$ cm water-level; $p<0.001$) respectively. Although all values were within the range (5–40 C m$^{-2}$ year$^{-1}$) found in the natural peatland [3], the results indicated that DOC export would increase by 69% and 102% annually if the water table at 0 cm was lowered by 10 cm and 20 cm, respectively. In contrast, no significant effect on DOC export was observed in the warming treatments when the peat temperature was raised by 1.35°C ($9.81\pm3.32$ g C m$^{-2}$ year$^{-1}$ vs. $8.30\pm1.82$ g C m$^{-2}$ year$^{-1}$, $p = 0.059$) throughout the growing season. Previous studies have demonstrated that DOC loss during the nongrowing season is similar between controlled and manipulated sites [41]. Therefore, our estimation of DOC export within the growing season did not suggest an alteration in its overall tendency throughout the year.

Eighty-seven percent of the variability in DOC annual export was explained by the combination of the water table level, temperature, and discharge. Among these variables, the level of the water table was the most important predictor, explaining more than 68% of the variation in DOC annual export. Furthermore, discharge was found to be significantly affected only by the water table treatment ($p = 0.028$, n = 18; Figure 2). The discharge

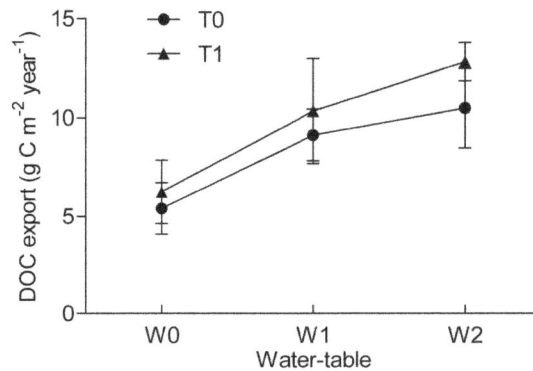

**Figure 1. Effect of water table levels and temperature on DOC annual export.** Data are means ± standard error. T0 and T1 correspond to ambient temperature and warming temperature, respectively, and W0, W1, and W2 indicate water table depths of 0 cm, −10 cm, and −20 cm, respectively.

volume was $265.08 \pm 1.88$ L year$^{-1}$ when the water table was at 0 cm, slightly lower than $293.58.04 \pm 6.71$ L year$^{-1}$ ($-10$ cm water-level; $p = 0.063$) and significantly lower than $294.92 \pm 10.84$ L year$^{-1}$ ($-20$ cm water-level; $p = 0.050$). In contrast, the discharge in the warming treatment was nonsignificantly smaller ($283.11 \pm 17.66$ L vs. $279.28 \pm 29.28$ L). Correlation analysis also showed that water table levels were significantly negatively correlated with discharge ($R^2 = 0.31$, $p = 0.008$; n = 18), while experimental warming was almost irrelevant to discharge ($R^2 = 0.004$, $p = 0.398$; n = 18).

## DOC Concentrations

DOC concentrations in discharged water varied significantly between the warmed and ambient temperature treatments as well as among the three water table treatments, despite the differences in their effects on DOC export. As shown in Figure 2, the DOC concentration was lower ($23.18 \pm 5.76$ mg L$^{-1}$) when the position of the water table level was at 0 cm, significantly lower than $32.81 \pm 2.88$ mg L$^{-1}$ ($-10$ cm water-level; $p = 0.028$) and $38.92 \pm 4.98$ mg L$^{-1}$ ($-20$ cm water-level; $p = 0.001$). Similarly, DOC concentration was higher in the warmed mesocosms ($34.90 \pm 5.25$ mg L$^{-1}$, $p = 0.005$) than in the ambient temperature mesocosms ($28.48 \pm 8.13$ mg L$^{-1}$). In addition, almost 78% of the variation in DOC concentrations was explained by water table and temperature treatments, and 61% of these could be contributed to the water table treatment alone.

## Qualities of DOC

There were similar trends for the effects of experimental warming and water table level on absorbance (at wavelengths of 254 nm and 400 nm) and specific absorbance (SUVA$_{254 \text{ nm}}$ and SUVA$_{400 \text{ nm}}$) of DOC in the filtered discharge. These four measures of the quality of DOC were all significantly higher under the warming treatment ($p = 0.004$, $p = 0.016$, $p = 0.004$, and $p = 0.015$, respectively; Table 1, Figure 3) than in the control. Similarly, values of the four measures for a lower water table were significantly higher than those observed at higher water table level ($p < 0.001$, $p = 0.010$, $p < 0.001$, $p = 0.006$, respectively; Table 1). We assessed the impact of the three positions of the water table level on the four DOC quality measures using a multiple comparison analysis as shown in Figure 3. Therefore, the results above showed that the water table and warming treatments clearly led to several changes in both the quality and absolute DOC concentrations of DOC, indicating a higher aromatic content and changes in the color of downstream water.

## Discussion

### Effect of Water Table Treatment on Export, Concentration, and Qualities of DOC

Water table manipulation had significant effects on the annual amount of DOC exported, DOC concentration, and the water discharge. Lower water tables were often accompanied by higher DOC exports and concentrations, and a larger discharge volume. These effects were significant when the water table at 0 cm was lowered to $-10$ cm and $-20$ cm. This supports the observed variation in the export of DOC, possibly due to site-specific characteristics, and is mainly derived from both fluctuations in the DOC concentration in runoff water and the quantity of water discharged, which disagrees with several previous studies [4,26,69]. Peatland has its own specific features such as plant community construction [70] and hydro-topographical characteristics [67]. Meanwhile, the export of DOC varies with catchment properties and hydrogeologic setting [71], such as precipitation, evapotranspiration [72] and annual runoff [73]. Therefore, it may have disparate performances in DOC production and export which may result in difficulty to draw a universal conclusion [67]. Actually, most reported studies on the response of DOC to water table were done in Europe [74] or North America [4,69]. In this study, when the water table at 0 cm was artificially lowered to $-10$ cm and $-20$ cm, DOC export increased by 69% and 102%, respectively. This suggests that an estimated extra $18.4 \times 10^9$ g C and $27.2 \times 10^9$ g C, respectively, would be transported downstream during the growing season in Zoige. Our ranges of estimated DOC exports with a water table level of $-10$ cm and $-20$ cm ($9.75$–$11.65$ g C m$^{-2}$ year$^{-1}$) were similar to those ($8.4$–$11.3$ C m$^{-2}$ year$^{-1}$) observed in Quebec, Canada, in two growing seasons following a water table drawdown [41]. This could result in a substantial DOC loading into downstream ecosystems, potentially altering the physical and chemical characteristics of aquatic ecosystems, such as acidity, light penetration, and metal and nutrient availability [75]. Such indirect effects on aquatic ecosystems may be the most severe consequence of the elevated DOC export from peatlands. Increases in DOC export may also indicate shifts in the carbon budget, suggesting either a decrease in carbon uptake and storage, or an increase in the turnover of organic carbon [39].

As expected, DOC concentrations were elevated when the water level declined, which agreed with the results of several previous studies [47–49], but disagreed with some other published observations [4,37,50,51]. These inconsistencies reflect the complicated mechanisms and processes involved in the production, consumption, and transport of DOC in peat [52]. Initially, water

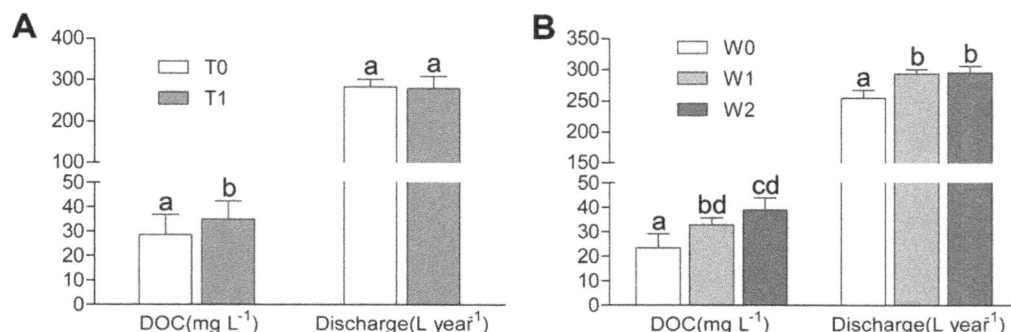

**Figure 2. Variations in DOC concentrations and discharge volumes under different treatments.** Data are means $\pm$ standard error. Same letter superscripts denote insignificant differences among the three water table levels from post hoc tests.

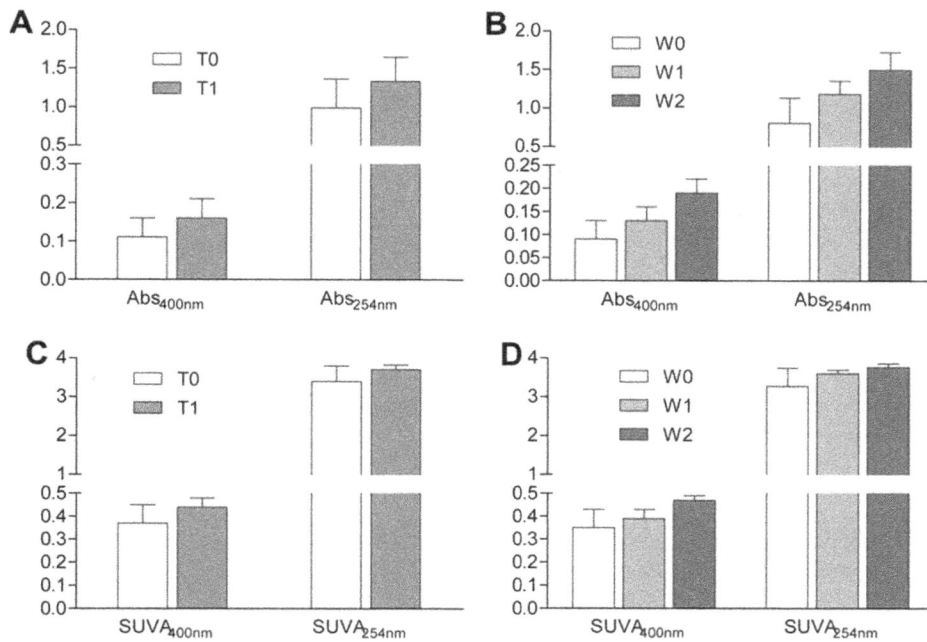

**Figure 3. Effects of water table levels and temperature treatments on $Abs_{254\ nm}$, $Abs_{400\ nm}$, $SUVA_{254\ nm}$ and $SUVA_{400\ nm}$.** Data are means ± standard error. Same letter superscripts denote nonsignificant differences among the three water table levels from post hoc tests.

table drawdown, which is closely associated with peat moisture content, can promote aerobic respiration [76,77] and release the activity of degrading hydrolase enzymes in peat, which is supported by reported changes in the specific absorbance of DOC [78]. Biomass is often related to water table, larger biomass being associated with lower water table [79]. Besides, water table could also affect decomposition indirectly through changes in plant community composition [80] or reduce productivity and even cause death in wetland species as the water table is lowered too far [79], thus it can't be determined water table effects on decomposition through changes in plant community or microbial activity within the results of our study. These two mechanisms described above lead to an increase in DOC production, which can then be flushed out from stagnant peat horizons [41] during rainfall. Furthermore, studies have shown that this could lead to peat subsidence and a lower porosity when the water level declines [42,81], likely resulting in slower interflow and a longer residence time for water moving through peat. This may also contribute to more DOC compounds from peat transferring into flowing water. There were 17.4 rainy days every month from May to September 2012 in Zoige, with a maximum of 21 in each of June and July. According to Harrison et al. [82], this high rainfall rate may have potential effect on DOC concentration in water by promoting DOC release from peat soil. Moreover, the water discharge also increased when the level of the water table was lowered in the mesocosm experiment. This suggests a decreased capacity for water storage, which might exacerbate severe water loss and peat erosion if artificial water drainage continues in Zoige peatland. Higher DOC concentrations could also create problems for the stable and sustainable development of peatland ecosystems because it would alter aquatic habitats through its effect on pH and various biological activities (e.g., transportation of nutrients). The weak correlation between precipitation and DOC concentration also indicates that precipitation could promote the export of DOC with a limited enhancement of the DOC concentration. It

also indicates that the elevated DOC in runoff water was the main contributor to the high discharge rather than the high concentration observed under higher rainfall.

The changes in DOC concentrations resulting from the water table treatment were accompanied by changes in UV absorbance characteristics. Our results showed that the aromatic and colored components of DOC increased when the water table was low, suggesting a possible increase in peat degradation. The colored components of DOC, as measured by absorbance at 400 nm and $SUVA_{400\ nm}$, increased sharply when the water table at 0 cm level was lowered to −20 cm, but showed only a slight decline when it was lowered to −10 cm. The increase in DOC aromaticity, measured by absorbance at 254 nm and $SUVA_{254\ nm}$, implied that more aromatic DOC substances should occur in discharge water following a decline in the water table level. Recent studies of Zoige peatland have demonstrated that aromaticity has substantially increased in sites that have experienced a long period of aerobic oxidation and water loss [11]. This is in accordance with our observation of a higher aromatic content when the water table was lower because more peat would be exposed to the air, resulting in increased aerobic respiration [76]. This would also support a reduction in the colored components and aromatic content when the water table rises, which has also been observed elsewhere [36,49,53]. Consequently, DOC in runoff consists of more colored and aromatic components when the water table was drawdown, making it less accessible to microbes within the fluvial ecosystem [54]. Finally, as our results suggested, it would lead to carrying more DOC compounds downstream.

## Effect of Warming on Export, Concentration, and Qualities of DOC

Experimental warming at a rate of $1.35°C·year^{-1}$ in peat had a significant effect on DOC concentration but limited effects on DOC export and discharge. DOC concentrations in warming mesocosms were higher than in normal mesocosms. However, the observed variability of DOC concentrations generally had a limited impact on the hydrological export of DOC which is explained by the relevance of the discharge volume to DOC exports and displayed a nonsignificant correlation with warming in our experiment. Some studies have observed high DOC concentrations at higher temperatures that increased the amount of DOC exported for as long as 12 years under natural warming conditions [25], whereas others have reported a decrease in DOC exports due to the lower discharge following a temperature increase of 1.6–4.1°C [4]. In the study, warming mesocosms have significantly higher DOC concentration but with insignificant lower discharge, which may result from its specific feature (such as evapotranspiration) [72] in Zoige that differs from anywhere else. Therefore, we assumed that temperature variation may have a complex influence on DOC export that is probably associated with both the rate of warming and the temporal scale. It is important to consider both present and future climate change when investigating the effects of experimental warming on peatland carbon turnover.

Temperature is the main factor influencing bacterial metabolism and the rate of decomposition of organic materials, and it also affects DOC dynamics in ecosystems [83]. High temperatures can not increase DOC production through enhanced phenol oxidase activity, but it also can increase the consumption of DOC [25,43]. Thus, determining the DOC concentration in specific regions is difficult without conducting practical experiments. Warming can decrease plant species richness but increase aboveground net primary production [84], thus it may influence inputs of carbon into peatland and lead to unstable DOC concentrations. Our correlation analysis in our experiment indicated that the monthly mean DOC concentration in the discharge increased with temperature. This suggests that the higher temperature increased the DOC concentration because the enhanced decomposition exceeded the gain in DOC consumption in the first year following a rise in temperature. Walker et al. [63] suggested that warming in OTC experiments increased the height and cover of deciduous shrubs and graminoids and decreased species diversity and evenness, implying that the increased DOC concentration we observed might also be attributable to the changes in plant primary productivity. However, temperature had a weak effect on the water discharge during the whole growing season, indicating a relatively stable discharge volume independent of the warming treatment, and differing slightly from the observations of Bridgham et al. [62]. However, this is understandable given the relatively plentiful precipitation at the study site as well as the relatively small change in peat temperature.

Similarly, all of the colored components displayed an upward trend with warming in the study, which was contrary to the results obtained in a laboratory experiment by Tang et al. [52]. But one recent study showed that warming could cause a shift in the composition of bacterial communities in the surface (1–3 cm) and middle layers (9–11 cm) of peat [85], which supported our observations of greater colored and aromatic content under warmer conditions. Therefore, our results indicated that the rising temperature could influence the composition of peat (especially color and aromatic content). Furthermore, peat degradation might arise following climatic warming according to the results, probably due to potential shifts in the function and structure of microbial communities in the peatland, a hypothesis that requires further investigation.

## Conclusion

We investigated the response of the hydrological export of DOC in Zoige peatland to changes in the water table level and temperature. Our one-year study provides a basis for understanding the rapid response of the carbon budget in Zoige peatland to climate change and/or artificial drainage, as well as the potential damage to downstream ecosystems, particularly in the Yellow River.

The differences between the water table and peat-temperature treatments implied that future short-duration water table drawdown events could have a greater impact than rising temperature on the export of DOC. In this study, water table affects DOC concentration and export as well as discharge, while temperature treatment only causes obvious effect on DOC concentration. It probably derives from that water table drawdown influenced temperature patterns in the decomposing litter [80] in the crossed factorial experiment. Meanwhile, the experimental warming may also not be high enough for changing the amount of DOC export, as warming could also increase evapotranspiration and therefore decrease discharge [62], which can be known from the relatively smaller discharge in warming mesocosm in the study. Thus, it supported the view that the influence of the water table or water content on peatland ecosystems (such as DOC loss in the study) is stronger than in variations in other environmental conditions, such as temperature [79]. The two experimental water table positions (−10 m and −20 m) resulted in increases in the annual export of DOC by 69% and 102%, respectively, through both a higher DOC concentration and larger discharge volume, indicating the potential release of both carbon and water from peatland after the level of the water table is lowered. The temperature treatment resulted in clear changes in the DOC concentration but had a limited effect on DOC export, probably indicating a shift in the turnover rate of organic carbon in peat because temperature is the main factor affecting bacterial metabolism and the rate of decomposition of organic materials. The nonsignificant effect of warming on DOC export and the notably positive relationship between mean monthly DOC export and peat temperature, which resulted from the shortage of recorded data for all mesocosms, appear contradictory, and suggest that further studies should be careful to consider these issues. Variable water levels and temperatures changed the absorbances of DOC in the year immediately after this experiment. This result suggests the varied nature or qualities of DOC that might influence fluvial systems, and also warns that using absorbance records as a proxy for DOC concentrations when studied in peatland should be done with caution.

Therefore, our observations in the first year immediately after the controlled experiment were helpful for understanding how the carbon budget might react to climate change and anthropogenic interference (i.e., drainage). This mesocosm experiment also provides useful information for local protection and sustainable development in Zoige peatland. Additional experiments and observations are required to achieve a comprehensive understanding of the carbon cycle (such as DOC fluxes) in peatlands facing changing environmental conditions.

## Supporting Information

**Figure S1** Topography and vegetation characteristics of the study area. (A) Drainage ditch. (B) Vegetation community growing in shallow water. (C) Scattered vegetation surrounded by surface water. All photographs were taken during May 2012 in Hongyuan County located in Zoige peatland.
(TIF)

**Figure S2** Peat temperature at −10 cm depth and precipitation during the growing season in 2012.
(TIF)

**Figure S3** Correlation analysis of temperature and precipitation with DOC. (A) Results of the correlation analysis between peat temperature recorded in four mesocosms and corresponding mean monthly export and concentration of DOC (DOC concentration: $y = 1.4573x + 9.9416$, $R^2 = 0.4025$, $p < 0.01$, n = 24; DOC export: $y = 0.2043x - 0.9177$, $R^2 = 0.4984$, $p < 0.01$, n = 24). (B) Results of the correlation analysis between mean monthly precipitation and mean monthly export and concentration of DOC in all mesocosms (DOC concentration: $y = 0.0379x + 26.409$, $R^2 = 0.3046$, $p = 0.128$, n = 6; DOC export: $y = 0.013x - 0.2987$, $R^2 = 0.8982$, $p < 0.01$, n = 6).
(TIF)

**Figure S4** The schematic drawing of the mesocosm in the study. The references of the number are shown as below: 1. Polycarbonate solar panels; 2. Water intake system; 3. Drainage system; 4. Water storage barrel; 201. Observation tube of water-level; 202. Sand filter pocket of inlet; 301. Sand filter pocket of outlet; 302. High pressure valves; 303. Water pipe; 304. Hanger loop; 305. Observation rule.
(TIF)

## Acknowledgments

The authors wish to acknowledge the assistance of Prof. Zhang Fen Chun at CRAES (Chinese Research Academy of Environmental Sciences) for his expert revision of the manuscript. The authors are also grateful to two anonymous reviewers for their valuable comments on the manuscript.

## Author Contributions

Conceived and designed the experiments: LLH BK. Performed the experiments: XDL SQZ. Analyzed the data: XDL BK YLH. Contributed reagents/materials/analysis tools: XDL SQZ BK LLH. Wrote the paper: XDL LLH.

## References

1. Aselmann I, Crutzen PJ (1989) Global distribution of natural fresh-water wetlands and rice paddies, their net primary productivity seasonality and possible methane emissions. Journal of Atmospheric Chemistry 8: 307–358.
2. Hobbie JE (1992) Microbial control of dissolved organic-carbon in lakes - research for the Future. Hydrobiologia 229: 169–180.
3. Moore TR, Roulet NT, Waddington JM (1998) Uncertainty in predicting the effect of climatic change on the carbon cycling of Canadian peatlands. Climatic Change 40: 229–245.
4. Pastor J, Solin J, Bridgham SD, Updegraff K, Harth C, et al. (2003) Global warming and the export of dissolved organic carbon from boreal peatlands. Oikos 100: 380–386.
5. Turunen J (2008) Development of Finnish peatland area and carbon storage 1950–2000. Helsinki, FINLANDE: Finnish Environment Institute. 16 p.
6. Yu Z, Beilman DW, Jones MC (2013) Sensitivity of northern peatland carbon dynamics to Holocene climate change. Carbon Cycling in Northern Peatlands: American Geophysical Union. pp. 55–69.
7. Moore TR, Dalva M (1993) The influence of temperature and water table position on carbon dioxide and methane emissions from laboratory columns of peatland soils. Journal of Soil Science 44: 651–664.
8. Price JS (2003) Role and character of seasonal peat soil deformation on the hydrology of undisturbed and cutover peatlands. Water Resources Research 39: 1241.
9. Xiang S, Guo R, Wu N, Sun S (2009) Current status and future prospects of Zoige Marsh in Eastern Qinghai-Tibet Plateau. Ecological Engineering 35: 553–562.
10. Shi C-c, Tu J (2009) Remote Sensing Monitory Study on Land Desertification in Ruoergai Plateau of Sichuan Province during 40 Years. Southwest China Journal of Agricultural Sciences 6: 035 (in Chinese).
11. Guo X, Du W, Wang X, Yang Z (2013) Degradation and structure change of humic acids corresponding to water decline in Zoige peatland, Qinghai-Tibet Plateau. Sci Total Environ 445–446: 231–236.
12. Zhang X, Lv X, Gu H (2005) To analyze threats, to describe present conservation situation and to provide management advices of the Ruoergai marshes. Wetland Sci 3: 292–297(in Chinese).
13. SAFS (Sichuan Academy of Forest Science) (2006) Scientific investigation report on Zoige marsh. Chengdu: Sichuan Science and Technology Press(in Chinese).
14. Wang M, Liu Z, Ma X, Wang G (2012) Division of organic carbon reserves of peatlands in China. Wetland Sci 10: 156–163(in Chinese).
15. Gao J (2006) Degradation factor analysis and solutions of Ruoergai Wetland in Sichuan. Sichuan Environ 25: 48–53(in Chinese).
16. Gorham E (1991) Northern peatlands- role in the carbon-cycle and probable responses to climatic warming. Ecological Applications 1: 182–195.
17. Trettin CC, Laiho R, Minkkinen K, Laine J (2006) Influence of climate change factors on carbon dynamics in northern forested peatlands. Canadian Journal of Soil Science 86: 269–280.
18. Bridgham SD, Pastor J, Dewey B, Weltzin JF, Updegraff K (2008) Rapid carbon response of peatlands to climate change. Ecology 89: 3041–3048.
19. Hogg EH, Lieffers VJ, Wein RW (1992) Potential carbon losses from peat profiles - effects of temperature, drought cycles, and fire. Ecological Applications 2: 298–306.
20. Freeman C, Lock MA, Reynolds B (1993) Fluxes of CO2, CH4 and N2O from a Welsh peatland following simulation of water-table draw-down - potential feedback to climatic-change. Biogeochemistry 19: 51–60.
21. Bohrer G, Chen H, Wu N, Wang Y, Zhu D, et al. (2013) Inter-Annual Variations of Methane Emission from an Open Fen on the Qinghai-Tibetan Plateau: A Three-Year Study. PLoS ONE 8: e53878.
22. Evans CD, Chapman PJ, Clark JM, Monteith DT, Cresser MS (2006) Alternative explanations for rising dissolved organic carbon export from organic soils. Global Change Biology 12: 2044–2053.
23. Zhang G, Tian J, Jiang NA, Guo X, Wang Y, et al. (2008) Methanogen community in Zoige wetland of Tibetan plateau and phenotypic characterization of a dominant uncultured methanogen cluster ZC-I. Environmental microbiology 10: 1850–1860.
24. Billett MF, Palmer SM, Hope D, Deacon C, Storeton-West R, et al. (2004) Linking land-atmosphere-stream carbon fluxes in a lowland peatland system. Global Biogeochemical Cycles 18: GB1024.
25. Freeman C, Evans CD, Monteith DT, Reynolds B, Fenner N (2001) Export of organic carbon from peat soils. Nature 412: 785–785.
26. Chen H, Yao S, Wu N, Wang Y, Luo P, et al. (2008) Determinants influencing seasonal variations of methane emissions from alpine wetlands in Zoige Plateau and their implications. Journal of Geophysical Research: Atmospheres 113: D12303.
27. Limpens J, Berendse F, Blodau C, Canadell JG, Freeman C, et al. (2008) Peatlands and the carbon cycle: from local processes to global implications – a synthesis. Biogeosciences 5: 1475–1491.
28. Dinsmore KJ, Billett MF, Dyson KE (2013) Temperature and precipitation drive temporal variability in aquatic carbon and GHG concentrations and fluxes in a peatland catchment. Global Change Biology 19: 2133–2148.
29. Aerts R, De Caluwe H (1999) Nitrogen deposition effects on carbon dioxide and methane emissions from temperate peatland soils. Oikos 84: 44–54.
30. Arnosti C, Holmer M (2003) Carbon cycling in a continental margin sediment: contrasts between organic matter characteristics and remineralization rates and pathways. Estuarine Coastal and Shelf Science 58: 197–208.
31. Wallage ZE, Holden J (2010) Spatial and temporal variability in the relationship between water colour and dissolved organic carbon in blanket peat pore waters. Science of The Total Environment 408: 6235–6242.
32. Chin W-C, Lennon JT, Hamilton SK, Muscarella ME, Grandy AS, et al. (2013) A Source of Terrestrial Organic Carbon to Investigate the Browning of Aquatic Ecosystems. PLoS ONE 8: e75771.
33. Evans CD, Monteith DT, Cooper DM (2005) Long-term increases in surface water dissolved organic carbon: Observations, possible causes and environmental impacts. Environmental Pollution 137: 55–71.
34. Carpenter SR, Pace ML (1997) Dystrophy and eutrophy in lake ecosystems: Implications of fluctuating inputs. Oikos 78: 3–14.
35. Wetzel RG (1992) Gradient-dominated ecosystems - sources and regulatory functions of dissolved organic-matter in fresh-water ecosystems. Hydrobiologia 229: 181–198.
36. Wallage ZE, Holden J, McDonald AT (2006) Drain blocking: An effective treatment for reducing dissolved organic carbon loss and water discolouration in a drained peatland. Science of the Total Environment 367: 811–821.
37. Grayson R, Holden J (2012) Continuous measurement of spectrophotometric absorbance in peatland streamwater in northern England: implications for understanding fluvial carbon fluxes. Hydrological Processes 26: 27–39.
38. Weishaar JL, Aiken GR, Bergamaschi BA, Fram MS, Fujii R, et al. (2003) Evaluation of specific ultraviolet absorbance as an indicator of the chemical composition and reactivity of dissolved organic carbon. Environmental Science & Technology 37: 4702–4708.
39. Worrall F, Armstrong A, Adamson JK (2007) The effects of burning and sheep-grazing on water table depth and soil water quality in a upland peat. Journal of Hydrology 339: 1–14.
40. Fraser CJD, Roulet NT, Moore TR (2001) Hydrology and dissolved organic carbon biogeochemistry in an ombrotrophic bog. Hydrological Processes 15: 3151–3166.
41. Strack M, Waddington JM, Bourbonniere RA, Buckton EL, Shaw K, et al. (2008) Effect of water table drawdown on peatland dissolved organic carbon export and dynamics. Hydrological Processes 22: 3373–3385.

42. Sommer M (2006) Influence of soil pattern on matter transport in and from terrestrial biogeosystems—A new concept for landscape pedology. Geoderma 133: 107–123.

43. Briggs J, Large DJ, Snape C, Drage T, Whittles D, et al. (2007) Influence of climate and hydrology on carbon in an early Miocene peatland. Earth and Planetary Science Letters 253: 445–454.

44. Strack M, Waddington JM, Tuittila ES (2004) Effect of water table drawdown on northern peatland methane dynamics: Implications for climate change. Global Biogeochemical Cycles 18: GB4003.

45. Scanlon D, Moore T (2000) Carbon dioxide production from peatland soil profiles: The influence of temperature, oxic/anoxic conditions and substrate. Soil Science 165: 153–160.

46. Morris PJ, Belyea LR, Baird AJ (2011) Ecohydrological feedbacks in peatland development: A theoretical modelling study. Journal of Ecology 99: 1190–1201.

47. Dai Y, Luo Y, Wang C, Shen Y, Ma Z, et al. (2010) Climate variation and abrupt change in wetland of Zoig Plateau during 1961 and 2008. Journal of Glaciology and Geocryology 32: 35–42(in Chinese).

48. Jager DF, Wilmking M, Kukkonen JVK (2009) The influence of summer seasonal extremes on dissolved organic carbon export from a boreal peatland catchment: Evidence from one dry and one wet growing season. Science of The Total Environment 407: 1373–1382.

49. Blodau C, Siems M (2012) Drainage-induced forest growth alters belowground carbon biogeochemistry in the Mer Bleue bog, Canada. Biogeochemistry 107: 107–123.

50. Sapek A, Sapek B, Chrzanowski S, Urbaniak M (2009) Nutrient mobilisation and losses related to the groundwater level in low peat soils. International Journal of Environment and Pollution 37: 398–408.

51. Ellis T, Hill PW, Fenner N, Williams GG, Godbold D, et al. (2009) The interactive effects of elevated carbon dioxide and water table draw-down on carbon cycling in a Welsh ombrotrophic bog. Ecological Engineering 35: 978–986.

52. Tang R, Clark JM, Bond T, Graham N, Hughes D, et al. (2013) Assessment of potential climate change impacts on peatland dissolved organic carbon release and drinking water treatment from laboratory experiments. Environmental Pollution 173: 270–277.

53. Watts CD, Naden PS, Machell J, Banks J (2001) Long term variation in water colour from Yorkshire catchments. Science of The Total Environment 278: 57–72.

54. Wilson L, Wilson J, Holden J, Johnstone I, Armstrong A, et al. (2011) Ditch blocking, water chemistry and organic carbon flux: Evidence that blanket bog restoration reduces erosion and fluvial carbon loss. Science of the Total Environment 409: 2010–2018.

55. Mitchell GN (1990) Natural discoloration of freshwater: Chemical composition and environmental genesis. Progress in Physical Geography 14: 317–334.

56. Mitchell G, McDonald AT (1992) Discolouration of water by peat following induced drought and rainfall simulation. Water Research 26: 321–326.

57. Li G, Liu Y, Frelich LE, Sun S (2011) Experimental warming induces degradation of a Tibetan alpine meadow through trophic interactions. Journal of Applied Ecology 48: 659–667.

58. Chen H, Wu N, Wang Y, Zhu D, Zhu Qa, et al. (2013) Inter-Annual Variations of Methane Emission from an Open Fen on the Qinghai-Tibetan Plateau: A Three-Year Study. PLoS ONE 8: e53878.

59. Yanbin H, Yanfen W, Xurong M, Xiangzhong H, Xiaoyong C, et al. (2008) CO2H2O and energy exchange of an Inner Mongolia steppe ecosystem during a dry and wet year. Acta Oecologica-international Journal Of Ecology 33: 133–143.

60. Luo C, Xu G, Wang Y, Wang S, Lin X, et al. (2009) Effects of grazing and experimental warming on DOC concentrations in the soil solution on the Qinghai-Tibet plateau. Soil Biology and Biochemistry 41: 2493–2500.

61. Zhang XH, Liu HY, Baker C, Graham S (2012) Restoration approaches used for degraded peatlands in Ruoergai (Zoige), Tibetan Plateau, China, for sustainable land management. Ecological Engineering 38: 86–92.

62. Bridgham SD, Pastor J, Updegraff K, Malterer TJ, Johnson K, et al. (1999) Ecosystem control over temperature and energy flux in Northern peatlands. Ecological Applications 9: 1345–1358.

63. Walker MD, Wahren CH, Hollister RD, Henry GH, Ahlquist LE, et al. (2006) Plant community responses to experimental warming across the tundra biome. Proc Natl Acad Sci U S A 103: 1342–1346.

64. Debevec EM, MacLean JrSF (1993) Design of greenhouses for the manipulation of temperature in tundra plant communities. Arctic and Alpine Research: 56–62.

65. Turetsky M, Treat C, Waldrop M, Waddington J, Harden J, et al. (2008) Short-term response of methane fluxes and methanogen activity to water table and soil warming manipulations in an Alaskan peatland. Journal of Geophysical Research: Biogeosciences (2005–2012) 113.

66. Chivers M, Turetsky M, Waddington J, Harden J, McGuire A (2009) Effects of experimental water table and temperature manipulations on ecosystem CO2 fluxes in an Alaskan rich fen. Ecosystems 12: 1329–1342.

67. Guo Y, Wan Z, Liu D (2010) Dynamics of dissolved organic carbon in the mires in the Sanjiang Plain, Northeast China. Journal of Environmental Sciences 22: 84–90.

68. Walling DE, Webb BW (1981) The reliability of suspended sediment load data: IAHS Publication.

69. Clair TA, Arp P, Moore TR, Dalva M, Meng FR (2002) Gaseous carbon dioxide and methane, as well as dissolved organic carbon losses from a small temperate wetland under a changing climate. Environmental Pollution 116, Supplement 1: S143–S148.

70. Weltzin JF, Pastor J, Harth C, Bridgham SD, Updegraff K, et al. (2000) Response of bog and fen plant communities to warming and water-table manipulations. Ecology 81: 3464–3478.

71. Fraser C, Roulet N, Moore T (2001) Hydrology and dissolved organic carbon biogeochemistry in an ombrotrophic bog. Hydrological Processes 15: 3151–3166.

72. Moore T (1989) Dynamics of dissolved organic carbon in forested and disturbed catchments, Westland, New Zealand: 1. Maimai. Water Resources Research 25: 1321–1330.

73. Urban N, Bayley S, Eisenreich S (1989) Export of dissolved organic carbon and acidity from peatlands. Water Resources Research 25: 1619–1628.

74. Worrall F, Reed M, Warburton J, Burt T (2003) Carbon budget for a British upland peat catchment. Science of The Total Environment 312: 133–146.

75. Steinberg (2003) Ecology of humic substances in freshwaters: determinants from geochemistry to ecological niches. Berlin: Springer.

76. Clymo RS (1984) The limits to peat bog growth. Philosophical Transactions of the Royal Society of London B, Biological Sciences 303: 605–654.

77. Mars H, Wassen MJ, Peeters WHM (1996) The effect of drainage and management on peat chemistry and nutrient deficiency in the former Jegrznia-floodplain (NE-Poland). Vegetatio 126: 59–72.

78. Freeman C, Ostle N, Kang H (2001) An enzymic 'latch' on a global carbon store. Nature 409: 149–149.

79. Moore PD (2002) The future of cool temperate bogs. Environmental Conservation 29: 3–20.

80. Strakova P, Penttilä T, Laine J, Laiho R (2012) Disentangling direct and indirect effects of water table drawdown on above-and belowground plant litter decomposition: consequences for accumulation of organic matter in boreal peatlands. Global Change Biology 18: 322–335.

81. Whittington PN, Price JS (2006) The effects of water table draw-down (as a surrogate for climate change) on the hydrology of a fen peatland, Canada. Hydrological Processes 20: 3589–3600.

82. Harrison AF, Taylor K, Scott A, Poskitt J, Benham D, et al. (2008) Potential effects of climate change on DOC release from three different soil types on the Northern Pennines UK: examination using field manipulation experiments. Global Change Biology 14: 687–702.

83. Froberg M, Berggren D, Bergkvist B, Bryant C, Mulder J (2006) Concentration and fluxes of dissolved organic carbon (DOC) in three norway spruce stands along a climatic gradient in sweden. Biogeochemistry 77: 1–23.

84. Lin X, Zhang Z, Wang S, Hu Y, Xu G, et al. (2011) Response of ecosystem respiration to warming and grazing during the growing seasons in the alpine meadow on the Tibetan plateau. Agricultural and Forest Meteorology 151: 792–802.

85. Kim SY, Freeman C, Fenner N, Kang H (2012) Functional and structural responses of bacterial and methanogen communities to 3-year warming incubation in different depths of peat mire. Applied Soil Ecology 57: 23–30.

# 15

# Incentivizing the Public to Support Invasive Species Management: Eurasian Milfoil Reduces Lakefront Property Values

**Julian D. Olden**[1]*, **Mariana Tamayo**[2]

**1** School of Aquatic and Fishery Sciences, University of Washington, Seattle, Washington, United States of America, **2** Faculty of Life and Environmental Sciences, University of Iceland, Reykjavík, Iceland

## Abstract

Economic evaluations of invasive species are essential for providing comprehensive assessments of the benefits and costs of publicly-funded management activities, yet many previous investigations have focused narrowly on expenditures to control spread and infestation. We use hedonic modeling to evaluate the economic effects of Eurasian milfoil (*Myriophyllum spicatum*) invasions on lakefront property values of single-family homes in an urban-suburban landscape. Milfoil often forms dense canopies at the water surface, diminishing the value of ecosystem services (e.g., recreation, fishing) and necessitating expensive control and management efforts. We compare 1,258 lakeshore property sale transactions (1995–2006) in 17 lakes with milfoil and 24 un-invaded lakes in King County, Washington (USA). After accounting for structural (e.g., house size), locational (e.g., boat launch), and environmental characteristics (e.g., water clarity) of lakes, we found that milfoil has a significant negative effect on property sales price ($94,385 USD lower price), corresponding to a 19% decline in mean property values. The aggregate cost of milfoil invading one additional lake in the study area is, on average, $377,542 USD per year. Our study illustrates that invasive aquatic plants can significantly impact property values (and associated losses in property taxes that reduce local government revenue), justifying the need for management strategies that prevent and control invasions. We recommend coordinated efforts across Lake Management Districts to focus institutional support, funding, and outreach to prevent the introduction and spread of milfoil. This effort will limit opportunities for re-introduction from neighboring lakes and incentivize private landowners and natural resource agencies to commit time and funding to invasive species management.

**Editor:** Bo Li, Fudan University, China

**Funding:** This work was supported by the Washington Department of Ecology Aquatic Weeds Management Program, University of Washington Royalty Research Fund, and the University of Washington - School of Aquatic and Fishery Sciences H. Mason Keeler Endowed Professorship. The funders had no role in study design, data collection and analysis, decision to publish, or preparation of the manuscript.

**Competing Interests:** The authors have declared that no competing interests exist.

* Email: olden@uw.edu

## Introduction

Despite the long history of investigating the ecology of nonindigenous species [1], the scope of economic damages associated with species invasions has only recently received greater attention [2,3]. Continental scale estimates suggest that thousands of invasive plants and animals have generated billions of dollars in economic losses [4–6]. These estimates, however, are conservative because they focus predominantly on expenditures to control the infestation and spread of invasive species. From an economic perspective, the full cost of biological invasions also includes the effects on host ecosystems and the human populations dependent on them [7]. The societal value that individuals give to both market (e.g., forestry) and nonmarket (e.g., landscape aesthetics) goods and services is also important to the economic valuation of damages incurred by invasive species. These values consider the market price of goods and services and people's willingness to pay and sell them [8–10].

In freshwater environments, previous studies have largely focused on the economic impacts of invasive species on fisheries, power plants, water treatment facilities, and recreation [11,12]. For example, the invasion of the rusty crayfish (*Orconectes rusticus*) into lakes in northern Wisconsin (USA) is estimated to generate damages of about $1.5 million USD annually to the panfish recreational fishery [13], and zebra mussels (*Dreissena polymorpha*) cost an estimated $267 million USD in lost power generation and drinking water treatment facilities in Lake St. Clair (USA) during the first 15 years of infestation [14]. However, a more complete understanding of the full spectrum of economic effects associated with aquatic invasive plants is needed to develop comprehensive policies and management strategies, as well as to incentivize the public to prevent future spread.

Eurasian milfoil (*Myriophyllum spicatum* L., herein referred to as milfoil) is an ideal study organism to enhance our knowledge regarding the economic effects of aquatic invasive species because extensive information is available on the ecology and management of this invasive plant [15]. Native to Europe, Asia, and northern

Africa, milfoil is now found on all continents except Australia and Antarctica, including almost all states and provinces of the United States and Canada [16]. This submersed perennial grows in a wide range of water temperatures, depths, and turbidities [15]. Milfoil can propagate through vegetative and sexual reproduction, although the former via stem fragments and runners provides the main mechanism of dispersal [17] by hitchhiking between waterbodies on trailered boats [18]. Milfoil invasions have become a major environmental nuisance in countless lakes across North American and globally, and many additional water bodies are susceptible to future invasions [19,20].

Freshwater ecosystems are often severely impacted by milfoil invasion. Milfoil form dense canopies in the water column (extending to the water surface) altering water chemistry, displacing native plants, and creating habitats that are unsuitable for wildlife [15,21,22]. The costs of controlling milfoil, which include mechanical harvesting, underwater cultivation, diver hand-pulling, water level manipulation, biological control, and aquatic herbicide application, exceed many millions USD annually [23]. For example, during a 15 year period (1985–2010) over $5 million USD was spent to control milfoil in Lake George (New York, USA) [24]. Moreover, milfoil can diminish the value of services like recreation, by hindering boating and swimming activities. In the Truckee River watershed (California and Nevada, USA), estimates of a potential decline in recreation values of only 1% due to the spread of milfoil were at least $500,000 USD annually [23]. Milfoil can also impact provisioning services such as agriculture and electricity generation, by reducing water circulation in irrigation projects and blocking water intakes in power plants.

In this study we evaluate the potential economic impacts of aquatic invasive plants on lakefront properties, using Eurasian milfoil as an illustrative example. Such impacts are largely unexplored (but see [25,26]), yet are critical to determine the benefits and costs of different strategies to manage invasive aquatic plants and to actively engage the public into management actions regarding the spread of non-native species. Specifically, we evaluate the economic effects of milfoil invasions on lakefront property values of single-family homes in a region of western Washington (USA) by applying a hedonic modeling framework. Furthermore, we assess the welfare effect of milfoil invading one additional lake in our study area in order to inform future prevention efforts.

## Methods

### Study region

Our study focused on lakefront properties in Pacific Northwest region of North America, specifically King County, Washington (USA). This county has the highest population density in the state (1,931,249 residents according to the 2010 census) and it is intersected (north-south) by the Interstate-5 highway, which serves as an invasion corridor for non-native plant species both in terms of high human populations (introduction via the aquarium trade) and movement of recreational boaters (introduction via entrainment on trailer boats). Lakes throughout King County are located along a distinct urban-rural land use gradient, and many have primary residences and support public recreation [27], making our study distinct from previous investigations examining milfoil impacts in rural landscapes. We assessed the economic effect of milfoil by comparing 1,258 lakeshore property sale transactions of single family homes in 41 small lakes (lake area <1 km$^2$) from 1995 to 2006 (Figure 1), prior to the 2007 decline of housing prices in the county and the state [28]. Although the county has >150

small lakes, we were limited to those containing complete datasets for sales transactions, structural, locational, and environmental characteristics (see below). The dataset consisted of 17 lakes with milfoil during the study period (611 total transactions) and 24 un-invaded lakes (647 total transactions) located in a predominantly urban-suburban landscape. The exact date when milfoil invaded each lake is unknown; however, based on county records [29] and personal communication with county officials 15 of the lakes were invaded prior to 1995 and two prior to 1999. Because the invasion dates of the latter two lakes were unclear, we treated them as being invaded throughout the 12-year study period. Unfortunately data on milfoil density is lacking for many lakes, therefore our analysis focused on presence/absence. Data sources were the King County Department of Assessments, King County Department of Natural Resources and Parks, and the Washington Department of Ecology.

### Statistical approach

We used hedonic modeling to quantify the effect of milfoil invasions on lake property values. This technique has proven useful in estimating the economic value of nonmarket amenities, for example, the effect of water quality on the recreational and aesthetic value of freshwater resources and shoreline properties [30–32]. We provide a brief description of this approach below, but refer the reader to Rosen [33] for further information. Hedonic modeling partitions a composite good (e.g., property value) into its defining characteristics and estimates the value (i.e., implicit price) of each characteristic. The relationship between the market price of the good and its attributes is the hedonic price function. We followed Halstead et al. [34] by defining the hedonic price function as $HP = f(S, L, E)$, where $HP$ represents home (property) price, $S$ are structural characteristics (e.g., house size), $L$ are locational characteristics (e.g., parcel density), and $E$ are environmental characteristics (e.g., water clarity).

We modeled $HP$ as a function of key property characteristics ($S$, $L$, $E$) to generate the value (i.e., marginal implicit price) consumers give to each characteristic. These estimated values were then used to evaluate the effect of milfoil presence on property value. A suite of independent variables (Table 1) that previous studies have identified as important in determining lakeshore property prices were analyzed [25,35]. We modeled $HP$ as a linear function of these variables for ease of interpretation and because this functional form has been used extensively in hedonic analyses [36,37]. Given that properties around a lake are influenced by the same lake-specific characteristics, we considered each lake a cluster of property sales and characteristics (see [25]). Unobserved lake characteristics can lead to endogeneity, whereby an independent variable is correlated with the error terms in the model, resulting in biased estimates of model coefficients. We used two-stage least squares regression to account for correlations between the error terms of the dependent variable and the independent variables. This regression uses instrumental variables that are uncorrelated with the error terms but are correlated to the endogenous variables to estimate the values of the endogenous variables (first stage), and then uses these estimated values to model the dependent variable (second stage) (see [38]).

### Hedonic model structure

The hedonic model comprised of lakefront property sales price as the dependent variable, which was deflated to 2006 property values (USD) using the house price index (purchase only) from the US Federal Housing Finance Agency. Independent variables used in the analysis included structural characteristics (i.e., house size, house age, lot size, frontage), locational attributes (i.e., presence of

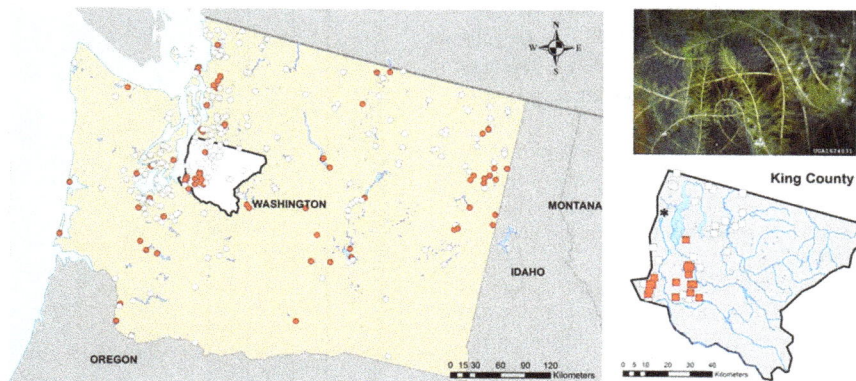

**Figure 1. Location of milfoil presences (red filled circle) and absences (white empty circles) in lakes of Washington, USA, including King County (bottom right) containing 17 invaded lakes (filled squares) and 24 uninvaded lakes (empty squares).** The city of Seattle, Washington (2010 population size of 608,660) is indicated as *.

a public boat launch, recreational fish stocking, lakefront parcel density) and environmental descriptors (i.e., presence of Eurasian milfoil, lake area, water temperature, water clarity) (Table 1).

The choice of endogenous and instrumental variables is influenced by geography and the specific characteristics of the focal property market (e.g., [25,26]). Our variable selection and model structure reflects a property market composed of primary residences in an urban-suburban landscape. Below we describe the endogenous and instrumental variables used in the hedonic model (independent variables are listed Table 1), noting that endogenous variables refer to factors whose values are determined by the state of other variables in the system and instrumental variables are hypothesized to be correlated to the endogenous variables but not to the dependent variable (property sales price).

Milfoil presence was treated as an endogenous variable; a choice supported by Horsch and Lewis [25] who showed the endogeneity of milfoil presence in a hedonic model. Recreational boaters commonly spread milfoil among lakes [18] and lake characteristics that increase the desirability for recreation are also attractive for homeownership. However, it is difficult to quantify many of these desirable characteristics, thus increasing the likelihood that milfoil presence is endogenous [25]. In our housing market, we used the occurrence of a public boat launch and fish stocking as instrumental variables because they are linked to recreational boating (i.e., the primary vector of milfoil introduction into lakes). Our choice is supported by the fact that all properties have direct dock assess to the lake and self-sustaining recreational fish populations exist in those lakes that are not stocked; therefore, these factors likely have little effect on a homeowner's willingness to pay in our housing market. We also used water clarity, lake area, and water temperature as instrumental variables for milfoil presence due to their influence on habitat suitability for milfoil establishment [15,20,27]. Although water quality is known to have an effect on property values of housing markets (e.g., [30,32,39]),

**Table 1.** Structural, locational, and environmental independent variables used in the hedonic analysis of property sales price (*).

| Variable | Description | Mean | S.E. |
|---|---|---|---|
| Sales price* | Selling price of the property (land + house; 2006 USD) | 502312.8 | 23942.4 |
| *Structural* | | | |
| Lot size | Size of a parcel (m$^2$) | 2394.5 | 216.2 |
| Frontage | Shoreline frontage of a property (m$^2$) | 22.0 | 1.1 |
| House size | Total living area (m$^2$) | 204.8 | 6.6 |
| House age | Age of a house (years) | 39.5 | 1.6 |
| *Locational* | | | |
| Boat launch | Presence of a public boat launch | 0.6 | 0.1 |
| Fish stocking | Presence of fish stocking for recreational angling | 0.7 | 0.1 |
| Parcel density | Number of parcels per km$^2$ | 512.1 | 35.6 |
| *Environmental* | | | |
| Milfoil presence | Presence of Eurasian milfoil | 0.4 | 0.1 |
| Lake area | Surface area of a lake adjacent to the property (km$^2$) | 0.2 | 0.03 |
| Temperature | Mean surface water temperature during the milfoil summer growing season (°C) | 19.8 | 0.3 |
| Water clarity | Mean Secchi depth of the lake during the milfoil growing season (m) | 3.4 | 0.2 |

this effect is unlikely to be manifested our property market where >90% of the study lakes had water clarity >2 m, with an average of 3.4 m and little variability (SD = 0.19 m). Given this, differences in water clarity are likely imperceptible to potential property buyers. Lake area and water temperature were also very similar among our lakes given their similar glacial history and elevation. Taken together, our housing market is characterized by similar sized lakes with good water clarity and similar water temperatures; therefore, it is unlikely that these attributes significantly affected a homeowner's willingness to pay.

A series of regression models were developed and compared using the modified Akaike's Information Criterion for small samples (AICc). AICc is a model selection technique based on the trade-off between model accuracy and parsimony [40]. Akaike weights were calculated with the AICc values to determine the relative likelihood that each model is the best model given the data and the other candidate models. Statistical analyses where conducted using PASW 18 (IBM SPSS).

We estimated the aggregate cost of milfoil invading one additional small lake in our study area by discounting a homeowner's willingness to pay for a property on a lake free of milfoil by 5% (same rate as in [25]) to estimate the average annual marginal willingness to pay, and then multiplying this annual average with the mean number of parcels for our study lakes (n = 80 parcels).

## Results

The presence of Eurasian milfoil had a significant negative effect on property values; mean reduction in property values was $94,385 USD, ranging from −$92,558 to −$94,670 USD according to the top three competitive models (Table 2). Based on an average sale price of $502,313 across all study lakes, the negative effect of milfoil presence corresponds to a 19% decline in mean property values.

The hedonic analysis revealed that larger homes located on lakes with larger surface areas had a significant positive effect on property values, on average selling for $2,600 per m$^2$ and $209,400 per km$^2$ more, respectively (Table 2). Parcel density showed a negligible effect on property value, whereas water clarity negatively influenced property sales prices, though not statistically significant. All model parameters, except for water clarity and parcel density, reflected the anticipated directional effect on property value (Table 1).

A homeowner's marginal willingness to pay for a waterfront property on a lake free of milfoil was on average $94,385 (model 1 in Table 2), resulting in an average annual marginal willingness to pay of $4,719 (using a 5% discount rate). The aggregate cost of milfoil invading one additional study lake was $377,542 per year ($4,719×80 lakefront parcels).

## Discussion

A broader understanding of the economic impacts of aquatic invasive plants is essential for promoting changes in policy and engaging more diverse stakeholder participation, such as lakefront property owners and recreational boaters, in the management of natural resources. Only until the full cost of biological invasions is considered (i.e. beyond control expenditures), will the optimal economical management of invasive species be possible [41]. Our study demonstrates that aquatic invasive plants can have dramatic economic impacts on the sale value of lakefront properties. The presence of milfoil in a lake results in an "invisible tax" on the real estate market by substantially reducing property values an average

**Table 2.** Hedonic analysis results for the two-stage least squares regression model predicting property price as a function of key independent variables describing structure, location, and the environment (see Table 1).

| Variable | Model 1 | | | Model 2 | | | Model 3 | | |
|---|---|---|---|---|---|---|---|---|---|
| | Coefficient | Sig. | S.E. | Coefficient | Sig. | S.E. | Coefficient | Sig. | S.E. |
| Constant | −63891.9 | | 97303.1 | −29052.8 | | 117533.5 | −95790.7 | | 149910.1 |
| Milfoil presence | −94385.4 | ** | 46712.9 | −94670.0 | * | 49174.7 | −92558.0 | * | 48255.6 |
| House size | 2608.1 | *** | 344.2 | 2474.0 | *** | 703.9 | 2681.6 | *** | 726.0 |
| Lot size | | | | −1.2 | | 13.2 | 3.1 | | 12.9 |
| Parcel density | 102.8 | | 70.3 | | | | 115.4 | | 78.6 |
| Lake area | 209407.8 | ** | 81848.0 | 154409.5 | * | 74073.6 | 215595.0 | ** | 84521.5 |
| Water clarity | −7430.3 | | 11160.2 | −6962.7 | | 12820.2 | −6954.9 | | 12600.7 |
| AICc | 23.931 | | | 23.986 | | | 23.998 | | |
| Relative likelihood | 1.000 | | | 0.969 | | | 0.963 | | |

See text for discussion of the endogenous variable (milfoil presence) and instrumental variables. Reported are the top three candidate models according to Akaike's Information Criterion for small samples (AICc) with their associated parameter coefficients and standard errors. The relative likelihood that the model is the best model given the data is denoted. Significant levels: *P<0.10, **P<0.05, ***P<0.01.

of over \$94 thousand USD, translating to 19% decline in value. We note that our estimates did not consider the level of infestation, the implementation of management actions, nor the losses to recreation.

Similar economic damages have been reported in northern Wisconsin, where waterfront property values in a popular recreational and rural area declined by approximately 8% after milfoil invaded a lake [25]. Furthermore, the process of milfoil infestation in five Vermont lakes (USA) resulted in property values that decreased by <1% to 16% depending on the level of infestation [26]. Both these studies examined rural properties containing mostly vacation homes (secondary residences) located in forested landscapes; our study adds to this understanding by demonstrating economic impacts to property values of primary residences in urban settings. Taken together, the negative effect of milfoil on property values of primary and secondary homes in different regions and landscape settings, suggests that the economic impacts of aquatic invasive plants are widespread and may be greater in urbanized landscapes. We recognize that milfoil presence/absence may overestimate economic impacts compared to plant density [26]. Additional studies that include detailed estimates of milfoil infestation (abundance) at the time of purchase, distance of property to nearest milfoil colony, and the level of property buyer knowledge of milfoil are warranted [34]. By contrast, the economic impacts of milfoil may be undervalued if those properties on highly infested lakes are the most difficult to sell and therefore remain on the market.

The costs of preventing new invasions of aquatic weeds are often thought to be greater than the benefits, thus leading to inaction. Our study, however, indicates there are benefits to preventing the spread of milfoil given that the invasion of one additional study lake leads to a high aggregate cost of over \$375 thousand USD annually. This aggregate cost represents a third of the amount spent annually (\$1 million USD) on managing milfoil across Washington State [42]. The knowledge that an invasion of milfoil can lead to a significant decline in property values provides the public an economic incentive to invest in prevention and/or control strategies [43]. Moreover, reductions in property values also translate directly to substantially losses in property taxes garnered by local governments. Thus the economic impacts of milfoil invasions may extend well beyond the infested lakefront properties by reducing local government revenue.

Lakefront property owners stand to benefit greatly (higher property values) from preventing milfoil invading their lake. In addition, it is necessary to engage recreational boaters in prevention efforts as well regardless whether or not they live on a lake, because they are an important dispersal vector of milfoil and other aquatic invasive species [18,19,44]. When recreational boaters spread milfoil into a new lake they are inadvertently creating hidden costs (negative externalities) to other lake users of the newly invaded lake; these costs include lower property values, reduction of biodiversity, and diminished recreational experience, among others.

Property owners could also benefit from aquatic weed control. Zhang and Boyle [26] showed that control efforts on a heavily infested lake that reduced milfoil areal coverage from 81–100% to 61–80% could offset losses to property values caused by the invasion. Similarly, we expect that properties on lakes where milfoil densities have been reduced will likely experience a reduced negative price effect. We did not consider milfoil management effects in our analysis because it requires a treatment to have taken place before the property transaction but within the same year. If the treatment were to take place after the transaction, the associated benefit to a selling property would not yet be capitalized into property price; ignoring expectations or knowledge of a pending treatment.

A key component for long-term management of invasive species is the participation of multiple groups representing ecological and socio-economic perspectives [45,46]. Often, however, engaging stakeholder groups is difficult because each entity may have different attitudes towards invasive species and resource allocation [47,48]. For example, a study of stakeholder perceptions about invasive species in the Doñana wetland (Spain), revealed remarkably different viewpoints among parties, which included local users, tourists, and conservation professionals [49]. People were more willing to support and pay for management of invasive species (including eradication) when they had a higher level of education, and a better understanding of the study. Therefore, to successfully manage Eurasian milfoil and other invasive species it is important to embrace the diversity of perceptions held by the stakeholders, by employing strategies (e.g., involving stakeholders at the beginning of the decision-making process) that promote cooperative participation and communication among parties [46,49].

Economic research on invasive species is essential for comprehensive assessments of the benefits and costs of management strategies aimed at increasing the effectiveness of publicly funded programs [50,51]. Prevention of future introductions and control of existing invasions are powerful management options [52], however, the ecological and economic benefits of these actions must be better illustrated. Individual costs of milfoil invasions (this study; [25,26]) coupled with local, regional and national costs associated with lost recreation, agriculture and power generation (e.g., [23]) make for a compelling case that even modest expenditures on prevention could help avoid substantial economic impacts and help preserve freshwater ecosystems. Public-derived funding for aquatic weed management in the United States is generally provided through state-derived sources and the creation of Lake Management Districts that allow lake property owners to tax themselves and other lake users to collect funds for various prevention and control activities. Only three of our study lakes were represented by a Lake Management District at the time of our analysis. We recommend coordinated efforts across the management mosaic (*sensu* [41]), whereby networks of Lake Management Districts operate together to focus institutional support, funding, and outreach to prevent the introduction and spread of milfoil. This effort will limit opportunities for re-introduction from neighboring lakes and incentivize homeowners to commit time and funding to invasive species management, including the education of transient boaters.

## Acknowledgments

We are very grateful to D. Kristófersson and C. Anderson for sharing their expertise in econometric valuation and Sudeep Chandra and Mark Sytsma for their constructive comments on the paper. We thank the King County's Department of Assessments, GIS Center, and Lake Stewardship Program, as well as the Washington Department of Ecology for providing the data for the study. We appreciate the assistance of S. Abella, M. Bell-McKinnon, B. Davíðsdóttir, H. Darin, M. Jenkins, J. Jóhannsson, K. Hamel, E. Larson, K. Messick, J. Ramos, S. Roe, and J. Withey.

## Author Contributions

Conceived and designed the experiments: JDO. Performed the experiments: JDO MT. Analyzed the data: MT. Contributed reagents/materials/analysis tools: MT JDO. Contributed to the writing of the manuscript: JDO MT.

## References

1. Richardson DM, Pyšek P (2008) Fifty years of invasion ecology-the legacy of Charles Elton. Diversity and Distributions 14: 161–168.

2. Keller RP, Lodge DM, Lewis MA, Shogren JF (2009) Bioeconomics of Invasive Species: Integrating Ecology, Economics, Policy, and Management. New York, New York: Oxford University Press.

3. Perrings C, Williamson M, Dalmazzone S (2001) The Economics of Biological Invasions. Cheltenham: Edward Elgar Publishing.

4. Colautti RI, Bailey SA, van Overdijk CDA, Amundsen K, MacIsaac HJ (2006) Characterised and projected costs of nonindigenous species in Canada. Biological Invasions 8: 45–59.

5. Pimentel D, Zuniga R, Morrison D (2005) Update on the environmental and economic costs associated with alien-invasive species in the United States. Ecological Economics 52: 273–288.

6. Vilà M, Basnou C, Pyšek P, Josefsson M, Genovesi P, et al. (2010) How well do we understand the impacts of alien species on ecosystem service? A pan-European, cross-taxa assessment. Frontiers in Ecology and the Environment 8: 135–144.

7. Pejchar L, Mooney HA (2009) Invasive species, ecosystem services and human well-being. Trend in Ecology and Evolution 24: 497–504.

8. Bockstael NE, Freeman AM, Koop RJ, Portney PR, Smith VK (2000) On measuring economic values for nature. Environmental Science and Technology 34: 1384–1389.

9. Corrigan JR, Egan KJ, Downing JA (2009) Aesthetic Values of Lakes and Rivers. In: Likens GE, editor. Encyclopedia of Inland Waters. Oxford: Academic Press. pp. 14–24.

10. McIntosh CR, Finnoff DC, Settle C, Shogren JF (2009) Economic valuation and invasive species. In: Keller RP, Lodge DM, Lewis MA, Shogren JF, editors. Bioeconomics of Invasive Species: Integrating Ecology, Economics, Policy, and Management. New York, New York: Oxford University Press. pp. 151–179.

11. Lovell SJ, Stone SF, Fernandez L (2006) The economic impacts of aquatic invasive species: a review of the literature. Agricultural and Resource Economics 35: 195–208.

12. Rockwell HW (2003) Summary of a survey of the literature on the economic impact of aquatic weeds. Aquatic Ecosystem Restoration Foundation. Available at: http://www.aquatics.org/pubs/economic_impact.pdf. Accessed 2012 Oct.

13. Keller RP, Frang K, Lodge DM (2008) Preventing the spread of invasive species: economic benefits of intervention guided by ecological predictions. Conservation Biology 22: 80–88.

14. Connelly NA, O'Neill CR, Knuth BA, Brown TL (2007) Economics impacts of zebra mussel on drinking water treatment and electric power generation facilities. Environmental Management 40: 105–112.

15. Smith CS, Barko JW (1990) Ecology of Eurasian watermilfoil. Journal of Aquatic Plant Management 28: 55–64.

16. Couch R, Nelson E (1985) *Myriophyllum spicatum* in North America. In: Anderson LWJ, editor. Proceedings first international symposium watermilfoil and related Haloragaceae species. Vicksburg, Mississippi: Aquatic Plant Management Society. pp. 8–18.

17. Madsen JD, Smith DH (1997) Vegetation spread of Eurasian watermilfoil colonies. Journal of Aquatic Plant Management 35: 63–68.

18. Rothlisberger JD, Chadderton WL, McNulty J, Lodge DM (2010) Aquatic invasive species transport via trailered boats: what is being moved, who is moving it, and what can be done. Fisheries 35: 121–132.

19. Johnson PTJ, Olden JD, Vander Zanden MJ (2008) Dam invaders: impoundments facilitate biological invasions into freshwaters. Frontiers in Ecology and the Environment 6: 357–363.

20. Madsen JD (1998) Predicting invasion success of Eurasian watermilfoil. Journal of Aquatic Plant Management 36: 28–32.

21. Boylen CW, Eichler LW, Madsen JD (1999) Loss of native aquatic plant species in a community dominated by Eurasian watermilfoil. Hydrobiologia 415: 207–211.

22. Madsen JD, Sutherland JW, Bloodfield JA, Eichler LW, Boylen CW (1991) The decline of native vegetation under dense Eurasian watermilfoil canopies. Journal of Aquatic Plant Management.

23. Eiswerth ME, Donaldson SG, Johnson WS (2000) Potential environmental impacts and economic damages of Eurasian watermilfoil (*Myriophyllum spicatum*) in western Nevada and northeaster California. Weed Technology 14: 511–518.

24. Boylen CW, Mueller N, Kishbaugh SA (2001) The costs of aquatic plant management in New York State. 51st Annual Meeting of the Aquatic Plant Management Society, Baltimore, MD.

25. Horsch EJ, Lewis DJ (2009) The effects of aquatic invasive species on property values: evidence from a quasi-experiment. Land Economics 85: 391–409.

26. Zhang C, Boyle KJ (2010) The effect of an aquatic invasive species (Eurasian watermilfoil) on lakefront property values. Ecological Economics 70: 394–404.

27. Tamayo M, Olden JD (2014) Forecasting the vulnerability of lakes to aquatic plant invasions. Invasive Plant Science and Management 7: 32–45.

28. Research WWCfRE (2009) Washington State's housing market: a supply/demand assessment - 4th quarter. http://www.wcrer.wsu.edu/WSHM/2008Q4/MKTRPT08d.pdf. Accessed 2012 Oct.

29. Walton SP (1996) Aquatic plant mapping for 36 King County lakes. Seattle, Washington: King County Surface Water Management Division.

30. Clapper J, Caudill SB (2014) Water quality and cottage prices in Ontario. Applied Economics 46: 1122–1126.

31. Lansford NH Jr, Jones LL (1995) Recreational and aesthetic value of water using hedonic price analysis. Journal of Agricultural and Resource Economics 20: 341–355.

32. Poor FJ, Pessagno KL, Paul RW (2007) Exploring the hedonic value of ambient water quality: a local watershed-based study. Ecological Economics 60: 797–806.

33. Rosen S (1974) Hedonic prices and implicit markets: product differentiation in pure competition. Journal of Political Economy 82: 34–55.

34. Halstead JM, Michaud J, Hallas-Burt S, Gibbs JP (2003) Hedonic analysis of effects of a nonnative invader (*Myriophyllum heterophyllum*) on New Hampshire (USA) lakefront properties. Environmental Management 32: 391–398.

35. Colwell PF, Dehring CA (2005) The pricing of lake lots. Journal of Real Estate Finance and Economics 30: 267–285.

36. Bao HXH, Wan ATK (2007) Improved estimators of hedonic housing price models. Journal of Real Estate Research 29: 267–301.

37. Griliches Z (1991) Hedonic price indexes and the measurement of capital and productivity: some historical reflections. In: E.R B, J.E T, editors. Fifty years of economic measurement: the jubilee of the conference on research in income and wealth. Chicago, Illinois: University of Chicago Press. pp. 185–206.

38. James LR, Singh BK (1978) An introduction to the logic, assumptions, and basic analytic prodcedures of two-stage least squares. Psychological Bulletin 85: 1104–1122.

39. Leggett CG, Bockstael NE (2000) Evidence of the effects of water quality on residential land prices. Journal of Environmental Economics and Management 39: 121–144.

40. Burnham KP, Anderson DR (2002) Model selection and multimodel inference: a practical information-theoretic approach. New York, New York: Springer.

41. Epancin-Niell RS, Hastings A (2010) Controlling established invaders: intergrating economics and spread dynamics to determine optimal management. Ecology Letters 13: 528–541.

42. Anonymous (2008) Invaders at the Gate. Available: http://www.invasivespecies.wa.gov/documents/InvasiveSpeciesStrategicPlan.pdf. Accessed 2012 Aug.

43. Provencher B, Lewis DJ, Anderson K (2012) Disentangling preferences and expectations in stated preference analysis with respondent uncertainty. The case of invasive species prevention. Journal of Environmental Economics and Management 64: 169–182.

44. Leung B, Bossenbroek JM, Lodge DM (2006) Boats, pathways, and aquatic biological invasions: estimating dispersal potential with gravity models. Biological Invasions 8: 241–254.

45. Bremner A, Park K (2007) Public attitudes to the managemnt of invasive non-native species in Scotland. Biological Conservation 139: 306–314.

46. Stoke KE, O'Neill KP, Montgomery WI, Dick JTA, Maggs CA, et al. (2006) The importance of stakeholder engagement in invasive species management: a cross-jurisdictional perspective in Ireland. Biodiversity and Conservation 15: 2829–2852.

47. Selge S, Fischer A, van der Wal R (2011) Public and professional views on invasive non-native species: a qualitative social scientific investigation. Biological Conservation 144: 3089–3097.

48. Verbrugge LN, Van den Born RJ, Lenders HJ (2013) Exploring public perception of non-native species from a visions of nature perspective. Environ Manage 52: 1562–1573.

49. García-Llorente M, Martín-López B, González JA, Alcorlo P, Montes C (2008) Social perceptions of the impacts and benefits of invasive alien species: implication for management. Biological Conservation 141: 2969–2983.

50. Homans FR, Smith DJ (2013) Evaluating management options for aquatic invasive species: concepts and methods. Biological Invasions 15: 7–16.

51. Larson DL, Phillips-Mao L, Quiram G, Sharpe L, Stark R, et al. (2011) A framework for sustainable invasive species management: Environmental, social, and economic objectives. Journal of Environmental Management 92: 14–22.

52. Vander Zanden MJ, Olden JD (2008) A management framework for preventing the secondary spread of aquatic invasive species. Canadian Journal of Fisheries and Aquatic Sciences 65: 1512–1522.

# 16

# Coexistence of Fish Species in a Large Lowland River: Food Niche Partitioning between Small-Sized Percids, Cyprinids and Sticklebacks in Submersed Macrophytes

**Małgorzata Dukowska\*, Maria Grzybkowska**

Department of Ecology and Vertebrate Zoology, Faculty of Biology and Environmental Protection, University of Łódź, Łódź, Poland

## Abstract

In the spring and summer of each year, large patches of submersed aquatic macrophytes overgrow the bottom of the alluvial Warta River downstream of a large dam reservoir owing to water management practices. Environmental variables, macroinvertebrates (zoobenthos and epiphytic fauna, zooplankton) and fish abundance and biomass were assessed at this biologically productive habitat to learn intraseasonal dynamics of food types, and their occurrence in the gut contents of small-sized roach, dace, perch, ruffe and three-spined stickleback. Gut fullness coefficient, niche breadth and niche overlap indicated how the fishes coexist in the macrophytes. Chironomidae dominated in the diet of the percids. However, ruffe consumed mostly benthic chironomids, while perch epiphytic chironomids and zooplankton. The diet of dace resembled that in fast flowing water although this rheophilic species occurred at unusual density there. The generalist roach displayed the lowest gut fullness coefficient values and widest niche breadth; consequently, intraspecific rather than interspecific competition decided the fate of roach. Three-spined stickleback differed from the other fishes by consuming epiphytic simuliids and fish eggs. The diet overlap between fishes reaching higher gut fullness coefficient values was rather low when the food associated with the submersed aquatic macrophytes was most abundant; this is congruent with the niche overlap hypothesis that maximal tolerable niche overlap can be higher in less intensely competitive conditions.

**Editor:** Tomoya Iwata, University of Yamanashi, Japan

**Funding:** This work was supported by a grant of the University of Lodz No 505/424. The funder had no role in study design, data collection and analysis, decision to publish, or preparation of the manuscript.

**Competing Interests:** The authors have declared that no competing interests exist.

\* Email: mdukow@biol.uni.lodz.pl

## Introduction

Dams usually affect downstream characteristics, like flow regimes, river-channel geomorphology, water and sediment quality, aquatic environment and biota [1–3]. The effects of the dam related river management practice may be evident for many kilometers of downstream reaches; one such effect is the development of submersed aquatic macrophytes (SAM) [4,5]. The main factors and processes controlling macrophyte status in lowland rivers are discharge and/or current velocity, light, substrate and nutrients, while the role of the first two is of most fundamental importance, as these hydrological parameters control instream macrophyte colonization, establishment and persistence [6]. The presence of SAM may be considered as a very important component of riverine biota, causing an increase in habitat structural complexity in alluvial lotic ecosystems. The water plants serve as a substrate for epiphyton, constituting a rich foraging habitat for macroinvertebrates [7], shelter against predation, heterogeneous substrate for co-existence and, to a small extent, a direct, food source [8–10]. Besides, the submersed plants' morphological characteristics (size, number and orientation of leaves and steams) influence both invertebrate [8,10] and fish distribution [11–13]. Thus SAM may support a high density of small fish individuals, because submersed plant beds offer protection from predators by hindering predators' foraging activities [14]. The foraging activity of vertebrate predators may decline monotonically with increasing habitat complexity [15].

At the river bed macrophyte patches, fish may exploit prey types from three ecological formations: zooplankton (especially numerous below dam reservoirs), fauna dwelling on the surface of vegetation (epiphytic fauna, mainly several taxa of Chironomidae and Simuliidae) and benthos. Submersed plants create favourable conditions for pelophilous forms like most of Oligochaeta and Chironomidae by extensive particle trapping and accumulation of a fine-grained, nutrient enriching sediment [16–18]. Thus the development of SAM on the alluvial bed river attracts many small fish individuals. Being a little competitive, the coexistence of these species is possible if there are differences in their responses to limiting resources. The species-specific differences that allow such coexistence can be considered as species' niche with four major axes: resources, predators, space and time [19,20].

Untypical, but abundant development of SAM has been observed in the large lowland alluvial Warta River downstream of the Jeziorsko Reservoir every year as an effect of a low discharge in late spring and summer [10]. Every early autumn large volumes of water start to be released through the reservoir dam sluices, and the SAM habitat is torn out of the bottom or gets inundated with the bottom substrate [3,10]. The trophic relationships among this rich but temporary habitat have been investigated with regard to the primary and secondary invertebrate consumers (gathering collectors, scrapers and predators) and the tertiary consumer of three-spined stickleback (*Gasterosteus aculeatus* L.), as well as percids [7,18,21,22].

For a long time every year, however, the Jeziorsko tailwater has been dominated by cyprinids [3,22–25], many small individuals of which were also foraging in the tailwater's SAM. The trophic impact of the fishes on this habitat has not been investigated, although we have long expected that all types of the rich food resources connected with the SAM (zoobenthos, epiphytic fauna and zooplankton) are exploited by these (and other) fishes. Consequently, the main objective of the present study was to identify patterns in the feeding of the five predominant fish species occurring in that area (roach, dace, perch, ruffe and three-spined stickleback), in order to evaluate their trophic niches' breadths and overlaps in relation to resources, time and space. To this end, we investigated in detail the gut contents and intraseasonal changes in the diet of these five fish species there.

## Study area

The Warta River rises 380 meters above sea level, is 808 km long and empties into the Oder River at 13 meters above sea level. Its catchment area is ca. 53 710 km$^2$ and its slope ranges from 2.0–1.0‰ in the upper course, and from 0.3–0.1‰ in the middle and lower courses [26]. The study site was established in this lowland alluvial river about 1.5 km downstream of the dam of the large Jeziorsko Reservoir, whose maximal surface area is 42.3 km$^2$ (Fig. 1). At the investigated site, the Warta River is about 60 m wide, with a maximum depth of 2.5 m in the erosion zone.

During sampling in 2004, similarly as in a few former years, the discharge of the Warta River below the dam was quite different in comparison with the natural, upstream reach, especially in summer, when its flow stabilized at a much lower level than in other seasons. One consequence of this phenomenon was the appearance of submersed macrophytes, which started to spread each year from the summer of 1992 along a short stretch of the tailwater [3,27]. Detailed site descriptions can be found in Grzybkowska et al. [10,28].

## Material and Methods

The sampling was conducted in 2004, 18 years after the reservoir started functioning routinely and 10 years after the construction of a hydroelectric plant. In the investigated reach, large patches of *Potamogeton pectinatus* L. and small patches of *Potamogeton lucens* L. gradually developed since late spring through late summer in the transitional riverbed zone, which is located between the depositional zone, close to a bank, and the midriver channel (Fig. 1). The percentage of river bottom covered by macrophytes, samples of particulate organic matter and inorganic substrate, macrophytes, zoobenthos and epiphytic fauna, zooplankton, and fish, were collected, at the same time, within an area sized 40 m by 2.5 m, and extending along the bank and along the transitional zone. The area was randomly selected within the zone. The samples were obtained twice a month from May through late August. As a result, nine samples of each biotic and abiotic component were collected.

On each sampling occasion each sample of benthos (containing benthic invertebrates, particulate organic and inorganic matter) consisted of five subsamples (each subsample covered 100 cm$^2$ of stream-bed area collected with a tabular sampler of a catching area of 10 cm$^2$). The sampling places were uniformly distributed

within the sampled area. The invertebrates of the samples were sorted from the detritus and benthic sediments by hand and preserved in 10% formalin. All invertebrates from these quantitative samples were counted and their wet weight (w.w.) assessed; these data were used to estimate the biomass of zoobenthos. Most invertebrates were classified to the lowest taxonomic level of the dominant macrobenthic group, while chironomids were identified to the species level when possible. As the exact identification on the basis of their larvae was often impossible, we reared their immature stages in the laboratory from additional qualitative samples taken each time in order to obtain larval and pupal skins, and imagines.

These samples were also used to determine the organic matter content in the bottom sediment. For this purpose a 1 mm mesh sieve was used to separate benthic particulate organic matter (BPOM) into:>1 mm (coarse – BCPOM) and <1 mm (fine – BFPOM) [29]. Next, the benthic organic matter was dried at 60°C for two days, weighed, ashed at 600°C for two hours and reweighed. A more detailed description of these methods can be found in Grzybkowska et al. [10], Grzybkowska & Dukowska [27].

To estimate the amount of dry weight of plants growing in the study site, a special frame (0.5×0.7 m) was placed on the riverine bottom and all the *Potamogeton* within the frame was collected. This procedure was repeated three times on each sampling occasion. In the laboratory, the pondweeds were dried for 24 hours at 65°C to estimate their dry weight (d.w.).

Five subsamples of the epiphytic fauna settled on *Potamogeton* were collected on each sampling occasion. Each of the subsamples consisted of three fragments of stems (about 20 cm long, on average) of the plants cut off under the water surface, stored in plastic containers, and preserved in 10% formalin in the fields. In the laboratory, the plant material was removed from the containers and the invertebrates were washed off the plants, sorted by hand, identified to the species level when possible, counted, had their wet weight assessed. The obtained data were recalculated to estimate the biomass of epiphytic invertebrates per 1 m$^2$ of *Potamogeton* covering the riverine bottom on given sampling occasions.

To evaluate the density of zooplankton (mainly Cladocera), 0.03 m$^3$ samples of river water were filtered through a planktonic net, of 50 μm mesh size, and preserved in formalin with riverine water in the fields. In the laboratory, individuals were identified to the species level and counted. The biomass of the zooplankton was estimated on the basis of suitable equations [30].

To assess fish density and gut contents, fish were caught in the sampled area using an electric current of 220 V and 3 A supplied from a backpack battery generator. A single pass CPUE sampling was carried out, which consisted in one person wading against the water current with an anode dipnet and another one with a bucket for collecting stunned fish along the longitudinal axis of the sampled area. The sampling period was 15 minutes on each occasion. As the equipment was battery-powered fish were not scared by noise, hence no barrier preventing fish from escaping was necessary. Immediately after the capture, fish were anaesthetised (MS-222, tricaine methanesulfonate) and then preserved in 4% formalin. The field studies did not involve endangered or protected species. The Polish Angling Association in Konin (a tenant of the water body, director Jerzy Olejnik, 1 Wyspiańskiego Str., 62–510 Konin) issued the permit to conduct the field study (see Fig. 1 for details of the study area location). Electrofishing was performed with the license No 1180/01 for the operation of electric fishing tools, and the other procedures were conducted under the permission No 219/2011 to perform experiments on animals, issued by the University of Lodz, and the individual

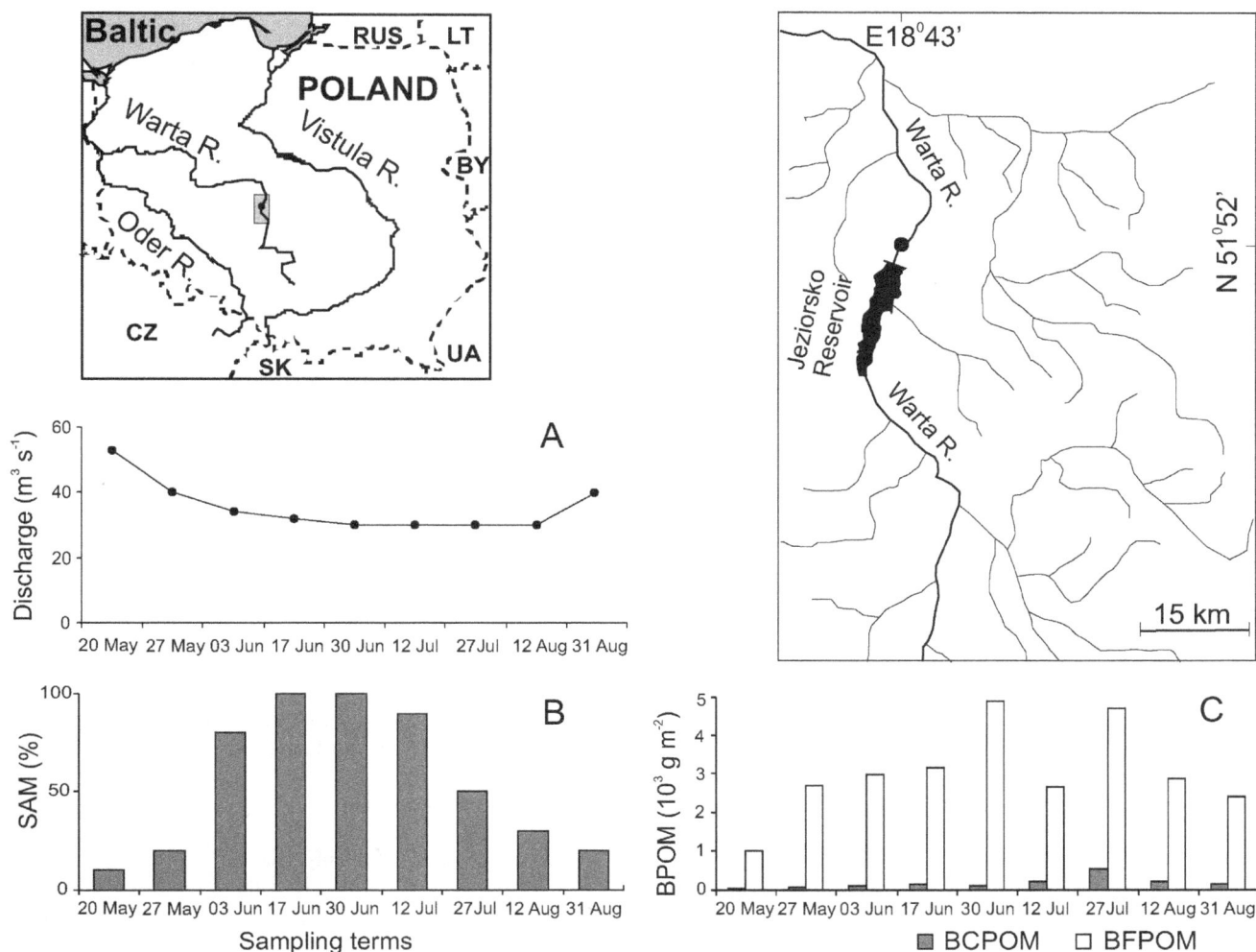

**Figure 1. Study area with the marked sampling site in the tailwater of the Jeziorsko Reservoir on the Warta River.** Parameters of the site over the study period: A. discharge; B. SAM -% of river bottom covered by hydrophytes; C. BFPOM and BCPOM, two fractions of benthic particulate organic matter (details in the text).

licence No 6/2006 to perform experiments on animals according to the Law on the protection of animals and the recommendations of the ICLAS.

In the laboratory, the total length (TL) of each analysed fish specimen was measured to the nearest 1 mm, and weighed to the nearest 0.1 g. The gut contents of all investigated individuals (n = 242) were analysed using a stereomicroscope and microscope. Prey types from the whole gut length, after identification to the lowest possible taxonomic category, were counted and weighed (w.w.), except zooplankton, the biomass of which was estimated in the same way as described above.

The gut fullness coefficient (FC) was calculated by the formula [31]:

$$FC = \frac{a}{b} \times 100$$

where:

$a$ – total gut content weight (g)

$b$ – weight of fish (g)

Niche breadth ($B_n$) was calculated using Levins' index:

$$B_n = \frac{1}{\sum p_j^2}$$

where:

$B_n$ – Levins' measure of niche breadth

$p_j$ – proportion of food type $j$

$n$ – number of possible food types.

We standardized the trophic niche values (ranging from 0 to 1) using the Hurlbert formula [32]:

$$B_A = \frac{B_n}{n-1}$$

where:

$B_A$ – Levins' standardized niche breadth

$B_n$ – Levins' measure of niche breadth

$n$ – number of possible food types.

The interspecific diet overlap among the investigated species was calculated using the Schoener overlap index [33]:

$$C_{xy} = 1 - 0.5 \left( \sum_i |P_{xi} - P_{yi}| \right)$$

where:

$C_{xy}$ – is the overlap index ranging from 0 (no overlap) to 1 (complete overlap)

$P_{xi}$ – is the proportion of food type $i$ of species $x$

$P_{yi}$ – is the proportion of food type $i$ of species $y$

It is worth noting that $C_{xy} > 0.6$ is considered biologically significant [34].

The cluster analysis was also applied for identifying fish species' groups with similar gut contents.

The Kruskal-Wallis test and the post-hoc Dunn test in the STATISTICA software package, v. 10 [35,36] were applied to compare the percentage of the most frequent food items in the diet, the gut fullness coefficient, as well as diet niche breadth, to determine whether significant changes during the investigated period occurred. The significance level of the tests was $\alpha = 0.05$.

## Results

### Intraseasonal dynamics of environmental parameters of the SAM habitat

The development of macrophytes started in late May. It gradually intensified the cumulation of benthic fine particulate organic matter on the bottom, which strengthened the possibilities of the development of pelophilous zoobenthos. The highest biomass values of *P. pectinatus* (over 210 g d.w. m$^{-2}$) and the highest density of BFPOM were recorded during the second half of June 2004; at that time the macrophytes were also covered with filaments of green algae *Cladophora glomerata* (L.) Kutz. The intraseasonal dynamics of the above characteristics are presented in Fig. 1. Inorganic substratum was mostly fine and coarse sand, as well as some gravel.

### Fish assemblage attributes

Over the investigated period the total of 14 fish species inhabiting *Potamogeton* patches in the SAM habitat were captured. The dominant species was perch (*Perca fluviatilis* L.), constituting 32.7% of the total fish density and 23.0% of their total biomass. In order of their decreasing importance in total fish biomass, the following species were: dace (*Leuciscus leuciscus* (L.) and roach (*Rutilus rutilus* (L.)), 38.2% and 12.1% of the biomass and 15.5% and 18.2% of the density, respectively. In turn ruffe (*Gymnocephalus cernuus* (L.)), accounted for 9.1% of the biomass and 11.8% of the density, while three-spined stickleback (*Gasterosteus aculeatus* L.) reached 4.0% of the biomass and 11.8% of the density. Spined loach (*Cobitis taenia* L.), gudgeon (*Gobio gobio* (L.)) and ide (*Leuciscus idus* (L.)), were captured less frequently, constituting with other fish species 13.6% of the biomass and 10.0% of the density (Fig. 2).

All guts of the analysed individuals of the five species were filled to various extent (n = 242). The characteristics of the fish are presented in Table 1.

### Food resources associated with SAM

**Microcrustaceans.** Over the whole investigated period Cladocera dominated in water column, reaching 99% of the zooplankton biomass of all investigated periods. Copepoda were very scarce, with the maximum in late August. The maximum values of cladoceran biomass were recorded in August due to the presence of large-sized predator *Leptodora kindtii* (Focke) (49.7%

of the total cladoceran biomass during the study period). The most numerous specimens in water column were the small sized species of *Chydorus sphaericus* (O. F. Müller) (which constituted only 15.1% of zooplankton biomass over the investigated period), *Bosmina* spp. (17.1%) and large size taxa, *Daphnia* spp. (17.2%) (Fig. 2).

**Epiphytic and benthic fauna connected with *Potamogeton*.** Simuliids dominated among the macrophytes, amounting to 84.7% of the biomass of the total epiphytic fauna, but were abundant only till late June. Different seasonal dynamics was showed by chironomids (14.1% of total epiphytic fauna biomass), which reached the highest biomass in June and July but occurred throughout the study period. Main chironomid species, on *Potamogeton*, were: *Cricotopus sylvestris* (Fabricius) (Orthocladiinae), and *Parachironomus gracilior* (Kieffer) (Chironominae-Chironomini). Other epiphytic invertebrates (including *Hydra* sp.) occurred rarely (1.2%) (Fig. 2).

Chironomidae dominated in the benthos over the whole investigated period, reaching 83.7% of total benthic invertebrate biomass, and their abundance decreased over time. The dominant chironomids in benthos were represented by *Chironomus riparius* Meigen, *Dicrotendipes nervosus* (Staeger), *Glyptotendipes cauliginellus* (Kieffer) and *Polypedilum* spp. (Chironominae-Chironomini), as well as *Cricotopus bicinctus* (Meigen) (Orthocladiinae) and *Paratanytarsus* (Tanytarsini). Oligochaeta (9.3%) were present at the same level over the whole studied period, while Trichoptera (4.1%, mainly *Hydropsyche* spp.) were present mostly in May. The macrophyte bed was also inhabited by *Hydra* sp. (1.2%), especially in May and June (Fig. 2).

### Trophic attributes of fish

Fish associated with the SAM exploited five main food categories: zooplankton (mainly Cladocera), benthos (Chironomidae during the whole studied period, and Trichoptera at the beginning of the investigated period), epiphyton (Chironomidae and Simuliidae), plants, detritus with algae, including the filamentous ones (Fig. 3).

In dace, Cladocera occurred on six of the seven sampling occasions on which the fish species was recorded. However, the biomass of Cladocera first rapidly increased to mid June, when they became dominant, and then decreased to none in late July. Both epiphytic and benthic chironomids became the first/second most important food category of dace in July. It is worth noting that dace also exploited imagines from the water surface. Three-spined stickleback consumed mainly Cladocera in May, which were replaced by Simuliidae in June. Benthic Chironomidae were the second/third most important food category in May and June but almost completely dominated the diet of this species in July. In ruffe, benthic chironomids were the main food category consumed on all sampling occasions, except late May and early July, when Trichoptera dominated instead. Epiphytic chironomids appeared in ruffe's diet since late June. Cladocerans and epiphytic chironomids dominated the diet of perch to a similar extent on most of the sampling occasions, although copepods became a subdominant and trichopterans a decisive dominant on one occasion each. In turn in roach, plants were the dominant food component on the first and a subdominant on all other sampling occasions. Cladocera were the main food type on three sampling occasions, in July and August. Detritus and algae were present during the whole study period (Fig. 3).

From the point of view of given food categories, in the course of the whole investigated period, the main component of roach diet were plants (about 28.2% of gut contents), while for other species it was the accessory food (Kruskal-Wallis test, H = 178.49, p<

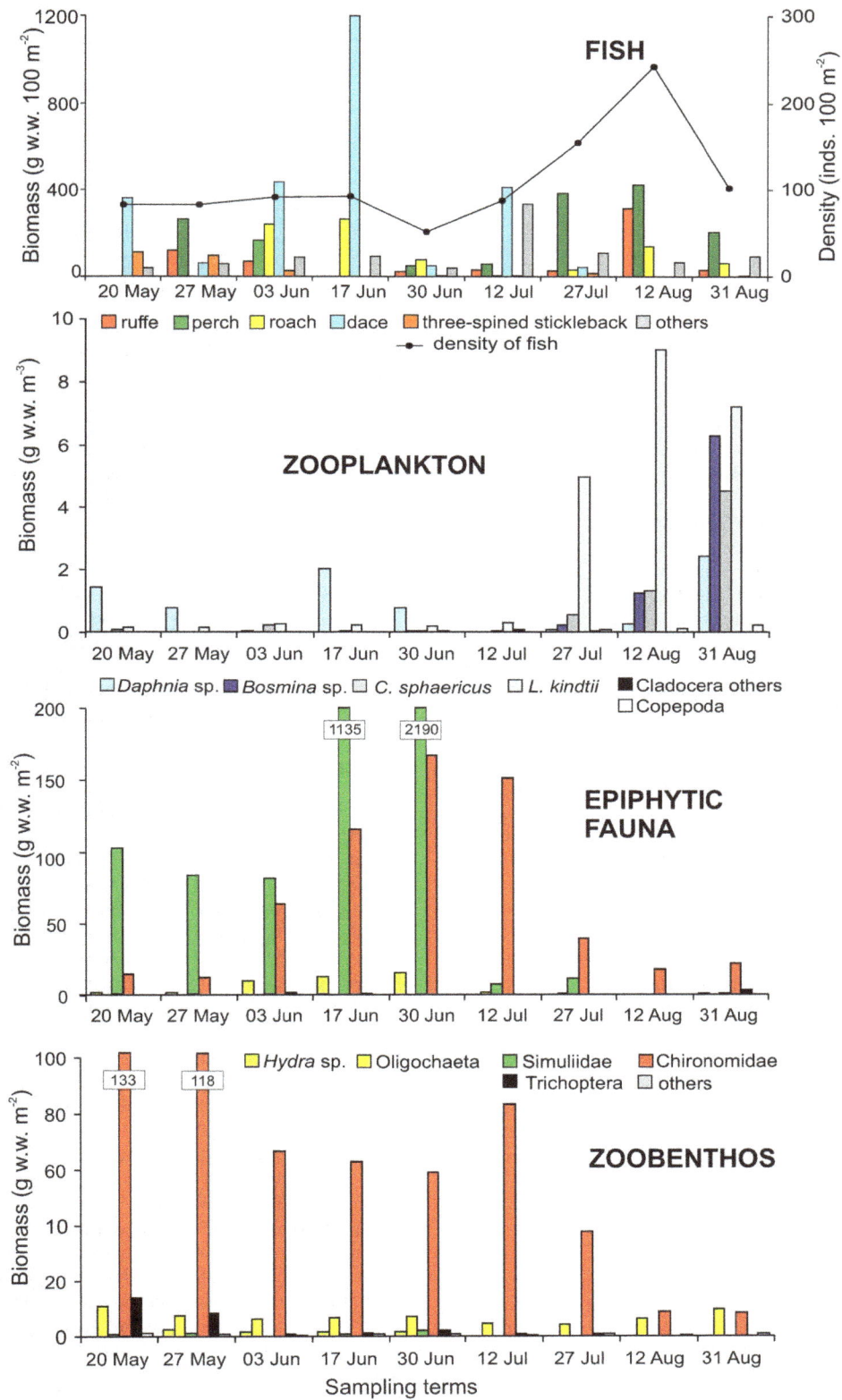

**Figure 2. Fish and their food resources.** Intraseasonal dynamics of biomass and density of fish and of their food resources (zooplankton, epiphytic fauna and zoobenthos) over the study period.

**Table 1.** Characteristics of analysed fish specimens in the SAM habitat of the Warta River, Poland.

| Fish species | N | | B (g) | TL (mm) | FC |
|---|---|---|---|---|---|
| Gymnocephalus cernuus (L.) (ruffe) | 41 | mean R | 2.7 0.8–16.0 | 63 44–113 | 2.4 0.6–7.3 |
| Perca fluviatilis L. (perch) | 55 | mean R | 3.3 0.8–14.2 | 66 41–102 | 0.8 0.1–4.9 |
| Rutilus rutilus (L.) (roach) | 76 | mean R | 1.8 0.13–44.2 | 56 24–147 | 0.4 0.03–1.5 |
| Leuciscus leuciscus (L.) (dace) | 37 | mean R | 16.6 10.8–29.3 | 116 100–136 | 1.0 0.1–2.1 |
| Gasterosteus aculeatus L. (three-spined stickleback) | 33 | mean R | 2.4 0.3–4.2 | 54 31–70 | 4.1 0.5–7.2 |

N – number of specimens, B – biomass, TL – total length, FC – gut fullness coefficient, R – range of values.

0.0001), among which higher percentages of SAM (about 4%) were recorded for dace.

Chironomidae were consumed by all fish during the whole studied period; these dipterans, living in the bottom, were the main food of ruffe and statistically more important than those of other fish species (Kruskal-Wallis test, H = 108.51, p<0.0001). In turn, epiphytic chironomid taxa were consumed by perch more than by other fishes (Kruskal-Wallis test, H = 57.87, p<0.0001). Other dipterans, Simuliidae, were the main food category (about 26.8% of gut contents) only for three-spined stickleback (Kruskal-Wallis test, H = 33.92, p<0.0001), and to a lesser extent, for dace and perch (about 6% of gut contents). Trichoptera (large size larvae represented by *Hydropsyche*) were exploited to a greater extent by dace, ruffe and perch, than by roach and three-spined stickleback (Kruskal-Wallis test, H = 61.16, p<0.0001); perch consumed *Hydropsyche* larvae at a lower level than dace (post-hoc Dunn test, p<0.036). Detritus with algae was the main component in the gut contents of roach and dace and statistically different than of the other studied fishes (Kruskal-Wallis test, H = 142.24, p<0.0001). Cladocerans were exploited by all fish; the lowest biomass of this prey was recorded in ruffe guts (Kruskal-Wallis test, H = 28.03, p<0.0001). Other microcrustaceans, Copepoda, were an important diet component for perch only (Kruskal-Wallis test, H = 60.08, p<0.0001). Complementary food types (included in the "others" category) were ephemeropterans, other insects and *Asellus aquaticus* (Isopoda), as well as ostracods.

Over the investigated season, the fullness coefficients of these five analysed fish species were statistically different (Kruskal-Wallis test, H = 130.58, p<0.0001). As showed by the post-hoc Dunn test the differences did not exist only between ruffe and three-spined stickleback and between perch and dace. Moreover, for the two last species the highest variability in FC values was noted: for perch from 0.25 at the end of August to 2.8 at the end of June, while for dace from 0.5 at the end of July to 2.8 at the end of May. However, the lowest variability was observed for roach: from 0.3 at the end of August to 0.6 in the middle of June. For the whole studied period, the highest FC values were noted for three-spined stickleback (at the beginning of June) (Fig. 3).

The diet compositions of the five fish species were analyzed using hierarchical cluster analysis, which distinguished three feeding groups: 1) three-spined stickleback which consumed epiphytic simuliids (main prey), and occasionally fish eggs, 2) roach, feeding mainly on plant materials, detritus and algae, zooplankton, and Chironomidae as an important complementary food category, 3) dace, perch and ruffe, which ate mainly animal materials (aquatic and terrestrial invertebrates) (Fig. 4).

Niche breadths (Levins' index) of these species were statistically different (Kruskal-Wallis test, H = 125.02, p<0.0001) and varied seasonally. On each sampling occasion when roach and ruffe occurred, the niche breadth of the former was always much wider

than of the latter (Fig. 5A). The evaluation of niche breadth revealed a less diverse diet of typical benthic fishes, such as ruffe, while the most diverse diet was in the case of roach. As regards niche breadths, three non-hierarchical groups of species (Kruskal-Wallis test, post hoc Dunn test, H = 77.015, p<0.0001) were distinguished: the first group was roach, the second group were dace and perch, and the third were ruffe and three-spined stickleback (Fig. 5A and B).

The Schoener interspecific diet overlap index did not differ between cyprinids, percids and three-spined stickleback over the studied period. However, this index varied seasonally, reaching the highest value between ruffe and dace (0.78 at the end of May when both species exerted pressure on the *Hydropsyche*) while the lowest ones were attained a few times: between roach and other fish (4 times lower or equal to 0.2) and between three-spined stickleback and other species (also 4 times lower, Fig. 6).

## Discussion

### Submersed aquatic macrophytes in large rivers

In the two recent decades, the bed zone between the bank and midriver channel of the tailwater stretch of the Warta River has undergone a seasonal regime shift from an alluvial to a macrophyte-dominated system. Such processes also took place in many other impacted large rivers. As this phenomenon is usually believed to exert negative environmental impacts, attempts to counteract it, using various methods, particularly artificial flooding, were applied in these rivers [37]. SAMs are rather undesirable elements in large rivers because they modify the in-river environment mainly by altering river flows, and by decreasing the depth of trapping sediments, which is also beneficial for large populations of pelophilous macrobenthic fauna. However, SAMs are the excellent surface for epiphyton. Some groups of invertebrates, represented by grazers, like snails, macrocrustaceans and cladoceran zooplankters, are able to protect aquatic macrophytes by removing epiphytes and phytoplanktonic algae. The plants themselves help the process of defense because they are a source of biochemical compounds that, on one hand, negatively affect the growth of algae (allelopathy), but, on the other hand, may attract grazers. Many fish species occupying this rich habitat also positively affect SAMs as main predators foraging on epiphytic invertebrates. Consequently, these vertebrates are an important part of a complex network of relations between nutrients, phytoplankton, epiphytes, herbivorous invertebrates and benthos [38–40].

### Fish foraging among vascular plants

From the beginning of the functioning of the Jeziorsko Reservoir (1986) in the Warta River downstream from its dam the diversity of fish declined regularly, while the number of these

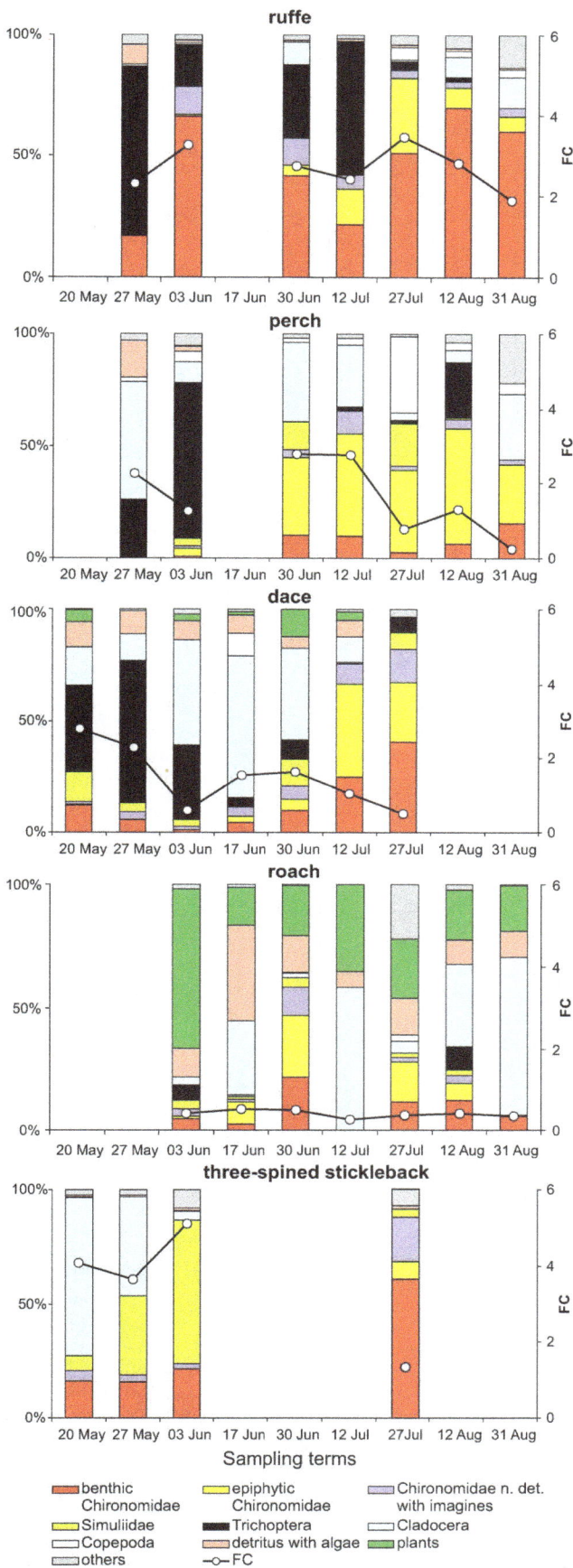

**Figure 3. Fish diet composition.** Food items and gut fullness coefficient (FC) of each fish species over the study period.

vertebrates showed quite opposite trends, mainly due to perch and roach, and to a lesser degree, to ruffe. This process of fish decline concerns, for example, dace, a rheophilic species (i.e. riverine specialist), which was present sporadically, and even absent in some years [3,24,25].

Dace, which exploits open water with fast water currents, usually suffer from human-induced changes in aquatic environment and is thus considered a good bioindicator [41–43]. The over-representation of dace among dense leaves and stems of *Potamogeton* recorded in the impacted stretch of the Warta River was rather unexpected. Dace showed a very high plasticity to food resources connected with SAM, feeding on larger, easily catchable benthic organisms, epiphytic fauna and/or zooplankton as well as imagines from water surface (mainly Chironomidae) which died after mating and/or rested on the surface above its pupal exuviae after eclosion. However, dace reached the highest value of FC when consuming large *Hydropsyche* larvae and at that time their niche overlapped that of ruffe. In June the diet of dace comprised also Cladocera, and the FC of dace was reduced. The presence of dace in SAM may testify to the greater attractiveness of this temporary habitat than the sandy mid-channel of the alluvial Warta River, where only tiny organisms, such as some Oligochaeta and Chironomidae, occasionally occur. Also, the potential migration of dace towards the depositional, SAM-less zone (see Material and methods), would not bring many benefits to the fish because the mud covering that zone is frequently flooded, or, to a high extent, exposed (to air) due to pulse water releases from the reservoir. Therefore, at the depositional zone benthic fauna was remarkably impoverished as compared with the natural river section (upstream of the reservoir); virtually no insects (except some families of Diptera), Hirudinea and Oligochaeta, were collected there [3,27,28,44]. Summing up, dace food spectrum in the very biologically productive SAM habitat was similar to those in other ecosystems [45–47] in which dace may also display daily

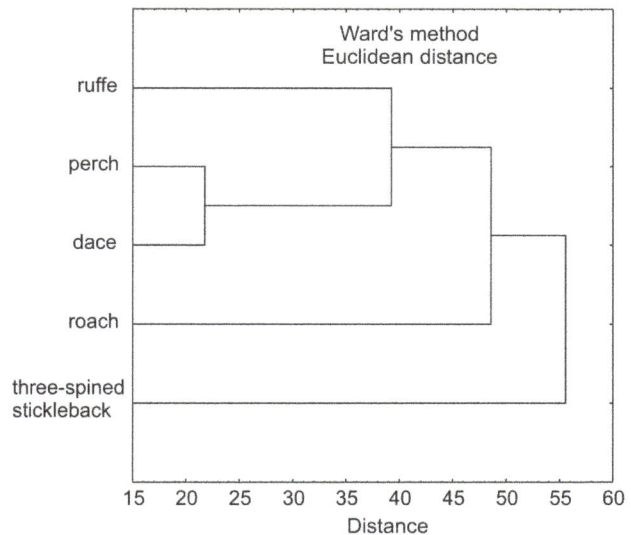

**Figure 4. Similarity of fish species in terms of diet.** Dendrogram resulting from the cluster analysis performed on gut contents of each species over the whole investigated period.

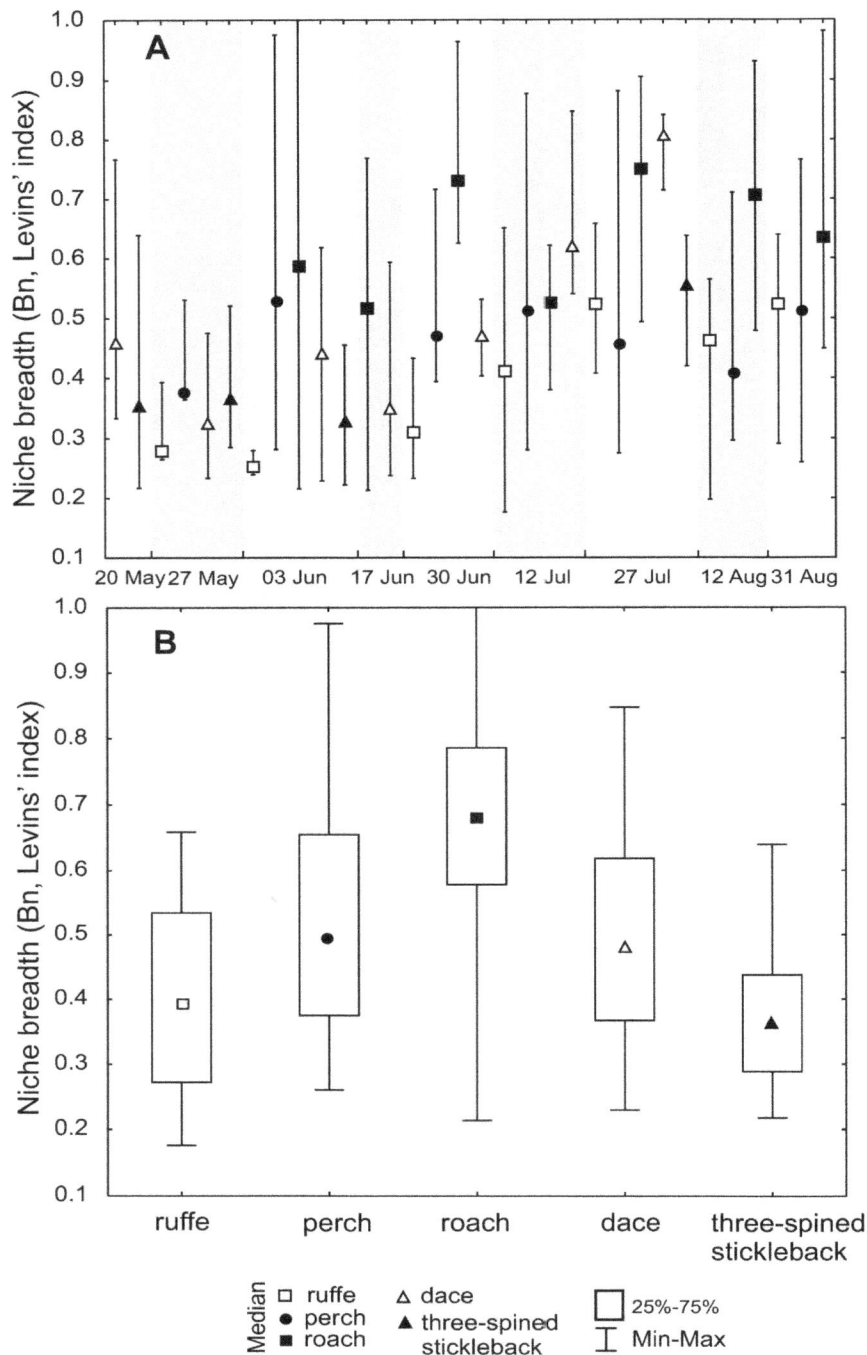

**Figure 5. Food niche breadth of fishes.** A. Intraseasonal variation in the niche breadth (medians of Levins' index, $B_n$) of each species. B. Median of the niche breadth for each species over the whole studied period. Vertical bars represent the interquartile range.

migrations to and from feeding places [48]. It is worth noting that this species also shows ontogenetic shifts in its food spectrum [49].

Three-spined stickleback was the second species for which over-representation in SAM as compared with its abundance in the whole tailwater reach was observed [50], but only at the beginning of the studied period. According to Bańbura [50], three-spined stickleback feed mainly on Copepoda, Cladocera, larvae of Chironomidae, and seasonally Mollusca, Oligochaeta and fish eggs, but none of these items exceeded 25% of the gut contents.

This finding is not in accordance with the present observations. In the macrophyte beds of the Warta River, the food of this species varied intraseasonally; the extraordinary development of the epiphytic fauna caused a foraging shift from Cladocera (constituting at the beginning of the study period about 70% of gut contents, and consisting mainly of large sized species of *Daphnia*, of reservoir origin, [51]) to epiphytic fauna, especially to the largest (oldest) larvae of the dominant dipterans, both simuliids and chironomids (the majority of gut contents in June). Low niche

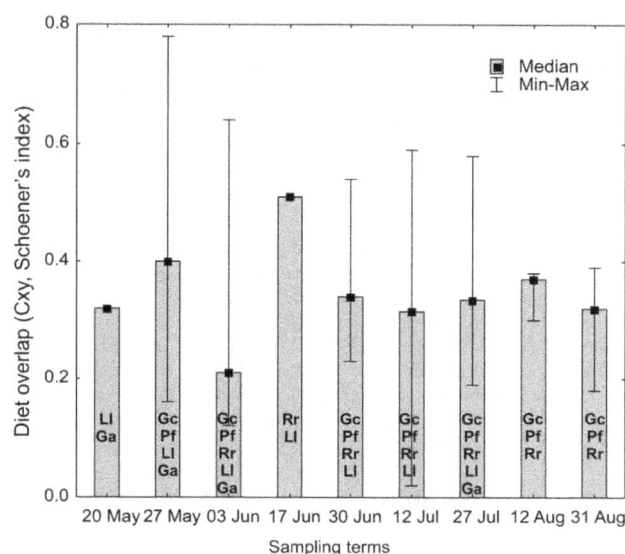

**Figure 6. Interspecific diet overlaps.** Interseasonal variation in the diet overlap (medians of the Schoener index, $C_{xy}$) between perch (Pf), ruffe (Gc), roach (Rr), dace (Ll) and three-spined stickleback (Ga); the above abbreviations in parentheses come from scientific names of fish species. Other explanations as in Fig. 5.

overlap between three-spined stickleback and other species was the effect of consuming epiphytic simuliids as the main prey and other prey types, such as fish eggs (the highest values of FC occurred when the species fed on these elements), rather untypical food items for other fish. This finding is confirmed by a dendrogram resulting from the cluster analysis.

The choice of older (larger) dipteran individuals by stickleback, as stated in our earlier paper [22], is consistent with the optimal foraging theory [52] - organisms forage in such a way as to maximize their net energy intake per unit time.

Ruffe, as a typical benthivorous fish, and partially three-spined stickleback and dace, play an important role in the internal nutrient cycling; firstly, during the searching for food (through bioturbation), secondly, during the translocation of nutrients from the bottom to the water column realized via their digesting and egestion. Ruffe, a small but aggressive fish native to Europe and Asia, may be classified rather among specialists [18,53–56] than generalists [57]. In lakes, ruffe may efficiently forage in deeper and darker bed patches due to its very sensitive lateral line system and the light-reflecting *tapetum lucidum* in its eye [58,59]. In the Warta River, this vertical gradient of food selection by ruffe was confirmed by Dukowska et al. [18], while its horizontal movement for searching food is rather unprofitable, similar to dace [3,27,28]. Overlaps between ruffe diet and those of other species were at the same level (from 0.21 to 0.34) regardless of ruffe's preference for benthic chironomids and *Hydropsyche*.

According to some studies, ruffe do not undergo ontogenetic dietary shifts, and remain the bottom feeders throughout the life cycle [57].

In rivers, high diversity and abundance values of fish are usually strongly associated with a high degree of trophic specialization (low amplitude of individual trophic niches) and a small degree of overlap between the diets of species constituting an assemblage [60]. But in the impacted stretch of the Warta River fish diversity has been low from the beginning of the functioning of the Jeziorsko Reservoir, although the abundance of several species was

very high [24,25]. Each year, when the amount of water released from the reservoir increases, a mass escapement of fish of eurytopic species, predominantly juvenile stages, gets intensified. This mainly concerns the species that have attained reproductive success in the reservoir in a given year (including mostly perch and roach, but also ruffe) and may lead to an increase in the abundance of these species (two or three orders of magnitude higher than in the pre-impoundment period or in the upstream river stretch) [3,24,61].

Perch, a potential competitor for ruffe [56,62], is included among generalists. It shows an ontogenetic shift in its diet during its life history, from zooplankton (at the onset of exogenous feeding) to macroinvertebrates, and from macroinvertebrates to fish [23,63–67]. However, the adults of this species may also be classified as trophic generalists switching frequently between piscivorous, zooplanktivorous, and benthivorous feeding style depending on the food resources in the environment. In the Warta River, in spite of this mode of foraging in areas drastically overcrowded by perch [61], the species starved also at the end of the present study period, despite the increased transport of zooplankton from the reservoir [18,51]. This is in accordance with assumptions of the theory of the ideal free distribution stating that an increased number of individuals in a given patch of resources reduces its quality, through either increased scramble competition or increased interference competition (including aggressive behavior) [68,69].

In the Warta River, diet overlap between perch and ruffe was rather low (the Schoener index, range 0.21–0.34) as perch preferred to eat epiphytic chironomids, simuliids and *Daphnia*, while ruffe chose large sized benthic chironomids and trichopterans (if they were present) and occasionally zooplankton [18]. It is worth stressing that the identification of prey, mainly Chironomidae, of these two fish species, to genus (mainly) or species level helped to determine the food resource partition between these two closely related species, because these fishes generally consumed different chironomid taxa [18].

In the Warta River, the second numerically most abundant species connected with SAM was roach. The interaction between perch and roach has been studied by many ecologists, especially in lentic ecosystems [67,70,71]. Both young roach, which fed mainly on detritus, chironomids and zooplankton, as well as older individuals, which consumed mostly detritus, filamentous algae, vascular plants, macroinvertebrates and molluscs, represented typically omnivorous feeding habitats, and are thus considered generalists [71–73]. Besides, roach may be classified into opportunists. If huge amounts of easily accessible food appear, roach use it immediately; for example, during chironomid emergences the alimentary tracts of roach are filled in with numerous pupae of these insects [74]. For each chironomid pupa being in a water column is a very dangerous experience (as the pupa may then end up as the prey of macroinvertebrates and/or fish), however, at that time zooplankton, mainly *Daphnia*, is released from fish predation [75].

In the present material, an insufficient amount of food available for roach (starvation?) is indicated by the low values of its FC. Its niche breadth was the largest one of those of all the five species during the whole studied period. The niche overlap between this species and other fishes was low, thus we may suppose that rather an intraspecific than interspecific competition decided the fate of small roach in this overcrowded environment. Even if certain individuals of this eurytopic species die because of competitive interactions, they will soon be replaced by juveniles migrating massively from the reservoir [25,61,76]. It is worth mentioning that the consumption of such food categories as algae and detritus may have resulted in a slow growth rate [73]. A large spectrum of food types consumed by roach, larger than that of perch, was recorded by other ecologists [67,70].

## Conclusions

In the SAM habitat the diet overlap was highest when the SAM patches, and the food resources associated with them, were most developed (twice during the investigated period). For the first time at the beginning of SAM development, when large-sized *Hydropsyche* were the main food category for ruffe, perch and dace, and for the second time throughout June and the first half of July, when benthic and/or epiphytic Chironomidae were the basic food items for these species. This finding is congruent with the niche overlap hypothesis saying that maximal tolerable niche overlap can be higher in less intensely competitive situations, i.e. in environments with lower demand/supply ratios [77].

One of the most important attributes of organisms to avoid direct overlap in the use of resources is diversification of body size of individuals of given species [60]. We put focus on three (time, space and resource) of the four niche axes of each of the five fish species living among the submersed macrophytes and we concluded that their diets only partly overlap, which allows them to coexist in this temporary, very rich habitat.

## References

1. Ward JS, Stanford JA (1980) Tailwater biota: ecological response to environmental alternations. In:Proceedings of the symposium on surface water impoundments ASCE.
2. Petts GE (1984) Impounded rivers. Perspectives for ecological management. Chichester: John Wiley and Sons. 326 p.
3. Głowacki Ł, Grzybkowska M, Dukowska M, Penczak T (2011) Effects of damming a large lowland river on chironomids and fish assessed with (multiplicative partitioning of) true/Hill biodiversity measures. River Res Appl 27: 612–629.
4. Bednarek AT (2001) Undamming rivers: a reviews of the ecological impacts of dam removal. Environ Manage 27: 803–814.
5. Feld CK, Birk S, Bradley DC, Hering D, Kail J, et al. (2011) From natural to degraded rivers and back again: A test of restoration ecology theory and practice. Adv Ecol Res 44: 119–209.
6. Franklin P, Dunbar M, Whitehead P (2008) Flow controls on lowland river macrophytes: A review. Sci Total Environ 400: 369–378.
7. Dukowska M, Grzybkowska M, Sitkowska M, Zelazna-Wieczorek J, Szeląg-Wasielewska E (1999) Food resource partitioning between chironomid species associated with submersed vegetations in the Warta River below the dam reservoir, Poland. Acta Hydrobiol 41: 219–229.
8. Tokeshi M, Pinder LCV (1985) Microhabitats of stream invertebrates on two submersed macrophytes with contrasting leaf morphology. Holarctic Ecol 8: 313–319.
9. Pinder LCV (1992) Biology of epiphytic Chironomidae (Diptera: Nematocera) in chalk streams. Hydrobiologia 248: 39–51.
10. Grzybkowska M, Dukowska M, Takeda M, Majecki J, Kucharski L (2003) Seasonal dynamics of macroinvertebrates associated with submersed macrophytes in a lowland river downstream of the dam reservoir. Ecohydrol Hydrobiol 3: 399–408.
11. Chick JH, McIvor CC (1997) Habitat selection by three littoral zone fishes: effects of predation pressure, plant density and macrophyte type. Ecol Freshw Fish 6: 27–35.
12. Grenouillet G, Pont D (2001) Juvenile fishes in macrophyte beds: influence of food resources, habitat structure and body size. J Fish Biol 59: 939–959.
13. Li J, Huang P, Zhang R (2010) Modeling the refuge effect of submerged macrophytes in ecological response of shallow lakes: A new model of fish functional response. Ecol Modell 221: 2076–2085.
14. Rozas LP, Odum WE (1988) Occupation of submerged aquatic vegetation by fishes: testing the roles of food and refuge. Oecologia 77: 101–106.
15. Manatunge J, Asaedaa T, Priyadarshana T (2000) The influence of structural complexity on fish–zooplankton interactions: a study using artificial submerged macrophytes. Environ Biol Fishes 58: 425–438.
16. Kleeberg A, Köchler J, Sukhodolova T, Sukhodolov A (2010) Effects of aquatic macrophytes on organic matter deposition, resuspension and phosphorus entrainment in a lowland river. Freshw Biol 55: 326–345.
17. Kohler J, Hachoł J, Hilt S (2010) Regulation of submersed macrophyte biomass in a temperate lowland river: Interactions between shading by bank vegetation, epiphyton and water turbidity. Aquat Bot 92: 129–136.
18. Dukowska M, Grzybkowska M, Kruk A, Szczerkowska-Majchrzak E (2013) Food niche partitioning between perch and ruffe: combined use of a self-organising map and the IndVal index for analysing fish diet. Ecol Modell 265: 221–229.
19. Chesson P (2000) Mechanisms of maintenance of species diversity. Annu Rev Ecol Syst 31: 343–366.
20. Schulze T, Dörner H, Baade U, Hölker F (2012) Dietary niche partitioning in a piscivorous fish guild in response to stocking of an additional competitor – The role of diet specialization. Limnologica 42: 56–64.
21. Dukowska M, Grzybkowska M, Folcholc I, Tszydel M, Szczerkowska E (2005) Predation of *Hydra* sp. on epiphytic fauna and zooplankton in a disturbed lowland river. Teka Commision of Protection and Formation of Natural Environment 2: 48–57.
22. Dukowska M, Grzybkowska M, Marszał L, Zięba G (2009) The food preferences of three-spined stickleback, *Gasterosteus aculeatus* L., downstream of a dam reservoir. Oceanol Hydrobiol Stud 38: 39–50.
23. Marszał L, Grzybkowska M, Penczak T, Galicka W (1996) Diet and feeding of dominant fish populations in the impounded Warta River, Poland. Polish Arch Hydrobiol 43: 185–202.
24. Penczak T, Głowacki Ł, Galicka W, Koszaliński H (1998) A long-term study (1985–1995) of fish populations in the impounded Warta River, Poland. Hydrobiologia 368: 157–173.
25. Penczak T, Głowacki Ł, Kruk A, Galicka W (2012) Implementation of a self-organizing map for investigation of impoundment impact on fish assemblages in a large, lowland river: Long-term study. Ecol Modell 227: 64–71.
26. EMPHP (Electronic Map of Hydrographic Partitions in Poland) (2007) Institiute of Meteorology and Water Management, Poland. Available: http://www.imgw.gov.pl. Accessed 10 January 2013.
27. Grzybkowska M, Dukowska M (2002) Communities of Chironomidae (Diptera) above and below a reservoir on a lowland river: long-term study. Annal Zool 52: 235–247.
28. Grzybkowska M, Hejduk J, Zieliński P (1990) Seasonal dynamics and production of Chironomidae in a large lowland river upstream and downstream from a new reservoir in Central Poland. Arch Hydrobiol 119: 439–455.
29. Petersen RC, Cummins KW, Ward GM (1989) Microbial and animal processing of detritus in a woodland stream. Ecol Monogr 59: 21–39.
30. Dumont HJ, Van de Velde I, Dumont S (1975) The dry weight estimate of biomass in a selection of Cladocera, Copepoda and Rotifera from the plankton, periphyton and benthos of continental waters. Oecologia 19: 75–97.
31. Opuszyński K (1983) The basics of fish biology. Warszawa: PWRiL. 590 p (in Polish).
32. Krebs CJ (1999) Ecological methodology. USA: Addison-Wesley Educational Publishers Inc. CA. 620 p.
33. Schoener TW (1970) Non-synchronous spatial overlap of lizards in patchy habitats. Ecology 51: 408–418.
34. Wallace RK (1981) An assessment of diet-overlap indexes. Trans Am Fish Soc 110: 72–76.
35. Zar JH (1984) Biostatistical analysis. Englewood Cliffs, New Jersey: Prentice-Hall Inc. 717 p.

## Supporting Information

**Table S1 Food items (% of biomass) in alimentary tracts of ruffe (Gc), perch (Pf), dace (Ll), roach (Rr) and three-spined stickleback (Ga).**
(PDF)

## Acknowledgments

We are greatly indebted to L. Kucharski for the identification of macrophytes, to W. Jurasz for the identification of cladocerans, and to our colleagues, L. Marszał, G. Zięba and J. Lik for assistance in the field and/or laboratory work, as well as to Ł. Głowacki and R. Reddy for revising the English. We also thank the Department of Hydrography and Morphology of River Beds, the Institute of Meteorology and Water Management, Poland, for granting us a licence to use the Electronic Map of Hydrographic Partitions in Poland. The study was carried out in the Field Station at Pęczniew.

## Author Contributions

Conceived and designed the experiments: MD MG. Performed the experiments: MD. Analyzed the data: MD. Wrote the paper: MD MG. Analyzed the food resources and gut contents of fish: MD MG. Prepared graphics: MD.

36. StatSoft Inc. (2011) STATISTICA (data analysis software system), version 10. Avaiable: http://www.statsoft.com. Accessed 2014 March 3.

37. Ibáñez C, Caiola N, Rovira A, Real M (2012) Monitoring the effects of floods on submerged macrophytes in a large river. Sci Total Environ 440: 132–139.

38. De Nie HW (1987) The decrease in aquatic vegetation in Europe and its consequences for fish populations. EIFAC/CECPI Occasional paper 19: 52 p.

39. Vanderstukken M, Mazzeo N, Van Colen W, Declerck SAJ, Muylaert K (2011) Biological control of phytoplankton by the subtropical submerged macrophytes *Egeria densa* and *Potamogeton illinoensis*: a mesocosm study. Freshw Biol 56: 1837–1849.

40. Gross EM (2003) Allelopathy of aquatic autotrophs. Crit Rev Plant Sci 22: 313–339.

41. Penczak T, Kruk A (2000) Threatened obligatory riverine fishes in human-modified Polish rivers. Ecol Freshw Fish 9: 109–117.

42. Raat AJP (2001) Ecological rehabilitation of the Dutch part of the River Rhine with special attention to the fish. River Res Appl 17: 131–144.

43. Kruk A (2007) Role of habitat degradation in determining fish distribution and abundance along the lowland Warta River, Poland. J Appl Ichthyol 23: 9–18.

44. Grzybkowska M (1991) Development and habitat selection of chironomid communities at long- and short-term water discharge fluctuation. Regul Riv 6: 257–264.

45. Tadajewska M (2000) Dace *Leuciscus leuciscus* (Linnaeus, 1758). In: Brylińska M, editor. Freshwater fishes of Poland.Warszawa: PWN.pp. 311–314 (in Polish).

46. Kucharczyk D, Mamcarz A, Kujawa R, Targońska K (2008) Dace *Leuciscus leuciscus* (Linnaeus, 1758) - monography Olsztyn: Mercurius78 p (in Polish).

47. Vlach P, Švátora M, Dušek J (2013) The food niche overlap of five fish species in the Úpoř brook (Central Bohemia). Knowl Manag Aquat Ecosyst. Available: http://dx.doi.org/10.1051/kmae/2013070. Accessed: 3 March 2014.

48. Clough S, Ladle M (1997) Diel migration and site fidelity in a stream-dwelling cyprinid *Leuciscus leuciscus*. J Fish Biol 50: 1117–1119.

49. Nunn AD, Harvey JP, Cowx IG (2007) The food and feeding relationships of larval and 0+ year juvenile fishes in lowland rivers and connected waterbodies. I. Ontogenetic shifts and interspecific diet similarity. J Fish Biol 70: 726–742.

50. Bańbura J (2000) Three-spined stickleback. In: Brylińska M, editor. Freshwater fishes of Poland.Warszawa: PWN.pp. 439–443 (in Polish).

51. Grzybkowska M, Temech A, Najwer I (1996) Seston, (particles>400 μm) of the Warta River downstream from the new reservoir of Jeziorsko. Acta Univ Lodz, Folia limnol 6: 47–61.

52. MacArthur RH, Pianka ER (1966) On optimal use of a patchy environment. Am Nat 100: 603–609.

53. Gunderson JL, Klepinger MR, Bonte CR, Marsden JE (1998) Overview of the international symposium on Eurasian ruffe (*Gymnocephalus cernuus*) biology, impacts and control. J Great Lakes Res 24: 165–169.

54. Tarvainen M, Ventelä A, Helminen H, Sarvala J (2005) Nutrient release and resuspension generated by ruffe (*Gymnocephalus cernuus*) and chironomids. Freshw Biol 50: 447–458.

55. Schleuter D, Eckmann R (2006) Competition between perch *Perca fluviatilis* and ruffe *Gymnocephalus cernuus*: the advantage of turning night into day. Freshw Biol 51: 287–297.

56. Schleuter D, Eckmann R (2008) Generalist versus specialist: the performances of perch and ruffe in lake of low productivity. Ecol of Freshw Fishes 17: 86–99.

57. Rösh R, Kangur A, Kangur K, Krämer A, Ráb P, et al. (1996) Ruffe (*Gymnocephalus cernuus*). Acta Zool Litu 19: 18–24.

58. Ogle DH, Selgeby JH, Newman RM, Henry MG (1995) Diet and feeding periodicity of ruffe in the St. Louis River Estuary, Lake Superior. Trans Am Fish Soc 124: 356–369.

59. Fullerton AH, Lamberti GA (2006) A comparison of habitat use and habitat-specific feeding efficiency by Eurasian ruffe (*Gymnocephalus cernuus*) and yellow perch (*Perca flavescens*). Ecol Freshw Fish 15: 1–9.

60. Barili E, Agostinho AA, Gomes LC, Latin JD (2011) The coexistence of fish species in streams: relationships between assemblage attributes and trophic and environmental variables. Environ Biol Fishes 92: 41–52.

61. Penczak T, Kruk A (2005) Patternizing of impoundment impact (1985–2002) on fish assemblages in a lowland river using the Kohonen algorithm. J Appl Ichthyol 21: 169–177.

62. Bergman E (1988) Foraging abilities and niche breadths of two percids, *Perca fluviatilis* and *Gymnocephalus cernua*, under different environmental conditions. J Animal Ecol 57: 443–453.

63. Kornijów R (1997) The impact of predation by perch on the size-structure of Chironomus larvae - the role of vertical distribution of the prey in the bottom sediments, and habitat complexity. Hydrobiologia 342/343: 207–213.

64. Dieterich A, Baumgärtner D, Eckmann R (2004) Competition for food between Eurasian perch *Perca fluviatilis* and ruffe *Gymnocephalus cernuus* over different substrate types. Ecol Freshw Fish 13: 236–244.

65. Rezsu E, Specziár A (2006) Ontogenetic diet profiles and size-dependent diet partitioning of ruffe *Gymnocephalus cernuus*, perch *Perca fluviatilis* and pumpkinseed *Lepomis gibbosus* in Lake Balaton. Ecol Freshw Fish 15: 339–349.

66. Borcherding J, Magnhagen C (2008) Food abundance affects both morphology and behaviour of juvenile perch. Ecol Freshw Fish 17: 207–218.

67. Nurminen L, Pekcan-Hekim Z, Horppila J (2010) Feeding efficiency of planktivorous perch *Perca fluviatilis* and roach *Rutilus rutilus* in varying turbidity: an individual-based approach. J Fish Biol 76: 1848–1855.

68. Kennedy M, Gray RD (1993) Can ecological theory predict the distribution of foraging animals? A critical analysis of experiments on the ideal free distribution. Oikos 68: 158–166.

69. Danchin E, Giraldeau LA, Cézilly F (2008) Behavioural ecology. Oxford: Oxford University Press. 912 p.

70. Persson L (1987) Effects of habitat and season on competitive interactions between roach (*Rutilus rutilus*) and perch (*Perca fluviatilis*). Oecologia 73: 170–177.

71. Kornijów R, Vakkilainen K, Horppila J, Luokkanen E, Kairesalo T (2005) Impacts of a submerged plant (*Elodea canadensis*) on interactions between roach (*Rutilus rutilus*) and its invertebrate prey communities in a lake littoral zone. Freshw Biol 50: 262–273.

72. Grzybkowska M (1988) Food of roach in the Widawka River. Acta Univ Lodz, Folia limnol 3: 85–100 (in Polish with English summary).

73. Horppila J, Nurminen L (2009) Food niche segregation between two herbivorous cyprinid species in a turbid lake. J Fish Biol 75: 1230–1243.

74. Grzybkowska M, Zalewski M (1983) The food of roach, bream and white bream in the Sulejów Dam Reservoir. In:Proceedings of the Conference of the Polish Hydrobiological Society. Lublin, pp 71–72 (in Polish).

75. Makino W, Kato H, Takamura N, Mizutani H, Katano N, et al. (2001) Did chironomid emergence release *Daphnia* from fish predation and lead to *Daphnia*-driven clear-water phase in Lake Towada, Japan? Hydrobiologia 442: 309–317.

76. Kruk A, Penczak T (2003) Impoundment impact on populations of facultative riverine fish. Ann Limnol 39: 197–210.

77. Pianka ER (1974) Niche overlap and diffuse competition. Proc Nat Acad Sci 71: 2141–2145.

# 17

# Dispersal Ability Determines the Role of Environmental, Spatial and Temporal Drivers of Metacommunity Structure

**André A. Padial**[1,2]\*, **Fernanda Ceschin**[2], **Steven A. J. Declerck**[3], **Luc De Meester**[4], **Cláudia C. Bonecker**[5], **Fabio A. Lansac-Tôha**[5], **Liliana Rodrigues**[5], **Luzia C. Rodrigues**[5], **Sueli Train**[5], **Luiz F. M. Velho**[5], **Luis M. Bini**[6]

1 Departamento de Botânica, Universidade Federal do Paraná, Curitiba, Paraná, Brazil, 2 Programa de Pós-graduação em Ecologia e Conservação, Universidade Federal do Paraná, Curitiba, Brazil, 3 Department of Aquatic Ecology, Netherlands Institute of Ecology (NIOO-KNAW), Wageningen, The Netherlands, 4 KU Leuven, University of Leuven, Laboratory of Aquatic Ecology, Evolution and Conservation, Leuven, Belgium, 5 Núcleo de Pesquisa em Limnologia, Ictiologia e Aqüicultura (Nupelia), Universidade Estadual de Maringá, Maringá, Brazil, 6 Departamento de Ecologia, Universidade Federal de Goiás, Goiânia, Brazil

## Abstract

Recently, community ecologists are focusing on the relative importance of local environmental factors and proxies to dispersal limitation to explain spatial variation in community structure. Albeit less explored, temporal processes may also be important in explaining species composition variation in metacommunities occupying dynamic systems. We aimed to evaluate the relative role of environmental, spatial and temporal variables on the metacommunity structure of different organism groups in the Upper Paraná River floodplain (Brazil). We used data on macrophytes, fish, benthic macroinvertebrates, zooplankton, periphyton, and phytoplankton collected in up to 36 habitats during a total of eight sampling campaigns over two years. According to variation partitioning results, the importance of predictors varied among biological groups. Spatial predictors were particularly important for organisms with comparatively lower dispersal ability, such as aquatic macrophytes and fish. On the other hand, environmental predictors were particularly important for organisms with high dispersal ability, such as microalgae, indicating the importance of species sorting processes in shaping the community structure of these organisms. The importance of watercourse distances increased when spatial variables were the main predictors of metacommunity structure. The contribution of temporal predictors was low. Our results emphasize the strength of a trait-based analysis and of better defining spatial variables. More importantly, they supported the view that "all-or-nothing" interpretations on the mechanisms structuring metacommunities are rather the exception than the rule.

**Editor:** Frédéric Guichard, McGill University, Canada

**Funding:** CNPq (http://www.cnpq.br/) and CAPES (http://www.capes.gov.br/) provided scholarships. KU Leuven (http://www.kuleuven.be/) provided research fund grant PF/2010/07. The funders had no role in study design, data collection and analysis, decision to publish, or preparation of the manuscript.

**Competing Interests:** The authors have declared that no competing interests exist.

\* Email: aapadial@gmail.com

## Introduction

The identification of the mechanisms driving variation in and among local communities is central to community ecology. The role of environmental and spatial processes operating in multiple scales to shape local community composition is explicit in the metacommunity framework [1–5]. If community composition is mainly predicted by environmental variables, then niche-related mechanisms are considered the primary drivers of metacommunities and species are sorted across habitats [6,7]. An alternative view has emphasized that the structure of local communities differ from each other mainly due to stochastic processes, including dispersal limitation and ecological drift [8]. In an effort to reveal the main mechanisms driving spatial variation in local communities, several studies have investigated the relative importance of environmental gradients and spatial processes in shaping meta-

community structure ([9,10] and references therein). Not uncommonly, studies indicate that both niche and spatial processes may account for variation in community structure [11].

One may expect that the relative importance of deterministic (e.g., species sorting) and stochastic processes (e.g., dispersal) will be dependent on the dispersal ability of the biological groups under study. Recently, studies have compared organism groups with different dispersal abilities in the same set of habitats to test the hypotheses that: i) niche related processes are important in structuring local communities for organisms with high dispersal ability, and ii) spatial structure is a better predictor of local community composition for biological groups with low dispersal ability [9,10,12]. Organisms with high dispersal ability may be less affected by spatial structure simply because they reach suitable patches more often than those with low dispersal ability [13]. In this case, species are sorted according to their environmental requirements. In freshwater ecosystems, dispersal ability is generally inversely related to body size [9,12,14,15]. Therefore, large-bodied organisms, such as fish and aquatic macrophytes, may have comparatively lower dispersal ability than small organisms, such as plankton and benthic invertebrates. This difference allows one to predict an increased role of environmental predictors in the structure of local communities from large to small organisms [9].

The structure of local communities also varies through time [16]. For instance, a recent study has found that temporal environmental variation is an important mechanism explaining zooplankton beta-diversity [17]. However, studies simultaneously testing the relative role of environmental, spatial and temporal processes on metacommunities are uncommon [18]. In this context, it is important to emphasize that in some ecosystems (e.g., floodplains), the magnitude of temporal variation in community structure may be as high or higher than the magnitude of spatial variation [19]. The test of this conjecture is increasingly relevant given the long list of environments changes caused by human activities [20].

For this study, we analyzed a dataset on different biological groups in the Upper Paraná River floodplain, Brazil. We tested the hypothesis that the relative role of environmental conditions in structuring local communities is high for communities composed by small organisms (with high dispersal ability). Conversely, the relative role of spatial variables in predicting community structure would increase for communities composed by large-bodied organisms. We also assessed the role of temporal dynamics on community structure in this highly dynamic system.

## Methods

### Study area

The Upper Paraná River and its floodplain (Figure 1) represent the last unregulated stretch of the Paraná River in the Brazilian territory. It is an important area for several migratory fish species and still has high species diversity [21]. Sampling sites in the Upper Paraná River floodplain were located along an environmental gradient of limnological, hydrological, and biological variables [22] within a protected area called "APA das Ilhas e Várzeas do Rio Paraná". All samplings were authorized by the Brazilian agency for environmental protection (Instituto Brasileiro do Meio Ambiente – IBAMA, https://www.ibama.gov.br). The hydrological regime is characterized by a dry season (June–September) and a wet season (October–February). However, due to hydrological control by recently built hydropower reservoirs, the frequency, amplitude, and duration of the floods have substantially changed [21].

### Sampling

We collected data on six biological communities: aquatic macrophytes, fish, benthic macroinvertebrates, zooplankton, periphyton, and phytoplankton. Sampling was carried out during February, May (wet season), August and November (dry season) in 2000 and 2001. Depending on sampling month, we sampled up to 36 sites spread throughout the Upper Paraná River floodplain (Figure 1). These sites included floodplain lakes permanently connected to rivers, floodplain lakes connected to rivers only during floods, and river channels.

We recorded the presence and absence of all aquatic macrophytes in the field from a boat, with the help of a grapnel to search for submerged vegetation. We determined fish abundance (individuals × 24 hours/1000 m$^2$ gillnet) by standardized fishing using gill nets with different mesh sizes. We used a Petersen's grab to collect benthic macroinvertebrates. The total number of individuals of each taxon per sample was used as the abundance data. We collected zooplankton samples by pumping 600 L of water through a 68 μm mesh net and, after laboratory procedures, data were expressed as individuals/m$^3$. We sampled periphyton from petioles of *Eichhornia azurea* Kunth in the mature stage, as this macrophyte was common in most of the environments in the Upper Paraná River floodplain. Abundance was expressed in individuals/cm$^2$. We used Van Dorn bottle to sample phytoplankton, and species densities were expressed as individual units (cells, coenobia, colonies, or filaments) per milliliter. Individuals of all biological groups were identified to the lowest taxonomic level possible. With the exception of benthic macroinvertebrates, identification reached species or genus level [22]. For benthic macroinvertebrates, some groups were identified as family, order or even class.

Although it is difficult to accurately classify organisms in terms of dispersal ability, we assumed, based on the body size and dispersal strategies [9], that microalgae (phytoplankton and periphyton) have the highest dispersal ability at the scale of the floodplain, whereas fish and macrophytes were expected to have the lowest. Zooplankton and macroinvertebrates were expected to have intermediate dispersal ability. Given the intricate spatial configuration of floodplain systems, with several dendritic watercourses temporarily or permanently connecting lakes, channels and main rivers, those groups dispersing directly through water may exhibit the lowest dispersal ability. Microorganisms may disperse via several vectors (e.g., air, watercourse, and attached to animals and plants).

We obtained the following environmental variables for each sampling site: depth (m), water temperature (°C), dissolved oxygen (mg/L), water transparency (m), pH, conductivity (μS/cm), total alkalinity (mEq/L), turbidity (NTU), total nitrogen concentration (μg/L), total phosphorus concentration (μg/L), chlorophyll-*a* (μg/L), total suspended matter (mg/L), and dissolved organic matter (mg/L). All environmental variables, except for pH, were log (*x*) transformed prior to the analyses described below. Details on sampling and laboratory procedures used to obtain the biological and environmental data can be found elsewhere [22].

### Data analysis

We used partial redundancy analysis (pRDA) to estimate the relative role of environmental, spatial and temporal predictors on the structure of the aquatic communities. As response matrices, we used abundance data for all biological groups (except for macrophytes because only presence/absence data are available for this group) and the total variance in the community data matrix was divided into unique and shared components of a set of environmental, spatial and temporal predictors [18]. As data are lacking for some groups in some sites and periods, we did not carry out an analysis with 36 sites and eight sampling periods for all biological groups (see Appendix S1).

The environmental matrix was composed of the limnological variables described above. We checked for collinearity among variables and removed variables that were strongly correlated with another variable before pRDA. Chlorophyll-*a* was not used as a predictor of periphyton and phytoplankton.

We used different strategies to generate spatial variables. Firstly, we calculated matrices of Euclidean ("overland") and watercourses distances between sites (**D** and **W** respectively). Four possible scenarios of spatial relationships between the sites are possible considering the unidirectional flow of the main rivers (i.e., Paraná, Baia and Ivinheima) and bidirectional flows of lateral channels (see Figure 1). For instance, sampling sites located in the Baia River

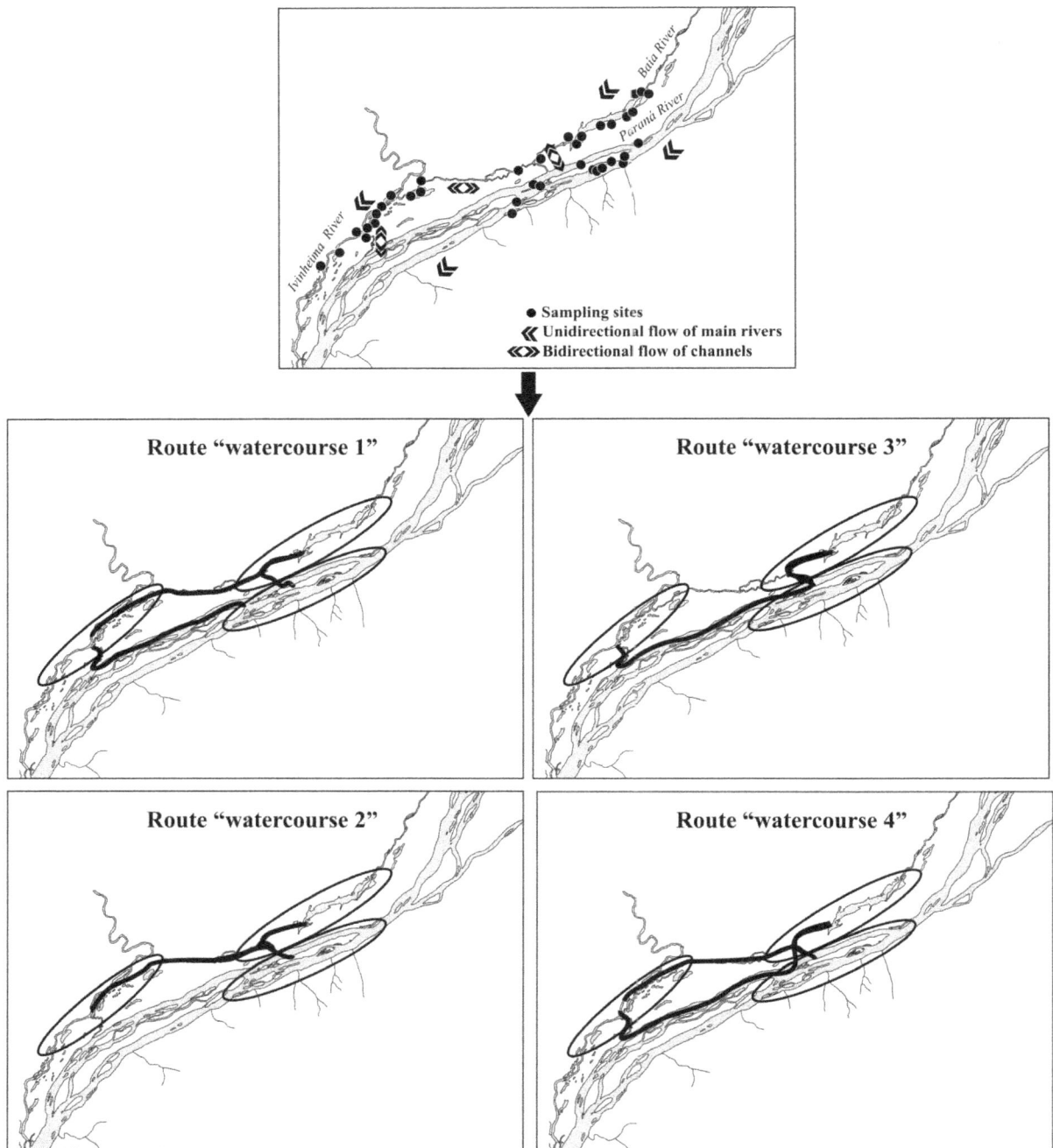

**Figure 1. Possible routes of dispersion among sampling sites in the Upper Paraná River floodplain.** These hypothetical routes were charted based on the unidirectional flow of main rivers and bidirectional flow of lateral channels.

subsystem can be connected to sampling sites from the Ivinheima River subsystem by a lateral channel (Figure 1, left schemes) or by the Paraná River main channel (Figure 1, right schemes). Also, Ivinheima River sampling sites and Paraná River sampling sites can be connected through a lateral channel downstream (Figure 1, upper schemes) or upstream (Figure 1, lower schemes). We generated four matrices **W** to represent the possible organism dispersion routes. Spatial variables based on the five distance matrices described above (one **D** and four **W**) were created using Moran's Eigenvector Maps [23]. Therefore, these spatial variables

(i.e., eigenvectors extracted from the distance matrices) are different representations of how sampling sites are spatially related [24,25]. We have also generated spatial predictors using asymmetric eigenvector maps (AEM) considering the directional flow of main rivers [26]. We selected only the eigenvectors with positive Moran's $I$ autocorrelation coefficients, assuming that these eigenvectors are proxies for dispersal processes or unmeasured environmental variables that are spatially structured.

The temporal matrix was composed by dummy variables differentiating sampling periods. For instance, a temporal matrix that had eight sampling periods was generated with seven vectors: each having "1" for a certain sampling period and "0" for the others. Thus, the temporal matrix was composed by dummy variables indicating that a group of sites was sampled at the same time. By using three explanatory matrices (environmental, spatial and temporal), eight variance components (or fractions of variation in canonical analysis; (see [27]) are generated in variation partitioning (see [18]): (1) Pure environmental, $E$: the fraction of variation in community structure explained by environmental variables that are neither spatially nor temporally structured; (2) Pure spatial, $S$: Spatial patterns in the species data that are independent of any temporal or environmental predictors included in the analysis; (3) Pure temporal, $T$: Temporal patterns in the species data that are independent of any spatial or environmental predictors included in the analysis; (4) $SE$: the variation in biological data explained by spatially structured environmental variables; (5) $TE$: the variation in biological data explained by temporally structured environmental variables; (6) $ST$: represents the explained variation that is co-structured in time and space, for instance in the case of temporally structured habitat connectivity; (7) $STE$: spatially and temporally structured environmental variation. This component, if important, indicates that the explanation of one predictor is correlated with the two others; (8) $U$: the unexplained variation in the community data - the fraction that cannot be explained by spatial, temporal or environmental predictors. These components were calculated using adjusted fractions, which take sample size and number of variables into account [27]. The significances of the fractions $E$, $S$ and $T$ were tested using 999 random permutations.

Before the analysis described above, presence-absence and abundance data were Hellinger transformed [27,28]. Results were similar after excluding rare species (those occurring in only one sampling site). Therefore, analyses were done with the total dataset. We used the R language and environment for statistical computing [29] with 'vegan' [30] and 'spdep' [31] packages for analyses.

Following previous studies [4,9,32], we assumed that fish and aquatic macrophytes are, comparatively, poor dispersers and that macroinvertebrates, zooplankton, phytoplankton and periphyton generally have higher dispersal abilities. In addition, the fish dataset was divided into a table of sedentary fish and a table of migratory fish [33]. To create a coarse quantitative measure of dispersal ability, these biological groups were ranked in the following order: phytoplankton (1), periphyton (2), zooplankton (3), macroinvertebrates (4), migratory fish (5), sedentary fish (6) and aquatic macrophytes (7). We used Spearman rank correlation to test the relationship between this crude ranking of dispersal ability and the relative importance of environmental and spatial variables in predicting community structure (difference between components $E$ and $S$). We recognize that the dispersal classification listed above is qualitative and not only reflects a general expected trend. It is, for instance, impossible to reliably assert that phytoplankton have higher dispersal than periphyton given both biological groups are comprised mainly by microalgae. Also, there is no doubt variation within groups. Yet, it remains that it is very likely that of all groups microalgae have the highest dispersal ability because of their abundances (sources of migrants) and small body size (see [9]). To take into account uncertainties in the way our measure of dispersal ability was created and thus increase robustness of our results, we repeated the Spearman rank test after considering different rank schemes (see Appendix S1).

## Results

Explanatory matrices explained up to 36.4% of the variation in biological datasets (Figure 2). Temporal variables (component $T$) significantly explained part of the variation in the structure of all groups except periphyton. The highest adjusted coefficient of determination associated with this fraction was obtained for migratory fish (6.7%). Spatial variables (component $S$) explained a significant proportion of the total variation in the structure of several groups. The lowest (and non-significant) component $S$ was obtained for planktonic communities, independently of the type of distance used to create spatial variables. On the other hand, the $S$ component was particularly high for aquatic macrophytes and sedentary fish when watercourse distances or AEM were used to generate spatial predictors. Environmental variables (component $E$) significantly accounted for part of the variation in the community structure of all groups. The highest shared fraction of variation was the spatially structured temporal variation (component $ST$), recorded for periphyton (Figure 2).

We found a negative correlation between the difference $E$–$S$ and dispersal ability (Figure 3 and Appendix S1), indicating that variation in community structure of groups with high dispersal ability (e.g., phytoplankton) were better predicted by environmental variables. Conversely, spatial variables were the main predictors of variation in groups with lower dispersal ability (e.g., aquatic macrophytes). The relative roles of environmental and spatial variables in structuring periphyton, zooplankton and macroinvertebrate communities were intermediate compared to phytoplankton and macrophytes (Figure 3). These patterns were nearly independent of the type of distance matrix, i.e. the hypothesized dispersal routes used to generate the spatial variables, and of the dispersal ability ranks (e.g. whether phytoplankton or periphyton are considered to be the group with the highest dispersal ability; see abscissa of Figure 3 and Appendix S1).

## Discussion

Our results suggest that there exists an overall association in the study area, the Upper Paraná River floodplain, between the dispersal ability of different organism groups and the relative roles that species sorting and neutral spatial dynamics play in structuring their metacommunities. Our results are in line with the existing studies on this theme [4,9,10,32,34], which also found that the variation in community structure of organism groups with high dispersal ability is mainly accounted for by environmental variables, while spatial predictors are more important in groups with low dispersal ability [9,10,32,35]. We assumed that macrophytes and fish, being the larger bodied-organisms in our study, have lower dispersal ability than the other taxa. Although fish actively search for habitats, dispersal in floodplains is not always evident given the complexity of channels and floodplain lakes and the fact that habitats can be temporarily isolated [19,21]. We indeed observed that spatial variables were important in explaining variation in community structure for fish and macrophytes, and that the spatial scenario that takes watercourse connectivity into account had the highest explanatory power [25]. Environmental variables were especially important in structuring the local communities of periphyton and phytoplankton, small organisms with typically large population sizes and high dispersal abilities [14,36]. In line with previous studies [4,9], environmental drivers had a stronger contribution to explaining community structure in phytoplankton than in zooplankton. Although our results are in line with previous studies, to the best of our knowledge, only De Bie et al. [9] so far were able to formally test the relationship

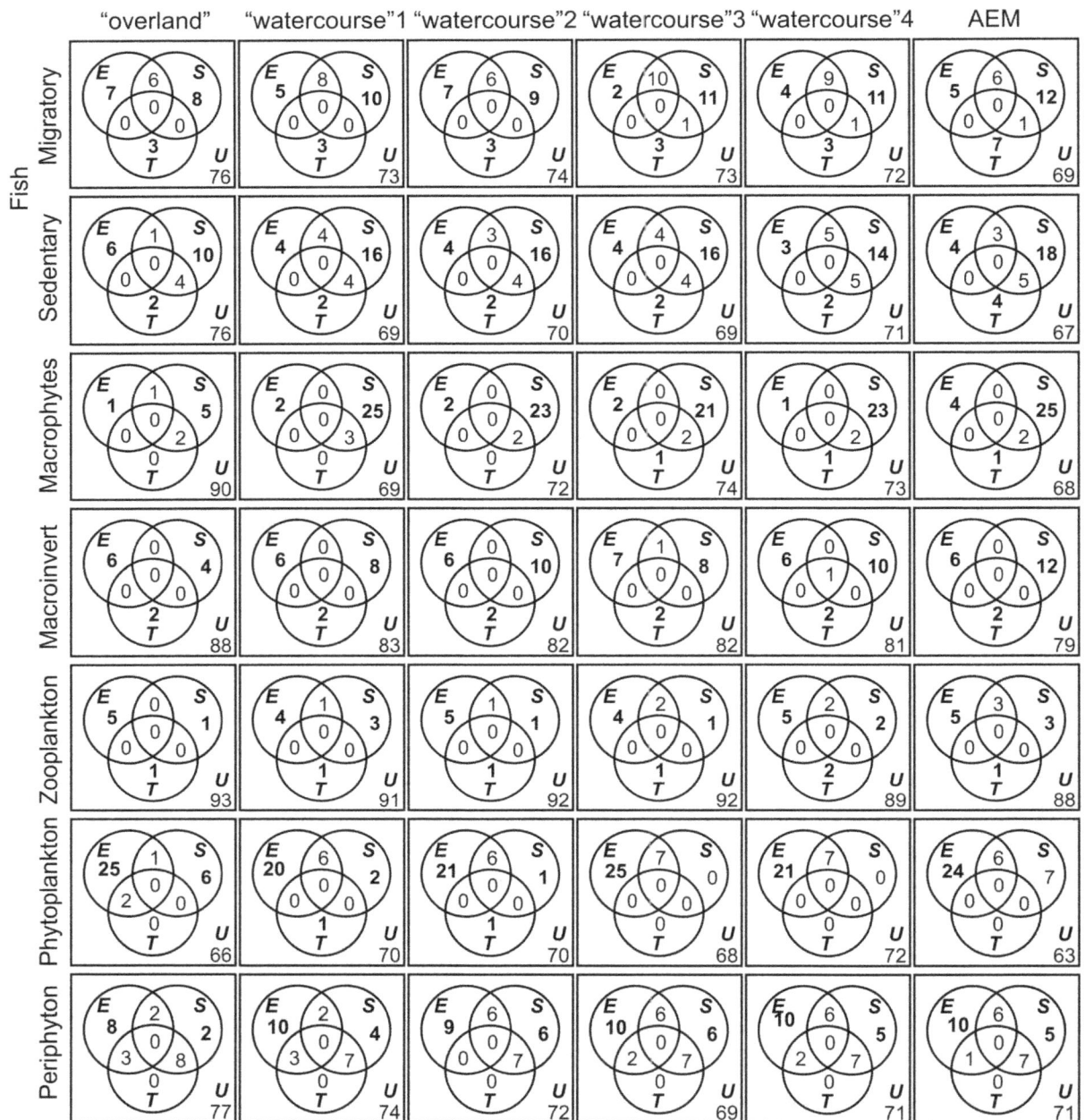

**Figure 2. Results from partial redundancy analysis.** Shown are the relative contributions (% of explanation) of environmental (*E*), spatial (*S*), and temporal (*T*) variables, as well as the shared components explaining variation in abundance of aquatic metacommunities (except for aquatic macrophytes, in which only presence/absence is available), using overland and four watercourse distances to generate spatial predictors. *U* = unexplained component. Zeros indicate values lower than 0.5%. The significance of the pure components (*E*, *S* and *T*) was tested using random permutations; bold numbers indicate significant values. Macroinvert = Benthic Macroinvertebrates.

between the relative roles of processes driving metacommunity structure and dispersal ability inferred by body size. Other studies with similar goals (e.g., [32]) were not able to formally test this relationship due to the lower number of organism groups that differ in dispersal abilities that could be compared. To test for generality of this pattern, we encourage further studies to test this relationship using data on multiple biological groups that differ in dispersal abilities surveyed in the same set of sampling sites.

The role of dispersal processes cannot be inferred from the mere observation of a significant spatial component *S*. Variance

component *S* may, for instance, also reflect the importance of unmeasured, spatially structured environmental variables [24,37,38]. Rather, it is a combination of results - significant component *S* combined with a negative relationship between the difference *E*–*S* and dispersal ability in a cross-group analysis - that constitutes strong evidence for the role of dispersal limitation. Although it is difficult to accurately rank the different study taxa with respect to their dispersal ability, we here work with a crude, robust ranking, with protists having the highest dispersal ability [39], and fish and macrophytes that generally need direct water

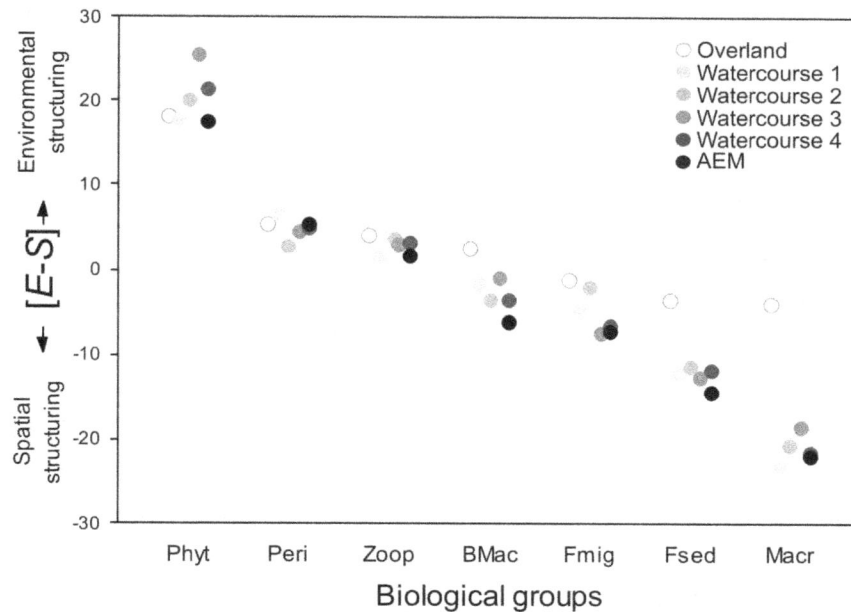

**Figure 3. Difference in the contribution of environmental (*E*) and spatial variables (*S*) for the different biological groups, ranked according to the presumed dispersal ability.** Different distance matrices were used to generate spatial predictors (one overland and four watercourse distances). *E* is the fraction of variation in community structure explained by environmental variables that are neither spatially nor temporally structured; *S* is the spatial patterns in the biological data that are independent of any temporal or environmental predictors. Phyt = phytoplankton; Peri = periphyton; Zoop = zooplankton; BMac = benthic macroinvertebrates; Fmig = migratory fish; Fsed = sedentary fish; Macr = macrophytes.

connections having the lowest dispersal ability. As in earlier studies, zooplankton are expected to have lower dispersal ability than microalgae, but higher than fish [4,9]. We recognize that macroinvertebrates could be split into flying dispersers in adult stage and those that have sedentary behavior (see [10,34]). However, taxonomical resolution of our data did not allow us to accurately split benthic macroinvertebrate community into weak and strong dispersers. Even considering these caveats, the main pattern of an increase in the importance of spatial variables (relative to the importance of environmental factors) with decreasing dispersal ability proved to be robust to the choice of ranks used to create our measure of dispersal ability (see Appendix S1). Also, the comparison between sedentary and migratory fishes highlights the importance of a trait-based metacommunity analysis (see [9]) and provides further support for the interpretation of the variance component *S* as reflecting dispersal limitation.

Our results do not support that communities would be structured by only one of the four different metacommunity paradigms proposed by Leibold et al. [1]. Instead, our results suggest that both niche-driven and spatial processes contributed to a varying degree to the structure of the local communities and that this variation is structured by dispersal ability (see also [11]). Thus, although our results (i.e., high frequency of a significant component *E*) reflect generally strong species sorting [15], they also indicate that "all-or-nothing" interpretations on the mechanisms structuring metacommunities are rather the exception than the rule. The view of high variation in the importance of mechanisms underlying metacommunity structure has been supported by experimental [40], temporal [41] and spatial studies [9,42], and our study adds to the evidence that this variation is in structuring mechanisms can be related to traits of the organisms [9]. In short, based on previous studies and on our results, we are

of the opinion that, most likely, there is no "silver-bullet" explanation for metacommunity patterns.

Our results also highlight the relevance of better defining the spatial variables used in variation partitioning analysis [4,25,26,35,43]. For instance, when watercourse distances or AEM were used instead of overland distances, we recorded a substantial increase in the magnitude of component *S* for sedentary fish (from 9.9% to an average of 16.1%) and macrophytes (from 5.4% to an average of 23.4%; see Figure 2). This reflects that direct hydrological connections are important for dispersal of many organisms that live in floodplain river systems, particularly fish and macrophytes. Surprisingly, AEM, accounting for directional flow of the main rivers [26], did not generate better spatial predictors than symmetric eigenvector maps (see Figures 2 and 3). A likely explanation is that some habitats are connected by channels that exhibit bidirectional flow (see Figure 1). In short, our results reinforce that, at least in floodplain systems and river networks, ecologists interested in quantifying the relative role of spatial and environmental variables on community structure should go beyond the evaluation of simplistic spatial proxies [26,35,44,45].

Floodplains are fundamentally seasonal systems, mainly due to the pervasive effects of the floods [46]. Indeed, studies carried out worldwide [47,48] and in the Upper Paraná River itself [49,50], indicate that floods account for important ecological patterns (e.g., species distribution) and processes (e.g., primary productivity and decomposition rates) in these systems. As a result, one could envisage that the role of spatial variables in explaining variation in community structure would be dependent on temporal predictors. It can, for instance, be expected that during high water periods, when there is a high level of connectivity within the floodplain, the role of spatial variables should be decreased. In this context, the lack of a strong temporal signal was a surprising result. Our data suggest that temporal dynamics do not massively change metacommunity structure, even in a floodplain setting, where the flood pulse is known to be a major structuring factor ([19,46] and references therein). Yet, we need to be cautious in generalizing our result, as the flood pulse in the study system is reduced because of the construction of dams [21]. Also, our study involved only two years of sampling, and thus may underestimate long-term effects. Irrespective of these caveats, we found that temporal predictors (component *T*) were generally low, and highest (up to 6.7%) for those groups were expected to show much temporal variability, such as migratory fish.

We tried to circumvent at least some of the caveats that have commonly been discussed in the metacommunity literature, such as, for instance, the limitations of snapshot sampling [4]. Moreover, the combination of a trait-based analysis (here through comparing patterns across different organism groups along a body size gradient and by comparing sedentary and migratory fish) and the inclusion of better spatial variables, allowed us to be more confident in the interpretation of the spatial variation component as reflecting the role of dispersal ability [9,24,37]. Yet, we cannot discard the possibility that relevant environmental predictors were missing in our dataset [38]. We did not, for instance, include sediment characteristics, which might be important for macroinvertebrates. In addition, long-term ecological studies are necessary to properly evaluate the explanatory power of temporal processes on shaping metacommunities.

In conclusion, by analyzing data on different biological groups, we supported the hypothesis that the relative role of environmental and spatial processes on structuring local communities depends on

the dispersal ability of these organisms [4,9,10]. We also demonstrated that spatial variables generated using watercourse distances and AEM (see also [4,25,26]) resulted in better estimates of the spatial drivers than geographical distances particularly when spatial structuring is the main mechanism in metacommunities. This is in line with a growing number of studies showing that one need to model connectedness of river networks. Finally, our study adds to the evidence that trait-based analyses [9] provide a deeper understanding of processes underlying metacommunity structure.

## Supporting Information

**Appendix S1.**
(DOCX)

## Acknowledgments

This research was supported by PELD-CNPq. A. A. Padial and F. Ceschin received CNPq and CAPES scholarships. We also acknowledge CNPq for research grants. LDM acknowledges KU Leuven research fund grant PF/2010/07. Finally, we acknowledge AA Agostinho, A. Takeda and SM Thomaz for providing data on fish, macroinvertebrates and aquatic plants.

## Author Contributions

Conceived and designed the experiments: AAP SAJD LDM LMB. Performed the experiments: CCB FALT LR LCR ST LFMV. Analyzed the data: AAP FC. Contributed reagents/materials/analysis tools: CCB FALT LR LCR ST LFMV. Contributed to the writing of the manuscript: AAP FC SAJD LDM LMB.

## References

1. Leibold MA, Holyoak M, Mouquet N, Amarasekare P, Chase JM, et al. (2004) The metacommunity concept: a framework for multi-scale community ecology. Ecol Lett 7: 601–613.
2. Chase JM (2005) Towards a really unified theory for metacommunities. Funct Ecol 19: 182–186.
3. Cottenie K (2005) Integrating environmental and spatial processes in ecological community dynamics. Ecol Lett 8: 1175–1182.
4. Beisner BE, Peres-Neto PR, Lindström ES, Barnett A, Longhi ML (2006) The role of environmental and spatial processes in structuring lake communities from bacteria to fish. Ecology 87: 2985–2991.
5. Stendera S, Adrian R, Bonada N, Cañedo-Argüelles M, Hugueny B, et al. (2012) Drivers and stressors of freshwater biodiversity patterns across different ecosystems and scales: a review. Hydrobiologia 696: 1–28.
6. Heino J, Mykrä H (2008) Control of stream insect assemblages: roles of spatial configuration and local environmental factors. Ecol Entomol 33: 614–622.
7. Vanormelingen P, Cottenie K, Michels E, Muylaert K, Vyverman W, et al. (2008) The relative importance of dispersal and local processes in structuring phytoplankton communities in a set of highly interconnected ponds. Freshw Biol 53: 2170–2183.
8. Hubbell SP (2001) A Unified Neutral Theory of Biodiversity and Biogeography. Princeton University Press, New Jersey, 375 p.
9. De Bie T, De Meester L, Brendonck L, Martens K, Goddeeris B, et al. (2012) Body size and dispersal mode as key traits determining metacommunity structure of aquatic organisms. Ecol Lett 15: 740–747.
10. Heino J (2013) Does dispersal ability affect the relative importance of environmental control and spatial structuring of littoral macroinvertebrate communities? Oecologia 171: 971–980.
11. Thompson R, Townsend C (2006) A truce with neutral theory: local deterministic factors, species traits and dispersal limitation together determine patterns of diversity in stream invertebrates. J Anim Ecol 75: 476–484.
12. Farjalla VF, Srivastava DS, Marino NAC, Azevedo FD, Dib V, et al. (2012) Ecological determinism increases with organism size. Ecology 93: 1752–1759.
13. Martiny JBH, Bohannan BJM, Brown JH, Colwell RK, Fuhrman JA, et al. (2006) Microbial biogeography: putting microorganisms on the map. Nature 4: 102–112.
14. Shurin JB, Cottenie K, Hillebrand H (2009) Spatial autocorrelation and dispersal limitation in freshwater organisms. Oecologia 159: 151–159.
15. Van der Gucht K, Cottenie K, Muylaert K, Vloemans N, Cousin S, et al. (2007) The power of species sorting: Local factors drive bacterial community composition over a wide range of spatial scales. Proc Natl Acad Sci USA 104: 20404–20409.
16. Bengtsson J, Baillie SR, Lawton J (1997) Community variability increases with time. Oikos 78: 249–256.
17. Bellier E, Grotan V, Engen S, Schartau AK, Herfindal I, et al. (2013) Distance decay of similarity, effects of environmental noise and ecological heterogeneity among species in the spatio-temporal dynamics of a dispersal-limited community. Ecography 36: 001–011.
18. Anderson MJ, Gribble NA (1998) Partitioning the variation among spatial, temporal and environmental components in a multivariate data set. Aust J Ecol 23: 158–167.
19. Thomaz SM, Bini LM, Bozelli RL (2007) Floods increase similarity among aquatic habitats in river-floodplain systems. Hydrobiologia 579: 1–13.
20. Legendre P, Cáceres M, Borcard D (2010) Community surveys through space and time: testing the space–time interaction in the absence of replication. Ecology 91: 262–272.
21. Agostinho AA, Gomes LC, Veríssimo S, Okada EK (2004) Flood regime, dam regulation and fish in the Upper Parana River: effects on assemblage attributes, reproduction and recruitment. Rev Fish Biol Fisher 14: 11–19.
22. Padial AA, Siqueira T, Heino J, Vieira LCG, Bonecker CC, et al. (2012) Relationships between multiple biological groups and classification schemes in a Neotropical floodplain. Ecol Indic 13: 55–65.
23. Dray S, Legendre P, Peres-Neto PR (2006) Spatial modeling: a comprehensive framework for principal coordinate analysis of neighbor matrices (PCNM). Ecol Model 196: 483–493.
24. Peres-Neto PR, Legendre P (2010) Estimating and controlling for spatial structure in the study of ecological communities. Global Ecol Biogeogr 19: 174–184.
25. Landeiro VL, Magnusson WE, Melo AS, Espírito-Santo HMV, Bini LM (2011) Spatial eigenfunction analyses in stream networks: do watercourse and overland distances produce different results? Freshw Biol 56: 1184–1192.
26. Blanchet FG, Legendre P, Borcard D (2008) Modelling directional spatial processes in ecological data. Ecol Model 215: 325–336.
27. Peres-Neto PR, Legendre P, Dray S, Borcard D (2006) Variation partitioning of species data matrices: estimation and comparison of fractions. Ecology 87: 2614–2625.
28. Legendre P, Gallagher E (2001) Ecologically meaningful transformations for ordination of species data. Oecologia 129: 271–280.
29. R Core Team (2013) R: A language and environment for statistical computing.R Foundation for Statistical Computing, Vienna, Austria. URL http://www.R-project.org/.
30. Oksanen J, Blanchet FG, Kindt R, Legendre P, Minchin PR, et al. (2013). vegan: Community Ecology Package. R package version 2.0–8. http://CRAN.R-project.org/package=vegan.

31. Bivand R (2013) spdep: Spatial dependence: weighting schemes, statistics and models. R package version 0.5–65. http://CRAN.R-project.org/package=spdep.

32. Hájek M, Rolecek J, Cottenie K, Kintrová K, Horsák M, et al. (2011) Environmental and spatial controls of biotic assemblages in a discrete semi-terrestrial habitat: comparison of organisms with different dispersal abilities sampled in the same plots. J Biogeogr 38: 1683–1693.

33. Graça WJ, Pavanelli CS (2007) Peixes da planície de inundação do Alto rio Paraná e áreas adjacentes, Eduem, Maringá, 241 p.

34. Grönroos M, Heino J, Siqueira T, Landeiro VL, Kotanen J, et al. (2013) Metacommunity structuring in stream networks: roles of dispersal mode, distance type, and regional environmental context. Ecol Evol 3: 4473–4487.

35. Liu J, Soininen J, Han BP, Declerck SAJ (2013) Effects of connectivity, dispersal directionality and functional traits on the metacommunity structure of river benthic diatoms. J Biogeogr 40: 2238–2248.

36. Finlay BJ (2002) Global dispersal of free-living microbial eukaryote species. Science 296: 1061–1063.

37. Diniz-Filho JAF, Siqueira T, Padial AA, Rangel TF, Landeiro VL, et al. (2012) Spatial autocorrelation analysis allows disentangling the balance between neutral and niche processes in metacommunities. Oikos 121: 201–210.

38. Chang LW, Zeleny D, Li CF, Chiu ST, Hsieh CF (2013) Better environmental data may reverse conclusions about niche- and dispersal-based processes in community assembly. Ecology 94: 2145–2151.

39. Wetzel CE, Bicudo DC, Ector L, Lobo EA, Soininen J, et al. (2012) Distance Decay of Similarity in Neotropical Diatom Communities. PLoS One 7: 1–8.

40. Chase JM (2010) Stochastic Community Assembly Causes Higher Biodiversity in More Productive Environments. Science 328: 1388–1391.

41. Langenheder S, Berga M, Östman Ö, Székely AS (2012) Temporal variation of β-diversity and assembly mechanisms in a bacterial metacommunity. Isme J 6: 1107–1114.

42. Bini LM, Landeiro VL, Padial AA, Siqueira T, Heino J (2014) Nutrient enrichment is related to two facets of beta diversity in stream invertebrates. Ecology 95: 1569–1578.

43. Landeiro VL, Bini LM, Melo AS, Pes AMO, Magnusson WE (2012) The roles of dispersal limitation and environmental conditions in controlling caddisfly (Trichoptera) assemblages. Fresh Biol 57: 1554–1564.

44. Rouquette JR, Dallimer M, Armsworth PR, Gaston KJ, Maltby L, et al. (2013) Species turnover and geographic distance in an urban river network. Divers Distrib 11: 1429–1439.

45. Peterson EE, Merton AA, Theobald DM, Urquhart S (2006) Patterns of spatial autocorrelation in stream water chemistry. Environ Monit Assess 121: 571–596.

46. Junk WJ, Bayley PB, Sparks RE (1989) The flood pulse concept in river-floodplain systems. In: Dodge DP editor. Proceedings of the International Large River Symposium. Canadian Special Publication of Fisheries and Aquatic Sciences. pp. 110–127.

47. Jellyman PG, Booker DG, McIntosh AR (2013) Quantifying the direct and indirect effects of flow-related disturbance on stream fish assemblages. Freshw Biol 58: 2614–2631.

48. Roach KA, Winemiller KO, Davis SE (2014) Autochthonous production in shallow littoral zones of five floodplain rivers: effects of flow, turbidity and nutrients. Freshw Biol 59: 1278–1293.

49. Thomaz SM, Pagioro TA, Bini LM, Roberto MC, Rocha RRA (2004) Limnological characterization of the aquatic environments and the influence of hydrometric levels. In: Thomaz SM, Agostinho AA, Hahn NS, editors. The Upper Paraná River and its floodplain: Physical aspects, Ecology and Conservation. Leiden: Backhuys Publishers. pp. 75–102.

50. Padial AA, Carvalho P, Thomaz SM, Boschilia SM, Rodrigues RB, et al. (2009) The role of an extreme flood disturbance on macrophyte assemblages in a Neotropical floodplain. Aquat Sci 71: 389–398.

# 18

# High Abundances of Potentially Active Ammonia-Oxidizing Bacteria and Archaea in Oligotrophic, High-Altitude Lakes of the Sierra Nevada, California, USA

**Curtis J. Hayden, J. Michael Beman***

Life and Environmental Sciences and Sierra Nevada Research Institute, University of California Merced, Merced, California, United States of America

## Abstract

Nitrification plays a central role in the nitrogen cycle by determining the oxidation state of nitrogen and its subsequent bioavailability and cycling. However, relatively little is known about the underlying ecology of the microbial communities that carry out nitrification in freshwater ecosystems—and particularly within high-altitude oligotrophic lakes, where nitrogen is frequently a limiting nutrient. We quantified ammonia-oxidizing archaea (AOA) and bacteria (AOB) in 9 high-altitude lakes (2289–3160 m) in the Sierra Nevada, California, USA, in relation to spatial and biogeochemical data. Based on their ammonia monooxygenase (*amoA*) genes, AOB and AOA were frequently detected. AOB were present in 88% of samples and were more abundant than AOA in all samples. Both groups showed >100 fold variation in abundance between different lakes, and were also variable through time within individual lakes. Nutrient concentrations (ammonium, nitrite, nitrate, and phosphate) were generally low but also varied across and within lakes, suggestive of active internal nutrient cycling; AOB abundance was significantly correlated with phosphate ($r^2 = 0.32$, $p < 0.1$), whereas AOA abundance was inversely correlated with lake elevation ($r^2 = 0.43$, $p < 0.05$). We also measured low rates of ammonia oxidation—indicating that AOB, AOA, or both, may be biogeochemically active in these oligotrophic ecosystems. Our data indicate that dynamic populations of AOB and AOA are found in oligotrophic, high-altitude, freshwater lakes.

**Editor:** Melanie R. Mormile, Missouri University of Science and Technology, United States of America

**Funding:** This work was supported University of California start-up funds (to JMB) and the UC Merced Sierra Nevada Research Institute (to JMB). The funders had no role in study design, data collection and analysis, decision to publish, or preparation of the manuscript.

**Competing Interests:** The authors have declared that no competing interests exist.

* Email: jmbeman@gmail.com

## Introduction

Nitrogen (N) is an essential nutrient for all life, and its availability serves as a critical factor for the growth of individual organisms, community composition, and ecosystem primary productivity in freshwater lakes [1,2]. In many ecosystems, N availability—both quantity and chemical form—is largely dictated by microbial communities, which transform inorganic N into bioavailable forms, and actively cycle N through oxidation-reduction (redox) processes. Phosphorus (P) typically limits primary production in freshwater [3], but both absolute amounts, and relative ratios, of N and P are highly variable due to variations in lake nutrient sources, as well as internal cycling by phytoplankton, zooplankton, and microbes [1,4,5]. In oligotrophic aquatic systems, in particular, differences in size, growth rate, and chemical form of available nutrients may favor microorganisms in competition with phytoplankton for N [6]. Microbial control of both N quantity and chemical form has important implications for the degree of eutrophication in these ecosystems, and the degree to which allochthonous N inputs (i.e. atmospheric pollutants) may affect oligotrophic lakes [7].

Within the microbial N cycle, nitrification is a two-step process that involves the aerobic oxidation of reduced inorganic N compounds (i.e. $NH_3/NH_4^+$) to nitrite ($NO_2^-$) and the subsequent oxidation of $NO_2^-$ to nitrate ($NO_3^-$). Nitrification links the mineralization of N to its eventual removal as dinitrogen gas ($N_2$) via either denitrification or anaerobic ammonium oxidation (anammox). The first step of nitrification is carried out by a few bacterial lineages within the *Beta-* and *Gamma-proteobacteria* and also by the archaeal phylum *Thaumarchaeota* (previously know as the group 1 *Crenarchaeota*) [8]. These ammonia-oxidizing bacteria (AOB) and ammonia-oxidizing archaea (AOA) use ammonia monoxygenase (AMO) to catalyze the oxidation of $NH_4^+$ to $NO_2^-$. As AOA were confirmed to be capable of ammonia oxidation only recently [9], the physical and chemical factors that control the abundance and function of these organisms, and their relative influence on nitrification rates, are not entirely understood—particularly in freshwater environments [10–13].

AOA, AOB, and nitrification have been examined within few freshwater lakes, yet AOA appear to be important and dynamic components of lake plankton and biogeochemical cycles: *Thaumarchaeota* are abundant [14], most appear to be AOA [15], and their populations fluctuate over time [15] and with depth [16]. AOB are found in lakes ranging from temperate eutrophic, to high-altitude oligotrophic, but in contrast to AOA, how AOB abundance varies in lakes through space and time is not well known [17–19]. The abundances of AOA and AOB can be controlled by differential sensitivities to temperature [20], pH [21,22], ammonium concentrations [23], and light [24]—all of which may be relevant in high elevation lakes, but have not been examined. Nitrification varies with depth and time, and is quantitatively important within lake water columns [10,25–27]—for example, Finlay et al. [25] showed that within-lake production of $NO_3^-$ through nitrification is the predominant source of $NO_3^-$ in Lake Superior. However, a lone study has measured both AOA and ammonia oxidation rates in freshwater lakes [10], and AOB have rarely been quantified in lakes [28]. We therefore know little about variations in AOA and AOB abundance and activity over time and across different lakes—let alone how AOA, AOB, and ammonia oxidation rates respond to changes in temperature, N availability, and other environmental factors within freshwater systems.

Of particular relevance are the potential inhibitory effects of light on nitrification: while these have been known for some time (reviewed by [29]), the relative effects of different wavelengths of light on AOA versus AOB, and in the field versus lab, are mixed. AOA appear highly sensitive to light in controlled experiments [24,30]: the AOA *Nitrosopumilus maritimus* and *Nitrosotalea devanaterra* were inhibited by lower light levels than AOB, and showed little recovery of ammonia oxidation over 8/16 hour light/dark cycles [30]. French et al. [24] likewise found that ammonia oxidation by three freshwater AOA isolates was strongly inhibited by white and blue light, whereas an AOB isolate was inhibited only by blue light and recovered partial oxidation ability in the dark. Notably, all of the AOA isolates used in these studies have been recovered from sediments or soil, and it is possible that pelagic AOA are less light-sensitive—for example, Auguet and Casamayor [14] proposed that surface waters of mountain lakes are an archaeal 'hotspot' based on high crenarchaeal abundance in the neuston. AOA also actively express *amoA*, and nitrification is known to occur at least transiently, in the upper ocean [31–33].

We quantified the abundance of AOA and AOB across a high-elevation lake transect in Yosemite National Park, in the Sierra Nevada mountain range, California, USA (Figure 1). In Sierra Nevada lakes, the differing susceptibility of AOA and AOB to photoinhibition could be a crucial factor in N cycling, as there is a strong natural increase in ultraviolet (UV) radiation with increasing elevation [26]. Moreover, these high-altitude lakes have relatively low light attenuation due to high transparency typical of oligotrophic aquatic ecosystems found at high elevations [26,34]. Freshwater lakes are traditionally limited by phosphorous availability [35], but the availability of N is also a critical factor for primary productivity in aquatic ecosystems of the Sierra Nevada, where biological activity in ~22% of lakes is strictly limited by N availability [36]. Internal N cycling may therefore play an important role in the overall productivity and structure of these freshwater ecosystems. We used natural variations in temperature, radiation, and N deposition, based on elevational and temporal variability between sampling sites, to examine the prevalence and abundance of AOA and AOB in high-altitude lakes.

## Materials and Methods

### Study Site

The Sierra Nevada (California) is a 400-mile long mountain range that gradually rises from the valley floor from west to east and reaches an apex of 3,000 to 4,200 meter peaks on its eastern edge (Figures 1 and 2). Vegetation along the mountain range is composed of grasslands and foothill woodlands at lower elevations, with a transition to mixed conifer forests, and then alpine meadows and lakes at higher elevations. Aquatic ecosystems in

the Sierra Nevada are located downwind of urban and agricultural areas that emit high levels of N [37] and so experience elevated levels of N deposition [38]. This N deposition is known to increase N concentrations in lakes [7], and in the case of the Sierra Nevada, represents a large fraction of the N input to high-elevation lakes: Baron et al. [39] established a critical load threshold of 1.5 kg N $ha^{-1}$ $year^{-1}$ for high-elevation lakes located in Rocky Mountain National Park, yet current annual N loading in the Sierra Nevada (i.e. Emerald Lake Watershed, Sequoia National Park) ranges from 2.0 to 4.9 kg N $ha^{-1}$ $year^{-1}$ [40]. Moreover, the watersheds are high in granitic parent material and generally have thin soils; both characteristics cause aquatic ecosystems in the Sierra Nevada to have limited buffering capacity in terms of their ability to neutralize foreign chemical species [41]. Clow et al. [42] suggested that this property, coupled with high precipitation at high elevations, leads to high N loading at high elevations despite greater distances from emission sources. Our data are relevant to this as our selected sites range from 2300 m (Harden Lake) to 3160 m (Upper Gaylor Lake) (Table 1 and Figure 2) and our transect terminus is adjacent to the steep Sierra escarpment (Figure 1).

### Sampling

Water samples were collected from lakes located in Yosemite National Park (YNP) during the 2012 summer. Nine lakes, which ranged in elevation from 2289 m to 3160 m and in depth from 4.5 m to 11 m (Table 1), were sampled in June, July, August, and September. Water samples were collected from the middle of each lake at 1 m depth using a Van Dorn water sampler (Lamotte); this depth was selected to include the effects of UV radiation and high photosynthetically active radiation (PAR). Duplicate water samples were filtered in the field onto 0.22 µm PVDF Membrane Filters (Millipore) using sterile 60 mL Polycarbonate syringes (Cole-Parmer). After filtration, filter membranes were packed into bead tubes (MP Biomedicals) containing 800 µL of Sucrose-Tris-EDTA. Filtrate was collected in 60 mL HDPE Bottles (Nalgene) for subsequent nutrient analysis. Samples were transported on ice to UC Merced's main campus or Yosemite Field Station (within hours of collection) and stored at $-80°$ or $-20°C$ until extraction of DNA and nutrient analysis. Samples were collected under USA National Park Service permit YOSE-2012-SCI-0111.

### DNA Extraction and quantification and real-time QPCR Analysis

DNA extraction followed Beman et al. [31] using Sucrose-Tris-EDTA (STE) lysis buffer, sodium dodecyl sulfate (SDS), and proteinase K with bead-beating. DNA was further purified using a DNeasy Blood & Tissue Extraction Kit (Qiagen) and resolubolized in 50 µL of ultra-pure DNA free water (Qiagen). DNA was quantified using a PicoGreen dsDNA quantification kit (Invitrogen) and an Mx3005P real-time thermocycler (Agilent Technologies). Total yield of DNA ranged from 23.5 to 922 ng.

Quantitative Polymerase Chain Reaction (QPCR) was used to quantify the abundance of *amoA* genes in lake samples. Primers, reaction chemistry, thermocycling, QPCR standards, quality control procedures, and data analysis exactly followed Beman et al. [31] and Beman et al. [43]. In brief, we used SYBR Green chemistry and the primers crenamoAF (5′-STAAT-GGTCTGGCTTAGACG-3′) and crenamoAR (5′-GCGGC-CATCCATCTGTATGT-3′) for archaeal *amoA* (originally Arch-amoAF and R; [44]) and beta-amoA1F (5′-GGGG-TTTCTACTGGTGGT-3′) and beta-amoA2R (5′-CCCCT-CKGSAAAGCCTTCTTC-3′) for betaproteobacterial *amoA* (originally *amoA*-1F and *amoA*-2R [45]). QCPR efficiencies

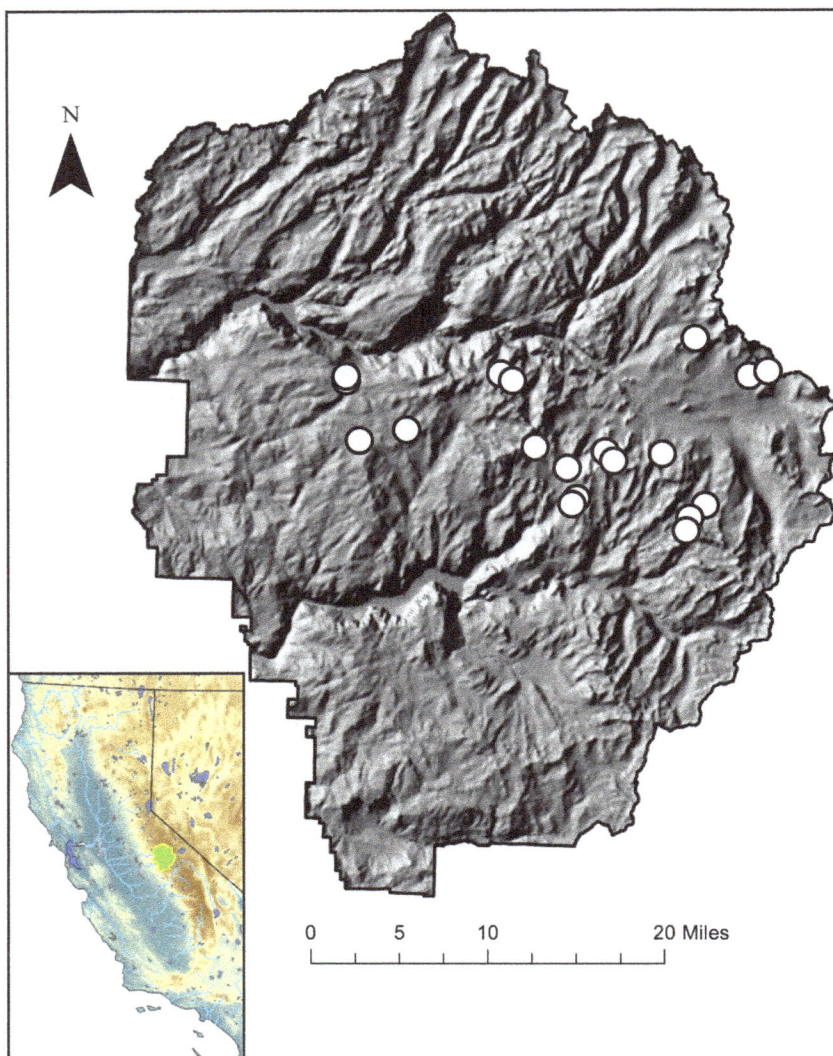

**Figure 1. Sampling locations (white circles) in Yosemite National Park displayed on 10 m resolution elevation data from the United States Geological Survey National Elevation Dataset** (http://nationalmap.gov/). Inset: location of Yosemite shown as the green shaded area within the state of California.

ranged from 89–91%, standard $r^2$ values ranged from 0.98 to 0.99, and we tested for inhibition (which was not detected) by 'spiking' standards with samples [31,43].

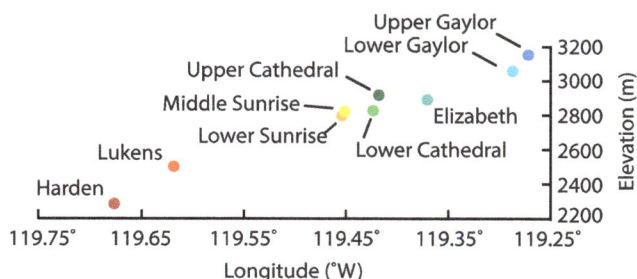

**Figure 2. Lake elevation plotted against longitude for the nine lakes sampled in this study.**

## Nutrient Analyses

Filtered lake water samples were analyzed for orthophosphate ($\mu mol\ l^{-1}$) nitrite ($\mu mol\ l^{-1}$) and nitrate ($\mu mol\ l^{-1}$) using flow-injection analysis on a QuikChem 8000 (Zellweger Analytics, Inc.) at the University of California, Santa Barbara Marine Sciences Institute Analytical Laboratory (standard curve $r^2 = 0.999$ for all assays). Filtered lake water samples were analyzed for ammonium ($NH_4^+$) following Holmes et al. [46]. 8 mL of filtered water was combined with 2 mL of reagent that consisted of 95% 0.1 M sodium tetraborate, 0.015 M O-pthaldialdehyde, 0.03 mM sodium sulfite, and 5% Ethanol. After aging, samples were measured in triplicate for fluorescence intensity using a fluorometer (Trilogy Laboratory Fluorometer, Turner Designs). Standards ranged from 31.2 to 186.8 nM and for different runs, standard curve $r^2 = 0.998$–0.999.

**Table 1.** Lake names, elevation, depth, and average nutrient concentrations.

| Lake | Elevatio (m) | Dept (m) | Average $PO_4^{3-}$ (nM) | Average $NH_4^+$ (nM) | Average $NO_2^-$ (nM) | Average $NO_3^-$ (nM) |
|---|---|---|---|---|---|---|
| Harden | 2289 | 4.5 | 80 | 15 | 44 | 297 |
| Lukens | 2506 | 6 | 98 | 32 | 41 | 374 |
| Lower Sunrise | 2801 | 5.5 | 74 | 13 | 38 | 432 |
| Middle Sunrise | 2826 | 5 | 116 | 25 | 52 | 212 |
| Lower Cathedral | 2832 | 10 | 73 | 13 | 28 | 281 |
| Upper Cathedral | 2923 | 3.5 | 91 | 20 | 67 | 419 |
| Elizabeth | 3050 | 9 | 84 | 10 | 36 | 232 |
| Lower Gaylor | 3064 | 11 | 81 | 41 | 33 | 398 |
| Upper Gaylor | 3160 | 7 | 79 | 20 | 26 | 307 |

## $^{15}NH_4^+$ oxidation rate measurements

Ammonia oxidation rates were measured by adding 99 atom percent (at%) $^{15}NH_4^+$ to a concentration of 200 nmol $L^{-1}$, and measuring the accumulation of $^{15}N$ label in the oxidized $NO_2^- + NO_3^-$ pool after incubation for ~24 hours [31,47]. All samples were incubated within lakes to mimic in situ conditions as accurately as possible. $\delta^{15}N$ of $NO_2^- + NO_3^-$ was measured at the UC Davis Stable Isotope Facility using the 'denitrifier method' [48], which produces $N_2O$ that can be analyzed on the mass spectrometer. Isotopic reference materials bracketed every 3–4 samples and coefficients of variation for these were 0.6%.

Initial at% enrichment of the substrate at the beginning of the experiment ($noNH_4^+$, see Eq. 1) was calculated by isotope mass balance based on $NH_4^+$ concentrations determined fluorometrically [46] assuming that the $^{15}N$ activity of unlabeled $NH_4^+$ was 0.3663 at% $^{15}N$. Rates of ammonia oxidation ($^{15}R_{ox}$) were calculated using equation 1 [31]:

$$^{15}R_{ox} = \frac{(n_t - n_{oNO_x^-}) \times [NO_3^- + NO_2^-]}{(n_{NH_4^+} - n_{oNH_4^+})} \quad \text{(eq.1)}$$

where $n_t$ is the at% $^{15}N$ in the $NO_3^- + NO_2^-$ pool measured at time $t$, $no_{NOx^-}$, is the measured at% $^{15}N$ of unlabeled $NO_3^- + NO_2^-$, $no_{NH4+}$ is the initial at% enrichment of $NH_4^+$ at the beginning of the experiment, $n_{NH4+}$ is at% $^{15}N$ of $NH_4^+$ at time $t$, and $[NO_3^- + NO_2^-]$ is the concentration of the $NO_x^-$ pool.

## Statistical analyses

Statistical analyses were performed using the R statistical environment (http://www.r-project.org/) and the vegan package.

## Results and Discussion

AOB $amoA$ genes were detected in all lakes from June to September, and in 88% of samples, whereas AOA were found in 46% of all samples. AOB $amoA$ gene copies ranged from $3.04 \times 10^2$ to $2.07 \times 10^5$ genes $mL^{-1}$, however the majority (73%) of AOB values fell below $2 \times 10^4$ genes $mL^{-1}$, with values exceeding this occurring in June, August and September at high elevations (Figure 3B). AOA ranged from 0 to $4.58 \times 10^3$ genes $mL^{-1}$ (See Figure 3A) and there were zero instances where AOA outnumbered AOB. When AOA were detected, they were outnumbered by AOB by 2.8- to 1080-fold. Highest average abundances of both AOB and AOA were found in Middle Sunrise Lake (2826 m),

while lowest average AOB abundances were found in Lukens Lake (2506 m), Lower Cathedral Lake (2832 m) and Lake Elizabeth (3019 m) (see Figure 3). For AOA, multiple lakes had low average values, including Lukens, Lower Cathedral, Elizabeth, and Lower Gaylor (3064 m). Over time, the highest numbers of $amoA$ genes were present in September for AOB (4 out of 6 lakes above 2800 m), and in June for AOA (5 out of 6 lakes above 2800 m). AOB and AOA therefore appear to be present, and sometimes abundant, within the water columns of oligotrophic lakes in Yosemite National Park.

Previous studies that have quantified AOA abundance in high-altitude oligotrophic lakes reported $amoA$ gene abundances as high as $3 \times 10^4$ genes $mL^{-1}$ [15], while AOB were below detection limits in the same lakes [49]. Our results indicate that AOB are more prevalent and abundant than AOA within oligotrophic lakes of the Sierra Nevada, CA, and AOB were dominant regardless of lake nutrient concentrations (see below), date of sampling, or lake elevation. This contrasts with earlier work that found AOA were more prevalent (detected via PCR but not quantified by QPCR) than AOB in oligotrophic lakes of the Tibetan plateau [18], and dominant in oligotrophic lakes of the Spanish Pyrenees [49]. In two contrasting lower elevation (231–825 m) lakes in France, Hugoni et al. [28] reported AOA dominance under low ammonium concentrations and oligotrophic conditions, whereas AOB were dominant in nutrient-rich waters [28]. The lakes sampled in Yosemite are uniformly nutrient-poor and have particularly low ammonium concentrations (all <75 nM), however our data indicate that AOB $amoA$ genes are more abundant and prevalent than those from AOA. One explanation for AOB dominance in these lakes is that the $amoA$ primers used to detect AOB and AOA may over- or under-estimate their abundances. However, the AOB primers used here are specific for betaproteobacterial AOB and are widely used [50]; the AOA primers amplify a wide range of AOA groups from water, sediments, and soils [44]. It is therefore unlikely that the AOB primers severely overestimate AOB $amoA$ genes or that the AOA primers severely underestimate AOA $amoA$ genes. Nor would this explain the prevalence of AOB, which were frequently detected. AOA were also detected in nearly half the samples—despite high light levels and oligotrophic conditions that are presumably hostile to nitrifiers in general.

Ultimately AOB and AOA must oxidize N to persist under oligotrophic conditions, and we suggest that their populations could be sustained by N fluxes that are not reflected in depleted nutrient pools: that is, $NH_4^+$ may be rapidly regenerated, assimilated, and/or oxidized, but because of high demand, does

A  AOA Abundance

B  AOB Abundance

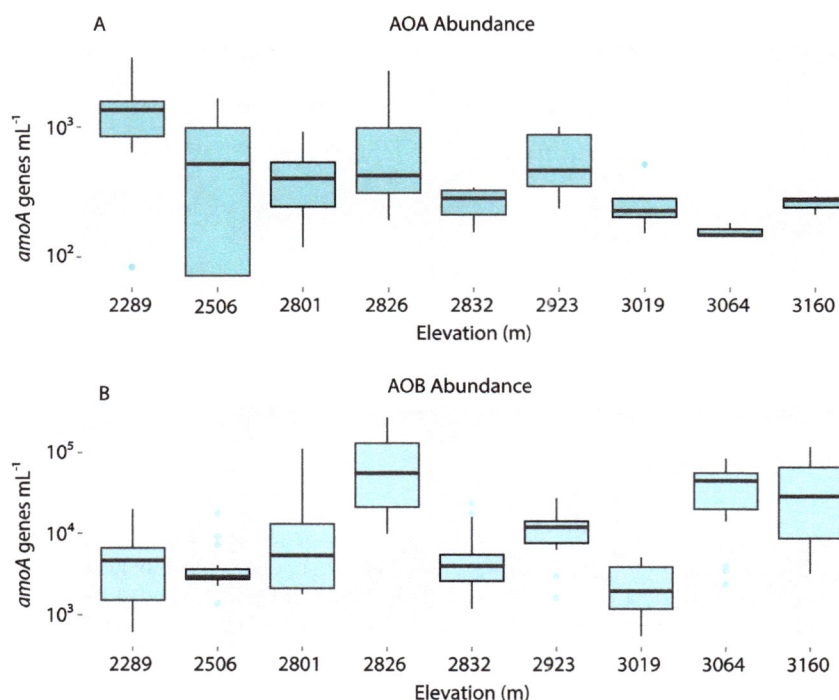

**Figure 3. Boxplot comparison of number of *amoA* genes per milliliter (genes mL$^{-1}$) for (A) ammonia-oxidizing archaea (AOA) and (B) ammonia-oxidizing bacteria (AOB).** The vertical axes (logarithimic scale) denote the number of genes mL$^{-1}$; the horizontal axes represent the elevation of the sampling sites, and are ordered from lowest elevation (Harden Lake) to highest elevation (Upper Gaylor Lake). In these plots, the box denotes the mean plus and minus one standard deviation; the line within the box represents the median value; and lines extending above and below the box span the full range of the data. Outliers were defined as sample values 1.5 times larger than the upper quartile and are represented by green and blue circles.

not accumulate in oligotrophic lakes. In fact, lakes are watershed 'integrators' [51] that can function as hotspots of N-cycling, including nitrification [14,25,42,52]. In Yosemite, lake N loading increases with elevation, and NO$_3^-$ concentrations are correlated with modeled N deposition rates [42]. N deposition to Sierra watersheds occurs primarily as NH$_4^+$ [37] and this flux of N— which can be comparatively large in these oligotrophic ecosystems [53]—must have been nitrified at least once if it accumulates as NO$_3^-$. Ammonium concentrations therefore may not be the sole predictor for AOA and AOB abundances—elevation and nitrate concentrations may also be relevant—and we analyzed relationships between AOA and AOB and several types of spatial and nutrient concentration data using a variety of statistical approaches.

During the sampling period NO$_3^-$ concentrations in all lakes ranged from 150 nM to 990 nM, PO$_4^{3-}$ concentrations from 60 nM to 120 nM, NO$_2^-$ concentrations from 20 nM to 120 nM, and NH$_4^+$ concentrations from 2.8 nM to 72 nM (Figure 4). As expected in aquatic ecosystems, NO$_3^-$ levels were higher than either PO$_4^{3-}$ or NO$_2^-$. We did not observe a significant trend with elevation (ANOVA $P > 0.05$), but this could emerge with additional sampling. In nearly all lakes, the molar concentration of PO$_4^{3-}$ was higher than that of nitrite, which is typical, as NO$_2^-$ is quickly oxidized to NO$_3^-$ in the presence of oxygen [54]. Across our samples, AOB and AOA were significantly correlated with a few individual variables, including nutrient concentrations. For

example, AOB abundance was most strongly correlated with PO$_4^{3-}$ concentrations ($r^2 = 0.32$, $p < 0.1$), consistent with work by Sundaweshar et al. [55] that showed P-limitation of N-cycling. AOB displayed an increasing trend with altitude, but this relationship was not significant ($r^2 = 0.21$, $p = 0.28$); in contrast, AOA abundance was inversely correlated with altitude ($r^2 = 0.43$, $p < 0.05$). Both groups showed wide variation in Middle Sunrise Lake at 2826 m. These patterns are evident in Figure 3, where AOB were notably more abundant—but also variable—in the Gaylor Lakes at >3000 m elevation, whereas AOA were more abundant and variable at lower elevations. For all samples collected in the three lakes >3000 m elevation, AOA were only detected four times. This inverse relationship between AOA abundance and increasing elevation could reflect the effect of increased UV radiation at higher elevations, or other factors that vary with elevation.

We therefore used redundancy analysis (RDA) to analyze multivariate relationships between AOA and AOB abundance and spatial (elevation and longitude), temporal (sampling date) and environmental (NO$_3^-$, NO$_2^-$, NH$_4^+$ and PO$_4^{3-}$) data (Figure 5). 33% of the variability in AOA and AOB abundance was explained by these data, with nutrient concentrations accounting for 24% of the variability in AOA and AOB abundance, and site location and sampling date accounting for 9% of the variability. Collectively, nutrients explain nearly a quarter of the variation in AOA and AOB abundance in these lakes, and this includes ammonium, as

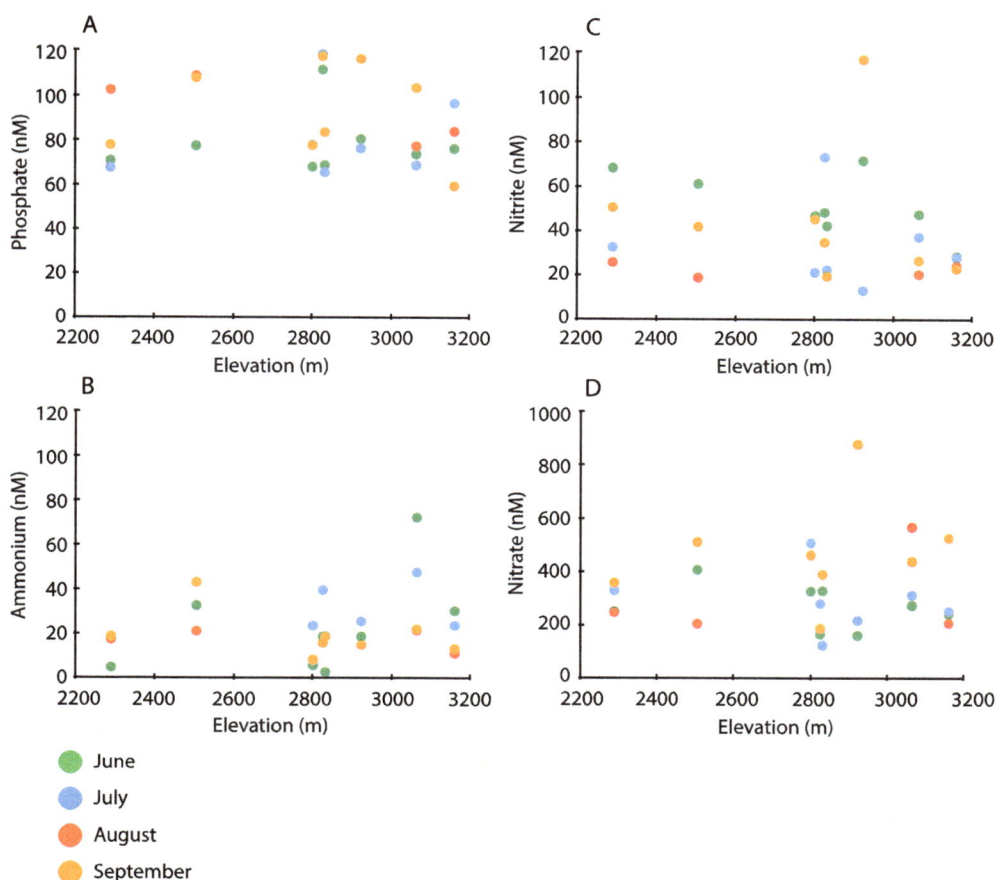

**Figure 4. Variation in lake nutrient concentrations versus elevation for (A) phosphate (PO$_4^{3-}$), (B) ammonium (NH$_4^+$), (C) nitrite (NO$_2^-$), and (D) nitrate (NO$_3^-$).** Vertical axes show nutrient concentrations in nanomolar (nM) plotted on a logarithmic scale, and colors denote month of sampling. Note that some samples have highly similar nutrient concentrations and fall nearly on top of one another.

well as nitrate, nitrite, and phosphate, concentrations. The modest percentage of constrained variance overall indicates that AOA and AOB populations are affected by other, un-measured factors, or this may reflect stochastic variation in populations. In lakes, AOB and AOA abundance may be modified by active growth, but also by transport of cells into lakes via air, water, or suspended particles; competition with other organisms for ammonium; and trophic interactions, such as grazing, viral infection, and lysis. None of these have been directly investigated.

The prevalence and abundances of AOA and AOB suggest that active N cycling and nitrification may be occurring in high-altitude lakes, but is surprising given that high-altitude oligotrophic lakes have low N concentrations and experience high light levels. To determine whether AOA and AOB may be active under these conditions, we performed $^{15}$NH$_4^+$ incubations to detect ammonia oxidation rates in three lakes that span a range of elevations and AOB/AOA abundances (Lukens Lake, Lower Cathedral Lake and Lower Gaylor Lake). These were conducted in both light and dark bottles that were incubated within the lakes under *in situ* conditions. At low N concentrations found in oligotrophic waters, measuring ammonia oxidation is extremely challenging due to multiple factors: (1) measuring low-level N concentrations is difficult, and uncertainties in basic nutrient concentration values

can strongly affect rate calculations; (2) addition of $^{15}$N label can significantly increase N concentrations and potentially introduce biases; (3) rates are expected to be low due to low N concentrations; and (4) measuring accurate isotopic values becomes difficult, which either introduces errors in the measurement or requires addition of N 'carrier'—yet this can make detection of low rates difficult or impossible. For these reasons, we propagated uncertainties in nutrient measurements and isotopic values through our calculations, and found that rates were above detection (limit = 10.2 pmol L$^{-1}$ d$^{-1}$) only in dark bottles in Lower Cathedral Lake and Lower Gaylor Lake—where they were extremely low (Figure 6). Ammonia oxidation was below the limits of detection in most light bottles in these lakes, as well as in Lukens Lake. That rates were undetectable under *in situ* light levels is consistent with light inhibition of ammonia oxidation, but more importantly indicates that AOB and AOA were inactive—or active at extremely low levels—at the time and location of sampling. Our data do suggest that they respond to changing light conditions and oxidize ammonia at low levels in completely darkened bottles. This expands the range of habitats in which ammonia oxidation can potentially occur to include surface waters of high-altitude, oligotrophic, freshwater lakes—but when and

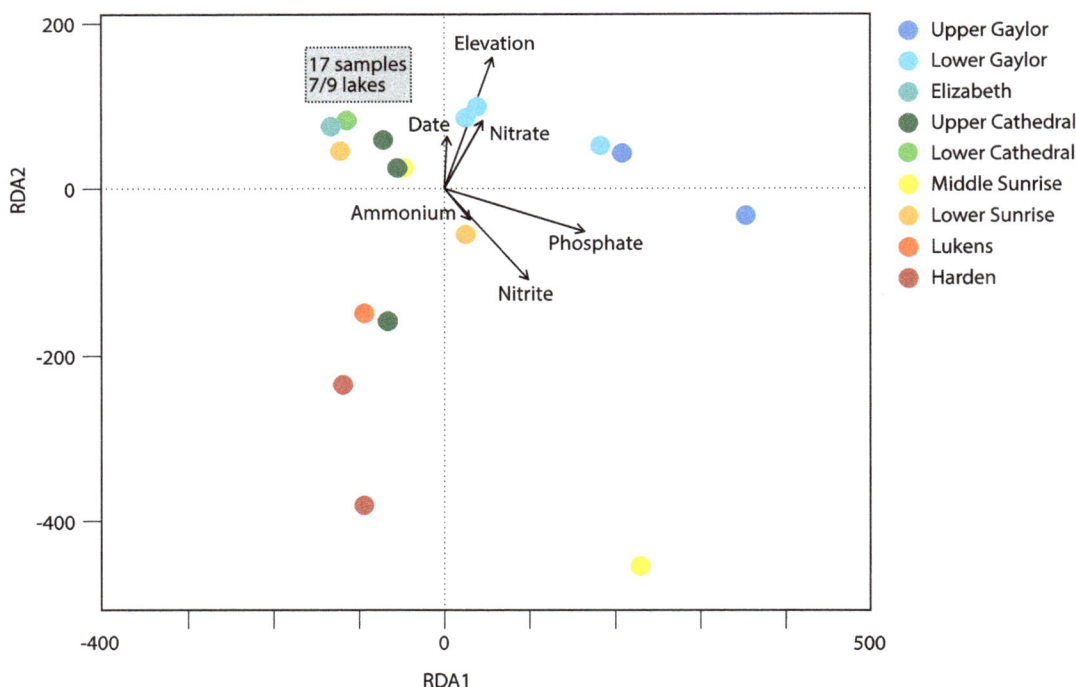

**Figure 5. Bi-plot of redundancy analysis of bacterial community composition.** Color key denotes different lakes, and arrows denote biplot scores for the constraining variables. The grey box encompasses 17 samples that fall within a narrow range of each other and are not visually distinguishable on the bi-plot; this includes 1–2 samples from every lake except Middle Sunrise and Upper Cathedral Lake. One sample from Middle Sunrise Lake with extremely high AOB *amoA* gene abundance is not shown, as it falls much farther along the RDA1 and RDA2 axes.

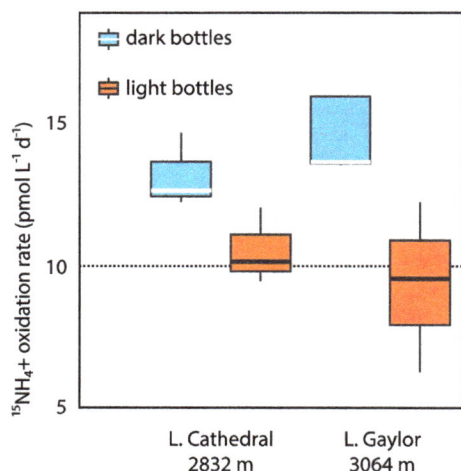

**Figure 6. Boxplot comparison of calculated $^{15}NH_4^+$ oxidation rates measured in Lower Cathedral and Lower Gaylor lakes.** The vertical axis shows $^{15}NH^+$ oxidation rates (pmol $L^{-1}$ $D^{-1}$). The blue bars represent dark bottle incubations that were incubated under zero-light conditions, while the orange bars represent samples that were exposed to ambient light over a 24-hour cycle during in-situ incubation at the sampling site. The dashed line notes the detection limit. In these plots, the box denotes the mean plus and minus one standard deviation; the line within the box represents the median value; and lines extending above and below the box span the full range of the data.

where these organisms are active at appreciable levels is not yet clear.

AOB and AOA were frequently present, varied in abundance between lakes, and fluctuated over time within individual lakes. The presence of AOA is consistent with the idea that they are adapted to the low-nutrient conditions that are characteristic of these lakes—for example, the marine AOA *Nitrosopumilus maritimus* has a remarkably high affinity for ammonia and appears to be adapted to life under extreme nutrient limitation [23]. However, AOB were more commonly detected and were more abundant than AOA under all sampled conditions; this fits with the recent idea that AOB are more light-tolerant than AOA [24,30]. High altitude lakes could also experience higher fluxes of N that would periodically favor AOB [42]. Both AOA and AOB may contribute to nitrification in freshwater oligotrophic lakes—as we detected low rates of ammonia oxidation in darkened bottles—but it is clear that they were inactive, or active only at low levels, during our incubations. Our data therefore add to the limited information available on microbial contributions to N cycling in lakes—particularly for AOA/AOB and nitrification—but additional work should expand these approaches to additional lakes and additional sampling periods. Nitrification under ice could be important during winter, for example, and nitrification may be especially significant following spring snowmelt, when pulses of N likely enter these lakes and may be metabolized by microbes. Altogether our findings are indicative of dynamic microbial communities and internal N cycling in high-altitude lakes.

## Acknowledgments

We thank Susan Alford, Julie Cline, Matt Meyerhof, Elizabeth Perkins, Sang Park, and Victoria Velez for help in the field and lab.

## Author Contributions

Conceived and designed the experiments: JMB. Performed the experiments: CJH. Analyzed the data: JMB CJH. Contributed reagents/materials/analysis tools: JMB CJH. Wrote the paper: CJH.

## References

1. Elser JJ, Bracken MES, Cleland EE, Gruner DS, Harpole WS, et al. (2007) Global analysis of nitrogen and phosphorus limitation of primary producers in freshwater, marine and terrestrial ecosystems. Ecology Letters 10: 1135–1142.
2. Vrede T, Ballantyne A, Mille-Lindblom C, Algesten G, Gudasz C, et al. (2009) Effects of N: P loading ratios on phytoplankton community composition, primary production and N fixation in a eutrophic lake. Freshwater Biology 54: 331–344.
3. Schindler DW (1977) Evolution of Phosphorus Limitation in Lakes. Science 195: 260–262.
4. Elser JJ, Andersen T, Baron JS, Bergstrom AK, Jansson M, et al. (2009) Shifts in Lake N: P Stoichiometry and Nutrient Limitation Driven by Atmospheric Nitrogen Deposition. Science 326: 835–837.
5. Harpole WS, Ngai JT, Cleland EE, Seabloom EW, Borer ET, et al. (2011) Nutrient co-limitation of primary producer communities. Ecology Letters 14: 852–862.
6. Cotner JB, Biddanda BA (2002) Small players, large role: Microbial influence on biogeochemical processes in pelagic aquatic ecosystems. Ecosystems 5: 105–121.
7. Fenn ME, Baron JS, Allen EB, Rueth HM, Nydick KR, et al. (2003) Ecological effects of nitrogen deposition in the western United States. Bioscience 53: 404–420.
8. Brochier-Armanet C, Boussau B, Gribaldo S, Forterre P (2008) Mesophilic crenarchaeota: proposal for a third archaeal phylum, the Thaumarchaeota. Nature Reviews Microbiology 6: 245–252.
9. Konneke M, Bernhard AE, de la Torre JR, Walker CB, Waterbury JB, et al. (2005) Isolation of an autotrophic ammonia-oxidizing marine archaeon. Nature 437: 543–546.
10. Small GE, Bullerjahn GS, Sterner RW, Beall BFN, Brovold S, et al. (2013) Rates and controls of nitrification in a large oligotrophic lake. Limnology and Oceanography 58: 276–286.
11. Erguder TH, Boon N, Wittebolle L, Marzorati M, Verstraete W (2009) Environmental factors shaping the ecological niches of ammonia-oxidizing archaea. Fems Microbiology Reviews 33: 855–869.
12. Francis CA, Beman JM, Kuypers MMM (2007) New processes and players in the nitrogen cycle: the microbial ecology of anaerobic and archaeal ammonia oxidation. Isme Journal 1: 19–27.
13. Hatzenpichler R (2012) Diversity, Physiology, and Niche Differentiation of Ammonia-Oxidizing Archaea. Applied and Environmental Microbiology 78: 7501–7510.
14. Auguet JC, Casamayor EO (2008) A hotspot for cold crenarchaeota in the neuston of high mountain lakes. Environmental Microbiology 10: 1080–1086.
15. Auguet JC, Nomokonova N, Camarero L, Casamayor EO (2011) Seasonal Changes of Freshwater Ammonia-Oxidizing Archaeal Assemblages and Nitrogen Species in Oligotrophic Alpine Lakes. Applied and Environmental Microbiology 77: 1937–1945.
16. Callieri C, Corno G, Caravati E, Rasconi S, Contesini M, et al. (2009) Bacteria, Archaea, and Crenarchaeota in the epilimnion and hypolimnion of a deep holo-oligomictic lake. Applied and Environmental Microbiology 75: 7298–7300.
17. Whitby CB, Saunders JR, Pickup RW, McCarthy AJ (2001) A comparison of ammonia-oxidiser populations in eutrophic and oligotrophic basins of a large freshwater lake. Antonie Van Leeuwenhoek 79: 179–188.
18. Hu AY, Yao TD, Jiao NZ, Liu YQ, Yang Z, et al. (2010) Community structures of ammonia-oxidising archaea and bacteria in high-altitude lakes on the Tibetan Plateau. Freshwater Biology 55: 2375–2390.
19. Cebron A, Coci M, Garnier J, Laanbroek HJ (2004) Denaturing gradient gel electrophoretic analysis of ammonia-oxidizing bacterial community structure in the lower Seine River: impact of Paris wastewater effluents. Appl Environ Microbiol 70: 6726–6737.
20. Tourna M, Freitag TE, Nicol GW, Prosser JI (2008) Growth, activity and temperature responses of ammonia'Äôoxidizing archaea and bacteria in soil microcosms. Environmental Microbiology 10: 1357–1364.
21. Nicol GW, Leininger S, Schleper C, Prosser JI (2008) The influence of soil pH on the diversity, abundance and transcriptional activity of ammonia oxidizing archaea and bacteria. Environmental Microbiology 10: 2966–2978.
22. Auguet JC, Casamayor EO (2013) Partitioning of Thaumarchaeota populations along environmental gradients in high mountain lakes. Fems Microbiology Ecology 84: 154–164.
23. Martens-Habbena W, Berube PM, Urakawa H, de la Torre JR, Stahl DA (2009) Ammonia oxidation kinetics determine niche separation of nitrifying Archaea and Bacteria. Nature 461: 976–U234.
24. French E, Kozlowski JA, Mukherjee M, Bullerjahn G, Bollmann A (2012) Ecophysiological Characterization of Ammonia-Oxidizing Archaea and Bacteria from Freshwater. Applied and Environmental Microbiology 78: 5773–5780.
25. Finlay JC, Sterner RW, Kumar S (2007) Isotopic evidence for in-lake production of accumulating nitrate in lake superior. Ecological Applications 17: 2323–2332.
26. Hall GH, Jeffries C (1984) The Contribution of Nitrification in the Water Column and Profundal Sediments to the Total Oxygen Deficit of the Hypolimnion of a Mesotrophic Lake (Grasmere, English Lake District). Microbial Ecology 10: 37–46.
27. Rudd JWM, Kelly CA, Schindler DW, Turner MA (1988) Disruption of the Nitrogen-Cycle in Acidified Lakes. Science 240: 1515–1517.
28. Hugoni M, Etien S, Bourges A, Lepere C, Domaizon I, et al. (2013) Dynamics of ammonia-oxidizing Archaea and Bacteria in contrasted freshwater ecosystems. Research in Microbiology 164: 360–370.
29. Lomas MW, Lipschultz F (2006) Forming the primary nitrite maximum: Nitrifiers or phytoplankton? Limnology and Oceanography 51: 2453–2467.
30. Merbt SN, Stahl DA, Casamayor EO, Marti E, Nicol GW, et al. (2012) Differential photoinhibition of bacterial and archaeal ammonia oxidation. FEMS Microbiology Letters 327: 41–46.
31. Beman JM, Popp BN, Alford SE (2012) Quantification of ammonia oxidation rates and ammonia-oxidizing archaea and bacteria at high resolution in the Gulf of California and eastern tropical North Pacific Ocean. Limnology and Oceanography 57: 711–726.
32. Church MJ, Wai B, Karl DM, DeLong EF (2010) Abundances of crenarchaeal amoA genes and transcripts in the Pacific Ocean. Environmental Microbiology 12: 679–688.
33. Santoro AE, Casciotti KL, Francis CA (2010) Activity, abundance and diversity of nitrifying archaea and bacteria in the central California Current. Environmental Microbiology 12: 1989–2006.
34. Sommaruga R (2001) The role of solar UV radiation in the ecology of alpine lakes. Journal of Photochemistry and Photobiology B-Biology 62: 35–42.
35. Sterner RW (2008) On the Phosphorus Limitation Paradigm for Lakes. International Review of Hydrobiology 93: 433–445.
36. Eilers JM, Kanciruk P, McCord RA, Overton WS, Hook L, et al. (1987) Characteristics of Lakes in the Western United States. Volume II: Data Compendium of Selected Physical and Chemical Variables. In: Agency UEP, editor.
37. Clarisse L, Shephard MW, Dentener F, Hurtmans D, Cady-Pereira K, et al. (2010) Satellite monitoring of ammonia: A case study of the San Joaquin Valley. Journal of Geophysical Research-Atmospheres 115.
38. Bytnerowicz A, Tausz M, Alonso R, Jones D, Johnson R, et al. (2002) Summer-time distribution of air pollutants in Sequoia National Park, california. Environmental Pollution 118: 187–203.
39. Baron JS (2006) Hindcasting nitrogen deposition to determine an ecological critical load (vol 16, pg 433, 2006). Ecological Applications 16: 1629–1629.
40. Sickman JO, Leydecker A, Melack JM (2001) Nitrogen mass balances and abiotic controls on N retention and yield in high-elevation catchments of the Sierra Nevada, California, United States. Water Resources Research 37: 1445–1461.
41. Meixner T, Gutmann C, Bales R, Leydecker A, Sickman J, et al. (2004) Multidecadal hydrochemical response of a Sierra Nevada watershed: sensitivity to weathering rate and changes in deposition. Journal of Hydrology 285: 272–285.
42. Clow DW, Nanus L, Huggett B (2010) Use of regression-based models to map sensitivity of aquatic resources to atmospheric deposition in Yosemite National Park, USA. Water Resources Research 46.
43. Beman JM, Sachdeva R, Fuhrman JA (2010) Population ecology of nitrifying Archaea and Bacteria in the Southern California Bight. Environmental Microbiology 12: 1282–1292.
44. Francis CA, Roberts KJ, Beman JM, Santoro AE, Oakley BB (2005) Ubiquity and diversity of ammonia-oxidizing archaea in water columns and sediments of the ocean. Proceedings of the National Academy of Sciences of the United States of America 102: 14683–14688.
45. Rotthauwe JH, Witzel KP, Liesack W (1997) The ammonia monooxygenase structural gene amoA as a functional marker: molecular fine-scale analysis of natural ammonia-oxidizing populations. Appl Environ Microbiol 63: 4704–4712.
46. Holmes RM, Aminot A, Kerouel R, Hooker BA, Peterson BJ (1999) A simple and precise method for measuring ammonium in marine and freshwater ecosystems. Canadian Journal of Fisheries and Aquatic Sciences 56: 1801–1808.
47. Ward BB (2008) Chapter 5 - Nitrification in Marine Systems. In: Capone DG, Bronk DA, Mulholland MR, Carpenter EJ, editors. Nitrogen in the Marine Environment (2nd Edition). San Diego: Academic Press.pp.199–261.
48. Sigman DM, Casciotti KL, Andreani M, Barford C, Galanter M, et al. (2001) A bacterial method for the nitrogen isotopic analysis of nitrate in seawater and freshwater. Analytical Chemistry 73: 4145–4153.
49. Auguet JC, Triado-Margarit X, Nomokonova N, Camarero L, Casamayor EO (2012) Vertical segregation and phylogenetic characterization of ammonia-oxidizing Archaea in a deep oligotrophic lake. Isme Journal 6: 1786–1797.

50. Junier P, Kim OS, Hadas O, Imhoff JF, Witzel KP (2008) Evaluation of PCR primer selectivity and phylogenetic specificity by using amplification of 16S rRNA genes from betaproteobacterial ammonia-oxidizing bacteria in environmental samples. Applied and Environmental Microbiology 74: 5231–5236.

51. Williamson CE, Saros JE, Vincent WF, Smol JP (2009) Lakes and reservoirs as sentinels, integrators, and regulators of climate change. Limnology and Oceanography 54: 2273–2282.

52. Baron JS, Hall EK, Nolan BT, Finlay JC, Bernhardt ES, et al. (2013) The interactive effects of excess reactive nitrogen and climate change on aquatic ecosystems and water resources of the United States. Biogeochemistry 114: 71–92.

53. Murphy DD, Knopp CM (2000) Lake Tahoe watershed assessment: volume I. General Technical Report-Pacific Southwest Research Station, USDA Forest Service.

54. Goldman CR, Horne AJ (1983) Limnology: McGraw-Hill.

55. Sundareshwar PV, Morris JT, Koepfler EK, Fornwalt B (2003) Phosphorus limitation of coastal ecosystem processes. Science 299: 563–565.

# 19

# Persistence of Environmental DNA in Freshwater Ecosystems

Tony Dejean[1,2,3], Alice Valentini[1,2], Antoine Duparc[2], Stéphanie Pellier-Cuit[4], François Pompanon[4], Pierre Taberlet[4], Claude Miaud[2]*

1 SPYGEN, Savoie Technolac - BP 274, Le Bourget-du-Lac, France, 2 Laboratoire d'Ecologie Alpine, UMR CNRS 5553, Université de Savoie, Le Bourget-du-Lac, France, 3 Parc Naturel Régional Périgord-Limousin, La Coquille, France, 4 Laboratoire d'Ecologie Alpine, UMR CNRS 5553, Université Grenoble I, Grenoble, France

## Abstract

The precise knowledge of species distribution is a key step in conservation biology. However, species detection can be extremely difficult in many environments, specific life stages and in populations at very low density. The aim of this study was to improve the knowledge on DNA persistence in water in order to confirm the presence of the focus species in freshwater ecosystems. Aquatic vertebrates (fish: Siberian sturgeon and amphibian: Bullfrog tadpoles) were used as target species. In control conditions (tanks) and in the field (ponds), the DNA detectability decreases with time after the removal of the species source of DNA. DNA was detectable for less than one month in both conditions. The density of individuals also influences the dynamics of DNA detectability in water samples. The dynamics of detectability reflects the persistence of DNA fragments in freshwater ecosystems. The short time persistence of detectable amounts of DNA opens perspectives in conservation biology, by allowing access to the presence or absence of species e.g. rare, secretive, potentially invasive, or at low density. This knowledge of DNA persistence will greatly influence planning of biodiversity inventories and biosecurity surveys.

**Editor:** Jack Anthony Gilbert, Argonne National Laboratory, United States of America

**Funding:** Tony Dejean was given a grant by the Ministry of Research (Grant CIFRE Number 966/2007), and Claude Miaud by the European Community (Marie Curie OIF grant Number 1018). The funders had no role in study design, data collection and analysis, decision to publish, or preparation of the manuscript.

\* E-mail: claude.miaud@univ-savoie.fr

## Introduction

The precise knowledge of species distribution is a key point for conservation strategies, especially when the focal species are invasive, threatened or endangered [1–4]. However, its detection may be extremely difficult in many environments, at specific life stages and in populations at very low densities [5,6]. To overcome this problem, DNA barcoding was recently used in order to detect species through extracellular DNA present in environmental samples, coming from cell lysis or living organism excretion or secretion [7]. This method allows species presence detection, without any contact (e.g. visual, auditory) when the only available indicators are hair, faeces or urine left behind by the organisms. For example, faeces and hair samples were used for monitoring the recent wolf range expansion in France and Switzerland [8]. In aquatic ecosystems, Ficetola et al. [5] proposed a new methodology for species detection using environmental DNA from freshwater samples. The aim was to detect the American bullfrog (*Rana catesbeiana = Lithobates catesbeianus*) in natural ponds in SW France where it was introduced about 40 years ago [9]. The method, efficient in detecting frogs even at very low density, can be integrated into the eradication strategy of this invasive species to estimate its distribution in ponds before and after frog removal. Environmental DNA could thus be used for a biodiversity inventory (e.g. introduced Asian carp in North America [10])but also to control the efficiency of eradication actions. In this context,

the precise assessment of the species presence requires knowledge of DNA persistence in water.

DNA persistence can be defined as the continuance of DNA after the removal of its source. However, any detection in the field is always imperfect and sampling is a stochastic process [11]. Detection probability depends on the species density and on the ratio between the DNA released by the organism and the DNA degraded by environmental factors.

In this study we estimated the time of DNA detection taking into account aquatic environment conditions and DNA concentrations. Experimentation was performed on two different species: the American bullfrog (*Rana catesbeiana = Lithobates catesbeianus*) and the Siberian sturgeon (*Acipenser baerii*).

## Materials and Methods

### Tests conditions and sampling

Two species were used for assessing the persistence of detectable amounts of DNA. Bullfrog tadpoles were studied to validate the possibility of integrating the approach of Ficetola et al. [5] in the species eradication strategy. Due to the risk of invasion and/or pathogen transmission to native populations (e.g. *Batrachochytrium dendrobatidis* [12] and Ranavirus [13]) bullfrog tadpoles cannot be used outdoors, and the experiments were performed in aquariums (as in Ficetola et al. [5]). The siberian sturgeon was used as DNA source species to test field conditions and placed in artificial ponds created about 20 years ago on the University campus, where this species has never been present.

For the bullfrog experiment, 3 different densities of tadpoles were used. One, 5 and 10 tadpoles were reared in 900 mL glass beakers for 5 days and each density was replicated 5 times. A 900 mL glass beaker without tadpoles was used as control. At the fifth day, the tadpoles were removed. At this time and every 24 h during 20 days, 15 mL of water were sampled from each glass beaker. Room temperature was maintained constant throughout the experimental period and the water temperature measured in the glass beakers was $17 \pm 1°C$.

For the sturgeon experiment, three ponds of dimensions $12 \text{ m}^2$ and 0.40 m deep were used. In each pond, a sturgeon (20 cm long) was housed for 10 days (from November the $04^{th}$ to $13^{th}$ 2009). On the tenth day, the sturgeons were removed and 15 mL of water were sampled from each pond. Water samples were collected every 24 h during 14 days. Water temperature fluctuated from 8 to 11°C during this period.

The duration of each experiment was determined after a preliminary test on the same condition without replication. In both experiments, water samples were added to a solution composed of 1.5 mL of sodium acetate 3 M, and 33 mL absolute ethanol immediately after collection, and then stored at $-20°C$ until the DNA extraction.

## DNA analysis

DNA extraction was adapted from Ficetola et al. [5]: we centrifuged the mixture at 9400 g for 1 h at 6°C to recover DNA and/or the cellular remains. The DNA from the pellet was extracted using QIAmp Blood and Tissue Extraction Kit (Qiagen, GmbH, Hilden, Germany), following manufactures' instructions. DNA extraction was performed in a dedicated room for degraded DNA samples. Control extractions were systematically performed to monitor possible contaminations.

Bullfrog DNA was amplified with primers described in Ficetola et al. [5]. Sturgeon DNA was amplified with primers designed to amplify a 98 bp fragment of the *Acipenser mt*DNA control region (5' – GACAGTAATTGTAGAGTTTC - 3'and 5' – CAGTAACAG-GCTGATTATG - 3'). *In silico* PCR, performed using the ecoPCR software [14] (http://www.grenoble.prabi.fr/trac/ecoPCR) on the whole GenBank dataset extracted on July 9 2009, showed the suitability of the primer pair. The only 4 species amplified were from the genus *Acipenser*. *A. persicus*, *A. brevirostrum*, *A. gueldenstaedtii* and *A. baerii*, the latter was the only species present in the ponds.

DNA amplifications were carried out in a final volume of 25 µL, using 3 µL of DNA extract as template. The amplification mixture contained 1 U of Ampli*Taq* Gold DNA Polymerase (Applied Biosystems, Foster City, CA), 10 mM Tris-HCl, 50 mm KCl, 2 mM of MgCl2, 0.2 mM of each dNTPs, 0.2 µm of each primer, and 0.005 mg of bovine serum albumin (BSA, Roche Diagnostics, Basel, Switzerland). After 10 min at 95°C (*Taq* activation), the PCR cycles were performed as follows: 55 cycles of 30 s at 95°C, 30 s at 54°C for *A. baerii* and 61°C for *R. catesbeiana* primer pair. The amplification for the sturgeon experiments was repeated 3 times using multi-tube approach [15]. PCR products were visualized using electrophoresis on 2% agarose gel.

For the bullfrog experiment, the DNA detectability was defined as the number of positive samples detected among the 15 samples collected per day (5 replicates and 3 densities). For the Sturgeon experiment, the DNA detectability was defined as the number of positive samples detected among the 9 samples analysed (3 samples collected and 3 PCR per sample).

## Statistical modelling

For the bullfrog experiment, the relationship between the DNA detectability, the time and density of tadpoles was inferred with a generalized linear model using binomial error. For the sturgeon experiment, the relationship between the DNA detectability and time was inferred with a linear mixed model with sites as random effect. In both experiments, a backward selection procedure was used starting with the full model containing all fixed explanatory components. Then, fixed variables were removed step by step. The best fitted model was selected based on AIC [16]. All analyses were done with R (R 2.10) [17].

## Ethics Statement

The research presented has been approved by the Animal Care and Use Committee (permit #CMLECA5553 05/19/05) of the Savoie University at Le Bourget du Lac (France).

## Results

In the bullfrog experiment, DNA was detected after tadpole removal at the three densities. DNA detectability was best explained by time and tadpole density factor. DNA detectability ($z = -8.032$ and $p < 0.001$) was negatively correlated with time. Tadpole density, although significant, showed no trends according to levels of density (no difference between 1 and 5 tadpoles, $z = -1.916$, $p = 0.0553$, while DNA detectability was higher with 10 tadpoles compared with 1 tadpole, $z = 2.091$, $p = 0.0365$). After the removal of tadpoles, the DNA was detected until day 25 with a detectability superior to 5% (all tadpole densities together; Figure 1a).

In the sturgeon experiment, DNA detectability was negatively correlated with time ($z = -6.136$ and $p < 0.001$, $R^2 = 0.5$). DNA was detected until day 14 with a detectability superior to 5% (Figure 1b). Using 3 replicates per pond, after 17 days there is a probability higher than 95% to not detect short DNA fragments (i.e. the probability that all 3 replicates are negative) or 21 days if a 99% threshold is considered.

## Discussion

Freshwater environments and oceans constitute a great reservoir of extracellular DNA [18]. Its detection in an aquatic environment depends on its release and degradation. The density of individuals influences the dynamics of DNA detectability in water samples, as shown by the bullfrog experiment in this study.

Once released from organisms, extracellular DNA in the environment may persist, adsorbed in organic or inorganic particles. It may also be transformed by competent soil microorganisms, or may be degraded (see [19] for a comprehensive review).

Several factors operate in DNA degradation. Endogenous nucleases, water, UV radiation and the action of bacteria and fungi in the environment contribute to DNA decay [20]. Different studies demonstrated that 300–400 bp fragments could be detected in water up to one week in controlled conditions [21–24]. Short DNA fragments are usually very slowly degraded and can be recovered from environmental samples [25]. They are well preserved in dry and cold environments and in the absence of light [20]. For example, the Greenland ancient communities of plants and animals was described using 450 000 year old silty ice samples extracted from the bottom of the Greenland ice cap [26]. In this study, using short fragments, DNA was detectable up to c. a. one month after the removal of its source, for both animal species used. This discrepancy in DNA persistence in for example soil and water and can be due to the action of endogenous nuclease and water

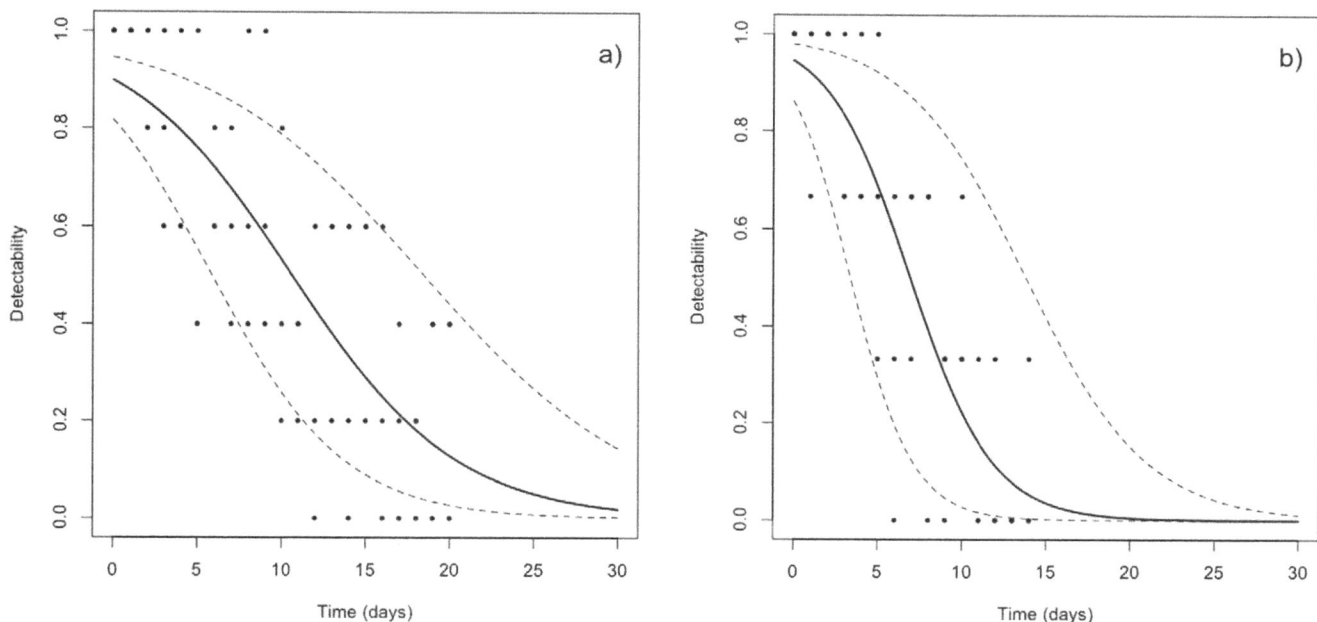

**Figure 1. DNA detectability in water according to time.** DNA detectability in water in control conditions (a) and in natural conditions (b) according to the time elapsed since the DNA source removal.

that hydrolyses DNA molecules and creates DNA strand breaks by direct cleavage of the DNA phosphodiester backbone or breakage of the sugar backbone after depurination [27]. UV radiation [28] and DNA uptake by micro-organisms, as source of nutriments (carbon, nitrogen and phosphorous) and to repair their own DNA damages [29], contribute also to damage and decrease DNA molecules density. Microorganisms' uptake varies with temperature; and as a consequence, DNA detectability can vary according to the period of the year. In fact low temperature can slow down enzymatic and microbial activity resulting in slower DNA degradation [24].

DNA detection and, as a consequence, DNA persistence estimation, is influenced by sampling and analysis strategy, other than environmental factors. The sampling and the analysis strategy must be extremely rigorous. Before any environmental DNA analysis, the reliability and the robustness of primers must be tested. First, the analysis must be performed *in silico* (e. g. using ecoPCR software [14]) in order to insure primer specificity (e. g. other species were not amplified at the same time as the species of interest) [30]. Once specific primers were found, reliabilty must be tested on very high quality DNA (e.g. extracted from tissue samples), and PCR conditions must be otpimized. Environmental DNA is rare and preacautions similar to those used for ancient DNA analysis must be taken [31]. DNA must be extracted in a dedicated room for rare DNA, mock samples without DNA have to be analysed in parallel, as well as positive samples. PCR cycles have to be increased and high attention must to be taken to avoid contamination. The optimum strategy to enhance the reliability is

to increase the number of analysed samples, i.e. more water samples in the field and more genetic replicates (multi-tube approach [15]) in the laboratory. However, all the sampling and analysis strategy must be adapted to the studied environments (e.g. large water bodies, marshes, etc) and species. In running waters, other sampling strategies will be developed, based e.g. on pumping water samples to increase DNA collection.

The dynamics of detectability reflects the persistence of DNA fragments in freshwater ecosystems. In this study we demonstrate that DNA persistence is less than one month. The short time persistence of detectable amounts of DNA opens new perspectives in conservation biology, by allowing access to the presence or absence of species e.g. rare, secretive, potentially invasive, or at low density. This knowledge of DNA persistence will greatly influence planning of biodiversity inventories and biosecurity surveys.

## Acknowledgments

We thank Christian Miquel, Delphine Rioux, Carole Poillot and Nathalie Tissot for the support and suggestions during the study.

## Author Contributions

Conceived and designed the experiments: TD CM. Performed the experiments: TD AV AD SP-C FP. Analyzed the data: TD AV AD FP. Lead writers of the paper: TD AV. Contributed to the writing of the paper: SP-C FP PT CM.

## References

1. Harvey CT, Qureshi SA, MacIsaac HJ (2009) Detection of a colonizing, aquatic, non-indigenous species. Divers Distrib 15: 429–437.
2. Margurran AE (2004) Measuring Biological Diversity. Malden: Blackwell Science.
3. Mehta SV, Haight RG, Homans FR, Polasky S, Venette RC (2007) Optimal detection and control strategies for invasive species management. Ecol Econ 61: 237–245.
4. Smith DR (2006) Survey design for detecting rare freshwater mussels. J N Am Benthol Soc 25: 701–711.
5. Ficetola GF, Miaud C, Pompanon F, Taberlet P (2008) Species detection using environmental DNA from water samples. Biol Letters 4: 423–425.
6. MacKenzie DI, Nichols JD, Sutton N, Kawanishi K, Bailey LL (2005) Improving inferences in popoulation studies of rare species that are detected imperfectly. Ecology 86: 1101–1113.

7. Valentini A, Pompanon F, Taberlet P (2009) DNA barcoding for ecologists. Trends in Ecol Evol 24: 110–117.

8. Valiere N, Taberlet P (2000) Urine collected in the field as a source of DNA for species and individual identification. Mol Ecol 9: 2150–2152.

9. Détaint M, Coïc C (2001) Invasion de la Grenouille taureau (*Rana catesbeiana* Shaw) en France: Synthèse bibliographique - Suivi 2000–2001 - Perspectives. Le Haillan (33): Association Cistude Nature. 30 p.

10. Jerde C, Mahon A, Chadderton W, Lodge D (2011) Sight-unseen detection of rare aquatic species using environmental DNA. Conserv Lett 4: 150–157.

11. Nichols JD, Thomas L, Conn PB (2009) Inferences About Landbird Abundance from Count Data: Recent Advances and Future Directions. In: Patil GP, ed. Modeling Demographic Processes In Marked Populations. New York: Springer. pp 201–235.

12. Garner TWJ, Perkins MW, Govindarajulu P, Seglie D, Walker S, et al. (2006) The emerging amphibian pathogen Batrachochytrium dendrobatidis globally infects introduced populations of the North American bullfrog, Rana catesbeiana. Biol Letters 2: 455–459.

13. Une Y, Sakuma A, Matsueda H, Nakai K, Murakami M (2009) Ranavirus Outbreak in North American Bullfrogs (Rana catesbeiana), Japan, 2008. Emerg Infect Dis 15: 1146–1147.

14. Taberlet P, Coissac E, Pompanon F, Gielly L, Miquel C, et al. (2007) Power and limitations of the chloroplast trnL (UAA) intron for plant DNA barcoding. Nucl Acids Res 35: e14.

15. Taberlet P, Griffin S, Goossens B, Questiau S, Manceau V, et al. (1996) Reliable genotyping of samples with very low DNA quantities using PCR. Nucl Acids Res 24: 3189–3194.

16. Burnham KP, Anderson DR (2002) Model Selection and Multimodel Inference: A Practical Information-Theoretic Approach. New York: Springer.

17. Team RDC (2010) R: A Language and Environment for Statistical Computing. Vienna, Austria: R Foundation for Statistical Computing.

18. Potè J, Ackermann R, Wildi W (2009) Plant leaf mass loss and DNA release in freshwater sediments. Ecotox Environ Safe 72: 1378–1383.

19. Levy-Booth DJ, Campbell RG, Gulden RH, Hart MM, Powell JR, et al. (2007) Cycling of extracellular DNA in the soil environment. Soil Biol Biochem 39: 2977–2991.

20. Shapiro B (2008) Engineered polymerases amplify the potential of ancient DNA. Trends Biotechnol 26: 285–287.

21. Alvarez AJ, Yumet GM, Santiago CL, Toranzos GA (1996) Stability of manipulated plasmid DNA in aquatic environments. Environ Toxicol Water Qual 11: 129–135.

22. Matsui K, Honjo M, Kawabata Z (2001) Estimation of the fate of dissolved DNA in thermally stratified lake water from the stability of exogenous plasmid DNA. Aquat Microb Ecol 26: 95–102.

23. Romanowski G, Lorenz MG, Sayler G, Wackernagel W (1992) Persistence of free plasmid DNA in soil monitored by various methods, inclding a transformation assay. Appl Environ Microb 58: 3012–3019.

24. Zhu B (2006) Degradation of plasmid and plant DNA in water microcosms monitored by natural transformation and real-time polymerase chain reaction (PCR). Water Res 40: 3231–3238.

25. Deagle B, Eveson JP, Jarman S (2006) Quantification of damage in DNA recovered from highly degraded samples - a case study on DNA in faeces. Front Zool 3: 11.

26. Willerslev E, Cappellini E, Boomsma W, Nielsen R, Hebsgaard MB, et al. (2007) Ancient biomolecules from deep ice cores reveal a forested Southern Greenland. Science 317: 111–114.

27. Willerslev E, Cooper A (2005) Ancient DNA. P Roy Soc B-Biol Sci 272: 3–16.

28. Ravanat JL, Douki T, Cadet J (2001) Direct and indirect effects of UV radiation on DNA and its components. J Photoch Photobio B 63: 88–102.

29. Chen I, Dubnau D (2004) DNA uptake during bacterial transformation. Nat Rev Microbiol 2: 241–249.

30. Ficetola GF, Coissac E, Zundel S, Riaz T, Shehzad W, et al. (2011) An In silico approach for the evaluation of DNA barcodes. BMC Genomics 11: 434.

31. Cooper A, Poinar HN (2000) Ancient DNA: Do it right or not at all. Science 289: 1139–1139.

# 20

# Climate Change Simulations Predict Altered Biotic Response in a Thermally Heterogeneous Stream System

**Jacob T. Westhoff**[1]*, **Craig P. Paukert**[2]

1 Missouri Cooperative Fish and Wildlife Research Unit, Department of Fisheries and Wildlife Sciences, University of Missouri, Columbia, Missouri, United States of America,
2 U.S. Geological Survey, Missouri Cooperative Fish and Wildlife Research Unit, University of Missouri, Columbia, Missouri, United States of America

## Abstract

Climate change is predicted to increase water temperatures in many lotic systems, but little is known about how changes in air temperature affect lotic systems heavily influenced by groundwater. Our objectives were to document spatial variation in temperature for spring-fed Ozark streams in Southern Missouri USA, create a spatially explicit model of mean daily water temperature, and use downscaled climate models to predict the number of days meeting suitable stream temperature for three aquatic species of concern to conservation and management. Longitudinal temperature transects and stationary temperature loggers were used in the Current and Jacks Fork Rivers during 2012 to determine spatial and temporal variability of water temperature. Groundwater spring influence affected river water temperatures in both winter and summer, but springs that contributed less than 5% of the main stem discharge did not affect river temperatures beyond a few hundred meters downstream. A multiple regression model using variables related to season, mean daily air temperature, and a spatial influence factor (metric to account for groundwater influence) was a strong predictor of mean daily water temperature ($r^2 = 0.98$; RMSE = 0.82). Data from two downscaled climate simulations under the A2 emissions scenario were used to predict daily water temperatures for time steps of 1995, 2040, 2060, and 2080. By 2080, peak numbers of optimal growth temperature days for smallmouth bass are expected to shift to areas with more spring influence, largemouth bass are expected to experience more optimal growth days (21 – 317% increase) regardless of spring influence, and Ozark hellbenders may experience a reduction in the number of optimal growth days in areas with the highest spring influence. Our results provide a framework for assessing fine-scale (10 s m) thermal heterogeneity and predict shifts in thermal conditions at the watershed and reach scale.

**Editor:** Michael Sears, Clemson University, United States of America

**Funding:** This project was funded through the United States Geological Survey Natural Resources Preservation Project through Research Work Order 116 of the Missouri Cooperative Fish and Wildlife Research Unit (http://www.nature.nps.gov/challenge/2011/nrpp.cfm). The funders had no role in study design, data collection and analysis, decision to publish, or preparation of the manuscript.

**Competing Interests:** The authors have declared that no competing interests exist.

* Email: westhoffj@missouri.edu

## Introduction

The ecological importance of water temperature to aquatic organisms has been the impetus for numerous studies that sought to develop predictive temperature models for various systems [1–3]. Many external drivers interact with the physical properties of rivers to determine water temperature and include air temperature, solar radiation, relative humidity, wind speed, riparian shade, cloud cover, solar angle, discharge, tributary contributions, and groundwater contributions [4–6]. However, it is often not feasible to obtain information on all of these factors for an aquatic system of interest, especially at the fine spatial scales required to document thermal heterogeneity. Lotic systems heavily influenced by groundwater inputs create spatially heterogeneous thermal environments and are difficult to explain with coarse-scale temperature models [7,8]. Progress has been made to address the heterogeneity of stream water temperatures at finer spatial scales [9–12], but collecting appropriate data to parameterize models that can be applied over long distances (100 s km) at a fine-scale spatial resolution (10 s m), while accounting for seasonal variation, is difficult.

Groundwater springs occur in patchy distributions around the globe and provide unique physical and chemical environments that support many biological assemblages [13,14]. Systems with significant groundwater input or cold-water tributaries serve as thermal refuges for aquatic species [15–20]. Further, certain species can exist in groundwater fed systems that may not be able to survive in geographically proximate systems lacking groundwater influence [21,22]. Species composition within and outside of springs is also known to differ, enhancing beta diversity in the system [14]. At a coarse spatial scale, groundwater can influence the distribution and abundance of aquatic organisms [23,24] and reduce the occurrence of temperature fluctuations that may result in reproductive failure of certain species [25]. At fine spatial scales, some fishes show behavioral responses to thermal refuges by selecting spawning locations [26,27], avoiding ice break up and frazil ice [28], or thermoregulating by occupying groundwater influenced areas during warm or cool water periods [16,19,20,29].

Climate change has the potential to alter environmental conditions in streams in many ways, but especially the physical properties associated with discharge and water temperature [30,31]. Altered environmental conditions in aquatic systems may result in physiological effects [32,33], behavioral or competitive effects [34,35], or shifts in the distribution and abundance of aquatic organisms [36–38]. Some of the effects of climate change may be buffered in thermally heterogeneous stream systems with high levels of groundwater influence [28,39]; however, little information exists linking predicted water temperatures to thermal requirements of aquatic organisms in these systems [40,41]. Climate change is frequently listed as a threat to groundwater-dependent biota, but direct quantification of potential effects is less common [42,43]. Efforts to predict climate change effects on thermal conditions in streams at a regional scale often do not include predictive variables that specifically account for groundwater influence, especially at fine-spatial scales, which can result in models with limited inference for heavily groundwater influenced systems [7,9,44]. Because of the importance of groundwater influenced systems, it is important to quantify their dynamics and how they might respond to a changing climate so that managers tasked with protecting biodiversity in the face of climate change proceed effectively [45,46].

Our goal was to develop an approach that could be used to model water temperature in groundwater dominated lotic systems that are of high conservation importance and do not conform with coarse-scale temperature modeling approaches. Further, we wished to explain thermal heterogeneity within a mainstem river system heavily influenced by groundwater inputs and predict how biota of high conservation concern may respond. To achieve these goals, we addressed four main objectives: 1) document longitudinal variation in stream temperature at the warmest and coldest times of the year, 2) create a spatially-explicit temperature model based on empirical data to predict daily average temperature, 3) apply the predictive model to forecast the effects of two climate change scenarios on water temperature, and 4) link predicted water temperatures based on climate change simulations to three aquatic organisms of concern to conservation and management but that have different temperature requirements.

## Methods

### Study Location

Our study occurred within the Ozark National Scenic Riverways (ONSR), which is a National Park Service Unit located in south-central Missouri, USA (37° N, 91° W) on the Ozark Plateau [47]. The park encompasses approximately 32,700 hectares, creating a narrow corridor along 215 km of the Current River and its largest tributary, the Jacks Fork River (Figure 1). The Current River is a southerly flowing stream which enters the ONSR as a 4th order [48] stream and reaches 6th order upon its exit, whereas the Jacks Fork River is an eastern flowing 5th order stream within the ONSR. Average (range) wetted channel width in the Current River was 47.6 m (17.5 – 127.3 m) and 26.3 m (12–49 m) in the Jacks Fork (J. Westhoff, unpublished data). The deepest pools in the Current and Jacks Fork Rivers rarely exceed 5 and 3 m, respectively (J. Westhoff, unpublished data). Substrate composition in the river channel was generally dominated by coarse chert gravel or large boulders associated with bluff pools or high gradient reaches [49]. The riparian zone was dominated by deciduous forest and was mostly intact along the entirety of the river contained with the ONSR. The overall catchment was primarily forested with 14% of the catchment in cleared land, only 2% of which is on areas with > 7° slopes [50].

The ONSR is characterized by deep valleys overlaying karst topography, which creates many caves and springs. Big Spring, one of the largest springs in the world, is located on the Current River and has an estimated annual mean discharge of 12.6 m³/sec [47]. Many other large springs exist along the Current and Jacks Fork Rivers (Appendix S1) and groundwater sources account for over 90% of the total discharge within the ONSR [47]. Throughout the remainder of the manuscript, the acronym "ONSR" will refer only to the mainstem Current and Jacks Fork Rivers within the park. Field research for this study was completed under permit OZAR-2011-SCI-0007 from the United States National Park Service.

### Data Collection

Longitudinal temperature surveys were conducted by boat during winter (Jan 18 – Feb 23, 2012) and summer (July 30 – Aug 15, 2012) over the entire ONSR. Temperatures were obtained over multiple days during daylight hours and in different sections of the ONSR from 10 – 25 km long each day, depending on river access locations. For the winter survey, temperature was recorded using an Aqua Troll 100 (In-Situ Inc., Fort Collins, CO; accuracy ±0.1°C) by recording temperature approximately 10 cm below the water surface every 30 seconds while moving in a downstream direction. Temperature values were spatially linked using a GPS (Archer Field PC with Hemisphere GPSXF101, Juniper Systems Inc., Logan UT) with time settings synced to the Aqua Troll 100 device to record UTM coordinates every 30 seconds. During summer, water temperatures were taken 10 cm below the water surface every 250 m along the ONSR using an YSI 550A (YSI Inc., Yellow Springs, OH; accuracy ±0.3°C) and linked with UTM coordinates.

Longitudinal surveys conducted in the winter and summer captured spatial variation in temperature over the entire ONSR, whereas temporal variation in temperature was captured using stationary temperature loggers (HOBO Pendant, Onset Computer Corp., Cape Cod, MA; accuracy ±0.53°C). Loggers were installed at 26 locations throughout the ONSR and were generally located within a few hundred meters of an established river access location or just above and below major groundwater inputs (Figure 1 and Appendix S2). The average (± Std. Dev) distance between loggers was 9.3 ± 5.6 km. Temperature logger installation and removal dates varied, but 25 loggers recorded data every 30 minutes throughout calendar year 2012 (Appendix S2).

### Longitudinal Water Temperature Variation Analysis

The use of three different temperature recording devices necessitated correction of temperature readings. Laboratory derived correction factors resulted in the addition of 0.1°C and 1.0°C to raw temperature values for the Aqua Troll 100 and YSI 550A, respectively, to achieve standardization with the HOBO Pendant loggers. Longitudinal temperature survey data were collected at various times of day and on different days, which required correction of values and standardization to the same moment in time for valid comparisons. Therefore, we used data collected from stationary temperature loggers to address issues of both temporal and spatial standardization. We collected information on the closest upstream and downstream temperature loggers for each longitudinal survey point, the distance to those loggers, and the temperature of those loggers at the time the survey point occurred. The distances from the survey point to each logger were used to create a weight for the logger specific to each survey point based on inverse distance weighting (IDW), where a factor of −2 was used for the exponent [51]. The difference in temperature between expected and observed values based on interpolation of

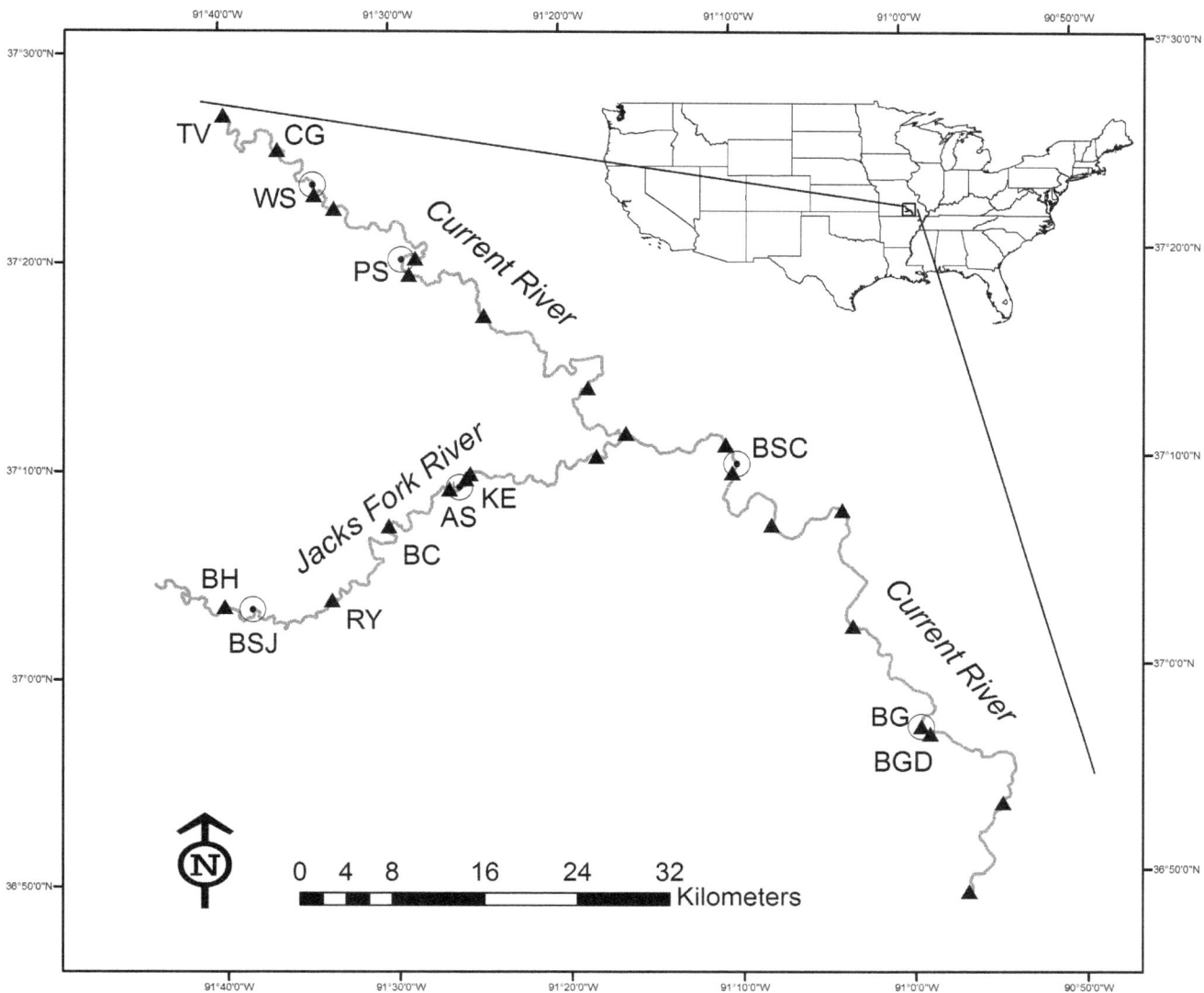

**Figure 1. Location of the Ozark National Scenic Riverways (Jacks Fork and Current Rivers).** Triangles indicate locations of stationary temperature loggers deployed for the entirety of 2012, with locations mentioned in the text noted by abbreviations (Tan Vat, TV; Cedar Grove, CG; Big Spring Downstream, BGD; Buck Hollow, BH; Rymers, RY; Bay Creek, BC, and Keatons, KE). Hollow circles surrounding a point indicate the locations of major springs referenced in the text (Welch Spring, WS; Pulltite Spring, PS; Blue Spring Current, BSC; Big Spring, BG; Blue Spring Jacks Fork, BSJ; and Alley Spring, AS).

temperature logger values was termed the spatial correction factor (SCF) and was calculated with the following equation,

$$SCF = T_{S_i} - \frac{(U * T_{U_i}) + (D * T_{D_i})}{U + D};$$

Where $Ts_i$ was the temperature recorded at survey point $i$, $U$ was the weight of the closest upstream logger to point $i$, $D$ was the weight of the closest downstream logger to point $i$, $Tu_i$ was the temperature recorded at the closest upstream logger at the time when temperature was recorded at survey point $i$, and $Td_i$ was the temperature recorded at the closest downstream logger at the time when temperature was recorded at survey point $i$.

Next, to standardize all the measured temperature readings to a single point in time, the weighting process was repeated using the extreme (either minimum or maximum, depending on season of survey) temperature values for the two loggers closest to the survey point. The mode of the coldest day and time observed (January 14, 2012 at 0830 hours) and the mode of the warmest day and time observed (July 25, 2012 at 1700 hours) were selected from the 25 stationary logger datasets and the values for those days served as the values to which all others would be standardized. This weighting provided a predicted high or low temperature value for that survey point location which was then added to the spatial correction factor. These final values represented temporally and spatially corrected temperature values for every survey point for the coldest and warmest time of the year. If a major groundwater feature entered the system, the IDW methodology varied in that the nearest logger may not have been used (e.g., a point just downstream of a major spring would have none of its weight based on a logger above the spring).

## Relation of Spring Magnitude to Spring Influence

Each spring in a system contributes a different percentage of discharge to the total river discharge [47] and affects the water temperature for various distances downstream [47,52]. This relationship between spring discharge and distance from spring is complex, especially when considered in conjunction with other spatially varying factors that affect water temperature [5]. Knowledge of individual thermal contribution from a given spring to the river was investigated based on observed longitudinal temperature variation. We selected nine springs that contributed at least 5% of the river discharge and used information obtained from our summer maximum and winter minimum estimates to determine the relationship between spring magnitude (M) and spring influence (I). Spring magnitude was defined as the discharge of the spring divided by the discharge of the river at its confluence with the spring and calculated with data from Mugel et al. [47]. Spring influence was calculated for summer ($SI_{Max}$) and winter calculated winter ($SI_{Min}$) conditions using the corrected temperature values described in the previous section (maximum or minimum corrected temperatures) following the equations;

$$SI_{Max} = \frac{(U_S - D_S)}{(U_S - G)};$$

and

$$SI_{Min} = \frac{(D_W - U_W)}{(G - U_W)};$$

where $U_S$ is the temperature upstream of a given spring in the summer, $U_W$ is the temperature upstream of a given spring in the winter, $D_S$ is the temperature downstream of a given spring in the summer, $D_W$ is the temperature downstream of a given spring in the winter, and G is the temperature of the spring water ($14°C$ was used in our analysis). This allowed us to display the spring influence relative to how much temperature change was possible given the difference between river temperature and groundwater temperature. Spring influence values of 0 indicated no influence and a value of 1 indicated complete spring influence. Linear regression was used to determine the relationship between spring magnitude and spring influence for both summer and winter conditions.

## Predictive Water Temperature Modeling

We used multiple regression to create a predictive water temperature model for the main stem Jacks Fork and Current Rivers within the ONSR. We chose a statistical approach because of data availability and concerns about addressing the spatial variation caused by groundwater influence. Statistical models are based on correlations among water temperature, air temperature, and other factors and often have less comprehensive data requirements than deterministic models [6,53–55]. Deterministic models rely on the physical properties of water and heat exchange and require large amounts of meteorological and hydrological data [56,57].

Because the system was greatly influenced by groundwater, we standardized all air and water temperatures based on the average temperature of groundwater entering the system. Our approach follows the equilibrium temperature concept described by Mohseni & Stefan [37] which identified the point in time when no heat is transferred between air and water; however, we instead focused on heat transfer between groundwater and river water. Instead of estimating heat flux using data on radiation, evapora-

tion, and other physical properties of heat exchange in water, we assumed a constant groundwater temperature of $14°C$ and subtracted that value from observed air and water temperature values to create a linear transformation of the response of water temperature to any groundwater input. Discharge was not included as a predictor in the model, but was controlled for by removing any observations from days when discharge exceeded the 75% percentile of records for the closest USGS gage station.

We accounted for spatial influences on water temperature by creating a spatial influence factor (SIF). Numerous spatial drivers of water temperature (e.g., stream size, land use, riparian coverage) were accounted for using the SIF, but the relative effects of any one driver were unknown. However, the primary spatial driver in this system captured by the SIF was groundwater influence. The SIF was calculated based on results from the longitudinal temperature survey conducted in the summer of 2012 and the resulting spatial correction values. The corrected temperature at each survey point along the longitudinal transect was subtracted from the warmest temperature that occurred in the ONSR ($31.9°C$). Resulting values were then divided by the greatest observed difference in temperature between any two points in the system ($13.7°C$) to standardize them for the ONSR. Resulting values ranged from 0 (no spring influence) to 1 (greatest spring influence). We determined the SIF values for every river reach in the Jacks Fork and Current Rivers to determine the composition of SIF values within the ONSR. The SIF value differs from the spring influence value because it incorporates all spatial drivers of water temperature and can be determined for any location on the river, not just below a spring.

Air temperature is a strong predictor of water temperature in lotic systems, but generally performs best when considered as a moving average as opposed to an instantaneous value congruent with the water temperature reading [58]. Thus, we used five day average daily air temperature (AirTemp) where the average daily air temperatures on the day of the reading and the four previous days were averaged using weighting factors. Daily air temperatures were averaged by multiplying the average temperature on the day of the observation by 0.3, the prior day by 0.3, the next previous day by 0.2, and the other two previous days by 0.1. All air temperature data were from the Round Spring weather station which was centrally located in the ONSR and within 70 km of the furthest location on the river to which the model applies.

Finally, we accounted for seasonal effects on water temperature with a metric (Season) based on climate normals and how they related to groundwater temperature. Climate normals [59] were obtained from the West Plains, MO weather station (40 – 80 km southwest of ONSR) as opposed to the Round Springs weather station because the data record had fewer missing values and a longer record. The average daily air temperature value for each calendar day was subtracted from $14°C$; thus, the metric had a value of zero when air temperatures were equal to groundwater temperatures (Mid April and Mid October). Days with average temperatures above $14°C$ had negative values and days below $14°C$ had positive values, which mimicked the effect of groundwater input on the ambient water temperatures. The interaction between season and the SIF was of interest because multiplying the SIF by the groundwater influence resulted in season values that were essentially weighted by the amount of spring influence (Figure 2). This interaction accounted for both the direction and magnitude of the effect of groundwater influence at a given air temperature.

Temperature data from each of 25 temperature loggers were summarized by determining the average daily water temperature for each calendar day of 2012. Those values were standardized to

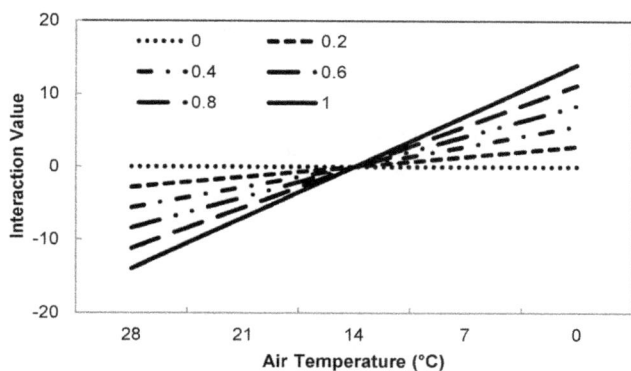

**Figure 2. Simulated values of the interaction term when spring influence is multiplied by season influence under six levels of spring influence (0, 0.2. 0.4. 0.6, 0.8, and 1; 0 = no spring influence, 1 = heavy spring influence).** Results are displayed by unadjusted air temperature for clarity, but air temperature values used in calculations were subtracted from 14°C.

groundwater temperature (as described above) and served as the response variable in our models. Data from all logger locations over all days in 2012 were compiled into one dataset and records were removed if discharge exceeded the 75% quartile during that day at that location, or if data were corrupt or missing. From the reduced dataset, 10% of the records were randomly selected to serve as training data for the models.

We used multiple regression techniques to develop a temperature model that predicted daily average water temperatures throughout the ONSR. Combinations of predictor variables were used to create six candidate models, which were compared based on their Akaike Information Criterion (AIC) values to select the best model (Table 1). These candidate models were chosen to examine the explanatory power of each of the variables by itself, and in combination. Model parameter estimates and intercepts were used along with the data withheld from the model creation set (90% of total observations) to validate the final model. All analyses were done in SAS 9.3 (Cary, NC).

## Climate Change Simulations and Predicted Biotic Implications

Simulated air temperature values for a central location within the ONSR were obtained from dynamically downscaled climate simulations [60] and incorporated into our predictive temperature

model. Air temperature data (2-m above surface) were obtained from the downscaled versions of the MPI ECHAM5 (EH5) and GENMOM climate simulations [60]. These models were chosen because they provided data at a 15-km grid scale and to show a range of potential conditions within the A2 emissions scenario [60]. Air temperature estimates were averaged for each day of the year across five years (e.g., 2040 − 2044) for four time steps of 1995, 2040, 2060, and 2080. We calculated 5-day moving averages of the simulated air temperature values as outlined above, and substituted those values in our predictive model of average daily temperature in place of the AirTemp variable. We did not incorporate predicted changes in precipitation, discharge, or groundwater temperature based on climate simulations in our model. All other model inputs were identical to the original predictive temperature model.

We summarized forecasted results by applying ecologically important thermal criteria for three aquatic species that occur in the ONSR. Smallmouth bass *Micropterus dolomieu* and largemouth bass *Micropterus salmoides* are competitors [61], economically important sportfish [62], and possess different thermal tolerances and optimal growth temperatures (Table 2). The Ozark hellbender *Cryptobranchus alleganiensis bishopi* is a rare species of salamander listed by the United States Endangered Species Act. Temperatures at which these organisms no longer exhibit growth due either to cold or warm water temperatures, along with the optimal temperature range for growth, and the range for potential growth (Table 2) were used in our models. The number of days per year with average daily temperature within these ranges (final predicted water temperatures were rounded to the nearest whole digit) was summed for each simulated climate change scenario and across the range of spatial influence factor values. These biologically relevant temperature estimations allowed us to examine potential change in thermal suitability and bioenergetic response associated with climate change while accounting for spatial heterogeneity of stream temperatures.

## Results

The greatest annual temperature variability (2.2 − 32.0°C) within the ONSR occurred at the upstream most location on the Jacks Fork River (Buck Hollow; Figure 1, Appendix S3), which is a location with very little groundwater influence. The lowest annual water temperature variation occurred at a site (Tan Vat) with a high degree of groundwater on the Current River, where water temperature ranged from 8.4 to 20.5°C. Alley Spring was the only location influenced entirely by groundwater with a temperature

**Table 1.** Candidate models used in multiple regression modeling to predict average daily water temperature in the Current and Jack's Fork Rivers, Missouri.

| Model Number | Model Structure | AIC | $w_i$ |
|---|---|---|---|
| 1 | $\beta_0 + \beta_1(\text{AirTemp})$ | 1245.2 | 0 |
| 2 | $\beta_0 + \beta_1(\text{Season})$ | 1470.0 | 0 |
| 3 | $\beta_0 + \beta_1(\text{SIF})$ | 2823.6 | 0 |
| 4 | $\beta_0 + \beta_1(\text{Season}) + \beta_2(\text{SIF}) + \beta_3(\text{Season*SIF})$ | 807.7 | 0 |
| 5 | $\beta_0 + \beta_1(\text{AirTemp}) + \beta_2(\text{Season}) + \beta_3(\text{SIF})$ | 879.0 | 0 |
| 6 | $\beta_0 + \beta_1(\text{AirTemp}) + \beta_2(\text{Season}) + \beta_3(\text{SIF}) + \beta_4(\text{Season*SIF})$ | −329.2 | 1 |

The variable AirTemp represents a five-day weighted moving average air temperature. The variable Season represents a value based on climate normal air temperatures, and the variable SIF (spatial influence factor) represents a spatial variation in water temperature caused by groundwater and other factors. $\beta_0$ represents intercept and $\beta_1$ represents slope. Akaike Information Criterion (AIC) values and model weights ($w_i$) are displayed for each candidate model.

**Table 2.** Ecologically important thermal criteria for smallmouth bass (*Micropterus dolomieu*), largemouth bass (*Micropterus salmoides*), and Ozark hellbender (*Cryptobranchus alleganiensis bishopi*) in the Ozark National Scenic Riverways.

| Species | Positive growth (°C) | Optimal growth (°C) | Sources |
|---|---|---|---|
| Smallmouth bass | 10–27 | 20–24 | [52,63] |
| Largemouth bass | 15–36 | 24–30 | [64–66] |
| Ozark hellbender | 3–27 | 10–16 | [67], consultation with experts |

logger and ranged from 13.4 to 15.7°C with an average annual temperature (± standard deviation) of 14.2±0.4°C. Overall, average annual temperature ranged from 14.2°C (Tan Vat) to 17°C (Rymers and Bay Creek) within the ONSR (Figure 1, Appendix S3).

## Longitudinal Water Temperature Variation

Spatial patterns in maximum and minimum temperatures observed from the longitudinal temperature surveys demonstrated the influence of groundwater input locations (Figure 3). The greatest effect of a single spring on the overall river temperature was observed downstream of Alley Spring on the Jacks Fork River, where summer maximum temperature decreased by approximately 10°C and winter minimum temperature increased by approximately 5°C (Figures 1 and 3). On the Current River, Welch Spring had the most influence on river temperature based on the 7°C decrease of maximum temperature and 3°C increase of minimum temperature (Figures 1 and 3). Other major springs that had greater than 1°C influence on stream temperature included Pulltite, Blue (Current), Big, and Blue Springs (Jacks Fork). Groundwater influence on river temperature directly downstream of springs was not as substantial in the winter as in the summer. Equipment malfunctions resulted in a loss of data for approximately 25 km of the Current River combined over two locations (River km 66 – 75 and 122 – 138) during the winter of 2012 (Figure 3).

## Relation of Spring Magnitude to Spring Influence

Spring influence on main stem river temperature was strongly related to spring magnitude during both summer ($R^2 = 0.93$, intercept ± standard error = 3.2 ± 3.3, slope ± standard error = 99.1 ± 9.9) and winter ($R^2 = 0.68$, intercept ± standard error = 5.2 ± 8.8, slope ± standard error = 121.4 ± 34.2); Figure 4). As spring magnitude increased, the influence of groundwater on river water temperature increased. Intercept values from the models indicated that a single spring contributing less than 3% of the total discharge in the summer or 5% in the winter would have no observable influence on river temperature at a location beyond 400 m downstream.

## Predictive Water Temperature Modeling

We removed 547 temperature logger records (6%) because they exceeded the 75% quartile of discharge, with 95% of the removed records occurring between January 1 and March 26th. An additional 236 records (3%) were removed because they were corrupt or missing, resulting in 8367 valid records. The distribution of SIF values was uneven across the ONSR and had a distance weighted average SIF value of 0.40 (Figure 5). Approximately 75% of the ONSR had SIF values less than or equal to 0.50 (Figure 5). No model selection uncertainty existed as candidate model 6 (all variables and the Season*SIF interaction) had a model weight of one and indicated that the variables measured were more important than any single variable alone (Table 1). The average daily temperature model performed well based on R-squared (0.98) and RMSE (0.82) values, and compared favorably to other studies using RMSE as a validation

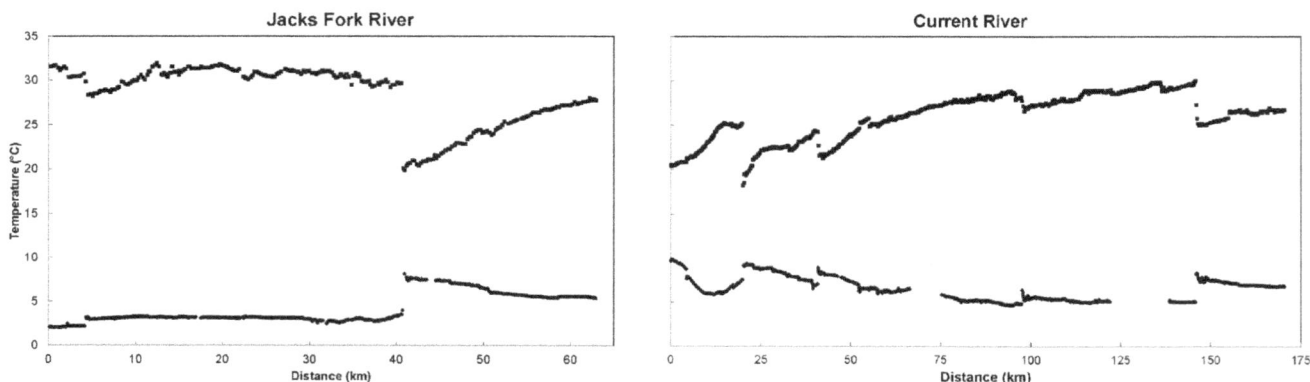

**Figure 3. Maximum (upper dots) and minimum (lower dots) predicted temperatures along the Jacks Fork (left panel) and Current Rivers (right panel) during the warmest (July 25 at 5 pm) and coldest (January 14 at 8 am) periods of 2012.** River distance for the Jacks Fork starts at the Buck Hollow access (0 km) and ends at the Current River confluence (63 km), with Blue Spring (4 km) and Alley Spring (41 km) accounting for major variation in temperature. River distance for the Current River starts at the Tan Vat access (0 km) and ends at the Gooseneck access (170 km), with Welch Spring (20 km), Pulltite Spring (40 km), Blue Spring (97 km), and Big Spring (146 km) accounting for major variation in temperature. Equipment malfunctions resulted in a loss of data during cold period sampling at two locations on the Current River (black boxes).

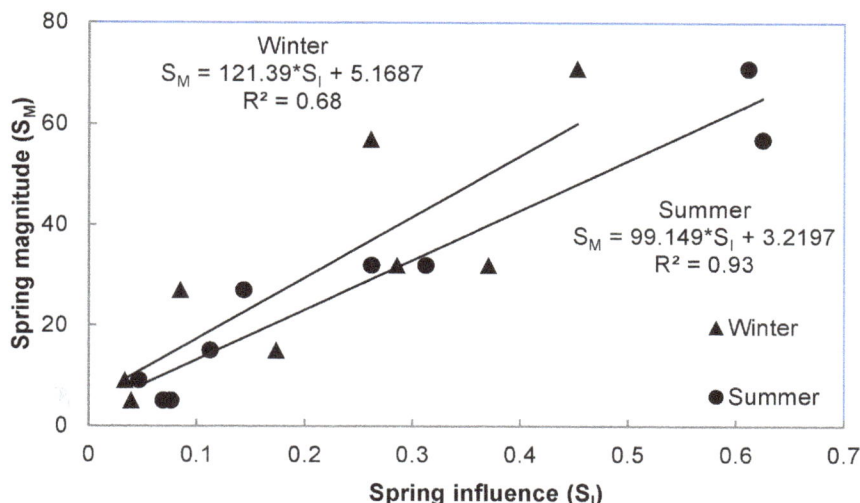

**Figure 4. Relationship between spring magnitude ($S_M$) and spring influence ($S_I$) for both winter (triangles) and summer (circles) observations.** Spring magnitude represents the percentage of discharge contributed to the river by the spring. Spring influence represents the percentage of change in water temperature from the upstream river temperature to groundwater temperature.

metric [68,69]. The final relationship to predict daily average water temperature (*WaterTemp*) was,

$$WaterTemp = 3.60 + 0.40(AirTemp) - 0.51(Season) - 3.61(SIF) + 0.66(Season \cdot SIF)$$

Associated standard error values for the parameter estimates were 0.05 (*Intercept*), 0.01 (*AirTemp*), 0.01 (*Season*), 0.11 (*SIF*), and 0.01 (*Season\*SIF*). We observed high Pearson Correlation Coefficients between AirTemp and Season ($-0.89$ to $-0.91$) and between Season and SIF (0.85). All other Pearson Correlation Coefficients were below 0.80. None of the predictor variables had variance inflation factor (VIF) values $>10$, indicating that multicollinearity did not produce problems in our regression coefficients [70].

Model validation indicated the greatest underestimate of temperature observed was 3.3°C and the greatest overestimate

was 3.4°C. The model predicted 98% of the observations within 2°C and 77% of the observations within 1°C. Estimated values 2°C warmer than observed came primarily (88%) from three sites (Buck Hollow, Cedar Grove, Tan Vat) and occurred in the summer (Figure 1). Estimates 2°C cooler than observed came primarily (82%) from two sites (Keatons and Big Spring Downstream) and occurred in the fall and winter (Figure 1).

## Climate Change Simulations and Predicted Biotic Implications

Average air temperature increased from the 1995 to 2080 time steps for all climate simulations (Table 3). The EH5 model predicted average air temperatures to increase by 2.8°C from 1995 to 2080, whereas the GENMOM model predicted a 2.1°C increase over the same time period. The average air temperatures at a given time for the EH5 model were commonly 2°C greater than those predicted by the GENMOM model. Mean yearly water temperatures were estimated to increase by 1.1°C from 1995 to

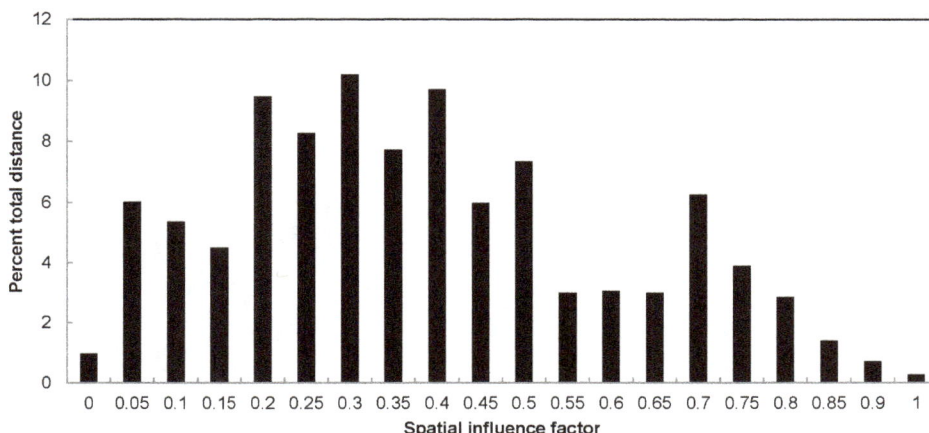

**Figure 5. Percent of Jacks Fork and Current River distance within the Ozark National Scenic Riverways comprised by each of the spatial influence factor (SIF) values ranging from 0 to 1, by increments of 0.05.**

2080 based on the EH5 model, versus only 0.8°C for the GENMOM model (Table 3). Areas that were highly buffered by groundwater inputs were predicted to have 3.5°C cooler yearly average temperatures than areas with minor groundwater influence (Table 3). The 2040 and 2060 simulations were intermediate between the 1995 and 2080 simulations, so further discussion of the 2040 and 2060 results was omitted. However, 2040 simulations were similar to 1995 results and noticeable shifts in temperature became apparent by 2060.

We observed similar trends among our simulated water temperature results regardless of climate scenario. The number of positive growth days for smallmouth bass increased with increasing SIF values (more groundwater) and plateaued near SIF values of 0.85 for most simulations (Figure 6A). Both climate models predicted an increase in the number of positive growth days from 1995 to 2080 for smallmouth bass at locations with SIF values greater than 0.15, with as many as 39 more days of growth per year (13% increase) at an SIF value of 0.50 in the EH5 simulation. These patterns were explained by the reduced number of days too cold for smallmouth bass growth at moderate to high (>0.15) SIF values and the increased number of days too warm for growth at low (<0.15) SIF values (Figure 7). The GENMOM simulations indicated a similar trend to the EH5 simulations, but there were on average 8% fewer positive growth days across SIF values above 0.15 predicted for the year 2080 than predicted by the EH5 simulations. Based on the 1995 simulations, smallmouth bass were expected to experience the maximum number of optimal growth days in areas of river with either a 0.25 (EH5) or 0.30 (GENMOM) SIF value (Figure 8A). However, by 2080, both climate models predicted that those maxima will shift to areas of the river with greater groundwater influence (0.40 SIF value).

The predicted largemouth bass response to climate change scenarios was less complex than that of smallmouth bass. The number of positive growth days for largemouth bass was predicted to increase from 1995 to 2080 due to fewer cold-limiting days, with the EH5 model again predicting the largest increase (Figure 6B). The GENMOM model predicted that the number of positive growth days for largemouth bass will increase by 2080 from 7 to 37% (11 to 40 days per year), depending on SIF value. The EH5 model predicted large gains (up to 70% more days) in the number of positive growth days for high SIF value areas and smaller gains (as low as 3%) for low SIF value areas (Figure 6B). The number of optimal growth days for largemouth bass may increase by 20 days (21%) in areas with SIF values of zero to as much as 57 days a year

(317%) in areas with SIF values near 0.3, based on the EH5 simulations for 1995 to 2080 (Figure 8B). The GENMOM simulations predicted a similar trend, but of lesser magnitude (5 to 33 days of additional optimal growth per year). The EH5 1995 model predictions indicate that in areas where SIF values are 0.15 or less, largemouth bass experience more days of optimal growth than do smallmouth bass. However, largemouth bass currently have three more days per year of optimal growth in areas where SIF = 0.15, and by 2080 that difference will increase to 23 days per year, which may have implications for interspecific competition (Figures 8A and B). All simulations indicated that areas with SIF values above 0.5 would not support any optimal growth days for largemouth bass due to water temperatures below 24°C.

The large range of temperatures at which hellbenders may experience positive growth resulted in negative growth days (< 3 or > 27°C) only in low spring influence area (SIF values < 0.15), and never for more than 48 days (13%) of a given year. The EH5 and GENMOM simulated trends were similar to each other for forecasted hellbender optimal growth days where both models predicted increasing optimal growth days with increasing SIF values (Figure 8C). Both models predicted fewer optimal growth days in 2080 than in 1995 for locations with either low (<0.25) or high (>0.85) SIF values, with up to a 20% reduction in optimal growth days in areas with SIF values of 1 (Figure 8C). The GENMOM and EH5 models predicted more optimal growth days in the future for mid-range SIF values (0.25–0.55 and 0.35 – 0.85; respectively) due to warmer winter temperatures; however, neither model predicted an increase of more than 27 days.

## Discussion

Our study provides a framework to document and predict fine-scale heterogeneous thermal conditions in lotic systems and complements other temperature modeling work typically conducted at larger spatial scales on surface-water systems. Our methods were based on easily obtainable data and accessible statistical techniques familiar to many biologists and managers. Characterization of thermal heterogeneity in the system allowed us to predict a probable shift in the spatial location of optimal growth temperatures available to two competing fish species under two climate change scenarios that otherwise might not have been identified by coarse-scale approaches.

**Table 3.** Summary temperature (°C) values for the Ozark National Scenic Riverways estimated using the MPI ECHAM5 (EH5) and GENMOM climate simulations at time steps of 1995, 2040, 2060, and 2080.

| Simulation | Time-step | Average Air Temperature | Average Water Temperature (High SIF, Low SIF) |
|---|---|---|---|
| EH5 | 1995 | 12.2 | 13.3, 16.8 |
| | 2040 | 13.0 | 13.6, 17.1 |
| | 2060 | 14.2 | 14.1, 17.6 |
| | 2080 | 15.0 | 14.4, 17.9 |
| GENMOM | 1995 | 10.9 | 12.8, 16.3 |
| | 2040 | 11.2 | 12.9, 16.4 |
| | 2060 | 12.1 | 13.2, 16.8 |
| | 2080 | 13.0 | 13.6, 17.1 |

Low spatial influence factor (SIF) indicates areas with little groundwater influence (SIF = 0) and high SIF indicates areas with high groundwater influence (SIF = 1).

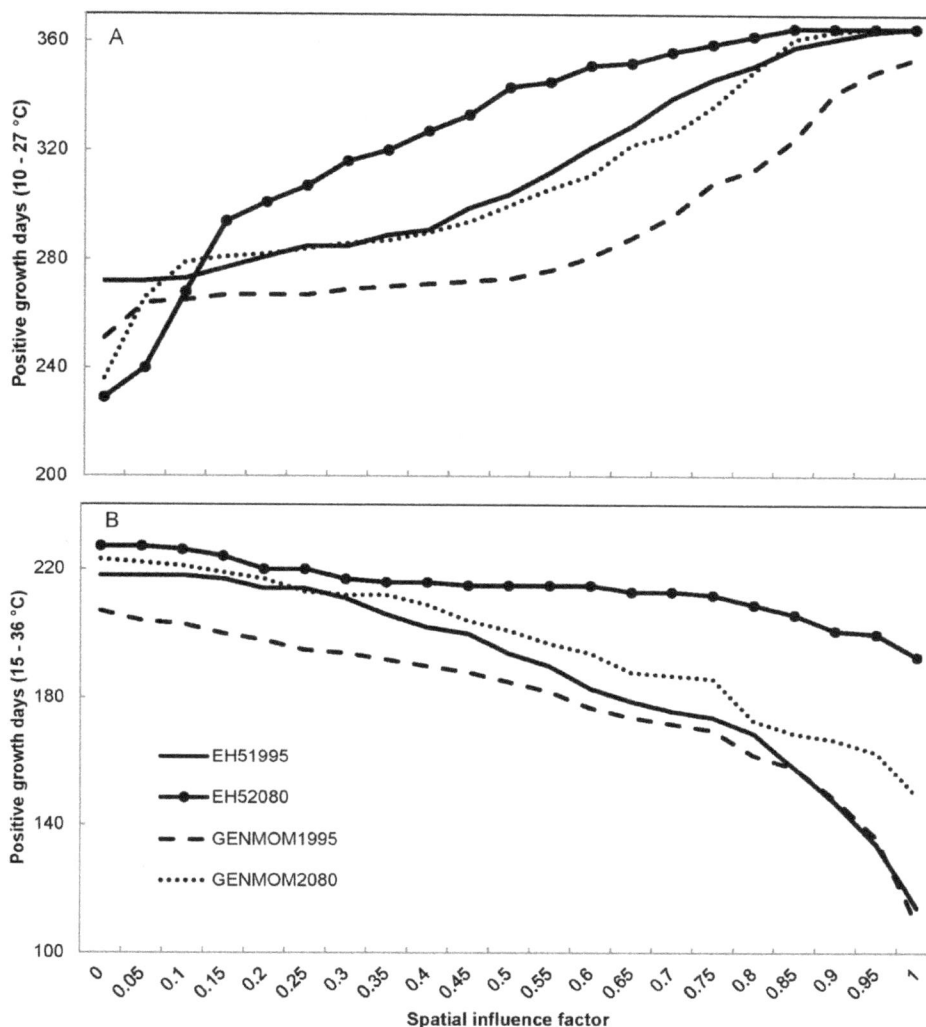

**Figure 6. Predicted number of positive growth days for smallmouth bass** *Micropterus dolomieu* **(Panel A) and largemouth bass** *Micropterus salmoides* **(Panel B) in the Ozark National Scenic Riverways displayed by spatial influence factor values for two climate scenarios (EH5 and GENMOM) during 1995 and 2080.**

## Temperature modeling approach

Scientists often consider multiple approaches to thermal modeling of stream networks and are faced with decisions regarding which predictor variables can and should be used [6]. These decisions sometimes result in the exclusion of groundwater streams from the dataset [7], post-hoc discussion of model limitations in relation to spatial variation caused by groundwater inputs [69,71], or no mention of potential groundwater influence [2,70]. We presented a novel approach to predict river temperatures using the spatial influence factor and readily available air temperature data to account for the combined effects of spatially important variables (e.g., groundwater input, stream size, amount of shading, tributary influence). Our modeling methods and the SIF could be easily adapted to use data from alternative methods for capturing broad-scale temperature variation data at fine resolutions such as distributed temperature sensing systems and thermal infrared imagery [72–76].

Our model indicated that groundwater had a significant effect on river water temperature and reduced water temperature variation; however, multiple linear regression models of water

temperature in other streams not influenced by groundwater often perform well using only air temperature and flow variables [77]. In a study of Pennsylvania streams, groundwater controlled the stream-air temperature relationship and reduced the coefficient of variation in water temperature relative to a stream with less groundwater influence [78]. More complex responses of groundwater inputs to river water temperatures were observed during the summer in a California stream, where a 3.7 km long shallow spring branch resulted in delivery of water to the stream that was warmer than the receiving water at night and cooler during the day [12]. This phenomenon was explained by solar radiation warming water in the spring branch during the day that did not arrive to the river until night [12]. The shorter (≤ 1 km) and heavily shaded spring branches in the ONSR are likely affected less by solar radiation and are more consistent with groundwater temperature when they enter the stream.

We believe the SIF approach could be scaled-up and applied across any system given concurrent temperature data are available from multiple locations for its creation. We applied it on two rivers, but the concept of capturing spatial variation in a single metric using empirical data could apply to a system of any size. Our use of the spatial correction factor to interpolate SIF values between permanent temperature logger locations was useful for our objectives, but the SIF could be based solely on stationary temperature logger data if fine-scale resolution is not required. The SIF was based only on summer temperature maximums and although we had information on winter temperature extremes, we chose to use only the summer values because they showed more spatial variation. Using a separate SIF based on data from each season may better account for temporal relations of spatially important factors. This is especially important if seasonal relations between spatial factors and water temperature depart significantly from linearity.

Our temperature modeling approach had several assumptions and limitations. First, we chose to use a static value for groundwater temperature of 14°C. The influence of groundwater inputs may vary regionally based on groundwater temperatures and may change with future climatic conditions [79]. We observed a groundwater temperature at Alley Spring of approximately 14.2°C, which is 0.8°C higher than mean annual air temperature [59]. This was consistent with Mohseni & Stefan [37] who found that groundwater temperature is $1-2$°C higher than mean annual air temperature in a given region. Temporal changes in groundwater temperature could be incorporated in the model to improve accuracy. Second, our model did not account for changes in river discharge and excluded temperature data from high water events ($>75^{th}$ quartile discharge). Discharge and water temperature are closely linked and are often both used in predictive temperature models [80], but without data on runoff water temperature it would be difficult to reliably predict the effect of high water events on river temperature. Further, we assumed that high water events would likely result in acute responses to thermal conditions by aquatic organisms in the system and would therefore be less informative related to growth predictions than the static conditions we modeled. Finally, including multiple years of data to train and validate the model would incorporate greater seasonal and yearly variation. All of the data used in this study came from the calendar year 2012, which was one of the warmest and driest on record [59]. However, the warmest 5-day average period and coldest 5-day average period were within the range observed during the previous 30 years.

Our approach to predicting the effects of future climate on river temperature and the effects of those temperatures on the biota also have limitations. We chose to use two downscaled climate models that represent extremes within the A2 emissions scenario [60]. This approach was intended to display a range in possible response, but neither model may provide the best approximation of future air temperatures. Multi-model ensemble approaches have become common and may provide future climate data that are more robust to individual model assumptions and more accurately reflect future conditions [81]. Further, we only used models from the A2 emissions scenario which predicts a medium-high level of carbon dioxide emissions relative to other scenarios [82]. Thus, our simulations are likely intermediate to what might be expected based on scenarios that differ in global population projections, carbon emissions, and other factors. We also did not include predicted changes in groundwater temperature, spring or river discharge, or precipitation based on climate change scenarios. Changes in precipitation are predicted and may result in altered flow regimes and groundwater discharges [83–85]. Reduced groundwater discharge would be expected to result in warmer water temperatures in the summer and colder water temperatures in winter due to less advective heat flux between groundwater and surface water. Thus, the importance of the SIF variable would be reduced and water temperatures would likely be driven more by air temperature. Streams with less groundwater inflow than our system may experience increased thermal stress from the synchrony of low flow conditions and high temperatures because they lack the more static discharge provided by springs [80]. Finally, the air temperature values used in our climate simulations were based on one, 15-km grid location from within the study area. Topographic variation and other factors may result in air temperatures that differ from those we used in both the temperature model and climate simulations which may affect model accuracy.

### Ecological Relevance of Thermal Heterogeneity

Groundwater inflows, such as those we documented on the ONSR, exemplify the patchy nature of environmental conditions in lotic systems [86,87]. Large magnitude springs in the ONSR altered water temperatures in the stream by up to 10°C during the warmest part of the year, effectively creating thermal patches

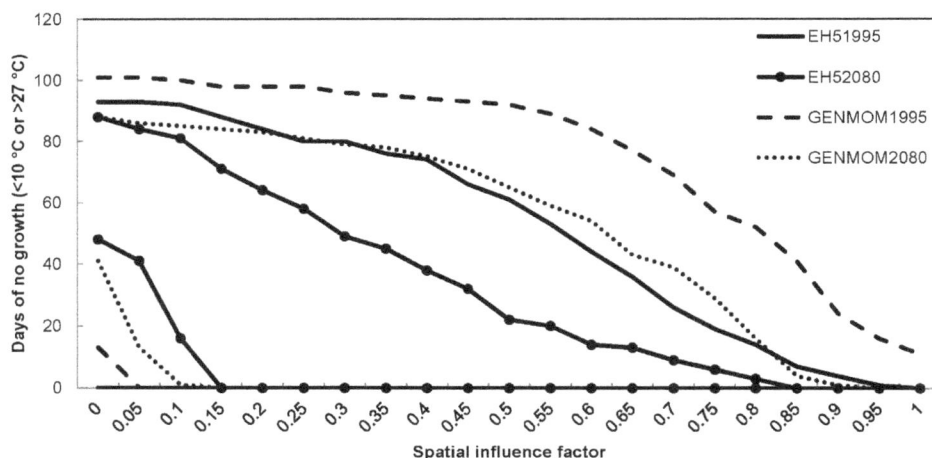

**Figure 7. Predicted number of non-growing days for smallmouth bass** *Micropterus dolomieu* **in the Ozark National Scenic Riverways displayed by spatial influence factor values for two climate scenarios (EH5 and GENMOM) during 1995 and 2080.** Days <10°C (too cold for growth) are displayed in the lower left corner and days >27°C (too warm for growth) are displayed as the top set of lines.

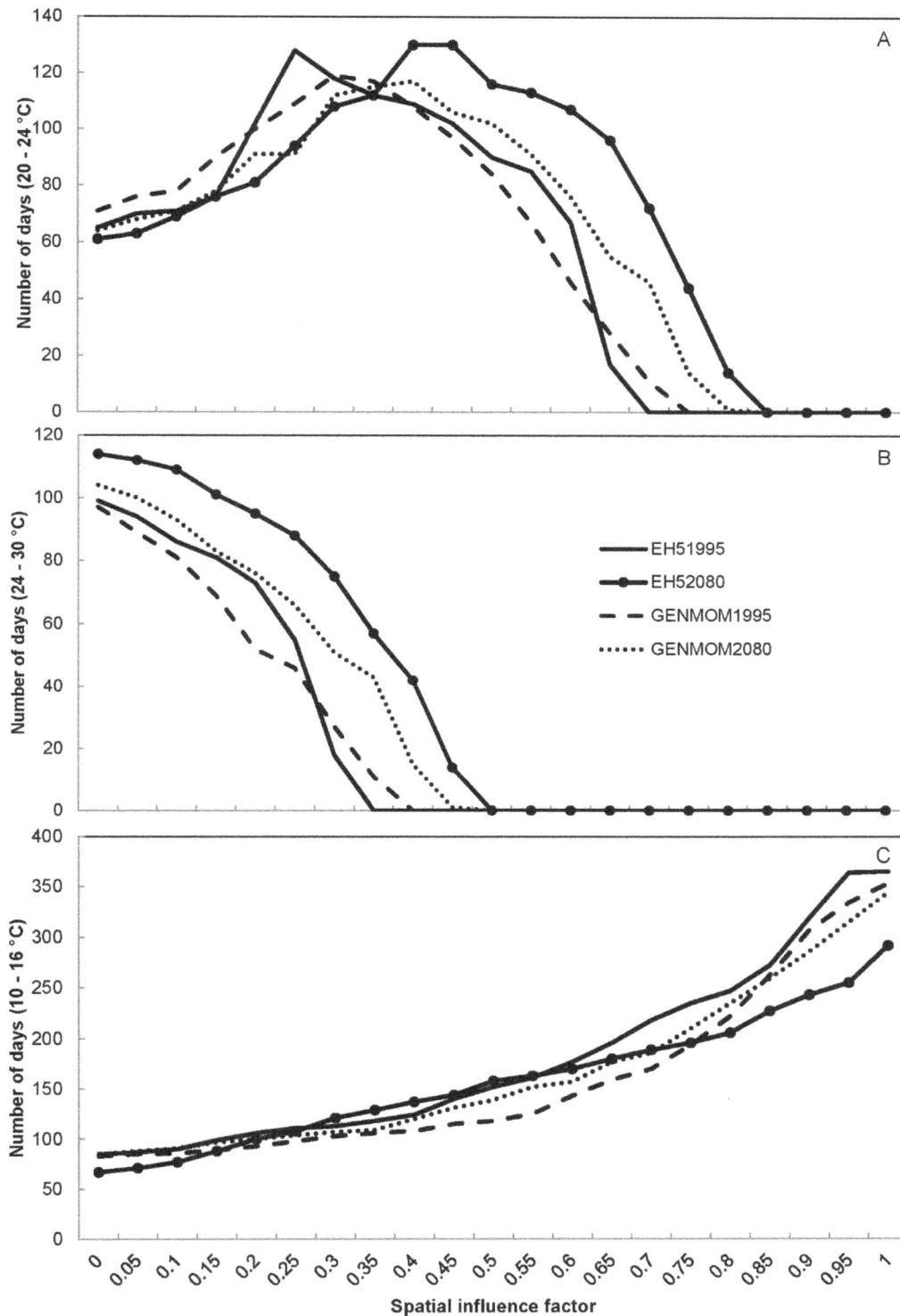

**Figure 8. Predicted number of optimal growth days for smallmouth bass** *Micropterus dolomieu* **(Panel A), largemouth bass** *Micropterus salmoides* **(Panel B), and Ozark hellbenders** *Cryptobranchus alleganiensis bishopi* **(Panel C) in the Ozark National Scenic Riverways displayed by spatial influence factor values for two climate scenarios (EH5 and GENMOM) during 1995 and 2080.**

which may affect the distribution of aquatic biota. For example in Tennessee, the Barrens topminnow *Fundulus julisia* is restricted to patchy environments created by groundwater springs [88]. Further, the physical processes that create patchy environments

are linked to patterns in biocomplexity and may influence assemblage patterns of aquatic organisms, meta-population dynamics, and biogeochemical processes [89–91]. Unlike patchy environments created by hydrodynamic processes, thermal patches are dynamic in a seasonally predictable way. Thus, the patches may disappear during certain times of the year when water temperature is near groundwater temperature. We observed a lower magnitude influence of groundwater on river temperatures in the winter as compared to the summer. This is important for modeling efforts as we suspect springs that contribute 3% (summer) or 5% (winter) of total discharge will have minimal effects on overall river temperatures beyond a few hundred meters.

Our results show that spring magnitude is positively related to spring influence on river temperature and is the primary spatial driver of water temperature. Whitledge et al. [52] examined groundwater influence on river temperature during the summer and noted a positive relationship between increasing spring discharge and the distance needed for water temperatures to warm to mean daily temperatures. They also concluded that riparian shading was important in Ozark Streams, but that shading alone did not affect water temperature more than 2°C and was less important than groundwater influences.

Stream water temperature can be modeled at spatial scales ranging from 1–10 m patches [19,92] to regional efforts covering thousands of km [9]. Our modeling efforts focused on a relatively fine-scale (10 s m grain size – 10 s km extent), which can provide area managers with detailed information on an entire system of interest and help identify the extent of potential thermal refugia for aquatic organisms. Our study did not identify thermal heterogeneity at very fine spatial scales ($1 - 10$ m$^2$ patches), but upwelling groundwater at that scale was used as thermal refuge by rainbow trout *Oncorhynchus mykiss* in Northwestern United States streams [19]. Despite the scale of observations, thermal refuges can also attract individual fish from great distances demonstrating the influence of groundwater beyond localized populations and single spatial scales, as evidenced by movements of >40 km by smallmouth bass (J. Westhoff, unpublished data).

## Thermal Heterogeneity and Biotic Response to Climate Change

Studies on the effects of climate change on fishes and other aquatic organisms have often focused on species distributional shifts at coarse spatial scales [36,93–95]. However, less is known about the potential fine-scale distributional responses by populations or individuals and it is expected that fish will experience population-level effects prior to coarse-scale range shifts [96,97]. Our study demonstrated that as the climate warms, smallmouth bass in groundwater-influenced rivers may need to occupy areas with greater spring influence to experience the maximum number of positive or optimal growth days. Failure to shift to optimal thermal conditions may result in decreased growth, as suggested for Great Lakes fishes [98]. Stream reaches with little spring-influence (SIF < 0.4) will become less thermally hospitable to smallmouth bass, while areas with moderate to high spring-influence (SIF > 0.4) will become more thermally suitable if air temperatures increase. We predicted that largemouth bass (a competitor of smallmouth bass) will also experience more optimal growth days in the future in areas of the river that were previously better thermally suited for smallmouth bass. Thus, as temperature regimes shift spatially, an increased competitive advantage may occur for species like largemouth bass in areas of low groundwater influence where resident smallmouth bass may experience more thermal stress and poorer growing conditions. Evidence for this already exists as largemouth bass were more successful than smallmouth bass in relatively warmer Ozarks streams [99]. Zweifel et al. [66] also predicted a similar trend for these species, but added that consumption demands for prey will favor largemouth bass over smallmouth bass in warmer systems. Species that are

normally dominant in an ideal thermal regime can experience reductions in growth when exposed to subdominant species in altered thermal regimes as observed in juvenile steelhead *Oncorhynchus mykiss* and Sacramento pikeminnow *Ptychocheilus grandis* [100]. Other organisms not buffered from climate change may be similarly tied to a combination of higher energetic costs from metabolic processes and altered interspecific community interactions [34]. Thus, aquatic organisms may need to adapt to changing water temperature or possibly face reduced fitness.

Climate change may have positive effects on certain species if water temperatures increase the number of days that occur within the optimal thermal range for a species [101,102]. Pease & Paukert [33] predicted that smallmouth bass in the midwest US (including Missouri Ozarks streams) would grow about 6% for every 1°C increase in water temperature, but will need 27% more food to reach that level of growth. Ideal growth conditions are maximized for hellbenders in areas with decreased temperature variability provided by large groundwater inputs. With warmer temperatures, the number of days too cold for hellbenders at moderate SIF values will decrease at a rate faster than the increase in the number of days that are too warm. This may provide marginal growth benefits for hellbenders occupying those sections of stream. However, we did not predict hellbenders in the ONSR to experience major shifts in distribution nor have significantly fewer optimal growth days (except in highly groundwater influenced habitats). Moreover, warmer temperatures may reduce the production of chytrid fungus *Batrachochytrium dendrobatidis* zoospores, which are produced in the greatest quantity at cold temperatures [103]. However, the thermal ranges we used to model hellbender response to climate change were not based on well-established and scientifically tested relationships, but rather our interpretation of available information. Thus, further study to refine the thermal ecology of hellbenders may better inform future models of hellbender response to climate change. .

Our study has implications for how climate change is assessed for fishes using thermal guilds at coarse-spatial scales. Fish thermal guilds are often used as a baseline or response variable to predict fish distributions based on current and future climate conditions [104,105]. However, stream systems with fine-scale heterogeneous thermal conditions may allow for species from multiple thermal guilds to occupy the same short (<10 km) reach of stream. For instance, following the methods of Wehrly et al. [104] to classify thermal conditions in streams, we noted three separate thermal categories (cold-, cool-, and warm-stable) present in the ONSR. Coarse-scale climate change predictions based on fewer temperature loggers would not likely detect this potential variation and could misrepresent the effects of climate change on fish communities.

## Conclusions

Understanding the spatial and temporal variation in water temperature in lotic systems provides an opportunity to better explore ecological phenomena and predict biotic response to change. We developed a methodology for assessing thermal heterogeneity and predicting water temperatures in the present and future with fine-scale spatial resolution. The SIF metric allowed us to predict daily average water temperature for over 200 km of stream heavily influenced by groundwater. The resulting model was then used to predict daily average water temperature based on future climate simulations. Those results demonstrated that smallmouth bass will likely need to shift their distribution closer to springs to experience maximum optimal growth conditions, largemouth bass will experience improved growing conditions in all sections of the stream, and hellbender salamanders will experience little change except near springs. Although we demonstrated potential uses of these data with three species, additional research avenues exist, including combining temperature predictions with habitat availability to determine suitability for various aquatic species and communities. We hope others will build on our results to better refine stream temperature models in groundwater systems.

## Supporting Information

**Appendix S1 Major springs located in the Ozark National Scenic Riverways, their discharges from Mugel *et al.*, (2009), and spatial locations.** UTM coordinates are in Zone 15 North, NAD 1983.
(DOCX)

**Appendix S2 Temperature logger locations and range of record for the Ozark National Scenic Riverways.** UTM coordinates are in Zone 15 North, NAD 1983.
(DOCX)

**Appendix S3 Summary data for temperature loggers in the Ozark National Scenic Riverways during the year**

**2012.** Bolded values should be interpreted with caution because they are based on less than the full year of data.
(DOCX)

## Acknowledgments

This research was a contribution of the Missouri Cooperative Fish and Wildlife Research Unit which is jointly sponsored by the Missouri Department of Conservation, the University of Missouri, the U.S. Geological Survey, the U.S. Fish and Wildlife Service, and the Wildlife Management Institute. The study was done in collaboration with the National Park Service. Any use of trade, product or firm name was for descriptive purposes only and does not imply endorsement by the U.S. Government. Data collection assistance was provided by T. Boersig III, J. Knerr, T. Schepker, N. Sievert, and J. Westhoff. Project design was aided by V. Grant, R. Jacobson, C. Rabeni, and J. Whittier. C. Jachowski, J. Briggler, C. Schuette provided hellbender thermal ecology information. The manuscript was improved by comments from Ty Wagner and an anonymous reviewer.

## Author Contributions

Conceived and designed the experiments: JW CP. Performed the experiments: JW. Analyzed the data: JW CP. Contributed reagents/materials/analysis tools: JW CP. Contributed to the writing of the manuscript: JW CP.

## References

1. Benyahya L, Caissie D, St Hilaire A, Ouarda TBMJ, Bobée B (2007) A Review of Statistical Water Temperature Models. Can Water Resour J 32: 179–192.
2. Jones LA, Muhlfeld CC, Marshall LC, McGlynn BL, Kershner JL (2013) Estimating thermal regimes of bull trout and assessing the potential effects of climate warming on critical habitats. River Res and Appl 30: 204–216.
3. Hague MJ, Patterson DA (2014) Evaluation of statistical river temperature forecast models for fisheries management. N Am J Fish Manage 34: 132–146.
4. Isaak DJ, Hubert WA (2001) A hypothesis about factors that affect maximum summer stream temperatures across montane landscapes. J Am Water Resour As 37: 351–366.
5. Poole GC, Berman CH (2001) An ecological perspective on in-stream temperature: natural heat dynamics and mechanisms of human-caused thermal degradation. Environ Manage 27: 787–802.
6. Cassie D (2006) The thermal regime of rivers: a review. Freshwater Bio 51: 1389–1406.
7. Al-Chokhachy R, Alder J, Hostetler S, Gresswell R, Shepard B (2013) Thermal controls of Yellowstone cutthroat trout and invasive fishes under climate change. Glob Change Biol 19: 3069–3081.
8. DeWeber JT, Wagner T (2014) A regional neural network ensemble for predicting daily river water temperature. J Hydrol 517: 187–200.
9. Isaak DJ, Luce CH, Rieman BE, Nagel DE, Peterson EE, et al. (2010) Effects of climate change and wildfire on stream temperatures and salmonid thermal habitat in a mountain river network. Ecol Appl 20: 1350–1371.
10. Herb WR, Stefan HG (2011) Modified equilibrium temperature models for cold-water streams. Water Resour Res 47: W06519.
11. Kanno Y, Vokoun JC, Letcher BH (2014) Paired stream-air temperature measurements revel fine-scale thermal heterogeneity within headwater brook trout stream networks. River Res and Appl 30: 745–755.
12. Nichols AL, Willis AD, Jeffres CA, Deas ML (2014) Water temperature patterns below large groundwater springs: management implications for coho salmon in the Shasta River, California. River Res and Appl 30: 442–455.
13. Meinzer OE (1927) Large springs of the United States. United States Geological Survey Water Supply Paper:557. Obtained online from: http://pubs.er.usgs.gov/publication/wsp557.
14. Smith H, Wood PJ (2002) Flow permanence and macroinvertebrate community variability in limestone spring systems. Hydrobiologia 487: 45–58.
15. Reynolds WW, Casterlin ME (1979) Behavioral thermoregulation and the "Final Preferendum" paradigm. Am Zool 19: 211–224.
16. Peterson JT, Rabeni CF (1996) Natural thermal refugia for temperate warmwater stream fishes. N Am J Fish Manage 16: 738–746.
17. Matthews KR, Berg NH (1997) Rainbow trout responses to water temperature and dissolved oxygen stress in two southern California stream pools. J Fish Biol 50: 50–67.
18. Torgersen CE, Price DM, Li HW, McIntosh BA (1999) Multiscale thermal refugia and stream habitat associations of Chinook salmon in northeastern Oregon. Ecol Appl 9: 301–319.
19. Ebersole JL, Liss WJ, Frissell CA (2001) Relationship between stream temperature, thermal refugia and rainbow trout *Oncorhyncnus mykiss* abundance in arid-land streams in the northwestern United States. Ecol Freshw Fish 10: 1–10.
20. Howell PJ, Dunham JB, Sankovich PM (2010) Relationships between water temperatures and upstream migration, cold water refuge use, and spawning of adult bull trout from the Lostine River, Oregon, USA. Ecol Freshw Fish 19: 96–106.
21. Sada DW, Fleishman E, Murphy DD (2005) Associations among spring-dependent aquatic assemblages and environmental and land use gradients in a Mojave Desert range. Divers Distrib 11: 91–99.
22. Martin RW, JT Petty (2009) Local stream temperature and drainage network topology interact to influence the distribution of smallmouth bass and brook trout in a central Appalachian watershed. J Freshwater Ecol 24: 497–508.
23. Labbe TR, Fausch KD (2000) Dynamics of intermittent stream habitat regulate persistence of a threatened fish at multiple scales. Ecol Appl 10: 1774–1791.
24. Brewer SK (2013) Groundwater influences on the distribution and abundance of riverine smallmouth bass, Micropterus dolomieu, in pasture landscapes of the Midwestern USA. River Res Appl 29: 269–278.
25. Shuter BJ, MacLean JA, Fry FEJ, Regier HA (1980) Stochastic simulation of temperature effects on first year survival of smallmouth bass. T Am Fish Soc 109: 1–34.
26. Garrett JW, Bennett DH, Frost FO, Thurow RF (1998) Enhanced incubation success for Kokanee spawning in groundwater upwelling sites in a small Idaho stream. N Am J Fish Manage 18: 925–930.
27. Baxter CV, Hauer FR (2000) Geomorphology, Hyporheic exchange, and selection of spawning habitat by bull trout (Salvelinus confluentus). Can J Fish Aquat Sci 57: 1470–1481.
28. Power G, Brown RS, Imhof JG (1999) Groundwater and fish – insights from northern North America. Hydrol Processes 13: 401–422.
29. Petty JT, Hansbarger JL, Hunstman BM, Mazik PM (2013) Brook trout movement in response to temperature, flow, and thermal refugia within a complex Appalachian riverscape. T Am Fish Soc 144: 1060–1073.
30. Hu Q, Wilson GD, Chen X, Akyuz A (2005) Effects of climate and landcover change on stream discharge in the Ozark Highlands, USA. Environ Model Assess 10: 9–19.
31. Palmer MA, Lettenmaier DP, Poff NL, Postel SL, Richter B, et al. (2009) Climate change and river ecosystems: protection and adaptation options. Environ Manage 44: 1053–1068.
32. Pörtner HO, Knust R (2007) Climate change affects marine fishes through the oxygen limitation of thermal tolerance. Science 315: 95–97.
33. Pease AA, Paukert CP (2014) Potential impacts of climate change on growth and prey consumption of stream-dwelling smallmouth bass in the central United States. Ecol Freshw Fish 23: 336–346.
34. Poloczanska ES, Hawkins SJ, Southward AJ, Burrows MT (2008) Modeling the response of populations of competing species to climate change. Ecology 89: 3138–3149.
35. Rahel FJ, Olden JD (2008) Assessing the effects of climate change on aquatic invasive species. Conserv Biol 22: 521–533.
36. Eaton JG, Scheller RM (1996) Effects of climate warming on fish thermal habitat in streams of the United States. Limnol Oceanogr 41: 1109–1115.
37. Mohseni O, Stefan HG (1999) Stream temperature/air temperature relationship: a physical interpretation. J Hydrol 218: 128–141.

38. Eby LA, Helmy O, Holsinger LM, Young MK (2014) Evidence of climate-induced range contractions in Bull Trout *Salvelinus confluentus* in a Rocky Mountain watershed, U.S.A.. PLOS ONE 9: e98812.

39. Chu C, Jones NE, Mandrak NE, Piggott AR, Minns CK (2008) The influence of air temperature, groundwater discharge, and climate on the thermal diversity of stream fishes in southern Ontario watersheds. Can J Fish Aquat Sci 65: 297–308.

40. Sinokrot BA, Stefan HG, McCormick JH, Eaton JG (1995) Modeling of climate change effects on stream temperatures and fish habitats below dams and near groundwater inputs. Climatic Change 30: 181–200.

41. Kundzewicz ZW, Mata LJ, Arnell NW, Döll P, Jimenez B, et al. (2008) The implications of projected climate change for freshwater resources and their management. Hydrolog Sci J 53: 3–10.

42. Lucas LK, Gompert Z, Ott JR, Nice CC (2009) Geographic and genetic isolation in spring-associated *Eurycea* salamanders endemic to the Edwards Plateau region of Texas. Conserv Genet 10: 1309–1319.

43. Carmona-Catot G, Magellan K, García-Berthou E (2013) Temperature-specific competition between invasive mosquitofish and an endangered Cyprinodontid fish. PLOS ONE 8: e54734.

44. Rieman BE, Isaak D, Adams S, Horan D, Nagel N, et al. (2007) Anticipated climate warming effects on bull trout habitats and populations across the interior Columbia River basin. T Am Fish Soc 136: 1552–1565.

45. Rahel FJ, Bierwagen B, Taniguchi Y (2008) Managing aquatic species of conservation concern in the face of climate change and invasive species. Conserv Biol 22: 551–561.

46. Ormerod SJ (2009) Climate change, river conservation and the adaptation challenge. Aquat Conserv 19: 609–613.

47. Mugel DN, Richards JM, Schumacher JG (2009) Geohydrologic investigations and landscape characteristics of areas contributing water to springs, the Current River, and Jacks Fork, Ozark National Scenic Riverways, Missouri. U. S. Geological Survey Scientific Investigations Report 2009–5138: 80 p.

48. Strahler AN (1957) Quantitative analysis of watershed geomorphology. EOS T Am Geophys Un 38: 913–920.

49. Peterson JT, Rabeni CF (2001) Evaluating the physical characteristics of channel units in an Ozark stream. T Am Fish Soc 130: 898–910.

50. Panfil MS, Jacobson RB (2001) Relations among geology, physiography, land use, and stream habitat conditions in the Buffalo and Current River systems, Missouri and Arkansas. USGS/BRD/BSR-2001-0005, 111 p.

51. Lloyd CD (2010) Spatial data analysis: an introduction for GIS users. Oxford: Oxford University Press. 206 p.

52. Whitledge GW, Rabeni CF, Annis G, Sowa SP (2006) Riparian shading and groundwater enhance growth potential for smallmouth bass in Ozark streams. Ecol Appl 16: 1461–1473.

53. Jeppesen E, Iversen TM (1987) Two simple models for estimating daily mean water temperatures and diel variations in a Danish low gradient stream. Oikos 49: 149–155.

54. Jourdonnais JH, Walsh RP, Prickett F, Goodman D (1992) Structure and calibration strategy for a water temperature model of the lower Madison River, Montana. Rivers 3: 153–169.

55. Pilgrim JM, Fang X, Stefan HG (1998) Stream temperature correlations with air temperatures in Minnesota: implications for climate warming. J Am Water Resour As 34: 1109–1121.

56. Bartholow JM (1989) Stream temperature investigations: field and analytic methods. Instream Flow Information Paper No. 13. U.S. Fish and Wildlife Service Biological Report 89(17). 139 p.

57. Bogan T, Mohseni O, Stefan HG (2003) Stream temperature-equilibrium temperature relationship. Water Resour Res 39(9): 1245.

58. Erickson TR, Stefan HG (2000) Linear air/water temperature correlations for streams during open water periods. J Hydrol Eng 5: 317–321.

59. NOAA, National Climatic Data Center. http://www.ncdc.noaa.gov/ Accessed 6/4/2014.

60. Hostetler SW, Alder JR, Allan AM (2011) Dynamically downscaled climate simulations over North America: methods, evaluation and supporting documentation for users. U. S. Geological Survey Open-File Report 2011–1238, 64 p.

61. Winemiller KO, Flecker AS, Hoeinghaus DJ (2010) Patch dynamics and environmental heterogeneity in lotic ecosystems. J N Am Benthol Soc 29: 84–99.

62. Schramm HL Jr, Armstrong ML, Fedler AJ, Funicelli NA, Green DM, et al. (1991) Sociological, economic, and biological aspects of competitive fishing. Fisheries 16(3): 13–21.

63. Whitledge GW, Hayward RS, Rabeni CF (2002) Effects of temperature on specific daily metabolic demand and growth scope of sub-adult and adult smallmouth bass. J Freshwater Ecol 17: 353–361.

64. Mohler HS (1966) Comparative seasonal growth of the largemouth, spotted, and smallmouth bass. Master's Thesis. University of Missouri, Columbia.

65. Struber RJ, Gebhart G, Maughan OE (1982) Habitat suitability index models: largemouth bass. United States Department of the Interior, Fish and Wildlife Service Report. FWS/OBS-82/10.16. 32 p.

66. Zweifel RD, Hayward RS, Rabeni CF (1999) Bioenergetics insight into black bass distribution shifts in Ozark Border Regions streams. N Am J Fish Manage 19: 192–197.

67. Hutchinson VH, Hill LG (1976) Thermal selection in the hellbender, *Cryptobranchus alleganiensis*, and the mudpuppy, *Necturus maculosus*. Herpetologica 32: 327–331.

68. Jeong DI, Daigle A, St-Hilaire A (2013) Development of a stochastic water temperature model and projection of future water temperature and extreme events in the Ouelle River basin in Québec, Canada. River Res Appl 29: 805–821.

69. Ouellet V, Secretan Y, St-Hilaire A, Morin J (2014) Daily averaged 2D water temperature model for the St. Lawrence River. River Res Appl 30: 733–744.

70. Ahmadi-Nedushan B, St.Hilaire A, Ouarda TBMJ, Bilodeau L, Robichaud E, et al. (2007) Predicting river water temperatures using stochastic models: case study of the Moisie River (Québec, Canada). Hydrol Processes 21: 21–34.

71. Imholt C, Soulsby C, Malcolm IA, Hrachowitz M, Gibbins CN, et al. (2013) Influence of scale on thermal characteristics in a large montane river basin. River Res Appl 29: 403–419.

72. Cherkauer KA, Burges SJ, Handcock RN, Kay JE, Kampf SK, et al. (2005) Assessing satellite-based and aircraft-based thermal infrared remote sensing for monitoring Pacific Northwest river temperature. J Am Water Resour As 41: 1149–1159.

73. Madej MA, Currens C, Ozaki V, Yee J, Anderson DG (2006) Assessing possible thermal rearing restrictions for juvenile coho salmon (*Oncorhynchus kisutch*) through thermal infrared imaging and in-stream monitoring, Redwood Creek, California. Can J Fish Aquat Sci 63: 1384–1396.

74. Selker J, van de Giesen N, Westhoff M, Luxemburg W, Parlange M (2006) Fiber optics opens window on stream dynamics. Geophys Res Lett 33: L24401.

75. Westhoff MC, Savenije HHG, Luxemburg WMJ, Stelling GS, van de Giesen NC, et al. (2007) A distributed stream temperature model using high resolution temperature observations. Hydrol Earth Syst Sc 11: 1469–1480.

76. Webb BW, Hannah DM, Moore RD, Brown LE, Nobilis F (2008) Recent advances in stream and river temperature research. Hydrol Processes 22: 902–918.

77. Neumann DW, Rajagopalan B, Zagona EA (2003) Regression model for daily maximum stream temperature. J Environ Eng-ASCE. 129: 667–674.

78. O'Driscoll MA, DeWalle DR (2006) Stream-air temperature relations to classify stream-groundwater interactions in a karst setting, central Pennsylvania, USA. J Hydro 329: 140–153.

79. Taylor CA, Stefan HG (2009) Shallow groundwater temperature response to climate change and urbanization. J Hydro 375: 601–612.

80. Arismendi I, Safeeq M, Johnson SL, Dunham JB, Haggerty R (2013) Increasing synchrony of high temperature and low flow in western North American streams: double trouble for coldwater biota? Hydrobiologia 712: 61–70.

81. Tebaldi C, Knutti R (2007) The use of multi-model ensemble in probabilistic climate projections. Philos T Roy Soc A 365: 2053–2075.

82. Intergovernmental Panel on Climate Change (2000) Summary for policy-makers: emissions scenarios. Special report of Working Group III on the Intergovernmental Panel on Climate Change. 27p.

83. Eckhardt K, Ulbrich U (2003) Potential impacts of climate change on groundwater recharge and streamflow in a central European low mountain range. J Hydro 284: 244–252.

84. Nohara D, Kitoh A, Hosaka M, Oki T (2006) Impact of climate change on river discharge projected by multimodel ensemble. J Hydrometeorol 7: 1076–1089.

85. Jyrkama MI, Sykes JF (2007) The impact of climate change on spatially varying groundwater recharge in the grand river watershed (Ontario). J Hydro 338: 237–250.

86. Townsend CR (1989) The patch dynamics concept of stream community ecology. J N Am Benthol Soc 8: 36–50.

87. Thorp JH, Thoms MC, Delong MD (2006) The riverine ecosystem synthesis: biocomplexity in river networks across space and time. River Res Appl 22: 123–147.

88. Williams JD, Etnier DA (1982) Description of a new species, *Fundulus julisia*, with a redescription of *Fundulus albolineatus* and a diagnosis of the subgenus Xenisma (Teleostei: Cyprinodontidae). Occas Pap Mus Nat Hist, University of Kansas 102: 1–20.

89. Winemiller KO, Taylor DH (1987) Predatory behavior and competition among laboratory-housed largemouth and smallmouth bass. Am Midl Nat 117: 148–166.

90. McClain ME, Boyer EW, Dent CL, Gergel SE, Grimm NB, et al. (2003) Biogeochemical hot spots and hot moments at the interface of terrestrial and aquatic systems. Ecosystems 6: 301–312.

91. Tonolla D, Wolter C, Ruhtz T, Tockner K (2012) Linking fish assemblages and spatiotemporal thermal heterogeneity in a river-floodplain landscape using high-resolution airborne thermal infrared remote sensing and in-situ measurements. Remote Sens Environ 125: 134–146.

92. Schmidt C, Bayer-Raich M, Schirmer M (2006) Characterization of spatial heterogeneity of groundwater-stream water interactions using multiple depth streambed temperature measurements at the reach scale. Hydrol Earth Syst Sci 10: 849–859.

93. Mohseni O, Stefan HG, Eaton JG (2003) Global warming and potential changes in fish habitat in U.S.streams. Climate Change 59: 389–409.

94. Ficke AD, Myrick CA, Hansen LJ (2007) Potential impacts of global climate change on freshwater fisheries. Rev Fish Biol Fisher 17: 581–613.

95. Comte L, Buisson LM, Daufresne M, Grenouillet G (2013) Climate-induced changes in the distribution of freshwater fish: observed and predicted trends. Freshwater Biol 58: 625–639.

96. King JR, Shuter BJ, Zimmerman AP (1999) Empirical links between thermal habitat, fish, and climate change. T Am Fish Soc 128: 656–665.

97. Ayllón D, Nicola GG, Elvira B, Parra I, Almodóvar A (2013) Thermal carrying capacity for a thermally-sensitive species at the warmest edge of its range. PLoS ONE 8: e81354.

98. Hill DK, Magnuson JJ (1990) Potential effects of global climate warming on the growth and prey consumption of Great Lakes fishes. T Am Fish Soc 119: 265–275.

99. Sowa SP, Rabeni CF (1995) Regional evaluation of the relation of habitat to distribution and abundance of smallmouth bass and largemouth bass in Missouri streams. T Am Fish Soc 124: 240–251.

100. Reese C, Harvey BC (2002) Temperature-dependent interactions between juvenile steelhead and Sacramento pikeminnow in laboratory streams. T Am Fish Soc 131: 599–606.

101. Lehtonen H (1996) Potential effects of global warming on northern European freshwater fish and fisheries. Fisheries Manag Ecol 3: 59–71.

102. Daufresne M, Lengfellner K, Sommer U (2009) Global warming benefits the small in aquatic ecosystems. P Nat Acad Sci USA 106: 12788–12793.

103. Woodhams DG, Alford RA, Briggs CJ, Johnson M, Rollins-Smith LA (2008) Life-history trade-offs influences disease in changing climates: strategies of an amphibian pathogen. Ecology 89: 1627–1639.

104. Wehrly KE, Wiley MJ, Seelbach PW (2003) Classifying regional variation in thermal regime based on stream fish community patterns. T Am Fish Soc 132: 18–38.

105. Busiion L, Thuiller W, Lek S, Lim P, Grenouillet G (2008) Climate change hastens the turnover of stream fish assemblages. Glob Change Biol 12: 2232–2248.

# 21

# The Morphometry of Lake Palmas, a Deep Natural Lake in Brazil

**Gilberto F. Barroso[1]\*, Monica A. Gonçalves[2], Fábio da C. Garcia[1]**

**1** Department of Oceanography and Ecology, Federal University of Espírito Santo, Vitória, Espírito Santo, Brazil, **2** Espírito Santo State Water Resources Agency, Vitória, Espírito Santo, Brazil

## Abstract

Lake Palmas ($A = 10.3$ km$^2$) is located in the Lower Doce River Valley (LDRV), on the southeastern coast of Brazil. The Lake District of the LDRV includes 90 lakes, whose basic geomorphology is associated with the alluvial valleys of the Barreiras Formation (Cenozoic, Neogene) and with the Holocene coastal plain. This study aimed to investigate the relationship of morphometry and thermal pattern of a LDRV deep lake, Lake Palmas. A bathymetric survey carried out in 2011 and the analysis of hydrographic and wind data with a geographic information system allowed the calculation of several metrics of lake morphometry. The vertical profiling of physical and chemical variables in the water column during the wet/warm and dry/mild cold seasons of 2011 to 2013 has furnished a better understanding of the influence of the lake morphometry on its structure and function. The overdeepened basin has a subrectangular elongated shape and is aligned in a NW-SE direction in an alluvial valley with a maximum depth ($Z_{max}$) of 50.7 m, a volume of $2.2 \times 10^8$ m$^3$ (0.22 km$^3$) and a mean depth ($Z_{mv}$) of 21.4 m. These metrics suggest Lake Palmas as the deepest natural lake in Brazil. Water column profiling has indicated strong physical and chemical stratification during the wet/warm season, with a hypoxic/anoxic layer occupying one-half of the lake volume. The warm monomictic pattern of Lake Palmas, which is in an accordance to deep tropical lakes, is determined by water column mixing during the dry and mild cold season, especially under the influence of a high effective fetch associated with the incidence of cold fronts. Lake Palmas has a very long theoretical retention time, with a mean of 19.4 years. The changes observed in the hydrological flows of the tributary rivers may disturb the ecological resilience of Lake Palmas.

**Editor:** Andrew C. Singer, NERC Centre for Ecology & Hydrology, United Kingdom

**Funding:** The authors acknowledge Espírito Santo State Agencies for Environment and Water Resources (IEMA) and Agriculture, Aquaculture (SEAG) as well as Linhares Aquaculture Association (AquaLin) for field work logistics. They also would like to acknowledge Espírito Santo State and Federal research supporting agencies FAPES and CNPq, respectively, for Dr. Fábio da Cunha Garcia's researcher fellowship grant. The funders had no role in study design, data collection and analysis, decision to publish, or preparation of the manuscript.

**Competing Interests:** The authors have declared that no competing interests exist.

\* Email: gilberto.barroso@ufes.br

## Introduction

Lake morphology has been recognized as a key factor for the understanding of lacustrine structure and function. Since the late 1930s, based on Rawson's diagram [1], lake area and depth contours have been viewed as factors controlling ecosystem productivity due to light penetration, heat balance, oxygen distribution, the input of allochthonous matter, the nature of the sediments and littoral zone development. The influence of the relative shape and size of lake basins on several lake processes has been investigated. These processes include mixing dynamics [2–4], hydrology [5], sedimentation [6], dissolved organic carbon content [7], the biomass of submersed macrophytes [8], primary productivity [9] and lake metabolism [10].

Lake morphology is quantified with morphometric metrics that are descriptors of the form and size of lake basins. This analysis provides crucial knowledge in support of approaches to lake management. Geographical information systems are becoming an important tool to process and analyze morphometric metrics of areas and volumes [11–13].

Geographical location of lakes (latitude, longitude and altitude) must also be considered due to the climatic drivers of insolation, wind and precipitation. Lake typologies for many different geographical settings have been established based on lake morphology and climate [14,15].

In Brazil, with the exception of the extensive study of 61 coastal lakes in the southern portion of the country [16], lake morphometric studies are relatively scarce. Morphometric data are generally available for artificial lakes, particularly in terms of reservoir engineering and management [17]. The deepest natural lake in the country for which data have been published is Dom Helvécio, in the Middle Doce River Valley - MDRV (State of Minas Gerais, Southeastern Brazil), with maximum and mean depths of 39.2 and 11.3 m, respectively [18]. This well-known lake, whose genesis is fluvial, shows a warm monomictic pattern. The lower Doce River Valley (LDRV) (State of Espírito Santo, Southeastern Brazil) has 90 lakes, comprising a valuable water resource that needs sound environmental management.

The present study aims to improve the ecological knowledge of moderate tropical deep lakes through the determination of several morphometric factors for Lake Palmas. A geographic information system (GIS) was developed as an environment for metrics calculation. Wind climate effects on lake stratification and mixing, based on wind direction and intensity, were also integrated in the GIS approach. Vertical water column profiles were developed to explore the relationships between lake morphology and water column stratification.

## Materials and Methods

No specific permissions were required to collect hydrographic data and temperature and dissolved oxygen data in Lake Palmas (19°23'S/40°17'W and 19°26'S/40°13'W). In addition, field and lab studies did not involve any biological species.

### Physiography of the study area

The LDRV is the location of a district including 90 lakes with areas ranging from 0.8 ha to 62.0 km$^2$, a total lake area of 165.5 km$^2$ (Figure 1a and 1b). The LDRV Lake District and its 'lakescape' comprise lakes located in dammed alluvial valleys and lakes located on the coastal plain. According to Bozelli et al. [19], the lakes of the LDRV show both intermittent and dynamic patterns of metabolism. The intermittent pattern is found in the lakes of the alluvial valleys of the Barreiras Formation (Cenozoic, Neogene Period), which are functionally deep and can be described by a seasonal metabolism model. The dynamic metabolism pattern is characteristic of the coastal plain lakes (Cenozoic, Quaternary Period, and Holocene Epoch), which are relatively shallow and are more efficient in processing organic matter.

In the easternmost part of the LDRV, neotectonic processes with patterns of alignment from the NW to the SE control the drainage system of the major river, the Doce River, and its tributary rivers and lakes in the alluvial valleys [20]. According to Martin et al. [21], the Doce River delta and the associated Holocene sedimentation of an ancient lagoon represent a breakthrough process of regional geomorphologic evolution. This process, which started approximately 5,100 yrs B.P., is associated with sea level transgression and regression on the paleodeltaic coastal plain and the damming of the alluvial valleys of the Barreiras Formation with fluvial sediments [21].

The regional climate is characterized by relatively wet and hot summers and dry/mild cold winters. Land use is dominated by pastureland, croplands and *Eucalyptus* forestry. The major areas of urbanization are located in the southeast portions of the District, in the vicinity of Juparanã (62.0 km$^2$), Meio (1.3 km$^2$), and Aviso (0.7 km$^2$) Lakes [22]. In general, lakes water resources have been used for crop irrigation, recreation, fishing and, more recently, for aquaculture. There are two intensive fish farming operations in Lake Palmas, which produce tilapia in floating cages.

### Hydrographic survey

The Lake Palmas (Figure 1c) shoreline was screen digitized with the geographic information system ArcGIS 10.1 ESRI (Redlands, California, USA), software licensing EFL615216336, using a digital aerial orthophotograph acquired on May 2008 at a scale of 1:15,000 and a spatial resolution of 1 m, georeferenced with Universal Transverse Mercator (UTM) projection zone 24 k and World Geodetic Datum (WGS) 1984. The polygon shapefile of the lake surface retained the projection and datum of the orthophoto. Echosounding survey lines were plotted on the polygon shapefile at a spacing of 200 m along the longitudinal axes of the lake (Figure 1d). A total of 100.7 km was selected for bathymetric sounding.

A hydrographic survey was performed in May 2011 with an Ohmex (Sway, Hampshire, UK) HydroLite XT echosounding system composed of a 210.0 kHz single beam transducer, a SonarMite V3 BT Bluetooth connection and a Trimble (Sunnyvale, California, USA) GeoXH DGPS receiver. Navigation along the survey lines was oriented with the shapefile of transects displayed with ArcPad 7.0 ESRI on a Trimble Juno GPS receiver. Spatial information was determined according to the Universal Transverse Mercator (UTM) projection and World Geodetic Datum (WGS) 1984. Lake hydrographic surveys were performed at a maximum boat speed of 5.0 km.h$^{-1}$ (2.7 knots).

### Bathymetric data processing

X, Y and Z (easting, northing and depth) data were downloaded to Ohmex SonarVista software and then exported as a *dxf* file. In ArcGIS 10.1, the *dxf* files were converted to point features in a shapefile format. The attribute table of the shapefile of depth values was edited to identify and erase depth spikes. The lake shoreline shapefile was converted from a polygon to a point file with a yield of 1,539 shoreline points, which were assigned to zero depth. This shapefile was later merged into the bathymetric survey shapefile.

The interpolation procedure for generating a surface model of the bathymetry data was conducted with Ordinary Kriging using the ESRI extension Geostatistical Analyst 10.1. The process is based on semivariogram modeling, neighborhood search and crossvalidation [23,24]. The resulting bathymetric map was presented with 5.0 m isobaths.

The intensity of the survey (L$_r$) is the ratio between the lake area in km$^2$ and the echosounding track length in km. The accuracy of the bathymetric map was assessed with the information value ($I$), which indicates a completely correct map when $I = 1$. $I$ was calculated as a product of correctly identified area ($I''$) and information number ($I''$). $I'$ and $I''$ also vary between 0 and 1, with a value of $I' = 1$ indicating that all contour lines are correct and a value of $I''$ indicating that the number of contour lines is optimal. The equations for $I$, $I'$ and $I''$, given by Håkanson [11,25], incorporate lake area (km$^2$), the distance between the sounding tracks (km), shoreline development and the number of bathymetric contour lines. The symbols for the morphometry metrics are based on Hutchinson [26].

### Lake size metrics

The maximum depth (Z$_{max}$) was determined from the echosounding points after editing to remove spike data. Lake perimeters, areas, and volumes were calculated with ArcGIS 10.1 routines to determine primary morphometric parameters for lake size, such as lake surface area - A (m$^2$), shoreline length - L$_0$ (m), maximum length - L$_{max}$ (m), maximum breadth - B$_{max}$ (m) and volume V (m$^3$).

### Lake form metrics

Lake form factors were calculated according to Håkanson [11] as follows: mean depth in m, $Z_{mv} = V/A$; relative depth in %, $Z_r = [(Z_{max} * \sqrt{\pi})/(20 * \sqrt{A})]$; shoreline development index, $L_d = \left\{ (L_0/2) * \left[ \sqrt{(\pi * A)} \right] \right\}$; volume development, $V_d = [(3 * Z_{mv})/Z_{max}]$; and mean basin slope in %, $S_{mv} = \left\{ [L_0 + (2 * L_{cot})] * [Z_{max}/(20 * n * A)] \right\}$.

Where, V = lake volume in m$^3$, A = lake water surface area m$^2$ in (in km$^2$ for S$_{mv}$), Z$_{max}$ = maximum depth in m, L$_0$ = normalized shoreline length in km, L$_{ctot}$ = total normalized length for all contour lines in km excluding the shoreline and $n$ = number of contour lines.

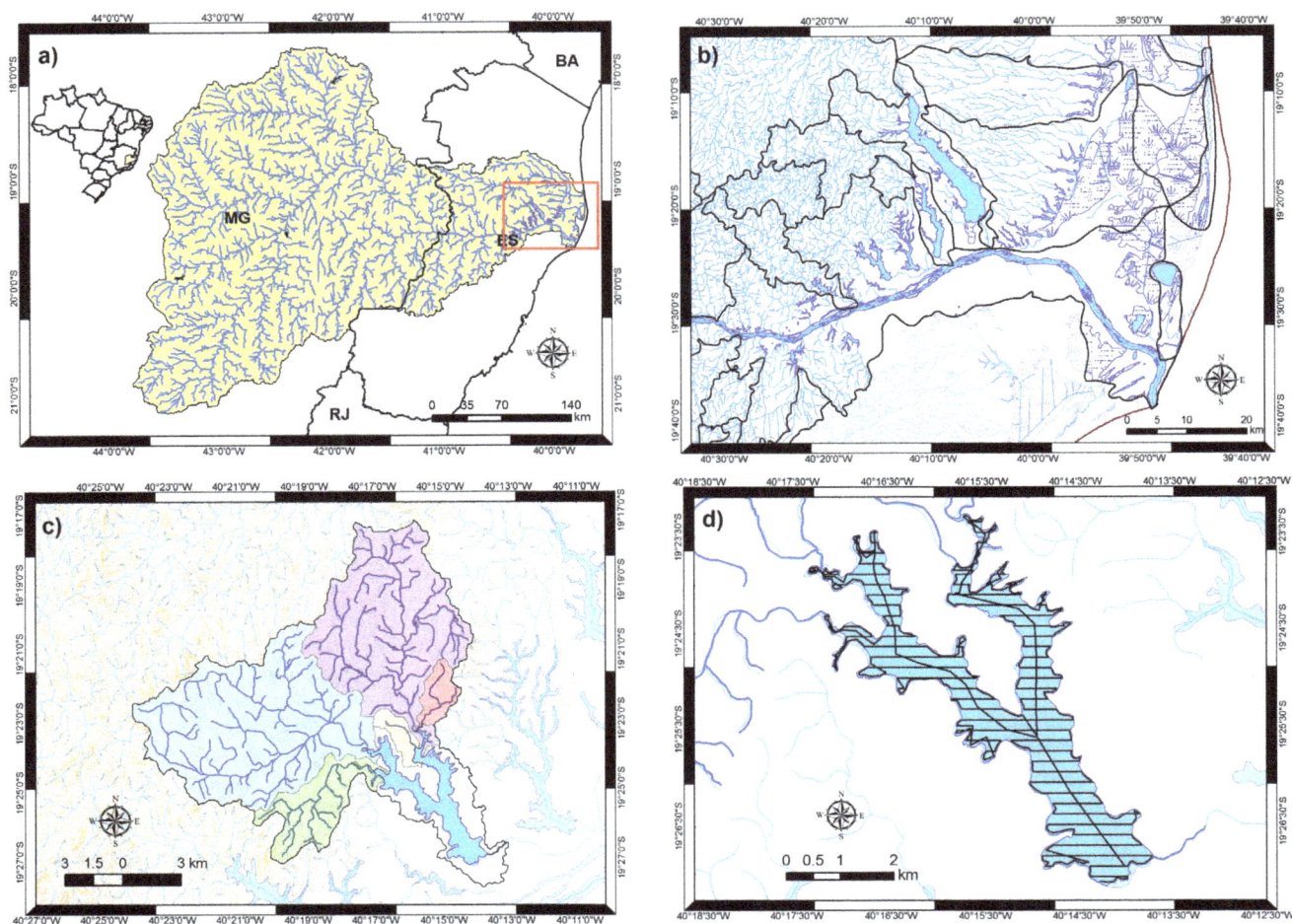

**Figure 1. Study area settings: a) LDRV location on the southeastern coast of Brazil in the Doce River Basin (State of Espírito Santo); b) the LDRV Lake District, with its "lakescape"; c) Lake Palmas and its watershed and height curves; d) echosounding transects for hydrographic survey of Lake Palmas.**

## Water column structure

Water column profiles for temperature (°C), photosynthetically active radiation (PAR) and dissolved oxygen (mg.L$^{-1}$) were recorded for the field samples from 2011 to 2013 in wet/warm months and in dry/mild cold months. Based on data from 1947 to 2013 from 13 meteorological stations (National Water Agency – ANA, hidroweb.ana.gov.br), the wet months show a regional mean monthly rainfall greater than 100 mm, whereas the regional mean monthly rainfall for the dry months is less than 50 mm (Figure 2). The wet/warm season extends from October to March, with a mean monthly rainfall of 167.6±32.2 mm and a mean air temperature of 24.8±3.25°C. The dry/mild cold season extends from May to August, with a mean monthly rainfall of 46.1±2.5 mm and a mean air temperature of 21.9±3.1°C.

Water transparency was estimated with the depth of Secchi disk. A Horiba (Minami-Ku, Kyoto. Japan) U-53G multiparameter water quality meter with a 30 m cable was used for vertical profiling at 4 sampling sites along the lake axes (Figure 3). Bottom water samples were taken with a Niskin bottle. The extent of the euphotic zone ($Z_{eu}$) was estimated with underwater light attenuation to a depth corresponding to 1% of subsurface PAR through vertical profiling with a LiCor (Lincoln, Nebraska USA) system with a LI-250A light meter and LI-193 spherical PAR quantum sensor.

The mixing depth ($Z_{mix}$) in m was calculated based on the maximum discontinuity in the relative thermal resistance (RTR) [27]. The thermal resilience of the water column, based on the Effective Wedderburn number ($W_e$) [28,29], was calculated for each sampling site at every field sampling event:

$$W_e = \left\{ \left[ \Delta \rho_\omega * g'(h_m)^2 \right] * \left[ \rho_\omega * (u^*) * L \right]^{-1} \right\}$$

where, $\Delta \rho_\omega$ = the difference between the water mass density at the upper and lower limits of the thermocline (kg.m$^{-3}$), $h_m$ = mixing depth (m), $L$ = effective fetch in m, g' = the reduced gravity and $u^*$ = the wind friction velocity, g' and u* are calculated with the following equations:

$$g' = [(g * \rho')/\rho_0]$$

where, g' is the reduced gravity, g is the normal gravitational acceleration, $\rho'$ is a density perturbation and $\rho_0$ is a standard reference density;

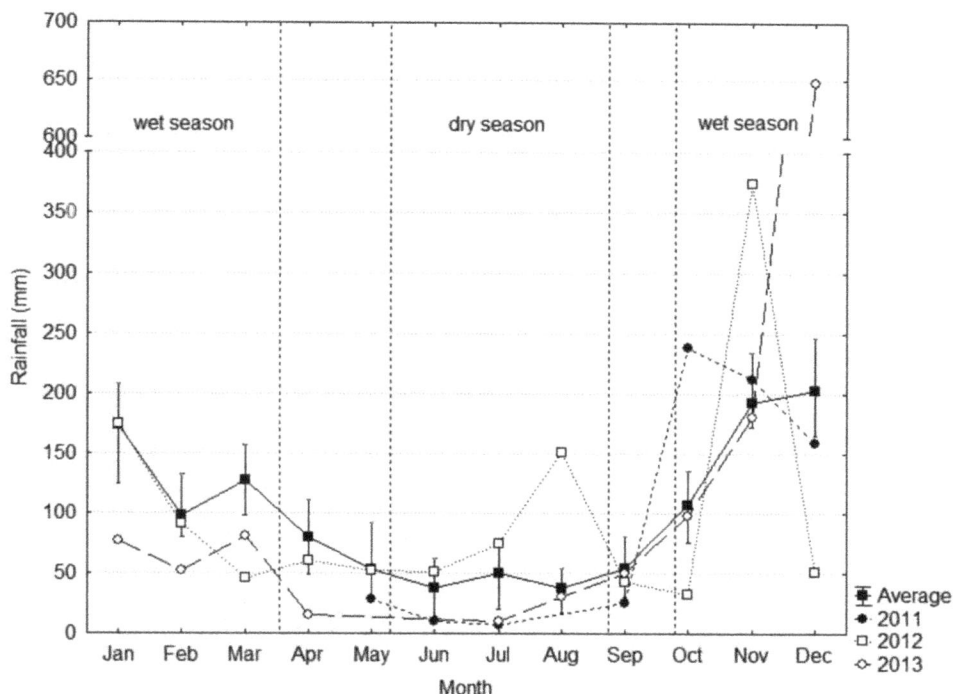

**Figure 2. Regional mean monthly rainfall (1947 to 2013) and mean monthly rainfall for the study period (2011 to 2013).** Data from 13 meteorological stations (National Water Agency – ANA).

$$u^* = [(\rho_{air}/\rho_s) * (C_d * u^2)]$$

Where, $\rho_{air}$ is the air specific mass in $kg.m^{-3}$, $\rho_s$ is the water specific mass at the lake surface in $kg.m^{-3}$; $C_d$ is the coefficient of friction (0.0015) and $u$ is the wind velocity in $m.s^{-1}$. If $W>1$, the thermal structure is stable; if $W<1$, the water column is susceptible to changes resulting from the effects of wind.

Hourly wind direction and intensity data were obtained from observations at the Linhares meteorological station (INMET - A614), approximately 18.0 km NE of Lake Palmas.

## Special metrics for lake morphometry

The A/V ratio was calculated to estimate the potential evaporation rate of lake water and the resistance of the water column to mixing. The slope of the lake basin was modeled in terms of the percent rise function using ArcGIS 10.1 with the function Slope, 3D Analyst Tool. The continuous surface model for the basin slope was reclassified in terms of gentle and steep slopes. According to Duarte and Kalff [8], a slope of 5.3% is the threshold value separating gentle and steep slopes in relationship to the development of submersed rooted aquatic vegetation in lakes.

The wave base depth ($Z_{wb}$) in meters was calculated from the $Z_{wb} = [(45.7 * \sqrt{A})/(21.4 * \sqrt{A})]$, with A in $km^2$ [11]. $Z_{wb}$ was the depth used to estimate the volume of the epilimnetic waters. The delimitation of the littoral and pelagic zones and their respective volumes was performed based on the mean depth of the euphotic zone ($Z_{eu}$) in meters. The area suitable for the development of rooted aquatic vegetation biomass was also based on the mean $Z_{eu}$ and within gentle slopes (<5.3%). Volumes in $m^3$

for hypoxic/anoxic bottom waters were determined based on the depth corresponding to hypoxia (<2.0 $mg.L^{-1}$) during the stratification season.

The effective fetch, $L_{ef}$ (km) and the wave heights (m) were estimated for 46 sites distributed along the lake surface with a grid with equal distances of 500.0 m. To estimate $L_{ef}$, distances from each site to the shoreline were measured according to the prevailing winds (defined as an angle of 0°) and every 6° on both sides of the 0° angle to 42° [11]. The below provides the integrated value of $L_{ef}$:

$$L_{ef} = [\Sigma x i \cos (ai)/(\Sigma x i \cos (ai) * SC')]$$

where, $\Sigma cos(a_i)$ is 13.5, a calculation constant, and $SC'$ is the map scale constant of 0.35.

Wave heights (H) in m were computed for the sites for which $L_{ef}$ was estimated, according to the Beach Erosion Board (1972) in Håkanson [11] using the following equation:

$$H = [0.105 * (\sqrt{L_{ef}})].$$

Both $L_{ef}$ and H were calculated based on two prevailing wind directions, one for the wet/warm season (NE) and one for the dry and wet season (SE, considering the major axis of the lake). Maps of $L_{ef}$ and H maps were created using GIS, interpolating the point data with a spline function with tension, a neighborhood of 5 points, a weight of 0.01 and a cell resolution of 5.0 m.

The basin permanence Index (BPI) ($m^3.km^{-1}$), which indicates the littoral effect on basin volume, is calculated according to the ratio of lake volume ($\times 10^6$ $m^3$) to shoreline length (km), $BPI = V/D_L$ [30]. The dynamic ratio (DR) was calculated according to the equation: $DR = [\sqrt{A}/Z_{mv}]$ with A in $km^2$ [11].

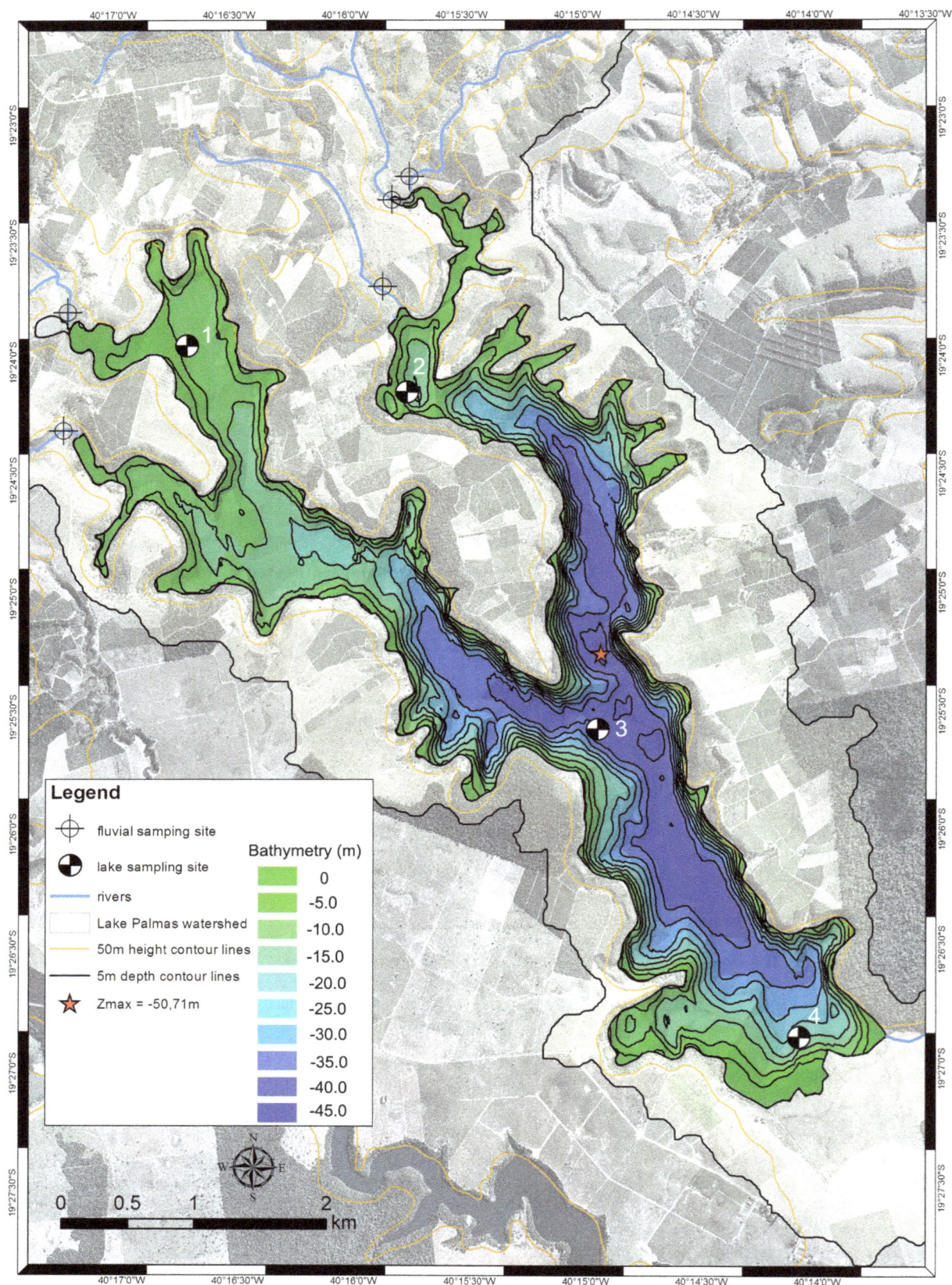

**Figure 3. Lake Palmas bathymetric map.**

To assess the cryptal depth ($Z_c$) and cryptal volume ($V_c$) of the lake, depth values were converted according to the altitude of the lake surface above sea level, using an altitude of 20 m as the reference value for the lake surface.

The theoretical lake water retention time was calculated according to the ratio of the lake volume ($m^3$) to the mean annual river tributary inflow ($m^3.s^{-1}$) $RT = V/Q_{mean}$ [41,42]. Discharge of the five tributary rivers (Figure 3) were measured during the wet/warm and dry/mild cold seasons (n = 8) with a SonTek (San Diego, California, USA) FlowTracker Handheld Acoustic Doppler Velocimeter (ADV). Mean annual river tributary discharges were then calculated, as well as discharge values for dry and wet seasons.

## Results

The total bathymetric sounding survey track was 122.9 km, yielding a survey intensity ($L_r$) of 0.08. A total of 46,941 valid depth points were computed. Ordinary kriging to obtain prediction results was applied as the interpolation method to yield a continuous surface of lake depths. A neighborhood search was used, considering a smooth type within an axis range between 100 to 2,000 points. Variogram modeling was based on 9 lags with a size of 280 and a spherical model with anisotropy (a direction of 125°). The regionalized variation of point data, optimized sampling and spatial pattern determination is addressed with the Semivariogram on Figure S1. The bathymetric map, with a cell size of $10 \times 10$ m (Figure 3).

Based on depth contour intervals of 5 m, the correctly identified area ($I'$) is 0.8571. This value means that 85.7% of the lake area was correctly identified and that 14.3% (1.5 $km^2$) of the lake area was incorrectly estimated. The information number ($I''$) for the 5 m contour lines was 0.9995, and the information value ($I$), indicating the overall map accuracy, was 0.8566.

The lake basin has a 'Y' shape aligned in a NW-SE direction, with a maximum length ($L_{max}$) of 7.1 km and an average breadth ($B_{mv}$) of 1.7 km. Other lake basin size metrics were a total surface area of 10.3 $km^2$, a shoreline length (L) of 51.9 km, a maximum depth ($Z_{max}$) of 50.7 m and a volume of $2.2 \times 10^8$ $m^3$ (0.2 $km^3$).

The lake form metrics were found to have the following values: the shoreline development index ($D_L$) was 4.5, the mean depth ($Z_{mv}$) was 21.4 m, the relative depth ($Z_r$) was 1.4%, the volume development ($V_d$) was 1.3, the mean slope ($S_{mv}$) was 15.8% and the $Z_{mv}:Z_{max}$ ratio was 0.42. These metrics indicate a flat-bottomed, overdeepened lake basin. Based on the value of $D_L$, the shoreline form is subrectangular elongate. The basin form is linear (L) according to the relative hypsographic area and volume curves (Figure S2a and S2b) as well as according to $V_d$.

There are 18 embayments along the lake axis, most of which are less than 15.0 m in depth. The area deeper than 40.0 m extends from the intersection of the two lake axes to the S shore and to the upper N axis. The three major deepest basins (>45.0 m) are located midway on the N-S axis next to the intersection of the two lake axes, representing an area of $1.5 \times 10^5$ $m^2$ (1.5% of the lake area) and a volume of $5.1 \times 10^5$ $m^3$ (0.02% of the lake volume).

The lake drainage is also oriented from NW to SE, with 5 tributary streams located along the upper NE and N shores. Lake Palmas discharges into the Doce River through a drainage river located along the SW shore.

The area/volume ratio is 0.05, indicating a deep basin with a small littoral zone. The mean basin slope ($S_{mv}$) of 15.8% represents an overall value for the shallow areas, central basin plain and steep lateral slopes. The basin slope GIS model (Figure 4a) shows steep lateral slopes, up to 112.8%, along the central E shore as well as

along the E and W shores of the promontory that separates the lake into two arms. The central basin plain is constrained with slopes lower than 20% and depths greater than 30 m. Shallow areas with a depth of less than 5 m and a slope of 5% are located at the mouths of the tributaries as well as the southernmost part of the lake. Very gentle slopes up to 2.0% are characteristic of deep basins (>40 m deep). Based on a threshold of 5.3%, in which fine sediments are retained [8,11], 9.1 $km^2$ (58.8% of the lake area) and 4.3 $km^2$ (41.3%) were classified as steep and gentle slopes, respectively (Figure 4b).

Data from field vertical profiling at 4 sampling sites (Table S1) show stratification during the warm months, with surface water temperatures reaching 31.3°C (Figure 5a), and a mixed water column during the mild cold months, with a mean water temperature of 23.2°C. The Effective Wedderburn ($W_e$) values, an overall indicator of the thermal stability of the water column, were 6.5±7.2 during the wet/warm season and zero in the dry/mild cold season. During the dry season, the mean $Z_{mix}$ value of 20.7±8.3 m indicated the presence of a deep mixing layer but with a very weak, i.e., unstable, stratification due to the zero W value and an RTR lower than 3.1.

During the season of stratification, $Z_{mix}$ was usually shallower or at least equal to $Z_{eu}$ (Figure 5a), yielding higher $Z_{eu}:Z_{mix}$ ratios. In contrast, the Secchi disk depth ($Z_{Sd}$) was higher during the mixing season. Based on a mean $Z_{eu}$ of 10.0 m, the volume of the euphotic layer ($V_{eu}$) was $9.2 \times 10^7$ $m^3$ (41.4% of the lake volume), whereas the volume of the aphotic zone was $1.3 \times 10^8$ $m^3$ (58.6%).

During the stratification season, hypoxic/anoxic conditions may develop below a depth of 13 m (Figure 5b). Under these conditions, the volume of anoxic waters may reach $1.1 \times 10^8$ $m^3$ or 48.6% of the lake volume. Bottom hypoxia/anoxia was recorded during the entire wet/warm season. In contrast, DO is well distributed in the water column throughout the mixing season (Figure 5c), even showing supersaturation at the surface (Figure 5d).

Special morphometry metrics show that the lake basin has a wave base depth ($Z_{wb}$), an indicator of the depth of turbulent mixing, of 6.0 m, with a surface layer volume (i.e., a mixing layer) of $5.8 \times 10^7$ $m^3$, or 26.1% of the lake volume. Thus, the volume of bottom waters was $1.6 \times 10^7$ $m^3$, or 73.9% of the lake volume. The dynamic ratio (DR) had a value of 0.15 indicating the predominance of slope processes over wind/wave processes in sediment resuspension.

The Basin Permanence Index (BPI), 4.3 $m^3.km^{-1}$, indicates that Lake Palmas is relatively less suitable for the development of the littoral zone and rooted aquatic plants. With the same threshold of 10.0 m for $Z_{eu}$, the littoral and pelagic zones are represented by $8.1 \times 10^6$ (3.7% of the lake volume) and $2.1 \times 10^8$ $m^3$ (96.3%), respectively (Figure 6a). The predicted potential areas for rooted submersed vegetation with nearshore gentle slopes comprise only $7.9 \times 10^5$ $m^2$, less than 1.0% of the bottom area of the lake (Figure 6b).

Wind pattern for the warm/wet months showed a dominance of 26% from the NE, with wind speeds up to 8.8 $m.s^{-1}$ (Figure 7a). During the dry/mild cold months, the wind was predominantly from the S, with speeds up to 11.1 $m.s^{-1}$ (Figure 7b).

The effective fetch ($L_{ef}$) model for wet/warm months with NE winds yielded values up to 0.8 km at the SW embayment (Figure 8a) and wave heights up to 0.5 m in the same embayment and at the confluence of the two axes (Figure 8c). For the dry/mild cold months, with SE winds, values up to 0.72 km were found at the lower section of the NW-SE axis and at the southern part of the land promontory (Figure 8b). Under SE winds, the wave heights were up to 0.6 m at the W shore next to the land

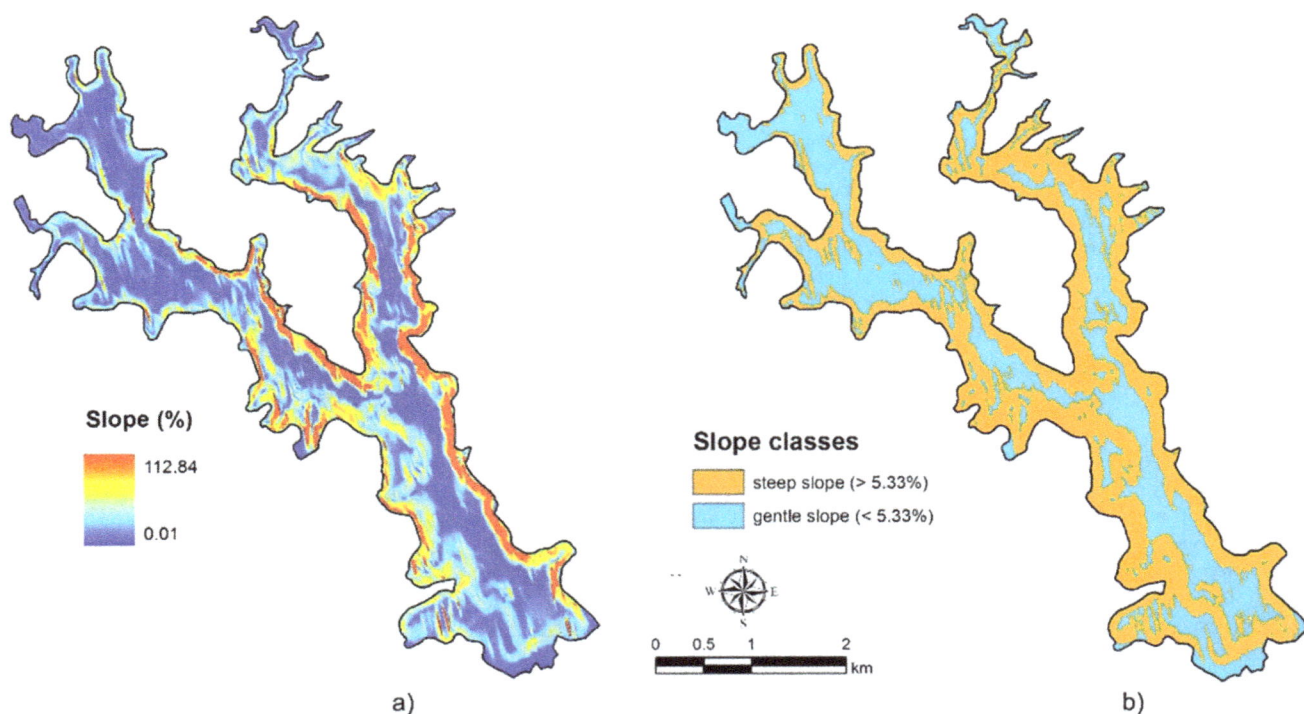

**Figure 4. Lake bottom slope (%): a) slope gradient in %; b) reclassified slope: gentle (<5.33%) and steep (>5.33%) slopes.**

promontory as well as in the central section of the NW-SE axis (Figure 8d).

As the lake surface is 20 m above mean sea level, the cryptal depth ($Z_c$) is 20.0 m, with a corresponding cryptodepression volume ($V_c$) of $2.8 \times 10^7$ m$^3$ (12.6% of the lake volume) (Figure 9). The water column is free from the influence of salt water.

The Lake Palmas watershed area ($W_A$) is 168.2 km$^2$, and the $W_A/A$ ratio is 16.3. The mean annual, dry/mild cold and wet/warm total tributary discharge values were 0.4±0.2, 0.3±0.3 and 0.4±0.04 m$^3$.s$^{-1}$, respectively. The river discharge during the wet/warm season was 10.0% higher than the annual mean. In contrast, the dry season discharge was 14.6% lower. Zero discharge was registered three times for tributary river 1 during the dry season, but tributary 5 dried up twice during the wet/warm season.

The theoretical retention time based on the mean annual tributary discharge was 19.4 years, which may increase or decrease up to 20.7 and 17.7 years, considering the low and high discharges of the dry and wet seasons, respectively.

## Discussion

According to the $D_L$ criteria proposed by Hutchison [26], $D_L >$ 2.5 and <5.0, the shoreline form of Lake Palmas is subrectangular elongated. Although this $D_L$ range was thought to designate lakes in overdeepened valleys associated with tectonic grabens or glaciated fjords, the geomorphology of Lake Palmas is associated with fluvial erosional and depositional processes in alluvial valleys and with Holocene sea level transgressions and regressions [21]. The relatively deep valley, from elevations up to 70 m at the Barreiras Formation plateaus down to – 30 m below sea level ($Z_{max} = -50.7$ m), may be associated with neotectonic processes, with valley alignments along the NW-SE axis [20]. The neotectonic hypothesis has also been supported by geophysical studies at Lake Juparanã (62.0 km$^2$) [32], the largest lake in the LDRV.

The high $V_d$ and $Z_r$ values indicate that the basin form of Lake Palmas is an overdeepened valley with a relatively flat bottom. The mean lake slope is moderate despite steep areas at the SE shore of the lake and around the S section of the land promontory at the confluence of the lake axes. In addition, the relative hypsographic area and volume curves indicate a linear basin, an intermediate profile between concave and convex basins. Nevertheless, the basin linear profile represents the major component of water storage for the pelagic volume (96.3%). This characteristic is supported by the low BPI value and the low A:V ratio, reinforcing the relatively deep morphology of Lake Palmas. These metrics also emphasize a low potential for lake water evaporation and a higher potential for water column resistance to mixing.

The basin slope influences the processes of sediment erosion, transport and deposition as well as macrophyte biomass. Based on the critical value of 5.3% used to differentiate gentle slopes from steep slopes, the areas with gentle slopes (below 5.3%) are characterized by the deposition of fine sediments and the thriving stands of rooted macrophytes. Nearshore steep slopes support erosion and transport processes and decrease macrophyte biomass [8]. The dynamic ratio (DR) of 0.15 indicates the predominance of slope processes in view of the threshold value of 0.25, with higher values indicating the predominance of resuspension from wind and wave action [11].

Another feature of deep basins is the significant volume of the aphotic zone (58.0% of lake volume), which may constrain the development and distribution of the photosynthetic biota. Consequently, the limited littoral zone (3.7% of lake volume) associated with gentle slopes (<5.3%) [8] restricts the habitat of rooted submersed macrophytes to less than 1.0% of the lake bottom area. The depth of 10.0 m, corresponding to 1.0 atm of hydrostatic

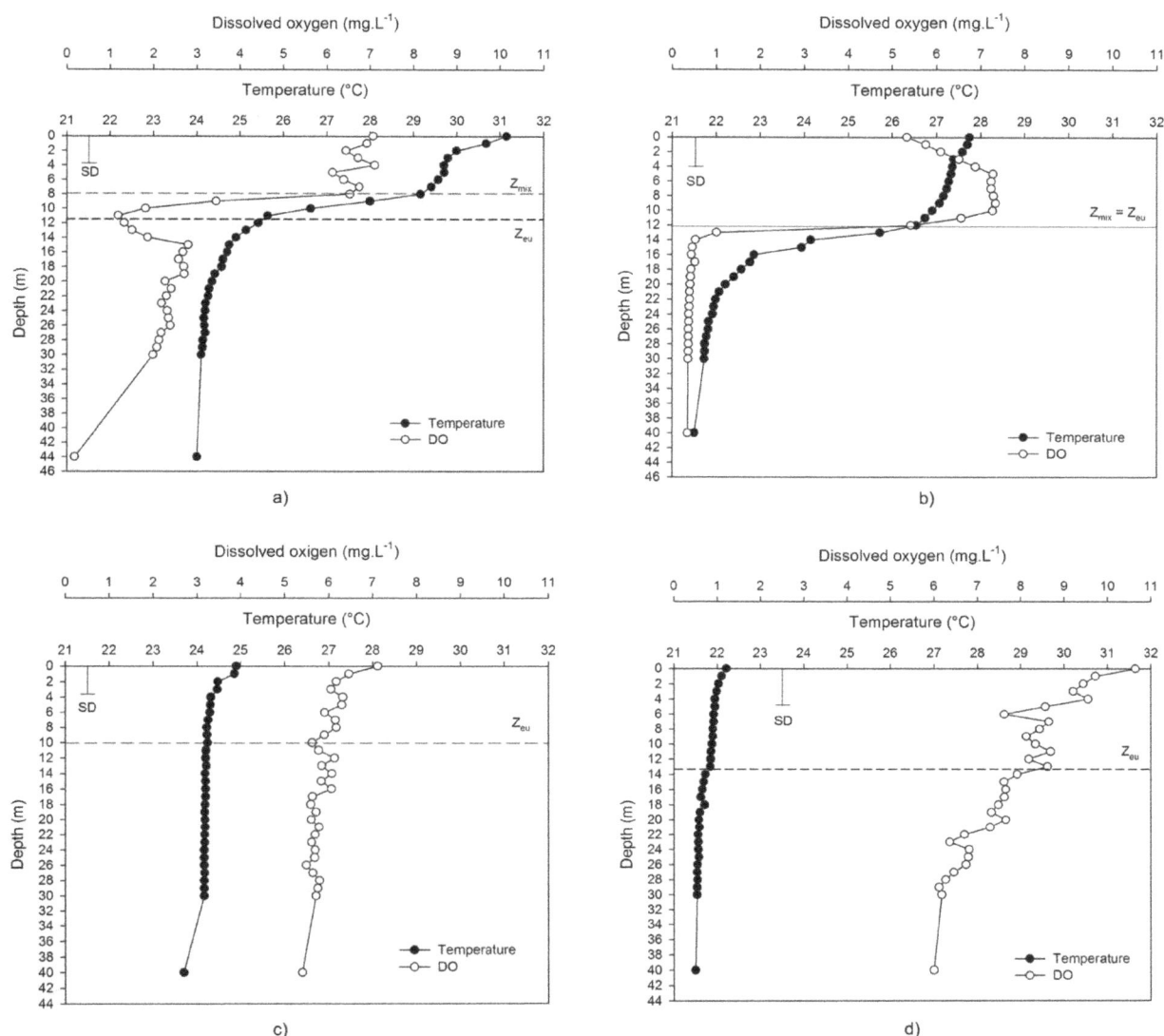

**Figure 5. Typical vertical profiles of temperature and dissolved oxygen with mixing ($Z_{mix}$) and euphotic ($Z_{eu}$) zones and Secchi disk depth (SD) at the deepest Lake Palmas sampling site during the wet/warm (a and b) and dry/mild cold (c and d) seasons: a) 03/07/12, b) 03/12/13, c) 07/24/12 and d) 09/04/13.**

pressure, is a threshold for the vascular system of angiosperms and defines the boundary of the lower infralittoral zone [33].

Based on published studies (Table S2), it seems that, Lake Palmas is the deepest natural lake in Brazil in terms of both $Z_{max}$ and $Z_{mv}$. In the light of the predominance of the fluvial and coastal geomorphological genesis of natural lakes in Brazil, these depths are remarkable. Lake Dom Helvécio (A = 5.3 km$^2$), in the MDRV (State of Minas Gerais), was formerly considered the deepest natural lake in Brazil, with the following metrics: $Z_{max}$ = 39.2 m, $Z_{mv}$ = 11.3 m and V = 59.6×10$^6$ m$^3$ [18]. In the same Lake District of Lake Dom Helvécio, Bezerra-Neto, Briguent and Pinto-Coelho [34] determined $Z_{max}$ and $Z_{mv}$ values of 11.8 and 4.7 m, respectively, for Lake Carioca, with an area of 0.14 km$^2$ and a volume of 6.7×10$^5$ m$^3$. Schwarzbold and Schäfer [16] conducted an extensive survey of 61 coastal lakes of southern Brazil and found a lake as large as 802 km$^2$, with $Z_{max}$ = 4.0 m and $Z_{mv}$ = 2.5 m (Lake Mangueira), but Lake Figueira (7.1 km$^2$) was found to have the highest $Z_{max}$ and $Z_{mv}$ values, 11.0 and 5.7 m,

respectively. Recent hydrographic surveys conducted in other LDRV lakes determined $Z_{max}$ values of 33.9, 31.6 and 22.1 m for Lakes Nova (A = 15.5 km$^2$, $D_L$ = 4.5 and $Z_{mv}$ = 14.7 m), Palminhas (A = 8.8 km$^2$, $D_L$ = 8.1 and $Z_{mv}$ = 14.2) and Terra Alta (A = 3.9 km$^2$, $D_L$ = 3.1 and $Z_{mv}$ = 9.0), respectively [35]. Although Lake Juparanã has a surface area 6 times greater than that of Lake Palmas, the estimated $Z_{max}$ is approximately 20.0 m. Even when considering Amazon lakes associated with fluvial processes in the floodplain, it seems that these lakes are shallower comparing to the ones of LDRV. For instance Lake Calado (A = 8.0 km$^2$) [36] and Tupé (A = 0.6 km$^2$) [51] show during Amazon River peak flooding $Z_{max}$ of 12.0 and 6.0 m, respectively.

$Z_{wb}$ is a functional depth that separates areas of sediment transport occurring through resuspension via wind turbulence from areas of sediment accumulation with no resuspension. The concept is also very useful for delimiting the boundary between surface (epilimnetic) and bottom (hypolimnetic) waters [11]. Considering the variability of $Z_{mix}$ during the stratification season,

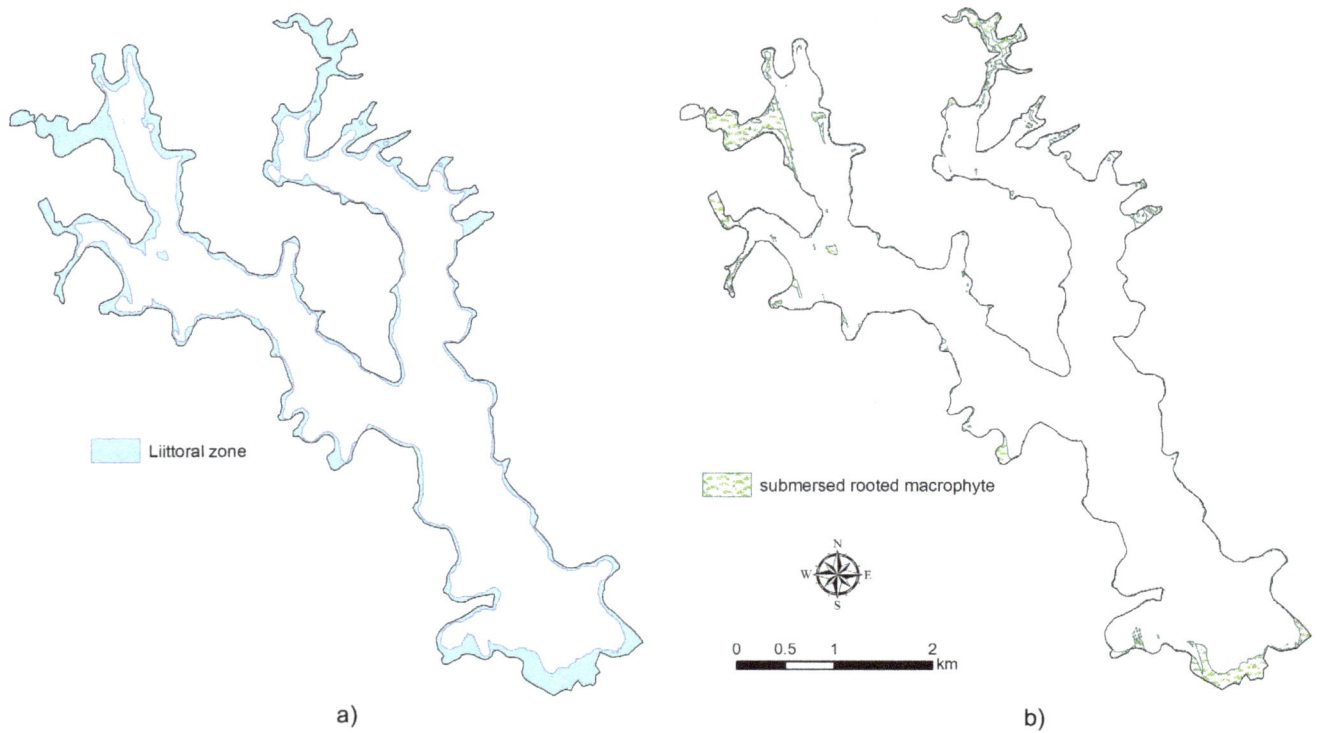

**Figure 6. Lake littoral zone: a) littoral zone with a threshold of a depth of 10.0 m for the euphotic zone; b) potential submersed macrophyte area with a depth of 10.0 m or less and gentle slopes (<5.33%).**

8.4±2.5 m, a $Z_{wb}$ value of 5.9 m can serve as an effective criterion to measure the significance of the physical and chemical stratification of Lake Palmas during the stratification season. Based on $Z_{wb}$, the epilimnion volume of Lake Palmas is only 26% of the lake volume. This value is consistent with the effects of the overdeepened basin on the resistance to mixing.

A moderate effective fetch ($L_{ef}$) may deepen the thermocline down to 12.0 m in the thermally stable water column of Lake Palmas. High $L_{ef}$ values usually occur with SE winds, which are characteristic of cold fronts. According to Marchioro [37], 3 and 16 cold fronts were recorded during summer and winter/spring 2011, respectively, for Vitória (ES), which is located 90 km south of Lake Palmas. These cold fronts, characterized by S, SW and

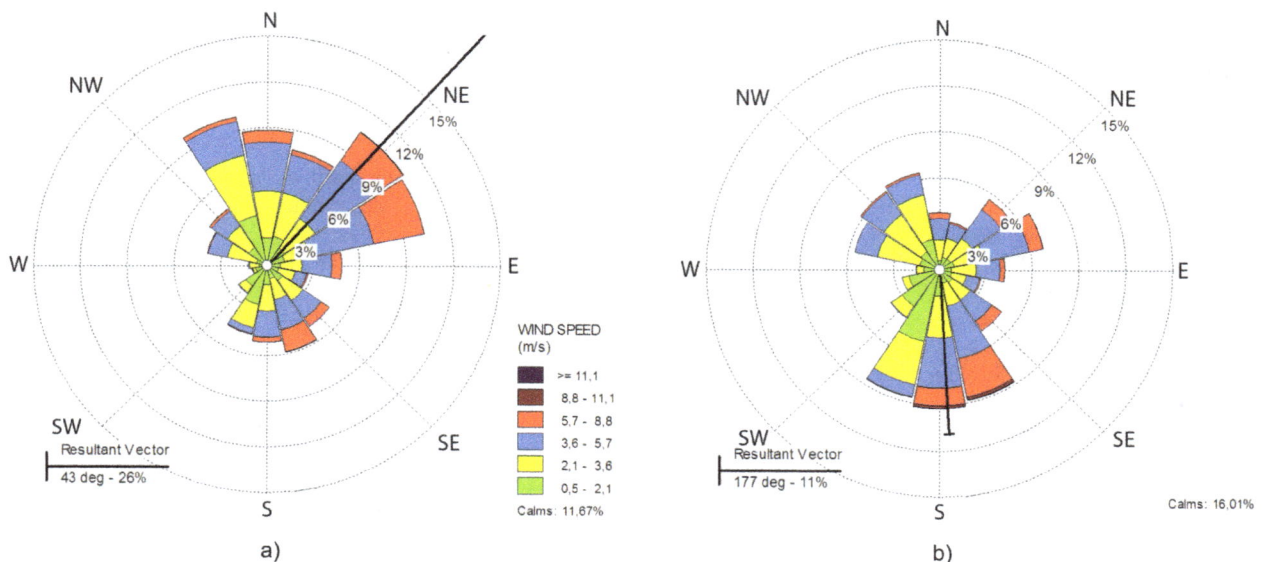

**Figure 7. Wind direction frequency (%) and intensity (m.s⁻¹) from Linhares meteorological station from 2007 to 2009: a) warm and wet months; b) dry/mild cold months.**

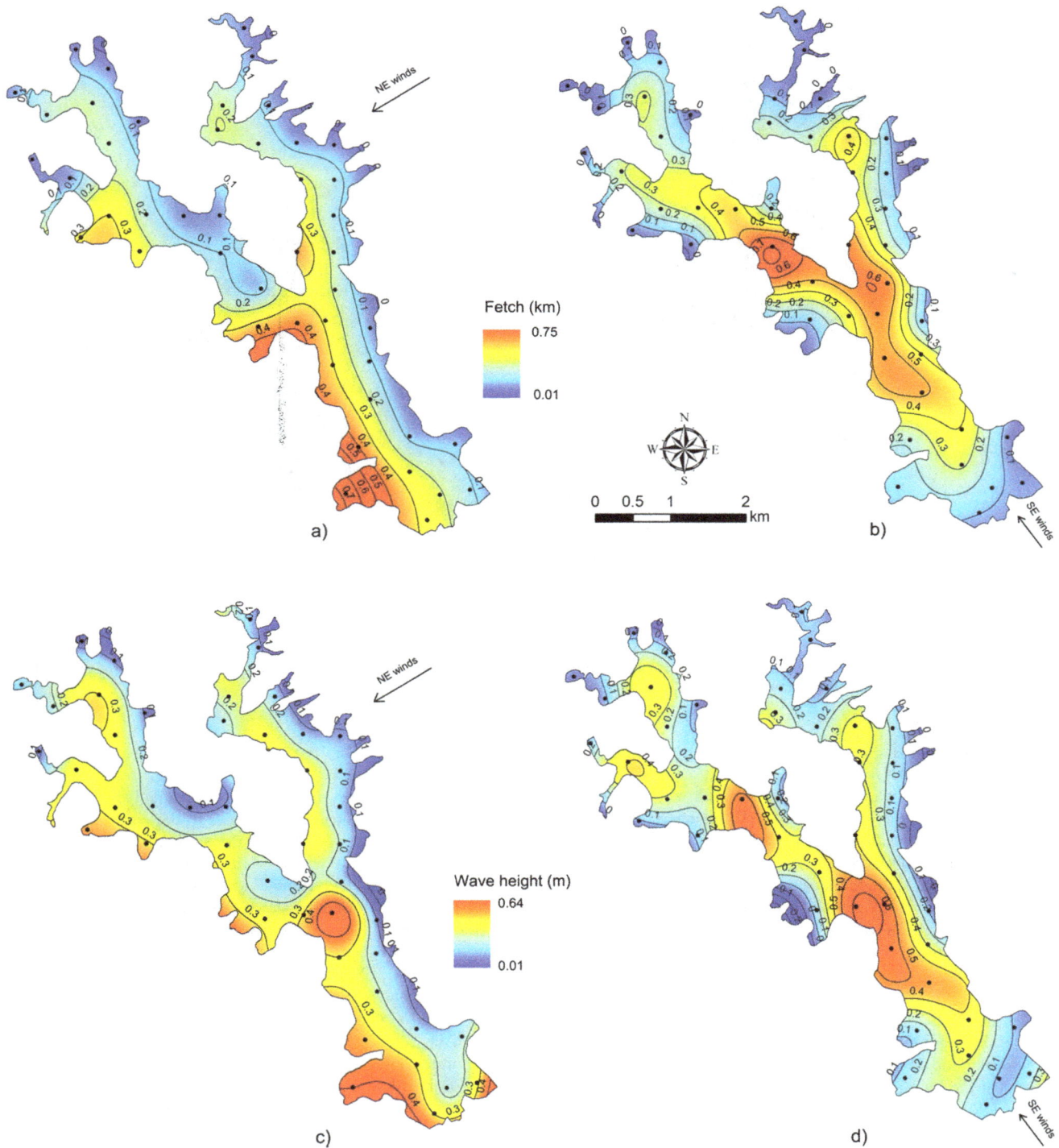

**Figure 8. Effective fetch (L$_{ef}$) and wave height (H) models.** a) L$_{ef}$ from NE winds; b) L$_{ef}$ from SE winds; c) wave heights from NE winds; and d) wave heights from SE winds. Contour lines show 0.1 m and km intervals.

SSE winds blowing up to 8.8 m.s$^{-1}$, may produce an average air temperature decrease of 7.1°C relative to the previous day's temperature. On average, cold fronts may persist up to 3.3 days. Tundisi et al. [38] have reported that the incidence of cold fronts may cause vertical mixing during the summer in reservoirs in Brazil.

The significant volume of hypoxic/anoxic bottom waters remains for the entire wet/warm season. This finding implies severe dissolved oxygen deficits and, as a consequence, the potential release of dissolved inorganic nutrients and heavy metals that are chemically bonded to the sediments. Nevertheless, Lake Palmas can be considered an oligotrophic ecosystem with a low

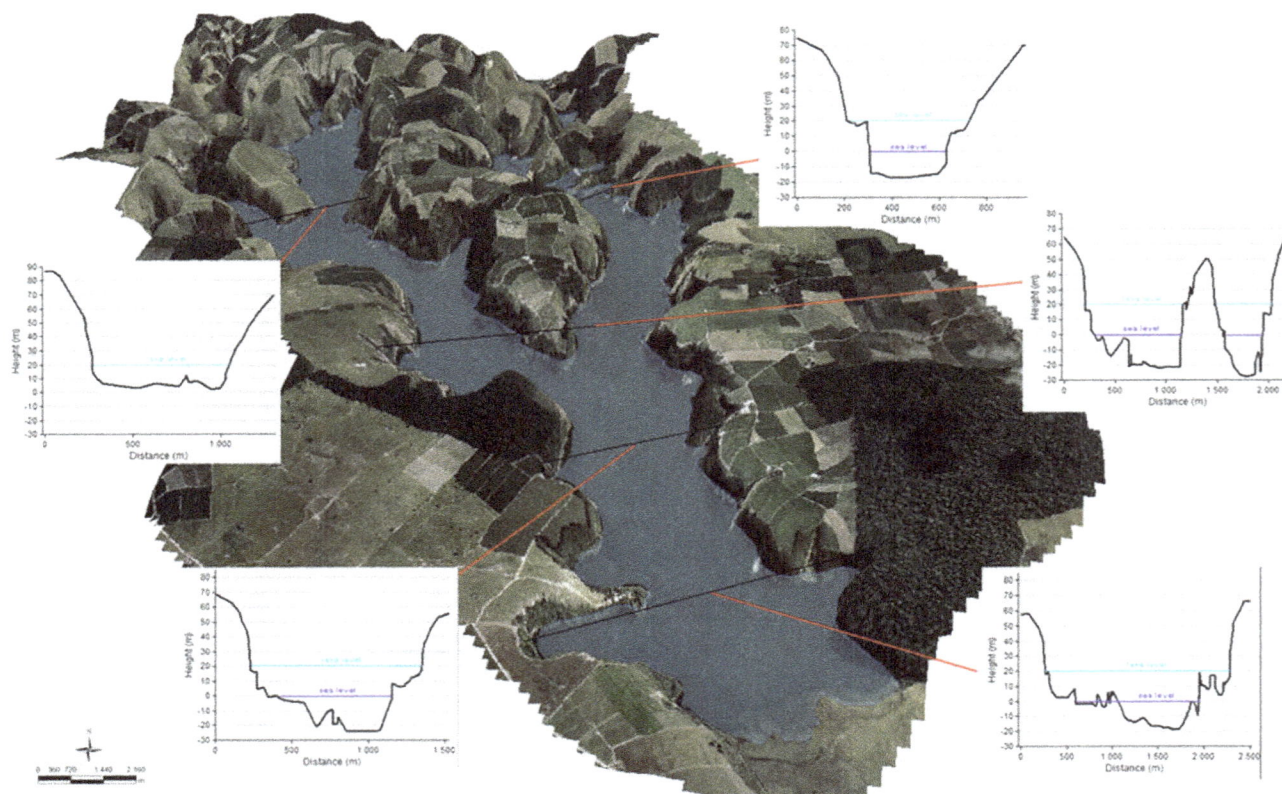

**Figure 9. 3D view of the terrain model and height profiles: the light blue line refers to the lake level, the dark blue line to sea level.** The cryptal volume is the volume below the dark blue line.

phytoplankton biomass, e.g., a mean chlorophyll $a$ value of less than $1.0 \ \mu g.L^{-1}$ [36]. Given that $Z_{eu}$ reaches the metalimnion and hypolimnion layers, it may produce a suitable climate with low light and rich nutrients, supporting the cyanobacteria community. This maximum in metalimnion phytoplankton pigments has been reported for Lake Dom Helvécio [39]. Lake Palmas may also exhibit maximum metalimnetic conditions with relative low concentrations of chlorophyll $a$. Despite the value of $Z_{eu}$:$Z_{mix}$ indicates that light is not a limiting factor for phytoplankton growth during the stratification season.

These metrics describe a relatively deep basin that promotes physical and chemical seasonal stratification with strong environmental gradients and a warm monomictic pattern. Stratification may inhibit phytoplankton biomass, producing an oligotrophic state despite anoxic bottom waters. These findings agree with the concept of the intermittent metabolism of the overdeepened lakes of the LDRV [19].

Wind climate is a key driving force in the deepening of the mixing depth during the stratified season. However, this process depends on the angle of incidence of the wind on the aquatic surface, i.e., the fetch. S winds associated with the occurrence of cold fronts approaching from the S of the continent are frequent during the dry/mild cold months. These winds blow along the major axis of the lake, which is aligned with the NW-SE direction of the drainage network. This condition, associated with lower thermal radiation, is effective in breaking down the thermal stability and mixing the entire water column. Under these circumstances, the effective Wedderburn number has low values, indicating low thermal stability.

The $W_A/A$ ratio of 16.27 indicates a relatively large drainage basin and a potentially significant discharge of river water into Lake Palmas, although the seasonality of rainfall and the potential negative effects of water and land uses in the watershed may halt the flow of tributary rivers during the dry season. The year-round unregulated use of the waters of the lake and tributary rivers for irrigation may also change the water balance in the lacustrine basin. Land use in the lake watershed is predominantly allocated to pasture, agriculture and forestry, representing 62.4% of the watershed area, whereas forested areas occupy only 32.8% of the watershed area [22]. If 30 and 100 m buffer areas are considered along the tributary river network and along the shoreline of Lake Palmas, agroecosystems represent up to 76.8 and 79.7%, respectively, of the buffer areas.

Accurate theoretical retention time (RT) estimates should be based on best knowledge of hydrological flows of tributary rivers inputs, evaporation rate, groundwater exchanges, and water consumption rates, instead of the simple ratio of the inflow to lake volume. Only 0.5% of the total number of natural lakes are known morphologically and hydrologically [40] in contrast, estimates of RT data are usually available for reservoirs construction and management. As rules of thumbs, RT is longer in deeper basin, reservoirs show shorter RT than lakes, and surface outflow from the lakes compared to deep outflow from the reservoirs. Reservoir RT in general is less than a year, with a threshold for reservoir limnology below 200 days [41,42]. The effect of flow is very significant for reservoir ecological structure and functioning, with RT >1 year show trend to stratification, eutrophication, anoxic bottom and recurrency/persistence of cyanobacteria bloom [42]. Blooms of cyanobacteria have been recorded in Funil reservoir, a tropical system in Brazil, despite the very short RT (annual mean of 41.5 days). High inputs of allochthonous nutrients from the reservoir watershed promote the eutrophic status of this artificial lake. The increase of residence time, up to 80 days during the dry season, promotes spatial variability in the ecological structure of the reservoir along the river-dam axis [43].

The RT of 19.4 years for Lake Palmas can be considered very long, particularly for the regional wet climate (annual mean of 1,027 mm.y$^{-1}$). In order to put it in perspective, and despite the lack of RT data for Brazilian natural lakes, Lakes in the Yunnan plateau, southwest of China, such as the oligotrophic Lakes Fuxianhu (A = 211.0 km$^2$, $Z_{mv}$ = 89.6 m and V = 18.9×10$^9$ m$^3$), Luguhu (A = 48.4 km$^2$, $Z_{mv}$ = 40.3 m and V = 1.9×10$^9$ m$^3$) and Cheghai (A = 77.2 km$^2$, $Z_{mv}$ = 25.7 m and V = 1.9×10$^9$ m$^3$) show RT of 35.5, 17.7 and 12.5 years, respectively [44]. These longer RTs imply in poorly flushed systems with negative correlations with total nitrogen and phosphorus and chlorophyll a. With such long RT of overdeepened lakes of Yunnan plateau are associated with oligotrophic systems, acting as a sink for inorganic and organic matter and with a delay response to additional nutrient inputs from lake watershed. In another hand, longer RT may imply in lack of resiliency after ecosystem distress from cultural eutrophication [44]. In addition to information about phosphorus and nitrogen loads input to the lake, knowledge of the theoretical residence time is a key factor for regulating the uses of Lake Palmas, such estimating lake carrying capacity for fish farming.

Hazards to water quantity and quality caused by pollution, silting and the introduction of exotic species may impair the ecological resilience of the lake. In addition, climate changes involving the intensification of extreme hydrological events (specifically, a predicted shortening of the rainfall season with fewer rainy days but with more intense and frequent storm events) [45] can be a major driver of shifts in lacustrine ecosystems in the LDRV. In December 2013, an extreme amount of regional precipitation, 650 mm of rainfall, 3 times the month mean rainfall at Linhares meteorological station, caused a major flood in the LDRV. Aditionaly, it must be considered a scenario of lake surface warming as a consequence of global warming. For deep lakes this scenario implies in an increasing loss of energy through evaporation, a deeper mixing layer, and an earlier summer stratification. These factors may lead monomitic lakes to turn into holo-oligomictic, with a complete vertical mixing occurring eventually in some years when stratification become weaker during winter [46].

Water uses in the lake watershed may also increase the stress on the theoretical retention time of Lake Palmas. Of the 5 tributary rivers, 2 showed no discharge at all at least 5 times. Rivers also became dry during the wet/warm season. These events might have resulted from river damming for irrigation purposes, given that 55 small reservoirs for irrigation purposes with an area up to 0.6 km$^2$ have been mapped in the Lake Palmas watershed [22].

Climate, lake morphology, and edaphic factors have been considered key drivers for trophic status of lakes, including the overdeepen basins. However, human impact factors have also been recognized, in some cases, as the leading driver for cultural eutrophication. The intensification of land and water uses in lake watersheds highlight the urgent need to regulate these uses in order to maintain a healthy lake ecosystems. In addition, climate change effects on water balance and related threats to ecosystem resilience and water security must be recognized.

## Conclusions

The subrectangular elongated shape and the relatively over-deepened basin of Lake Palmas place most of the lake's volume in the pelagic compartment. Approximately one-half of the lake's volume is within the aphotic zone. The overdeepened basin promotes the physical and chemical stratification of the water column during the wet/warm months of the year. Under these conditions, only a small part of the lake volume is prone to mixing effects, and a large volume remains hypoxic/anoxic. During the dry/mild cold months, the predominance of S-SE winds, characteristic of the arrival of cold fronts, and the high effective fetch of these winds on the basin aligned along a NW-SE axis effectively promote the mixing of the lake's water column. Thus, the thermal pattern of Lake Palmas is warm monomictic. This finding is consistent with the hypothesis of a pattern of intermittent metabolism in the overdeepened lakes of the LDRV. Based on published data, Lake Palmas seems to be considered the deepest natural lake in Brazil in terms of both its maximum and mean depths. Given the very long theoretical retention time of Lake Palmas, hydrological changes in tributary rivers may increase the retention time and foster water quantity and quality problems. There are warning signs that the water balance in the basin is under pressure due to the unregulated uses of water for the year-round irrigation of croplands.

## Supporting Information

**Figure S1   Semivariogram for kriging interpolation of point data to generate a continuous surface describing the lake depth measurements.**
(TIF)

**Figure S2   Hypsographic curves of percent total surface (a) and total volume (b).**
(TIF)

**Table S1   Descriptive statistics of limnological variables in wet/warm and dry/mild cold seasons (2011 to 2013).** Data from field vertical profiling at 4 sampling sites" from surface to the bottom.
(DOCX)

**Table S2   Morphometry of natural lakes in Brazil deeper than 6.0 m.**
(DOCX)

## Author Contributions

Conceived and designed the experiments: GFB. Performed the experiments: GFB MAG FCG. Analyzed the data: GFB MAG FCG. Contributed reagents/materials/analysis tools: GFB. Contributed to the writing of the manuscript: GFB MAG FCG.

## References

1. Cole GA (1994) Textbook of limnology. 4$^{th}$ ed., Prospect Heights: Waveland Press, Inc. 412p.
2. Kling GW (1988) Comparative transparency, depth of mixing, and stability of stratification in Lakes of Cameroon, West Africa. Limnology and Oceanography 33(1): 27–40.
3. Ambrosetti W, Barbanti L (2002a) Physical limnology of Italian lakes. 1. Relationship between morphometry and heat content. Journal of Limnology 61(2): 147–157. DOI:10.4081/jlimnol.2002.147.
4. Ambrosetti W, Barbanti L (2002b) Physical limnology of Italian lakes. 2. Relationships between morphometric parameters, stability and Birgean work. Journal of Limnology 61(2): 159–167. DOI:10.4081/jlimnol.2002.159.

5.  Kvarnäs H (2001) Morphometry and hydrology of the four large lakes of Sweden. Ambio 30(8): 469–474. DOI:10.1579/0044-7447-30.8.467.

6.  Blais JM, Kalff J (1995) The influence of lake morphometry on sediment focusing. Limnology and Oceanography 40(3): 582–588.

7.  Rasmussen JB, Godbout L, Schallenberg M (1989) The humic content of lake water and its relationship to watershed and lake morphometry. Limnology and Oceanography 34(7): 1336–1343.

8.  Duarte CM, Kalff J (1986) Littoral slope as a predictor of the maximum biomass of submersed macrophyte communities. Limnology and Oceanography 1(5): 1072–1080.

9.  Fee EJA (1980) A relation between lake morphometry and primary productivity and its use in interpreting whole-lake eutrophication experiments. Limnology and Oceanography 24(3): 401–416.

10. Staehr PA, Baastrup-Spohr L, Sand-Jensen K, Stedmon C (2012) Lake metabolism scales with lake morphometry and catchment conditions. Aquatic Sciences 74(1): 155–169. DOI:10.1007/s00027-011-0207-6.

11. Håkanson L (2004) Lakes: form and function. Cladwell: The Blackburn Press. 201p.

12. Hollister J, Milstead WB (2009) Using GIS to estimate lake volume from limited data. Lake and Reservoir Research and Management 26(3): 194–199. DOI:10.1080/07438141.2010.504321.

13. Hollister JW, Milstead WB, Urrutia MA (2011) Predicting maximum lake depth from surrounding topography. PLoS ONE 6(9): e25764. DOI:10.1371/journal.pone.0025764.

14. Nõges T (2009) Relationships between morphometry, geographic location and water quality parameters of European lakes. Hydobiologia 633(1): 33–43. DOI:10.1007/s10750-009-9874-x.

15. Kosten S, Huszar V, Mazzeo N, Scheffer M, Sternberg LSL, et al. (2009) Lake and watershed characteristics rather than climate influence nutrient limitation in shallow lakes. Ecological Applications 19(7): 1791–1804. DOI:10.1890/08-0906.1.

16. Schwarzbold A, Schäfer A (1984) Genesis and morphology of Rio Grande do Sul (Brazil) costal lakes. Amzoniana. 9(1): 87–104 (In Portuguese).

17. von Sperling E (1999) Morphology of lakes and reservoirs. Belo Horizonte: Ed. Universidade Federal de Minas Gerais – UFMG, 137p (In Portuguese).

18. Bezerra-Neto JF, Pinto-Coelho RM (2008) Morphometric study of Lake Dom Helvécio, Parque Estadual do Rio Doce (PERD), Minas Gerais, Brazil: a re-evaluation. Acta Limnologica Brasiliensia 22(2): 161–167 (In Portuguese).

19. Bozelli RL, Esteves FA, Roland F, Suzuki MS (1992) Lakes of the lower Doce River: abiotic variables and chlorophyll a (Espírito Santo – Brazil). Acta Limnologica Brasiliensia. 4: 13–21 (In Portuguese).

20. Bricalli LL, Mello CL (2013) Lineament patterns related to lithostructural and neotectonic fracturing (State of Espírito Santo, Southeastern Brazil). Revista Brasileira de Geomorfologia, v. 14, p. 301–311 (In Portuguese).

21. Martin L, Suguio K, Flexor JM, Archanjo JL (1996) Coastal quaternary formations of the southern part of the State of Espírito Santo (Brazil). Anais da Academia Brasileira de Ciências 68(3): 389–404.

22. Barroso GF, Mello FA de O (2013) Landscape compartments and indicators of environmental pressures on fluvial and lacustrine ecosystems of the Lower Doce River Valley. Proceedings of the 15th Brazilian Symposium of Applied Physical Geography. Vitória, UFES, 158–165p. Available: http://www.xvsbgfa2013.com.br/anais/ (In Portuguese).

23. Isaaks EH, Srivastava RM (1989) Applied geostatistics. Oxford: Oxford University Press.

24. Burrough PA, MacDonnell RA (1998) Principles of geographical informations systems. Oxford: Oxford University Press.

25. Håkanson L (1978) Optimization of lake hydrographic surveys. Water Resources Research 14(4): 545–560.

26. Hutchinson GE (1957) A treatise on limnology. Volume I: Geography, physics and chemistry. New York: John Wiley & Sons, Inc.

27. Dadon JR (1995) Calor y temperatura en cuerpos lenticos. In: Lopretto EC, Tell G. Ecosistemas de aguas continentals: metodologias para su studio. Vo. 1, La Plata: Ediciones SUR, 47–56.

28. Imberger J, Hambling PF (1982) Dynamics of lakes, reservoirs and cooling ponds. Annual Review of Fluid Mechanichs 14, 153–87. DOI:10.1146/annurev.fl.14.010182.001101.

29. Reynolds CS (2006) The ecology of phytoplankton. Ecology, Biodiversity and Conservation. Cambridge: Cambridge University Press, 537 p.

30. Kerekes J (1977) The index of lake basin permanence. Internationale Revue der Gesamten Hydrobiologie und Hydrographie. 62(2): 291–293.

31. Håkanson L (1983) Principles of lakes sedimentology. Cladwell, The Blackburn Press.

32. Hatushika RS, Silva CG, Mello CL (2007) High resolution seismic stratigraphy of Lake Juparanã, Linhares (ES, Brazil) as study aid for sedimentation and quaternary tectonics. Revista Brasileira de Geofísica. 25(4): 433–442. DOI:10.1590/S0102-261X2007000400007 (In Portuguese).

33. Wetzel RG (2001) Limnology: lake and river ecosystems. 2nd ed., San Diego, Academic Press, 1006p.

34. Bezerra-Neto JF, Briguent LS, Pinto-Coelho RM (2010) A new morphometric study of Carioca Lake, Parque Estadual do Rio Doce (PERD), Minas Gerais State, Brazil. Acta Scientiarum. Biological Sciences 32(1): 49–54. DOI:10.4025/actascibiolsci.v32i1.4990 (In Portuguese).

35. Barroso GF, Garcia F da C, Gonçalves MA, Martins, FC de O, Venturini JC, et al. (2012) Integrated studies on the lacustrine system of the Lower Doce River Valley (Espírito Santo). Proceedings of the I National Seminar of Sustainable Management of Aquatic Ecosystems: Complexity, interactivity and Ecodevelopment. Arraial do Cabo, COPPE/UFRJ, 21 a 23 de março. 7p. Available: http://gestaoecossistemas.files.wordpress.com/2012/11/1-i-1-estudos-integrados-no-sistema-lacustre-do-baixo-rio-doce-espc3adrito-santo.pdf (In Portuguese).

36. Melack JM, Fisher TR (1990) Comparative limnology of tropical lakes with emphasis on the central Amazon. Acta Limnologica Brasiliensia 3: 1–48.

37. Marchioro E (2012) Incidence of cold fronts in Vitória, ES, Brazil. Acta Geografica, v. 7, p. 49–60. DOI:10.5654/actageo2012.0002.0003 (In Portuguese).

38. Tundisi JG, Matsumura-Tundisi T, Pereira KC, Luzia AP, Passerini MD, et al. (2010) Cold fronts and reservoir limnology: an integrated approach towards the ecological dynamics of freshwater ecosystems. Brazilian Journal of Biology, 70(3 Suppl): 815–824. DOI:0.1590/S1519-69842010000400012.

39. Souza MBG, Barros CFA, Barbosa F, Hajnal É, Padisák J (2008) Role of atelomixis in replacement of phytoplankton assemblages in Dom Helvécio Lake, South-East Brazil. Hydrobiologia 607: 211–224. DOI:10.1007/s10750-008-9392-2.

40. Ryanzhin SV (1999) Fundamental limnological processes and relevant indicators used in lake studies. Kondratyev, K. Y. and Filatov, N. Limnology and Remote Sensing: A Contemporary Approach, Springer: 53–78.

41. Straškraba M (1998) Coupling of hydrobiology and hydrodynamics: lakes and reservoirs. Coastal and Estuarine Studies 288: 601–622.

42. Straškraba M (1999) Retention time as a key variable of reservoir limnology. In: Tundisi JG, Straškraba M. Theoretical reservoir ecology and its applications. São Carlos, Brazilian Academy of Sciences/International Institute of Ecology/Backhuys Publishers: 385–410.

43. Soares MCS, Marinho MM, Azevedo SMOF, Branco CWC, Huszar VLM (2012) Eutrophication and retention time affecting spatial heterogeneity in a tropical reservoir. Limnologica 42(3): 197–203. DOI:10.1016/j.limno.2011.11.002.

44. Liu W, Zhang Q, Liu G (2011) Lake eutrophication associated with geographic location, lake morphology and climate in China. Hydrobiologia 644(1): 289–299.

45. PBMC (2013) Contributions of the Working Group 2 to the First National Assessment Report of National Climate Change Panel. Executive Summary GT2: Impacts, vulnerability and adapting, PBMC, Rio de Janeiro (In Portuguese).

46. Ambrosetti W, Barbantti L, Sala N (2003) Residence time and physical processes in lakes. Journal of Limnology 62(1): 1–15.

47. Henry R, Tundisi JG, Calijuri M do C, Ibañez M do SR (1997) A comparative study of thermal structure, heat content and stability of stratification in three lakes. In Tundisi JG, Saijo Y (eds.) Limnological studies on the Rio Doce Lakes, Brazil. Brazilian Academy of Sciences. University of São Paulo. School of Engineering at São Carlos. Center for Water Resources and Applied Ecology. 528p.

48. Barros CF de A, dos Santos AMM, Barbosa FAR (2013) Phytoplankton diversity in the middle Rio Doce lake system of southeastern Brazil. Acta Botanica Brasilica 27(2): 327–346.

49. Brighenti LS, Pinto-Coelho RM, Bezerra-Neto JF, Gonzaga AV (2011) Morphometric features of Lake Central (Lagoa Santa, Minas Gerais State): comparison of two methodologies. Acta Scientiarum. Biological Sciences 33(3): 281–287 DOI:10.4025/actascibiolsci.v33i3.5545 (In Portuguese).

50. Nogueira F, Souza MD, Bachega I, Silva RL (2002) Seasonal and diel limnological differences in a tropical floodplain lake (Pantanal of Mato Grosso, Brazil). Acta Limnologica Brasiliensia 14(3): 17–25.

51. Aprile FM, Darwich AJ (2005) Geomorphological models for Lake Tupé. In: dos Santos-Silva E.N., Aprile FM, de Melo S, Scudeller VV (eds.). Biotupé: Physical, biodiversity and sociocultural dimensions of the Lower Negro River, Central Amazon. Manaus, AM: Editora INPA, p. 03–17. (In Protuguese).

52. Panosso R (2000) Geographical and gemorphological considerations. In: Bozelli RL, Esteves FA, Roland F (eds). Lake Batata: impacts and restoration of an Amazon ecosystem. IB-UFRJ/SBL. Rio de Janeiro. 342 p. (In Protuguese).

53. Herdendorf CE (1984) Inventory of the morphometric and limnologic characteristics of the large lakes of the world. Technical Bulletin OHSU-TB-017, Columbus, The Ohio State University, 78p.

54. Niencheski LFH, Baraj E, Windom HL, França RG (2004) Natural background assessment and Its anthropogenic contamination of Cd, Pb, Cu, Cr, Zn, Al and Fe in the sediments of the southern area of Patos Lagoon. Journal of Coastal Research SI39: 1040–1043.

55. Llames ME, Zagarese HE (2010) Lakes and reservoirs of the world: South America. In: Likens GE. (ed.). Lake ecosystem ecology: a global perspective. San Diego, Academic Press: 332–340.

56. Hennemann MC, Petrucio MM (2011) Spatial and temporal dynamic of trophic relevant parameters in a subtropical coastal lagoon in Brazil. Environmental Monitoring and Assessment (Print) 181: 347–361.

57. Kjerfve B, Schettini CAF, Knoppers B, Lessa G, Ferreira HO (1996) Hydrology and salt balance in a large, hypersaline coastal lagoon: Lagoa de Araruama, Brazil. Estuarine, Coastal and Shelf Science 42: 701–725.

# 22

# Emissions from Pre-Hispanic Metallurgy in the South American Atmosphere

**François De Vleeschouwer**[1,2]*, **Heleen Vanneste**[1,2], **Dmitri Mauquoy**[3], **Natalia Piotrowska**[4], **Fernando Torrejón**[5], **Thomas Roland**[6,7], **Ariel Stein**[8], **Gaël Le Roux**[1,2]

1 Université de Toulouse, INP, UPS, EcoLab (Laboratoire Ecologie Fonctionnelle et Environnement), ENSAT, Castanet Tolosan, France, 2 CNRS, EcoLab, Castanet Tolosan, France, 3 School of Geosciences, University of Aberdeen, Aberdeen, United Kingdom, 4 Department of Radioisotopes, Institute of Physics, Silesian University of Technology, Gliwice, Poland, 5 Environmental Sciences Center EULA-Chile, University of Concepción, Concepción, Chile, 6 Geography, College of Life and Environmental Sciences, University of Exeter, Exeter, United Kingdom, 7 Palaeoenvironmental Laboratory (PLUS), Geography and Environment, University of Southampton, Southampton, United Kingdom, 8 NOAA/Air Resources Laboratory, R/ARL - NCWCP, College Park, Maryland, United States of America

## Abstract

Metallurgical activities have been undertaken in northern South America (NSA) for millennia. However, it is still unknown how far atmospheric emissions from these activities have been transported. Since the timing of metallurgical activities is currently estimated from scarce archaeological discoveries, the availability of reliable and continuous records to refine the timing of past metal deposition in South America is essential, as it provides an alternative to discontinuous archives, as well as evidence for global trace metal transport. We show in a peat record from Tierra del Fuego that anthropogenic metals likely have been emitted into the atmosphere and transported from NSA to southern South America (SSA) over the last 4200 yrs. These findings are supported by modern time back-trajectories from NSA to SSA. We further show that apparent anthropogenic Cu and Sb emissions predate any archaeological evidence for metallurgical activities. Lead and Sn were also emitted into the atmosphere as by-products of Inca and Spanish metallurgy, whereas local coal-gold rushes and the industrial revolution contributed to local contamination. We suggest that the onset of pre-Hispanic metallurgical activities is earlier than previously reported from archaeological records and that atmospheric emissions of metals were transported from NSA to SSA.

**Editor:** John P. Hart, New York State Museum, United States of America

**Funding:** This research is supported by a Young Researcher Grant of the Agence Nationale de la Recherche (ANR) to F. De Vleeschouwer (Project ANR-2011-JS56-006-01 "PARAD"). The funders had no role in study design, data collection and analysis, decision to publish, or preparation of the manuscript.

**Competing Interests:** The authors have declared that no competing interests exist.

* Email: francois.devleeschouwer@ensat.fr

## Introduction

### 1. Background

Population expansion and territorial colonisation increased over the course of the last 5000 years in South America. Pre-Colombian civilizations flourished first in the Northern Andes and populations progressively migrated to the South. South American animal domestication and agriculture was followed by several periods of population expansion and metallurgical activities. In particular, copper extraction and smelting started in northwestern South America as well as in Argentina around 1400 BC and spread with the various South American civilizations (Chavin, Nasca, Tiwanaku, Inca). Incas also increasingly used silver, mainly around Peru, the Titicaca basin and Potosi (Bolivia). These changes in land use as well as the extraction and processing of metals up to the present day have released detectable amounts of trace elements into the atmosphere during the second half of the Holocene (*ca.* 4 kyrs cal BP to present).

Whereas South American metallurgy has been documented by archaeological finds, there is no record of how the resulting trace metals dispersed into the South American atmosphere through time and how far they have been transported away from production sites. To estimate the extent of anthropogenic trace metal emissions in South America, continuous and well-dated records of long-term metal deposition away from production centres are needed to highlight ancient metallurgical activities, given the absence/lack of archaeological evidence.

Our study of an ombrotrophic peat profile from Tierra del Fuego provides a continuous history of metallurgical activities in South America, and is therefore a valuable alternative to discontinuous archaeological findings. Andean trace metals of anthropogenic origin are found in our peat record from Tierra del Fuego, which suggests that they have been dispersed widely from their source areas. The data suggest for the first time that trace metals from pre-Hispanic and Hispanic Andean metallurgical activities were transported from North to South America, due to occasional North to South wind trajectories.

## 2. Scope

Continental archives, in particular raised/ombrotrophic peatlands or bogs have proven to be very useful in reconstructing past environmental changes and human activities [1–3]. Peat bogs are exclusively fed by atmospheric inputs and can therefore provide key archives of atmospheric metal deposition through time [3]. Long distance transport of trace metals is possible and hence suitable peat archive records can be used to identify the history of metallurgy even in remote areas [4–6]. In this study, we reconstruct the timing of metal deposition in SSA during the last 4200 years using a high-resolution record of metal/La ratios as well as Pb isotopes in a radiocarbon-dated peat core from Karukinka bog, located in the central western part of Isla Grande de Tierra del Fuego (Fig. 1). The ombrotrophic section of the core provides a reliable record of trace metallurgical activities throughout the Andes because 1/peat is mostly (>95%) composed of organic matter, hence trace to ultra-trace amounts of metals are detectable, 2/Modern-time particle back-trajectories supported by studies showing analogies between past and present climate [7–10] prove the feasibility of atmospheric transport from NSA to SSA and 3/the low level of local environmental and atmospheric contamination (no metallurgical activities before the 20th century) provides ideal background conditions to record long-range trace metal deposition.

## Materials and Methods

### 1. Location

Karukinka Park is a protected area managed by the Chilean branch of the Wildlife Conservation Society (WCS) and located at the southwestern edge of Isla Grande de Tierra del Fuego. (Fig. 1). Access and coring permits were obtained by contacting Ricardo Muza at the WCS office in Punta Arenas (http://www.karukinkanatural.cl/). The mean annual rainfall is 400 mm.yr$^{-1}$ and the mean annual temperature is 5°C. The reserve contains numerous peat bogs dominated by *Sphagnum magellanicum* with a sporadic cover of *Marsippospermum grandiflorum* and *Empetrum rubrum*. Occasional pools on the peatlands contain *Sphagnum falcatulum* and are fringed by *Tetroncium magellanicum*. Dense stands of deciduous forests surround the peat bog ecosystems and are dominated by *Nothofagus pumilio*. Peat deposits have accumulated continuously in this zone since the early Holocene (*ca.* 10kyrs cal BP) [*11,12*] Several bogs were identified as ideal dust traps because they are located on high altitude points in open areas solely dominated by *Sphagnum magellanicum* mosses. The site we sampled is a small 1-km diameter peat bog (S 53.86002°, N 69.57639°, 245 m a.s.l.) located at a relatively high altitude, 200 m aside from the main valley and 20 m above the main river. Given this, significant minerogenic input of river sediments is highly unlikely.

### 2. Coring

Peat profiles were recovered from Karukinka bog in February 2012 using a stainless steel Russian corer of 10 cm internal diameter and 50 cm length [13]. The barrel of the corer was cleaned between each section using deionized water. The cores were photographed, described, wrapped in plastic film and packed into PVC tubes, which were subsequently stored in wooden boxes and shipped to France. An overlapping core were collected in order to avoid any possible disturbance at the top and bottom of each master core section, which can occur because the tip of the corer can slightly disturb the directly underlying peat layer. However, this core was not analyzed because we did not detect any contamination in the master core. The maximum depth

(bottom of the peat bog) was reached at 4.5 m. In this study, we have only analysed the ombrotrophic part of the section (0–344 cm, see section 4.1).

## 3. Subsampling, depth correction and density calculation

A clean slicing and sub-sampling procedure was applied following published guidelines [14,15] to ensure minimal contamination and disturbance, whilst maintaining the highest sampling resolution. Samples were handled with powder free latex gloves. The cores were frozen at −20°C, unpacked and sliced contiguously at 1 cm resolution using a stainless steel band saw in a room specifically designed to process pristine samples. Each slice was cleaned with mQ (18 mega Ohm clean) water. The edges of each semi-cylindrical slice were then removed and the slices were subsampled using ceramic knives and plastic cutting boards. The dimensions of the still-frozen samples dedicated to geochemistry were measured using a vernier caliper in order to know the exact thickness of the sample and the loss of material due to each cut. The thickness of each sample ranges from 0.8 to 1.3 cm with 72% of the samples being 1.0±0.1 cm thick. The average loss on each cut (i.e. [total core length - cumulative sample thickness]/Nr of samples) is 0.18 cm. We have taken this loss due to cutting into account and reassessed the exact mid-point depth of each sample. All the subsamples were finally stored in a freezer at −20°C.

According to literature recommendations [15], the frozen samples dedicated for geochemistry were also measured for their lateral dimensions in order to accurately calculate their volume. These frozen samples were then freeze-dried to remove their water. The resulting dry samples were subsequently weighed and their densities calculated by dividing their dry weight by their respective volume.

## 4. Macrofossil analysis and selection for radiocarbon measurement

Plant macrofossil samples (5 cm$^3$) were boiled with 5% NaOH and sieved (mesh size 180 μm). Macrofossils were scanned using a binocular microscope (×10–×50), and identified using an extensive reference collection of type material collected during fieldwork in the study region in 2012. Plant macrofossils were assessed using the Quadrat and Leaf Count (QLC) method [16]. Volume percentages were estimated for all components with the exception of seeds, macroscopic charcoal particles, fungal fruit bodies and minerogenic grains, which were counted and expressed as the number (n) present in each sub-sample. The selection of plant macrofossils in 8 samples for AMS radiocarbon dating was made following established protocols [17]. The selection of aboveground plant macrofossils ensures that accurate dates are assigned (i.e. free of rootlets and therefore of contamination).

## 5. Radiocarbon sample preparation and measurement

The macrofossils from the 8 samples selected according to literature recommendations [15,18] for radiocarbon dating mainly consisted of *Sphagnum magellanicum* leaves and stems except the basal sample, which consisted of unidentified graminoids. The deepest dated sample is at 440.6 cm depth (below this depth and down to the base of the core at 450 cm, no datable macrofossils could be retrieved). Enough macrofossils (15–20 *Sphagnum* stems or graminoid fragments) were picked to ensure a minimum carbon mass of ca. 1 mg. All samples were prepared at the GADAM Centre (Gliwice, Poland) using acid-alkali-acid washing (to remove carbonate, bacterial $CO_2$ and humic/fulvic acid), drying, combustion and graphitisation [19]. Radiocarbon concentrations were measured at the Rafter Radiocarbon Laboratory (Lower Hutt,

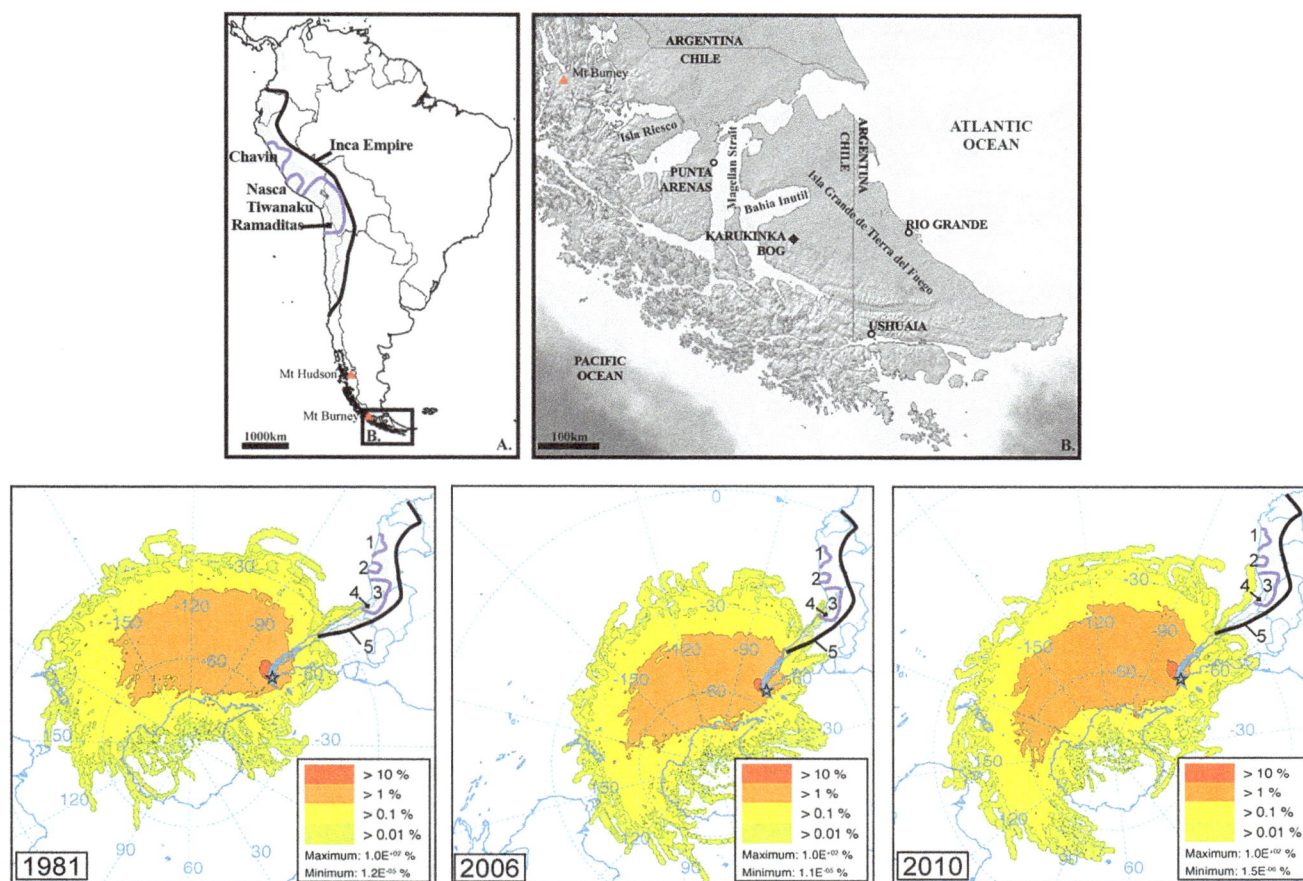

**Figure 1.** Upper Panel: A. Main polymetallic ores in South America (grey shaded area) and the extent of pre-Colombian civilisations and cultures discussed in the text (dark blue lines: 1-Chavin, 2-Nazca, 3-Tiwanaku; black square: 4-Ramaditas site black line: 5-Incas). B. Location of Karukinka bog in Tierra del Fuego. Lower Panel: back-trajectories for years when significant parcels of air masses travelled from NSA, specifically from the Inca, Tiwanaku and Nazca territories as well as from the Ramaditas site. Only the three most representative years are shown (see figure S1 for all years from 1948 to 2012). Other years (1997, 1999, 2000, 2008) are displayed in Fig. 3.

New Zealand) and at DirectAMS Laboratory (Bothell, USA) following established protocols [20]. Results are reported in Table 1.

## 6. Acid digestion in clean lab and dilutions

Dried samples were manually crushed using a clean agate pestle and mortar, which was rinsed between each sample using mQ water and *pro analysis* grade ethanol. They were then stored in 20-ml plastic vials. Approximately 100 mg of dried and homogenized peat from each interval was weighed out and digested in Teflon beakers in a clean room with class 100 flow benches. All acids used were of ultrapure (purchased Optima grade) quality. The digestion procedure consisted of three steps each succeeded by a slow evaporation at 55°C [21]: (1) a mixture of 0.5 ml HF and 2 ml 16M $HNO_3$ was added and left on the hotplate at 110°C for 2 days; (2) 1 ml of $H_2O_2$ was added to react for 6 h at room temperature; (3) 2 ml of 16M $HNO_3$ was added and left at 90°C for 2 days to finalize the digestion. Although peat is mainly composed of organic matter, a strong acid digestion is needed to dissolve any silica-based mineral residue that forms part of the remaining inorganic content (for peat bogs this is likely to be atmospheric dust) [21,22]. The samples were subsequently dissolved in 2 ml of 35% $HNO_3$, transferred into 15 ml polypropylene tubes (Falcon®), and further diluted with Milli-Q water up 14 ml. All samples were stored in a fridge for future analyses.

## 7. Q-ICP-MS analyses

After proper dilution Cu, Pb and La concentrations were measured on a Quadrupole ICP-MS (Agilent Technologie 7500 ce) at the Observatoire Midi Pyrénée, Toulouse. The ICP-MS was calibrated using a synthetic multi-element standard, which was run every 8 samples, while an In-Re solution was used as an internal standard. The accuracy of the analyses was assessed by analysis of 3 international certified reference materials (SRM1947-peach leaves; SRM1515-apple leaves and GBW-07063-bush branches and leaves) and is reported in Table 2. The reproducibility, determined by repeat analyses (n = 3) of 1 CRM and 2 peat samples, is better <5% for Cu and Pb, and < 12% for Sb, Sn and La. The blanks for all elements considered here, are <0.01%. Final dilution factors were approximately 700.

## 8. HR-ICP-MS

After the measurements of the Pb concentration by Q-ICP-MS, mother solutions of selected samples were sub-sampled and diluted to adjust the Pb concentration to 500 mg.kg$^{-1}$ prior to analyses by HR-ICP-MS at the *Observatoire Midi-Pyrénées*, Toulouse [23]. A 500-mg.kg$^{-1}$ SRM 981 standard solution was used to control and

**Table 1.** Results of the AMS radiocarbon dating of 8 selected samples from Karukinka.

| Sample Name | Sample Composition | Sample ID | AMS Lab ID | Age<br>14C BP | Error<br>yr | Depth<br>cm |
|---|---|---|---|---|---|---|
| KAR12-PB01/111 | *S. magellanicum* stems, leaves & branches | GdA-2764 | NZA-51256 | 921 | 20 | 113.6 |
| KAR12/PB01B/131 | *S. magellanicum* leaves & stems | GdA-3034 | D-AMS 002878 | 1528 | 25 | 141.1 |
| KAR12/PB01B/166 | *S. magellanicum* leaves & stems | GdA-3035 | D-AMS 002879 | 1962 | 26 | 185.1 |
| KAR12/PB01/209 | *S. magellanicum* leaves & stems | GdA-2870 | NZA-52239 | 2181 | 23 | 241.5 |
| KAR12/PB01B/241 | *S. magellanicum* leaves & stems | GdA-3036 | D-AMS 002880 | 2850 | 27 | 285.2 |
| KAR12-PB01/287 | *S. magellanicum* leaves | GdA-2765 | NZA-51257 | 3759 | 24 | 343.9 |
| KAR12-PB01/315 | *S. magellanicum* leaves & stems | GdA-2871 | NZA-52232 | 4482 | 26 | 380.2 |
| KAR12-PB01/363 | Unidentified graminoids | GdA-2766 | NZA-51258 | 7078 | 32 | 440.6 |

correct mass bias by standard bracketing of every sample. $^{206}Pb/^{207}Pb$ ratios in the NIMT peat standard were $1.175 \pm 0.001$ (n = 4) compared to the reported value of $1.1763 \pm 0.0004$ [24].

## 9. Sample preparation for tephra analysis

Based on the visual examination of the core as well as the lead isotope profile, 3 samples (237.5 cm, 34–301 yrs cal BC; 347 cm, 2354–2130 yrs cal BC; and 428 cm, 5745–5327 yrs cal BC) were processed to determine their potential volcanic origin by identifying glass shards. Dry samples were burned in a muffle furnace (550°C, 4 h) and ash residues were fine-sieved with the 10–125 μm fractions mounted in Canada Balsam on glass slides. Plane-polarised light, at ×100–400 magnification, was used to differentiate between tephra and other minerogenic material and 300 shards/grains were counted to determine their respective proportion. Fresh glass shards were identified and photographed.

## 10. Generation of the age-depth model

The 8 radiocarbon dates as well as two eruptions from Mt. Hudson (428 cm depth, 5745–5327 yrs cal BC) and Mt. Burney (347 cm depth, 2354–2130 yrs cal BC) were used as chronological time markers with the top of the core set to year 2012 AD (when it was recovered). CLAM software [25] was used to construct a cubic smooth spline age-depth model using the SHCal04 calibration curve [26]. The age-depth model is presented in Fig. 2. Based on 10000 iterations minimum and maximum ages for 2 sigma confidence interval were determined for each sample, as well as maximum likelihood age point estimates based on the weighted average of all generated age-depth curves. Maximum likelihood ages are given with the geochemical data in Table 3.

## 11. Back trajectory calculation

The Hybrid Single Particle Lagrangian Integrated Trajectory (HYSPLIT) model [27] has been used to run a series of back-trajectories from January 1, 1948 until December 31, 2012. The model is driven by meteorological data fields from the NCEP/ NCAR Reanalysis Project (http://ready.arl.noaa.gov/ gbl_reanalysis.php and ftp://arlftp.arlhq.noaa.gov/archives/ reanalysis/) [28]. The tools (http://www.meteozone.com/home/ tutorial/html/traj_freq.html) used to generate the trajectories are publicly available for the scientific community to reproduce the results. Previous observational and modelling palaoclimate studies have shown strong analogies between present and past climate and wind conditions [7–10], and that climate variability is partly controlled by modes or short-term pluriannual to decadal-scale oscillations [7]. Therefore, our calculations, which cover more than 60 years, encompass these climatological drivers. The calculation includes back-trajectories initialized from the measurement site every six hours and the duration of each trajectory is set to 15 days. The trajectory starting heights are set to 500 m above ground level to represent air parcels within the planetary boundary layer reaching the measurement site. A total of 93440 trajectories were computed for this study. Once all the trajectories were calculated, the model counts the number of trajectories that fall within each grid cell from an arbitrarily set horizontal grid that covers the Southern Hemisphere. The grid resolution is 1.0 degree. The trajectory frequency (F) is just the sum of the number of trajectories (T) that passed through each (i, j) grid cell divided by the total number (N) of trajectories analyzed:

$$F_{i,j} = 100 \ \Sigma \ T_{i,j}/N$$

Note that all trajectories are counted once in the source location grid cell and once per intersecting grid cell.

Out of the 64 years investigated in this model (from 1948 to 2012, figure S1 for more information), air parcels coming partially from NSA and the pre-Hispanic metallurgical territories 7 times (Fig. 3). This represents more than 10% of the total period. As each of the peat slice from Karukinka represents several decades of peat accumulation, and assuming that the 64-year climatology used to drive the back-trajectories is representative of the atmospheric circulation of the past 4500 years, each slice from Karukinka is likely to capture several years where small but significant amounts of metals coming from Pre-Hispanic metallurgical territories are incorporated in the air parcels travelling to Tierra del Fuego.

## Results and Interpretation

### 1. Determination of the bog trophic status

Karukinka bog is located on a small hill beside the main Karukinka Valley. It consists of an open area where the surface vegetation is dominated by *Sphagnum magellanicum* mosses. The 5-cm resolution macrofossil diagram records a poor-fen stage (zone KAR12-PB01B-A from the base of the core (450 cm) to approximately 415 cm depth (4555–4215 yrs cal BC) and is composed of graminoid remains (Monocots undifferentiated). The presence of abundant macrofossil charcoal fragments indicates that local fires may have occurred during this period. The fen

**Table 2.** Cu, Pb and La determined by Q-ICP-MS compared to certified values of SRM1947, SRM1515 and GBW-07063 (mean $\pm$ standard deviation, mg.kg$^{-1}$).

| | Measured | Certified |
|---|---|---|
| **GBW-07063 (n = 2)** | | |
| Cu | 6.4±1.0 | 6.6±0.8 |
| La | 1.1±0.6 | 1.3±0.06 |
| Pb | 49±1.4 | 47±3.0 |
| Sb | 0.1±3.7 | 0.1±0.01 |
| **SRM1947 (n = 2)** | | |
| Cu | 3.6±0.6 | 3.7±0.4 |
| Pb | 0.7±0.4 | 0.87±0,03 |
| **SRM1515 (n = 3)** | | |
| Cu | 6.3±0.8 | 5.6±0.24 |
| Pb | 0.45±1.5 | 0.47±0.02 |

La results are not certified for SRM1947 and SRM1515 hence not reported. Sn has no certified values. However, the good analytical reproducibility certifies the analytical quality of Sn measurement.

stage (Fig. 4 is followed by a transition to a bog (zone KAR12-PB01B-B) at 344 cm depth (2269–2057 yrs cal BC). This transition stage is however largely dominated by *Sphagnum magellanicum*. The ombrotrophic stage at 344 cm persists to the present, with a dominance of *Sphagnum magellanicum* (up to 95% of the total macrofossil assemblage). Moreover, from 344 cm to the surface, the peat bulk density is low (<0.2 g.cm$^{-2}$), which also suggests ombrotrophic conditions spanning this depth interval. These low bulk density values indicate that minimal inwash of mineral material from lateral streams has occurred and therefore the peat deposits are likely to record predominantly low atmospheric dust inputs. Under these conditions, previous studies

have shown that the elements analysed here (Pb, Cu, Sn, Sb) are not prone to post-depositional mobility and should therefore effectively record the history of metallurgy and other human activities [5,29–32].

## 2. Tephra identification

Of the 3 samples submitted for petrographic analysis, tephra horizons were identifiable in two (347 cm depth, 2354–2130 yrs cal BC and 428 cm depth, 5745–5327 yrs cal BC). The remaining sample contained small numbers of tephra shards but did not exceed baseline levels, which are typically present in such close proximity to numerous volcanic sources.

The tephrochronology of SSA is relatively well known [33]. Subsequently, the two tephra horizons identified in the Karukinka peat core could be confidently attributed to known eruptions based on the characteristics of the shards present and their respective ages, as calculated from the age-depth model. The Mt. Hudson eruption was identified at 428 cm depth (5745–5327 yrs cal BC), with shard colour and morphology consistent with an eruption from the predominantly andesitic southern Volcanic Zone in which Mt. Hudson is located (Fig. 5). Conversely, the transparent, rhyolitic shards found at 347 cm were consistent with the Mt. Burney eruption (2354–2130 yrs cal BC) and the Austral Volcanic Zone in which the volcano is situated [33]. These tephra horizons have been identified in various peat cores from Southern South America [33–36]. The presence of a younger eruption from Mt. Burney reported in the literature at 358 cal BC –201 cal AD in Southern Patagonia [33] is not clear in the Karukinka peat sequence, since no significant glass shard concentrations were found in the layer corresponding to this time range.

## 3. Brief description of the raw geochemical results (Cu, Pb, La, $^{206}Pb/^{207}Pb$) and calculation of Metal/La$_{UCC}$ ratio

Copper and Sb concentrations remain below 7 mg.kg$^{-1}$ from the base of the core to approximately 250 cm depth (Table 3). From 250 cm to 150 cm depth, the Cu concentration increases periodically, reaching more than 24 mg.kg$^{-1}$ to finally decrease rapidly and remain low from 150 cm depth to the surface. Lead and Sn concentrations remain mostly below 1 mg.kg$^{-1}$ except around 250 cm–150 cm where concentrations double and peak twice. Lanthanum concentration are stable between ca. 1 mg.kg$^{-1}$

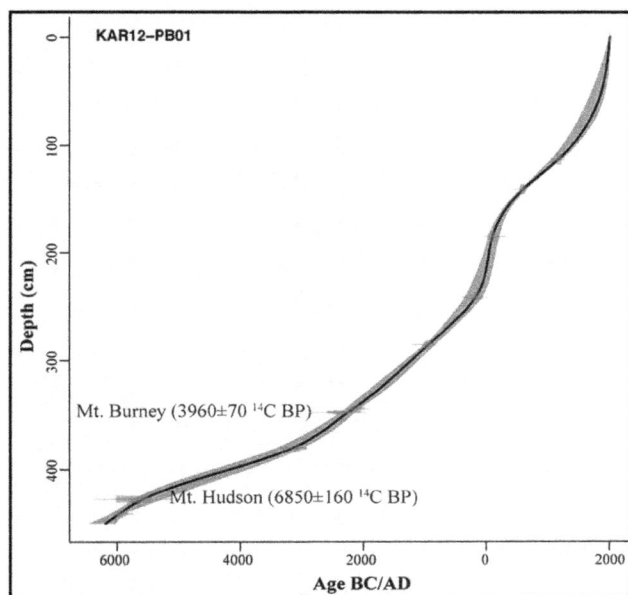

**Figure 2. Age-depth model for the Karukinka peat core.** Maximum likelihood ages are expressed in years Anno Domini (AD) and Before Christ (BC). The Hudson and Burney tephras are also used as chronological markers.

**Table 3.** Cu, Sn, Sb, Pb and La concentrations as well as $^{206}Pb/^{207}Pb$ isotopic ratios in the Karukinka core.

| Depth | Density | Calendar Age | Cu | Sn | Sb | Pb | La | $^{206}Pb/^{207}Pb$ | 2σ |
|---|---|---|---|---|---|---|---|---|---|
| cm | (g.cm$^{-3}$) | (BC/AD) | (mg kg$^{-1}$) | (mg kg$^{-1}$) | (mg kg$^{-1}$) | (mg kg$^{-1}$) | (mg kg$^{-1}$) | | |
| 1.3 | 0.04 | 2010 | 1.59 | 0.04 | 0.02 | 0.14 | 0.14 | | |
| 8.7 | 0.05 | 1998 | 0.99 | 0.02 | 0.02 | 0.14 | 0.07 | | |
| 15.9 | 0.05 | 1985 | 0.84 | 0.03 | 0.02 | 0.28 | 0.25 | 1.181 | 0.001 |
| 22.1 | 0.05 | 1972 | 0.84 | 0.02 | 0.02 | 0.27 | 0.08 | | |
| 28.0 | 0.10 | 1957 | 0.55 | 0.04 | 0.01 | 0.31 | 0.08 | | |
| 30.8 | 0.14 | 1949 | 0.84 | 0.02 | 0.01 | 0.48 | 0.55 | 1.182 | 0.003 |
| 34.8 | 0.19 | 1936 | 0.89 | 0.02 | 0.01 | 0.20 | 0.20 | 1.179 | 0.0001 |
| 36.0 | 0.15 | 1932 | 0.40 | 0.03 | 0.01 | 0.21 | 0.07 | | |
| 42.0 | 0.15 | 1910 | 0.46 | 0.01 | 0.01 | 0.19 | 0.06 | | |
| 50.4 | 0.07 | 1874 | 0.85 | 0.01 | 0.01 | 0.12 | 0.15 | 1.195 | 0.0003 |
| 54.3 | 0.14 | 1853 | 1.14 | 0.06 | 0.02 | 0.23 | 0.30 | | |
| 58.0 | 0.10 | 1831 | 0.81 | 0.01 | 0.01 | 0.15 | 0.15 | 1.185 | 0.004 |
| 62.0 | 0.09 | 1805 | 1.20 | 0.02 | 0.01 | 0.17 | 0.16 | | |
| 66.0 | 0.06 | 1776 | 1.91 | 0.01 | 0.01 | 0.11 | 0.11 | | |
| 72.6 | 0.06 | 1722 | 0.69 | 0.13 | 0.04 | 0.10 | 0.10 | 1.191 | 0.001 |
| 79.9 | 0.08 | 1653 | 0.99 | 0.01 | 0.01 | 0.15 | 0.20 | 1.193 | 0.001 |
| 84.1 | 0.09 | 1607 | 1.39 | 0.03 | 0.01 | 0.16 | 0.20 | 1.188 | 0.003 |
| 92.1 | 0.08 | 1509 | 2.24 | 0.02 | 0.01 | 0.17 | 0.33 | 1.190 | 0.002 |
| 96.1 | 0.10 | 1454 | 2.38 | 0.02 | 0.01 | 0.13 | 0.46 | 1.198 | 0.004 |
| 102.7 | 0.06 | 1355 | 2.12 | 0.03 | 0.02 | 0.14 | 0.63 | | |
| 109.2 | 0.07 | 1244 | 2.48 | 0.03 | 0.02 | 0.22 | 0.63 | | |
| 117.1 | 0.05 | 1093 | 3.63 | 0.02 | 0.02 | 0.12 | 0.54 | | |
| 122.2 | 0.08 | 988 | 4.56 | 0.07 | 0.02 | 0.20 | 0.87 | 1.194 | 0.002 |
| 127.7 | 0.09 | 871 | 3.43 | 0.05 | 0.02 | 0.19 | 0.64 | | |
| 135.3 | 0.09 | 713 | 4.03 | 0.08 | 0.05 | 0.19 | 1.14 | | |
| 142.7 | 0.08 | 571 | 3.62 | 0.04 | 0.01 | 0.37 | 3.13 | 1.193 | 0.001 |
| 146.9 | 0.11 | 499 | 6.71 | 0.06 | 0.02 | 0.85 | 7.30 | 1.197 | 0.0002 |
| 158.0 | 0.13 | 339 | 15.28 | 0.15 | 0.07 | 1.59 | 15.19 | 1.198 | 0.001 |
| 161.8 | 0.12 | 293 | 21.53 | 0.11 | 0.10 | 2.00 | 14.76 | | |
| 164.4 | 0.11 | 264 | 21.55 | 0.15 | 0.12 | 1.90 | 14.72 | | |
| 168.0 | 0.10 | 228 | 17.10 | 0.12 | 0.08 | 1.56 | 11.37 | | |
| 171.8 | 0.11 | 194 | 24.03 | 0.13 | 0.08 | 1.36 | 10.32 | 1.197 | 0.002 |
| 175.4 | 0.08 | 165 | 22.19 | 0.06 | 0.07 | 1.28 | 9.35 | | |
| 179.0 | 0.07 | 140 | 24.72 | 0.05 | 0.08 | 1.23 | 7.62 | | |

**Table 3.** Cont.

| Depth | Density | Calendar Age | Cu | Sn | Sb | Pb | La | $^{206}Pb/^{207}Pb$ | $2\sigma$ |
|---|---|---|---|---|---|---|---|---|---|
| cm | (g.cm$^{-3}$) | (BC/AD) | (mg kg$^{-1}$) | (mg kg$^{-1}$) | (mg kg$^{-1}$) | (mg kg$^{-1}$) | (mg kg$^{-1}$) | | |
| 182.8 | 0.09 | 117 | 10.69 | 0.07 | 0.06 | 0.80 | 6.26 | | |
| 190.0 | 0.08 | 83 | 18.05 | 0.05 | 0.06 | 0.91 | 5.37 | | |
| 195.4 | 0.07 | 64 | 14.49 | 0.05 | 0.05 | 0.93 | 5.73 | 1.190 | 0.001 |
| 201.3 | 0.06 | 46 | 12.46 | 0.05 | 0.07 | 0.68 | 4.81 | | |
| 207.8 | 0.08 | 29 | 12.91 | 0.05 | 0.05 | 0.87 | 6.66 | | |
| 215.2 | 0.07 | 6 | 9.1 | 0.1 | 0.1 | 1.4 | 5.4 | | |
| 221.9 | 0.08 | −21 | 13.86 | 0.07 | 0.04 | 1.39 | 6.85 | | |
| 227.4 | 0.06 | −50 | 21.63 | 0.26 | 0.07 | 1.63 | 7.68 | | |
| 231.5 | 0.06 | −77 | 20.04 | 0.09 | 0.06 | 1.33 | 6.21 | 1.196 | 0.002 |
| 237.5 | 0.16 | −128 | 24.53 | 0.09 | 0.06 | 1.50 | 9.00 | 1.198 | 0.003 |
| 244.3 | 0.13 | −204 | 4.49 | 0.04 | 0.03 | 0.44 | 5.64 | | |
| 250.2 | 0.09 | −285 | 9.26 | 0.04 | 0.04 | 0.80 | 6.35 | | |
| 253.1 | 0.14 | −330 | 4.68 | 0.03 | 0.03 | 0.40 | 4.78 | 1.199 | 0.001 |
| 258.7 | 0.12 | −424 | 5.30 | 0.04 | 0.03 | 0.41 | 4.36 | | |
| 264.8 | 0.10 | −535 | 5.19 | 0.03 | 0.03 | 0.33 | 2.96 | | |
| 270.2 | 0.10 | −639 | 5.78 | 0.05 | 0.03 | 0.41 | 3.79 | 1.196 | 0.001 |
| 274.0 | 0.09 | −714 | 5.64 | 0.04 | 0.03 | 0.39 | 3.98 | | |
| 279.7 | 0.10 | −828 | 3.0 | 0.1 | 0.0 | 0.4 | 2.9 | | |
| 285.2 | 0.09 | −937 | 2.35 | 0.02 | 0.02 | 0.32 | 2.03 | 1.198 | 0.001 |
| 289.3 | 0.09 | −1017 | 2.52 | 0.03 | 0.02 | 0.31 | 1.82 | | |
| 293.4 | 0.14 | −1096 | 3.49 | 0.08 | 0.04 | 0.25 | 2.04 | | |
| 300.9 | 0.10 | −1239 | 5.88 | 0.03 | 0.04 | 0.41 | 2.15 | 1.196 | 0.003 |
| 304.9 | 0.09 | −1316 | 6.28 | 0.04 | 0.03 | 0.21 | 1.41 | | |
| 308.6 | 0.08 | −1387 | 4.44 | 0.02 | 0.02 | 0.26 | 1.45 | | |
| 312.2 | 0.13 | −1457 | 4.63 | 0.07 | 0.06 | 0.58 | 2.87 | 1.195 | 0.002 |
| 320.5 | 0.10 | −1623 | 6.46 | 0.04 | 0.04 | 0.27 | 1.38 | | |
| 325.6 | 0.13 | −1729 | 9.90 | 0.07 | 0.06 | 0.54 | 2.47 | | |
| 330.8 | 0.11 | −1842 | 6.31 | 0.06 | 0.04 | 0.54 | 2.78 | 1.199 | 0.002 |
| 334.8 | 0.12 | −1932 | 6.25 | 0.06 | 0.05 | 0.68 | 3.29 | | |
| 339.8 | 0.12 | −2050 | 6.60 | 0.04 | 0.05 | 0.48 | 2.79 | | |
| 343.9 | 0.12 | −2152 | 6.62 | 0.09 | 0.04 | 0.85 | 3.56 | | |

Sample depths and maximum likelihood calendar ages (positive values are ages Anno Domini; negative values are ages Before Christ) are also reported.

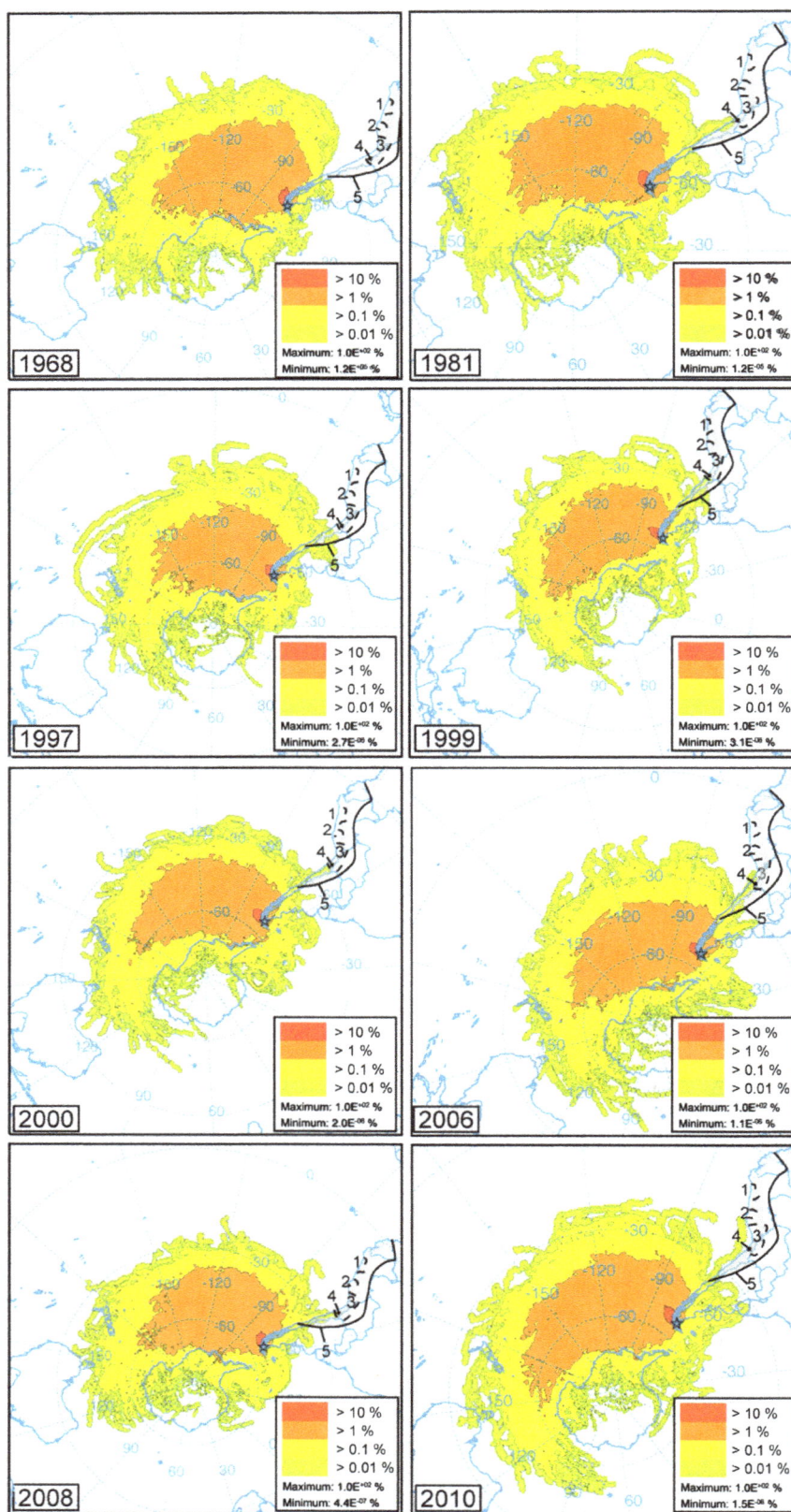

**Figure 3. Back trajectory frequency corresponding to calendar years where air parcels are partially coming from NSA.** The year 1968 is given as a comparison to show a year where no air back trajectory is coming from NSA (see figure S1 for all years from 1948 to 2012).

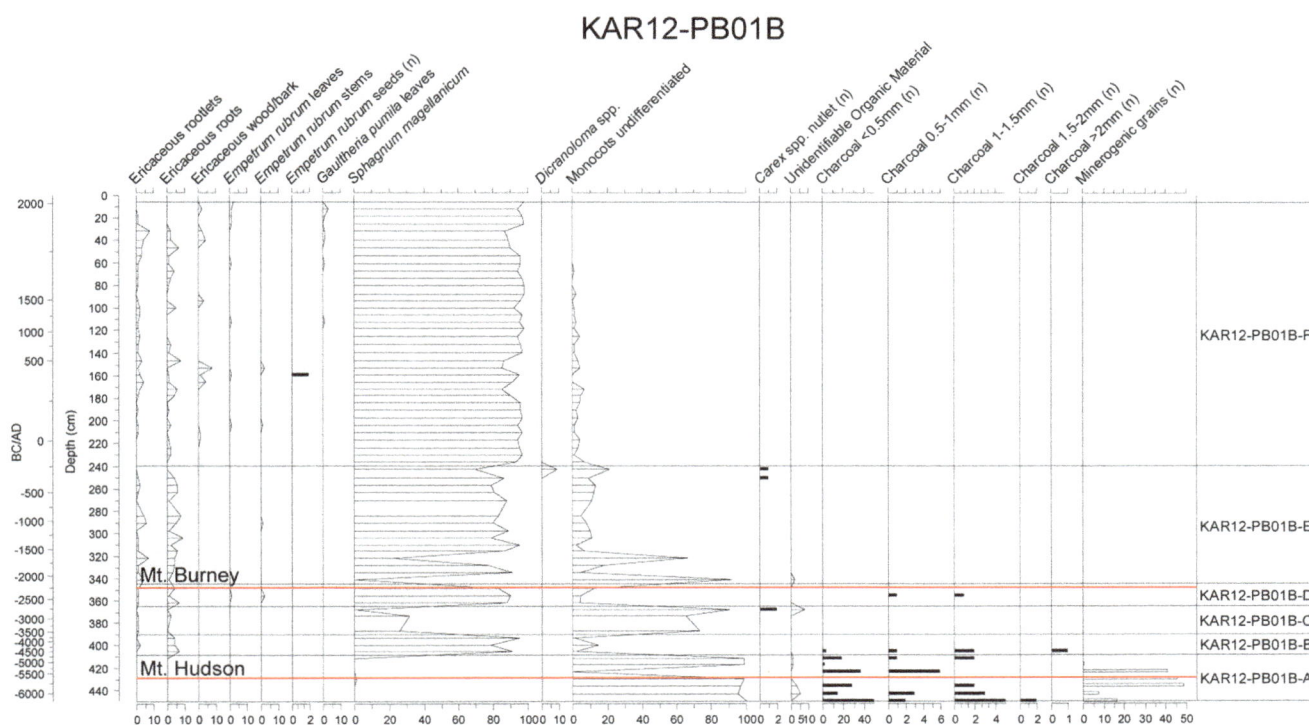

**Figure 4. Detailed macrofossil age and depth profile for the Karukinka peat profile.** Mineral grains are from the Hudson and Burney tephras (red lines).

and 6.5 mg.kg$^{-1}$ from 344 cm to 250 cm depth, and then increase twice from 3 mg.kg$^{-1}$ to 15 mg.kg$^{-1}$ between 250 cm to 150 cm, to then decrease below 1 mg.kg$^{-1}$. These increases in metals and La between 150 and 250 cm are not interpreted here since the concentration profiles are prone to changes in the accumulation rates or in the influx of dust to the bog. In order to accurately interpret changes in metal inputs to the bog, we will be using Me/La ratios in the rest of the manuscript. Upper Continental Crust (UCC)-normalized metal/La ratios were calculated by dividing the metal/La ratio by the same ratio in the UCC, with 25 mg.kg$^{-1}$, 17 mg.kg$^{-1}$, 0.2 mg.kg$^{-1}$, 5.5 mg.kg$^{-1}$ and 30 mg.kg$^{-1}$ as the UCC concentrations for Cu, Pb, Sb, Sn and La, respectively [37].

Lead isotopes fluctuate within a range typical for the Upper Continental Crust ($1.18 < {}^{206}Pb/{}^{207}Pb < 1.21$) with three slight shifts towards more radiogenic values around 425 cm, 351 cm and 195 cm depth. In shallower samples (above 60 cm depth), the ${}^{206}Pb/{}^{207}Pb$ shifts towards more radiogenic values, which are typical for a mix between anthropogenic and natural sources.

## 4. History of metallurgy as recorded in Karukinka bog

Whereas metallurgical activities occurred throughout South America since ca 2000 BC, there is no evidence that pre-Hispanic metal objects or artifacts reached Tierra del Fuego [38]. Hunters and gatherers as well as canoe-borne nomadic people from Tierra

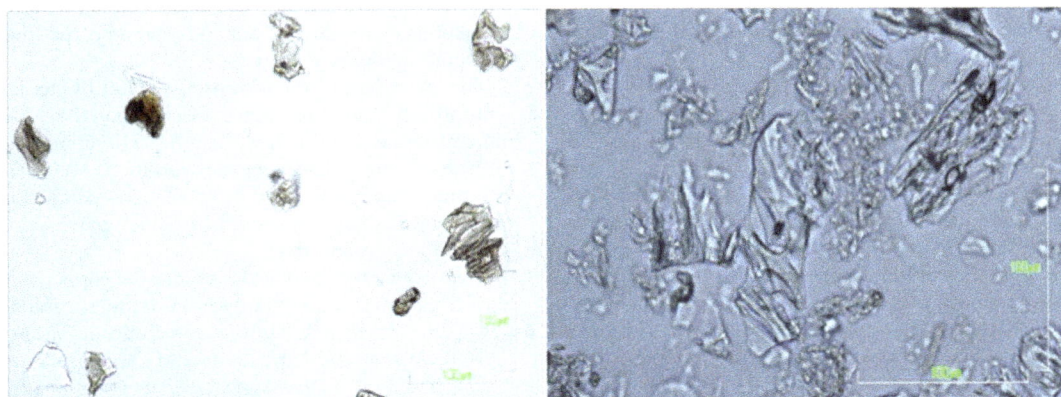

**Figure 5. Picture of Mt. Hudson 5745–5327 yrs cal BC (428 cm depth, left) and Mt. Burney 2354–2130 yrs cal BC (347 cm depth right) shards found in Karukinka peat core.**

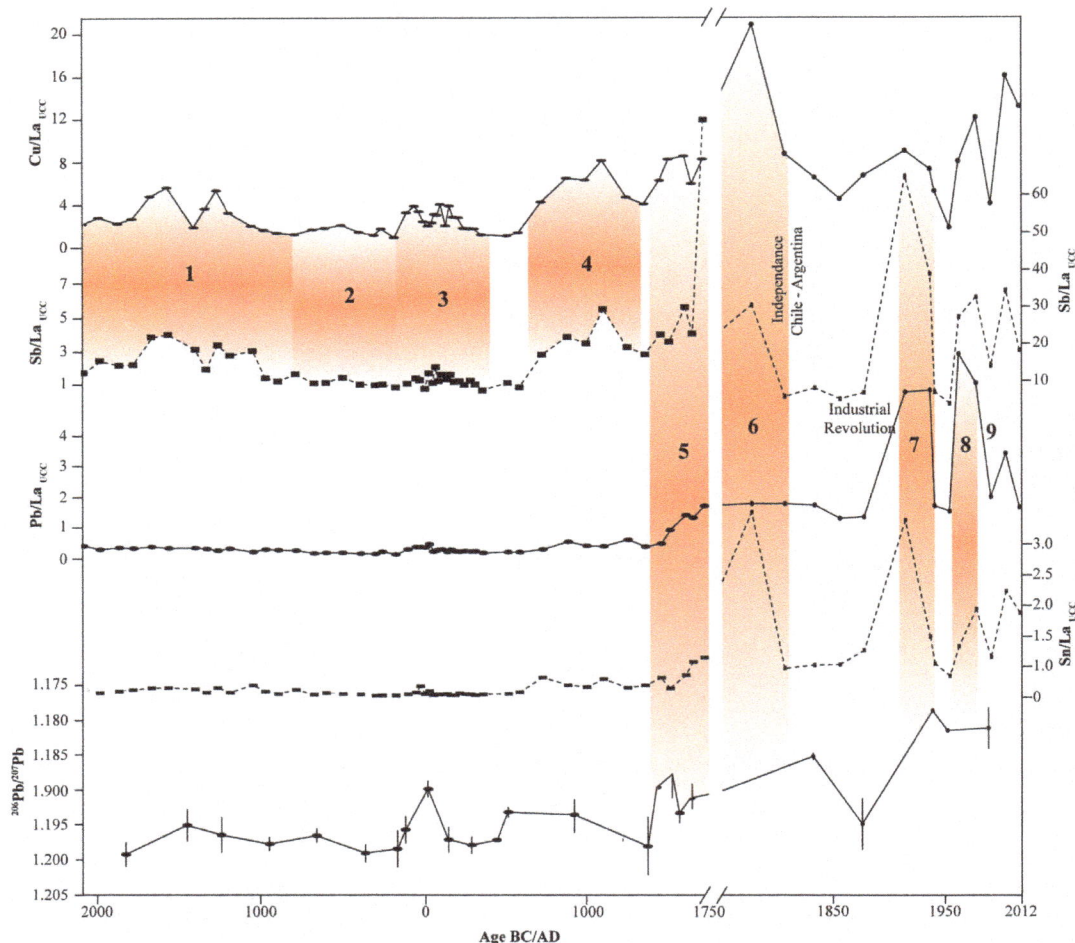

**Figure 6. UCC-normalized Cu, Sb, Pb and Sn vs. La ratios [37] and $^{206}Pb/^{207}Pb$ isotopic ratios in the Karukinka peat core during the last 4200 yrs (ombrotrophic part of the core).** Maximum likelihood ages are expressed in years Anno Domini (AD) and Before Christ (BC). The age scale is cut at 1750 AD and the Sb/La$_{UCC}$ scale changes at 1750AD. 1. Onset of Cu metallurgy, 2. Chavin, 3. Nazca and Ramaditas, 4. Tiwanaku, 5. Inca (transition from Cu to Ag mining), 6. European Colonization, 7. Gold rush and bronze industry, 8. Leaded gasoline period, 9. Unleaded gasoline period (see text for details).

del Fuego were seemingly unaware of purified metals before the Argentinean and Chilean colonization of these territories occurred in the 19th century [39,40]. Hence, compared to the Northern Hemisphere and mainly Europe, Tierra del Fuego remained almost pristine and free of any local metal contamination until the last two centuries.

The upper part of the Karukinka peat bog (4200 years) is ombrotrophic as demonstrated by the plant macrofossil succession (see Figure 4) and therefore is a suitable archive of atmospheric metal deposition. By accurately analysing Pb, Cu, Sb, Sn, La, and Pb isotopes in peat samples using Q-ICP-MS and HR-ICP-MS, supported by a $^{14}C$ age-depth model, we provide a unique history of anthropogenic activities and metallurgical developments in South America since the mid-Holocene (from *ca.* 4 kyrs cal BP) (Fig. 6). Modern-time back trajectory calculations further demonstrate that air parcels can be transported from the main metallurgical areas in NSA to SSA. The trajectory frequency plot confirms that this is the case for 7 times (1981, 1997, 1999, 2000, 2006, 2008, 2010) during the 64 investigated years. Accordingly, NSA, and especially northern Chile, southern Bolivia and southern Peru are potential source regions of metals found in Karukinka (Fig. 1). Assuming that the 64-year climatology is

representative of the atmospheric circulation of the past 4200 years and as each peat slice from Karukinka represents several decades, each single peat slice is likely to capture several years where small but significant amounts of metals coming from pre-Hispanic metallurgical territories are incorporated in the air parcels travelling to Tierra del Fuego.

The lower part of the peat profile (6500 BC to 2200 BC) is not ombrotrophic and may have been exposed to local in-washed mineral inputs and changes in redox conditions. As we want to investigate atmospheric metal deposition, we therefore do not interpret this part of the profile in detail but focus our interpretations on the ombrotrophic part of the sequence (2200 BC to present day).

The oldest archaeological evidence for copper metallurgy dates back to ca. 1400 BC in Southern Peru and the following centuries saw its development mainly in the Bolivian Altiplano, Northern Chile but also in the Antofagasta and northeast Argentina regions. This is based on archaeological evidence for metal ore extraction and smelting, various ritual objects, artifacts and weapons [41–47]. In the Karukinka core, the Upper Continental Crust (UCC)-normalized Cu/La$_{UCC}$ ratio increases starting at 2000 BC (ca. 600 yrs before the first archaeological evidence), suggesting that small but detectable amounts of Cu would be transported to SSA and that Andean Cu metallurgy may be older than previously reported. Antimony, associated with Cu ores such as tetrahedrite, also displays an increase in the Sb/La$_{UCC}$ ratio. A shift in Pb isotopes can also be observed around the BC-AD transition. Pb metallurgy was not yet developed in South America, therefore this shift may be an early manifestation of metallurgy of unknown origin or a shift in Pb isotopes from a natural origin. An eruption from Mt. Burney with a large associated age range (358 BC–201 AD) has been reported in various Patagonian sites [33,34] and could explain the shift observed in Karukinka. While volcanic eruptions can shift isotopic ratios [48], we found no volcanic material in this sample. Therefore, we interpret this shift as of an unknown anthropogenic origin, or caused by trace levels of Pb contained in the Cu ores that were processed by the Nazca populations at that time.

Civilizations such as the Chavin and Nazca are mainly known to have worked textiles and ceramics but they also processed copper [38,42]. At the BC-AD transition, the Atacama Desert was the cradle of a new Cu metallurgical technique in the Ramaditas area, northern Chile, with wind being directed into partly opened furnaces to increase the temperature and melt the ores [49]. The subsequent increase in metal production, as well as to a certain extent Chavin and Nazca metallurgical activities, also emitted metals into the atmosphere. Increases in Cu/La$_{UCC}$ and Sb/La$_{UCC}$ between 80 BC and 200 AD in Karukinka and the modern-time back-trajectories indicate that some of these metals may have been transported south as far as Tierra del Fuego. Between 500 and 1000 AD, the Tiwanaku civilization spread through and dominated the South-Central Andes, the Titicaca Basin and northern Chile [50] and most likely increased the metallurgical activities in these areas [51]. This constitutes a turning point in South American metallurgy, with the development of smelting techniques (wax moulds and tin-copper alloys) as recorded by a significant increase in metal objects and artefacts at this time in northern Chile, particularly in San Pedro de Atacama [48]. Between 650 and 850 AD, metallurgical activities also developed in northwest Argentina [42,52]. This boom in Cu metallurgy is shown by a drastic increase in the Cu/La$_{UCC}$ and Sb/La$_{UCC}$ ratios in Karukinka, starting around 550 AD and culminating at 1100 AD. Interestingly, an increase in Cu/La$_{UCC}$ during this period was reported in an ice core collected in Bolivia [1], but the resolution was too low to make a conclusive interpretation.

Copper production intensified under the control of the Incas (1480–1532 AD) [54]. During the Inca domination and during the Hispanic colonization, northwest Argentina and northern Chile remained important copper production areas [52,53]. The Karukinka peat record highlights the period of Inca copper metallurgy with an increase of Cu/La$_{UCC}$ and Sb/La$_{UCC}$ starting at 1400 AD and persisting until ca. 1600 AD. At the same time, the mining of silver in the Andes triggered a substitution of copper to silver smelting from ca. 1450–1533 AD (i.e. during the Inca Period) around Peru, the Titicaca basin and Potosi, Bolivia [51]. In order to reconstruct the Ag mining history, we report Pb and Sn data as these metals have been shown to be associated with Ag ores, for example in Potosi [51]. An increase in Pb/La$_{UCC}$ and Sn/La$_{UCC}$ from ca. 1400 to ca. 1800 AD indicates that Pb and Sn were dispersed into the atmosphere and transported to SSA during the silver mining era. This is consistent with similar increases in Pb and Sn in NSA lake sediments [51,53] as well as the Inca and European Ag production, especially at the famous Potosi mine. The increase in Ag and Pb smelting is also clearly marked in Karukinka by a shift in the $^{206}$Pb/$^{207}$Pb ratio towards values of ca.

1.19, corresponding to isotopic signatures found in Bolivian and Peruvian ores [55,56].

Although the most recent section of the Karukinka core is not as well chronologically constrained as the pre-Hispanic section, we can identify excursions in the metals/La$_{UCC}$ ratios and Pb isotopic profiles, which match historical data from the 18$^{th}$ to 20$^{th}$ century. We therefore present a tentative interpretation for this section of the profile. The early 18$^{th}$ century sees a depletion in Ag resources and a transition to Sn mining in the Potosi area [51], which could explain the increase in Sn/La$_{UCC}$ but also Cu/La$_{UCC}$ in Karukinka, reflecting the increase of bronze alloy production. The 19$^{th}$ century is marked by a period of unrest associated with the independence of South American countries, especially Chile and Argentina and is associated with a decrease of Cu and Ag production. In Karukinka, the decrease of Cu/La$_{UCC}$ and Sn/La$_{UCC}$ around 1750–1820 AD, the slight decrease in Pb/La$_{UCC}$ and Sb/La$_{UCC}$ around 1800 AD, and the shift in Pb isotopic values could reflected this period of unrest. Copper and Pb production remained low in South America until 1920–1930 AD [1]. The Fuegian gold rush in 1903–1908 AD [57] and two phases of coal exploitation – during the first half of the 19$^{th}$ century and from 1980 onwards [58] - lead to the first emissions of local anthropogenic metal particles in Tierra del Fuego. These events seem to be recorded in the Pb/La$_{UCC}$ profile, which increased drastically around 1900 AD. This regional 'industrial revolution' persisted into the 20$^{th}$ century and saw the introduction of leaded gasoline as well as the intensification of lead metallurgy until 1980 [1]. This would be recorded by an increase in the Pb/La$_{UCC}$ at the beginning of the second half of the 20$^{th}$ century to a maximum around 1970–1980 and by a shift in the Pb isotopic signatures towards less radiogenic values. The last period of atmospheric metal deposition recorded in Karukinka could be the introduction of unleaded gasoline in the late 1980's [59], which is seen by a drastic decrease in Pb/La$_{UCC}$ as well as a shift in Pb isotopes towards more radiogenic compositions, while the Cu/La$_{UCC}$ does not decrease, as Cu production is still flourishing in South America [1].

## Conclusions

The Karukinka peat bog sequence records past metallurgical activities in South America, as the area was isolated from any local anthropogenic sources until the 19$^{th}$ century. Based upon the excellent agreement between variations in UCC-normalized Cu, Pb, Sb and Sn to La ratios, modern-time back-trajectories and historical data, we conclude that Cu metallurgy predates archaeological evidence and that NSA atmospheric emissions from pre-Hispanic civilizations through to the Industrial Revolution are recorded in SSA.

## Supporting Information

**Figure S1** Back trajectory frequency corresponding over Tierra del Fuego 1948 to 2012.
(PDF)

## Acknowledgments

This research is supported by a Young Researcher Grant of the *Agence Nationale de la Recherche* (ANR) to F. De Vleeschouwer (Project ANR-2011-JS56-006-01 "PARAD"). We thank Aurélie Lanzanova from *Geoscience-Environnement-Toulouse* (France), Maxime Enrico (EcoLab, Toulouse, France) and Adrien Claustres (EcoLab, Toulouse, France) for their help in the careful measurements of trace metals and lead isotopes. Barbara Saavedra, Susan Arismendi, Ricardo Muza and the Karukinka Park rangers from the Wildlife Conservation Society (WCS) are warmly thanked for facilitating access to the Karukinka Park. Sébastien Bertrand and Zakaria Ghazoui from RCMG (UGent, Belgium) and Jean-Yves De Vleeschouwer (ULg, Belgium) are thanked for their help in the field. Nelson Bahamonde (INIA, Punta Arenas, Chile) and Ernesto Teneb (UMag, Punta Arenas, Chile) are thanked for their logistical support in preparing the field mission. The authors gratefully acknowledge the NOAA Air Resources Laboratory (ARL) for the provision of the HYSPLIT transport and dispersion model and READY website (http://www.ready. noaa.gov) used in this publication. Supporting data are available in Figure

S1. We thank one anonymous reviewers and Xabier Pontevedra Pombal for their valuable comments on an earlier version of this manuscript.

## Author Contributions

Conceived and designed the experiments: FDV GLR HV. Performed the experiments: HV GLR DM NP TR AS. Analyzed the data: HV GLR DM NP TR AS FT. Contributed reagents/materials/analysis tools: HV GLR DM NP TR. Contributed to the writing of the manuscript: HV GLR DM NP TR AS FT.

## References

1. Hong S, Barbante C, Boutron C, Gabrielli P, Gaspari V et al. (2004) Atmospheric heavy metals in tropical South America during the past 22 000 years recorded in a high altitude ice core from Sajama, Bolivia. J Environ Monit 6: 322–326.
2. Renberg I Persson MW, Emteryd O (1994) Preindustrial atmospheric lead contamination detected in Swedish lake sediments. Nature 368: 323–326.
3. Shotyk W, Weiss D, Appleby PG, Cheburkin AK, Frei R et al. (1998) History of atmospheric lead deposition since 12,370 [14]C yr BP from a peat bog, Jura Mountains, Switzerland. Science 281: 1635–1640.
4. De Vleeschouwer F, Le Roux G, Shotyk W (2010) Peat as an archive of atmospheric pollution and environmental change: A case study of lead in Europe. PAGES News 18: 20–22. Available at: http://www.pages-igbp.org/download/docs/newsletter/2010-1/Special%20Section/DeVleeschouwer_2010-1(20-22).pdf. Accessed 20 January 2014.
5. Shotyk W, Le Roux G (2005) Biogeochemistry and cyclcing of lead. In: Sigel A, Sigel H, Sigel RKO, editors. Metal ions in biological systems Vol. 43: Biogeochemical cycles of elements. Taylor & Francis, London. pp. 239–275.
6. Rosman KJR Chisholm W, Hong S, Candelone JP, Boutron C (1997) Lead from Carthaginian and Roman Spanish mines isotopically identified in Greenland ice dated from 600 B.C. to 300 A.D. Environ Sci Technol 31: 3413–3416.
7. Fagel N, Boës X, Loutre MF (2008) Climate oscillations evidenced by spectral analysis of Southern Chilean lacustrine sediments: the assessment of ENSO over the last 600 years. J Paleolimm 39: 253–266.
8. Haug GH, Hughen KA, Sigman DM, Peterson L, Ursula Röhl U (2001) Southward migration of the Intertropical Convergence Zone through the Holocene. Science 293: 1304–1308.
9. Markgraf V, Baumgartner TR, Bradbury JP, Diaz HF, Dunbar RB et al. (2000) Paleoclimate reconstruction along the Pole-Equator-Pole transect of the Americas (PEP 1). Quat Sci Rev 19: 125–140.
10. Killian K, Lamy F (2012) A review of Glacial and Holocene paleoclimate records from southernmost Patagonia (49–55°S) Quat Sci Rev 53: 1–23.
11. Auer V (1965) The Pleistocene of Fuego-Patagonia. Part 4: Bog profiles. Ann Acad Sci Fenn Ser A III Geol.-Geogr. 80: 1–60.
12. Rabassa J, Coronato A, Heusser CJ, Juñent FR, Borromei A, et al. (2006). The peatlands of Argentine Tierra del Fuego as a source for paleoclimatic and paleoenvironmental information. In: Martini IP, Martínez-Cortizas A, Chesworth W, editors. Peatlands: Evolution and Records of Environmental and Climate Changes. Elsevier, The Netherlands. pp. 129–144.
13. Belokopytov IE, Beresnevich VV (1955) Giktorf's peat borers Torfânaâ promyslennost' 8: 9–10.
14. Givelet N, Le Roux G, Cheburkin A, Chen B, Frank J, et al. (2004) Suggested protocol for collecting, handling and preparing peat cores and peat samples for physical, chemical, mineralogical and isotopic analyses. J Environ Monitor 6: 481–492.
15. De Vleeschouwer F, Chambers FM, Swindles G (2010) Coring and sub-sampling of peatlands for palaeoenvironmental research. Mires and Peat 7: article 1. Available at: http://pixelrauschen.de/wbmp/media/map07/map_07_01.pdf. Accessed 15 January 2014.
16. Barber KE, Chambers FM, Dumayne L, Haslam CJ, Maddy D, et al. (1994) Climatic change and human impact in north Cumbria: peat stratigraphic and pollen evidence from Bolton Fell Moss and Walton Moss. In: Boardman J, Walden J., editors. The Quaternary of Cumbria: Field Guide, Quaternary Research Association, Oxford. pp. 20–54.
17. Mauquoy D, Blaauw M, van Geel B, Borromei A, Quattrocchio M, et al. (2004) Late Holocene climatic changes in Tierra del Fuego based on multiproxy analyses of peat deposits. Quaternary Research 61: 148–158.
18. Piotrowska N, Blaauw M, Mauquoy D, Chambers FM (2010) Constructing deposition chronologies in peat deposits using radiocarbon dating. Mires and Peat 7: article 12. Available at: http://pixelrauschen.de/wbmp/media/map07/map_07_10.pdf. Accessed 20 March 2014.
19. Piotrowska N (2013) Status report of AMS sample preparation laboratory at GADAM Centre, Gliwice, Poland. Nucl Instrum Meth B 294: 176–181.
20. Donahue D, Linick TW, Jull AJT (1990) Isotope-Ratio and Background Corrections for Accelerator Mass Spectrometry Radiocarbon Measurements. Radiocarbon 32: 135–142.
21. Le Roux G, De Vleeschouwer F (2010) Preparation of peat samples for inorganic geochemistry used as palaeoenvironmental proxies. Mires and Peat 7: article 4. Available: http://pixelrauschen.de/wbmp/media/map07/map_07_04.pdf. Accessed 20 January 2014.
22. Le Roux G, Shotyk W (2006) Weathering of inorganic matter in bogs. In: Martini IP, Martínez-Cortizas A, Chesworth W, editors. Peatlands: Evolution and Records of Environmental and Climate Changes. Elsevier, The Netherlands. pp. 197–215.
23. Krachler M, Le Roux G, Kober B, Shotyk W (2004) Optimising accuracy and precision of lead isotope measurement ([206]Pb, [207]Pb, [208]Pb) in acid digests of peat with ICP-SMS using individual mass discrimination correction. J Anal Atom Spectrom 19: 354–361.
24. Yafa C, Farmer JG, Graham MC, Bacon JR, Barbante C et al. (2004) Development of an ombrotrophic peat bog (low ash) reference material for the determination of elemental concentrations. J Environ Monitor 6: 493–501.
25. Blaauw M (2010) Methods and code for classical' age-modelling of radiocarbon sequences. Quat Geochronol 5: 512–518.
26. McCormac FG, Hogg AG, Blackwell PG, Buck CE, Higham TFG, et al. (2004) SHCal04 Southern Hemisphere Calibration, 0–11.0 cal kyr BP. Radiocarbon 46: 1087–1092.
27. Draxler RR, Hess GD. (1997). Description of the HYSPLIT_4 modeling system. NOAA Technical Memo. Available: http://www.arl.noaa.gov/documents/reports/arl-224.pdf. Accessed 6 december 2013.
28. Kalnay E, Kanamitsu M, Kistler R, Collins W, Deaven D et al. (1996) The NCEP/NCAR 40-Year Reanalysis Project. Bull Amer Meteor Soc 77: 437–471.
29. Rausch N, Nieminen T, Ukonmaanaho L, Le Roux G, Krachler M, et al. (2005) Comparison of atmospheric deposition of copper, nickel, cobalt, zinc, and cadmium recorded by Finnish peat cores with monitoring data and emission eecords. Environ Sci Technol 39: 5989–5998.
30. Shotyk W, Krachler M, Chen B (2004) Antimony in recent peat from Switzerland and Scotland: comparison with natural background values (5,320 to 8,020 14C yr BP), correlation with Pb, and implications for the global atmospheric Sb cycle. Global Biogeochem Cycles, doi:10.1029/2003GB002113.
31. Allan M, Le Roux G, De Vleeschouwer F, Bindler R, Blaauw M, et al. (2013) High-resolution reconstruction of atmospheric deposition of trace metals and metalloids since AD 1400 recorded by ombrotrophic peat cores in Hautes-Fagnes, Belgium. Env Poll 178: 381–394.
32. Mearg A, Edwards KJ, Schofield JE, Raab A, Feldmann J et al. (2012) First comprehensive peat depositional records for tin, lead and copper associated with the antiquity of Europe's largest cassiterite deposits. J Archaeol Sci 39: 717–727.
33. Stern CR (2008) Holocene tephrochronology record of large explosive eruptions in the southernmost Patagonian Andes. Bull Volcanol 70: 435–454.
34. Stern CR (2000) The Holocene tephrochronology of southernmost Patagonia and Tierra del Fuego. Actas 9th Congreso Geológico Chileno 2: 77–80.
35. Heusser CJ, Heusser LE, Lowell TV, Moreira A, Moreira S (2000) Deglacial palaeoclimate at Puerto del Hambre, subantarctic Patagonia, Chile J. Quat Sci 15: 101–114.
36. McCulloch RD, Davies SJ (2001) ate-glacial and Holocene palaeoenvironmental change in the central Strait of Magellan, southern Patagonia. Palaeogeogr, Paleocean Palaeocl 173: 143–173.
37. Taylor SR, McLennan SM (1995) The geochemical evolution of the continental crust. Rev Geophys 33: 241–265.
38. Palacios T (2011) Metalurgia Prehispánica de Sudamérica. Revista SAM 8: 15–28.
39. Martinić M (1982) La Tierra de los Fuegos. Artegraf Ltda., Punta Arenas, Chile. 221 pp.
40. Gusinde M (1951) Fueginos. Hombres primitivos en la Tierra del Fuego. Publicaciones de la escuela de Estudios Hispano-Americanos de Sevilla, Sevilla. 398 pp.
41. Eerkens J, Vaughn K, Linares Grados M (2009) Pre-Inca mining in the Southern Nasca Region, Peru. Antiquity 83: 738–750.
42. Lleras R (2010) Una revisión crítica de las evidencias sobre metalurgia temprana en Suramérica. Maguaré 24: 297–312.
43. Núñez L, Cartajena I, Carrasco C, de Souza P, Grosjean M (2006) Emergencia de comunidades pastoralistas formativas en el sureste de la Puna de Atacama. Estud. Atacam. 32: 93–117.
44. Salazar D, Castro V, Michelow J, Salinas H, Figueroa V et al. (2010) Minería y Metalurgia en la Costa Arreica de la Región de Antofagasta, Norte de Chile. Bol Museo Chil Arte Precol 15: 9–23.
45. Salazar D, Figueroa V, Mille B, Morata D, Salinas H (2010) Metalurgia prehispánica en las sociedades costeras del norte de Chile (quebrada Mamilla, Tocopilla). Estud. Atacam. 40: 23–42.

46. Salazar D, Figueroa V, Morata D, Milleiv B, Manríquez G, et al. (2011) Metalurgia en San Pedro de Atacama durante el Período Medio: Nuevos Datos, Nuevas Preguntas. Rev Antropol Chil 23: 123–148.

47. Scattolin M, Bugliani M, Cortés L, Pereyra L, Calo M, et al. (2010) Una máscara de cobre de tres mil años. Estudios arqueometalúrgicos y comparaciones regionales. Bol Museo Chil Arte Precol 15: 25–46.

48. Hinkley T (2007) Lead (Pb) in old Antarctic ice: some from dust, some from other sources. Geophys Res Lett 34: GL028736.

49. Graffam G, Rivera M, Carevič A (1996) Ancient Metallurgy in the Atacama: Evidence for Copper Smelting during Chile's Early Ceramic Period. Lat Am Antiq 7: 101–113.

50. Augustyniak S (2004) Dating the Tiwanaku State. Rev Antropol Chil 36: 19–35.

51. Abbott M, Wolfe A (2003) Intensive Pre-Incan Metallurgy Recorder by Lake Sediments from the Bolivian Andes. Science 301: 1893–1895.

52. Scattolin S, Williams V (1992) Actividades minero metalúrgicas prehispánicas en el noroeste argentino.nuevas evidencias y su significación. Bul Inst Fran Etu And 21: 59–97.

53. González L (2002) A sangre y fuego. Nuevos datos sobre la metalurgia Aguada. Estud Atacam 24: 21–37.

54. Zori C, Tropper P, Scott D (2013) Copper production in late prehispanic northern Chile. J Archaeol Sci 40: 1165–1175.

55. Cooke C, Abbott M, Wolfe A, Kittleson J (2007) A Millennium of Metallurgy Recorded by Lake Sediments from Morococha, Peruvian Andes. Environ Sci Technol 41: 3469–3474.

56. Desaulty AM, Telouk P, Albalat E, Albarède F (2011) Isotopic Ag-Cu-Pb record of silver circulation through16[th]–18th century Spain, Proc Natl. Acad. Sci. USA 108: 9002–9007.

57. Martinić M (2003) La mineria aurifera en la Region Austral Americana (1869–1950). Historia 36: 219–254.

58. Martinić M (2004) La mineria del carbon en Magallanes entre 1868–2003. Historia 37: 129–167.

59. Onursal B, Gautam SP (1997) Contaminación atmosférica por vehículos automotores. Experiencias recogidas en siete centros urbanos de América latina. Banco Mundial, Oficina Regional de América Latina y el Caribe, Departamento Técnico, Washington D.C. 306pp.

# 23

# Biodegradation of Microcystins during Gravity-Driven Membrane (GDM) Ultrafiltration

Esther Kohler[1], Jörg Villiger[1], Thomas Posch[1], Nicolas Derlon[2], Tanja Shabarova[1], Eberhard Morgenroth[2,3], Jakob Pernthaler[1], Judith F. Blom[1]*

1 Limnological Station, Institute of Plant Biology, University of Zurich, Kilchberg, Switzerland, 2 Eawag: Swiss Federal Institute of Aquatic Science and Technology, Dübendorf, Switzerland, 3 Institute of Environmental Engineering, ETH Zurich, Zurich, Switzerland

## Abstract

Gravity-driven membrane (GDM) ultrafiltration systems require little maintenance: they operate without electricity at ultra-low pressure in dead-end mode and without control of the biofilm formation. These systems are already in use for water purification in some regions of the world where adequate treatment and distribution of drinking water is not readily available. However, many water bodies worldwide exhibit harmful blooms of cyanobacteria that severely lower the water quality due to the production of toxic microcystins (MCs). We studied the performance of a GDM system during an artificial *Microcystis aeruginosa* bloom in lake water and its simulated collapse (i.e., the massive release of microcystins) over a period of 21 days. Presence of live or destroyed cyanobacterial cells in the feed water decreased the permeate flux in the *Microcystis* treatments considerably. At the same time, the microbial biofilms on the filter membranes could successfully reduce the amount of microcystins in the filtrate below the critical threshold concentration of 1 µg L$^{-1}$ MC for human consumption in three out of four replicates after 15 days. We found pronounced differences in the composition of bacterial communities of the biofilms on the filter membranes. Bacterial genera that could be related to microcystin degradation substantially enriched in the biofilms amended with microcystin-containing cyanobacteria. In addition to bacteria previously characterized as microcystin degraders, members of other bacterial clades potentially involved in MC degradation could be identified.

**Editor:** Rajeev Misra, Arizona State University, United States of America

**Funding:** This research was supported by the ProDoc program "Predictive Toxicology" (PDFMP3_132466), funded by the Swiss National Science Foundation. J. Villiger was supported by the Swiss Federal Department of Foreign Affairs through Polish-Swiss Research Program, project PSPB-036/2010: Diversity and Ecology of Mixotrophic Nanoflagellates in the Gulf of Gdańsk (DEMONA). The funders had no role in study design, data collection and analysis, decision to publish, or preparation of the manuscript.

**Competing Interests:** The authors have declared that no competing interests exist.

* Email: blom@limnol.uzh.ch

## Introduction

During the last century, the anthropogenic input of nutrients into freshwaters has resulted in a distinct increase of cyanobacterial biomass in many water bodies worldwide [1]. Climate change and global warming may even increase the frequency and intensity of cyanobacterial blooms in the future [2]. Some cyanobacteria represent a major challenge for drinking water usage due to their production of microcystins (MCs), toxic secondary metabolites that affect a wide range of animals and humans [3]. The major route of human exposure to MCs is via oral ingestion, mainly due to the consumption of drinking water [4]. Even a subchronic dose of MCs in drinking water may elevate the rate of liver cancer, as was shown in China, where prevalent incidences of liver cancer correlated with MC-contaminations in drinking water [5]. Consequently, the WHO has developed a guideline for MCs in drinking water stating that an average exposure generally should be below the level of 1 µg L$^{-1}$ [4]. *Microcystis aeruginosa* is known to form massive blooms in many lakes worldwide, and it produces MCs in high amounts. The concentrations of the intracellular MCs range between 0.3 to 15 µg L$^{-1}$ [6] and up to 400 µg L$^{-1}$ [7] in cyanobacterial blooms, however, occasionally high concentrations of up to 1400 µg L$^{-1}$ were found [8]. Elimination of the MCs from drinking water is therefore highly desirable.

The largely cell-bound MCs are eliminated by removing the intact cyanobacterial cells by conventional water treatment procedures such as coagulation or flocculation [4]. However, cell damage (e.g. during the collapse of a bloom) will release toxins into water, and the above mentioned procedures will not sufficiently remove MCs from drinking water. Strategies such as powdered activated carbon [9], sediment sorption [10], or ozonisation [11] have been suggested to effectively eliminate dissolved MCs. However, these treatments are costly in terms of development and management (energy, need of chemicals) and thus not suitable for the application in developing and transient countries.

Gravity driven membrane (GDM) ultrafiltration is considered for drinking water production as a relevant alternative to common appliances [12]. GDM uses a simple set-up [13], which is inexpensive, electricity-free, easy to use, and it is already known to provide an effective barrier against pathogens, disease vectors and suspended solids [14]. Microbial activity as well as the total organic carbon content in the feed water have been shown to affect the performance of the GDM ultrafiltration without control of the biofilm formation (no backwashing or chemical cleaning) [13]. However, nothing is currently known about the possible degradation processes of intact cyanobacterial cells or of toxins such as MCs in these point-of-use membrane systems. In recent years, biodegradation by heterotrophic bacteria has been recognized as an alternative way to eliminate MCs [15]. A few bacterial isolates capable of MC degradation have been already characterized [16]. The best studied MC degrading bacteria are belonging to the *Alphaproteobacteria* such as *Sphingomonas* sp. [17], *Sphingopyxis* sp. [18], or *Novosphingobium* sp. [19]. However, only a few studies have tried to link the composition of bacterial communities in plankton [20] or in biofilms of biological drinking water treatment facilities [21] with the ability of these systems to degrade microcystins.

The objective of our study was to examine possible degradation of MCs by the microbial biofilm of a GDM ultrafiltration system. We simulated cyanobacterial blooms and their collapse (and thus the release of the cell-bound MCs into the surrounding water) and determined the ability of the microbial biofilm to remove MCs during drinking water production. We also analysed the composition of microbial assemblages of the biofilms of the GDM ultrafiltration systems by next generation sequence analyses in order to obtain information about the microorganisms that might potentially be involved in this process.

## Materials and Methods

### Cyanobacterial cultures and quantification of microcystins

Axenic cultures of Microcystis aeruginosa PCC 7806 were kept at 20°C in Cyano-medium in several Erlenmeyer flasks under constant light at 5 µmol quanta $m^{-2} s^{-1}$ from fluorescent tubes. Fresh Microcystis aeruginosa PCC 7806 cultures were taken every three to four days from the cultivation for the ongoing experiment. The cell number of the cyanobacterial culture was determined by flow cytometry (described below), and the MC concentration was quantified by high-performance liquid-chromatography (HPLC) as followed: A volume of 5 mL of the culture was frozen at −23°C for three hours. After thawing, 7.5 mL of 100% methanol (MeOH) were added to achieve a 60% aqueous methanolic solution. The extract was centrifuged for 15 min at 25′700 g. HPLC analysis was performed on a Shimadzu 10 AVP system with photodiode array detector (PDA) and a Hydrosphere C18 column (YMC, 4.6×250 mm, Stagroma, Switzerland), using solvent A: UV-treated $H_2O$ containing 0.05% trifluoroacetic acid (TFA, Merck) and solvent B: acetonitrile and 0.05% TFA. A gradient was achieved by applying linear increases in two steps (solvent B from 35% to 70% in 30 min, 70% to 100% in 2 min). For the quantification procedure, calibration curves for MC-LR and [D-Asp$^3$] MC-LR, the two MCs of M. aeruginosa PCC 7806 had to be established: The two MCs were isolated in high purity (>99%, HPLC) from Microcystis aeruginosa PCC 7806, and their specific molar absorption coefficient was used to prepare accurate standard solutions between 1 and 10 µg $mL^{-1}$. The calibration curves were based on the peak area recorded at a wavelength of 239 nm. The microcystin quantification was done in duplicate and is referred to as the sum of the concentrations of MC-LR and [D-Asp$^3$] MC-LR.

### Experimental setup of the Gravity-Driven-Membrane (GDM) system

Water from a depth of 5 m of Lake Zurich was continuously pumped by a fountain pump (Nautilus 450, Oase GmbH, Hörstel, Germany) to a storage tank (6 l volume, kept in the dark at room temperature). This storage tank was connected by silicon tubes (Saint-Gobin) to six parallel membrane modules consisting of filter holders of 48 mm inner diameter (Whatman, Maidstone, Kent, UK) and polyethersulfone ultrafiltration membranes with a 150 kDa nominal cut-off (PBHK, Biomax Millipore, Billerica, MA, USA). A hydrostatic pressure of 0.65 mbar was received by keeping the storage tank 0.65 m above the membrane surface. Overflow conditions at the storage tank guaranteed constant transmembrane pressure. Filter holders, silicon tubes and glass bottles for collection of filtrate water were autoclaved prior to experiment. Ultrafiltration membranes have been soaked in nanopure water (Bearnstead, Thermo Scientific, Basel, Switzerland) for 24 h before starting the experiment. The six membrane systems were split into three different treatments, each with two replicates (Figure 1): the control treatment (referred to as CON) received lake water only. A Microcystis bloom was simulated in two replicates: the lake water was enriched with the cyanobacterium Microcystis aeruginosa PCC 7806 (about $2 \times 10^8$ cells) once every 24 h (treatment subsequently referred to as LMA: living Microcystis aeruginosa). A collapsing cyanobacterial bloom was simulated in the last treatment. A culture of Microcystis aeruginosa PCC 7806 was first frozen for 3 h at −20°C, and thawed before adding to the membrane system once every 24 h (referred to as DMA: dead Microcystis aeruginosa). The cyanobacterial cells were directly added above the filtration membrane module to avoid MC degradation processes in the storage tank or during tubing passages. All filter systems were kept in the dark. The filtrate water of all treatments was collected every 24 h, and the volume was determined to quantify the permeate flux (as L $m^{-2} h^{-1}$). A subsample of 1 mL of each filtrate was fixed with 50 µl glutaraldehyde (2.5% final concentration) for cell enumeration at the flow cytometer. The rest was stored at 4°C for microcystin (MC) quantification (usually done within 24 to 48 h). The experiment was running for 21 days. At the end of the experiment, the ultrafiltration membranes were cut into three equal parts that were subjected to the following analyses: (i) One part of the filter was used to quantify MCs by HPLC that possibly remained in or attached to the biofilm on the filters. (ii) Phylogenetic analyses (454 tag pyrosequencing) were performed with the biofilm on another filter part, and (iii) the last part was used to take a closer look on biofilm structures by non-invasive methods such as Optical Coherence Tomography (OCT) (model 930 nm Spectral Domain, Thorlabs GmbH, Dachau, Germany). OCT images were analysed for average thickness and relative roughness using image analysis software developed under Matlab (MathWorks, Natick, US) [12]. To determine the exact membrane area of the pieces image analysis (Zeiss, AxioVision 4.7) was applied using a microscope (AxioImager.Z1, 1 x EC Plan-Neofluar, Zeiss) and a CCD camera (AxioCam MRm, 12 bit grayscale, 1388×1040 px, Zeiss).

### Flow cytometric enumeration of cells in the filtrate water

Samples for flow cytometry were stained with DAPI (4′,6-diamidino-2-phenylindole, 1 µg $mL^{-1}$ final concentration) for 15 min in the dark. Subsequently, samples were analysed using an Influx V-GS cell sorter (Becton Dickinson, Inc., San Jose, CA) equipped with a UV laser (60 mW, 355 nm; CY-PS; Lightwave Electronics) for detection of DAPI fluorescence, and a blue laser (200 mW, 488 nm; Sapphire; Coherent Inc.) for scattered light

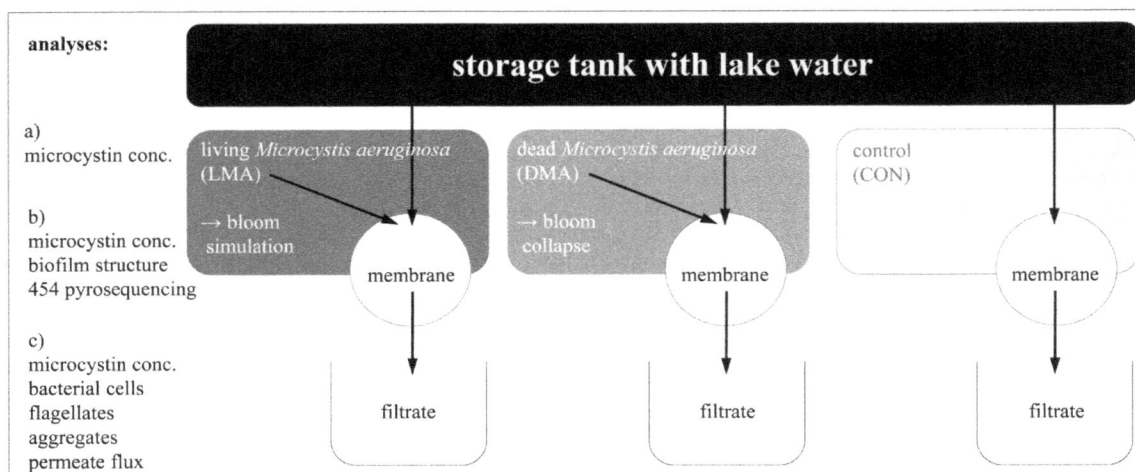

**Figure 1. Schematic view of the gravity-driven membrane (GDM) system.** Depicted are the three different operating treatments (LMA, DMA, and CON), the ultrafiltration membrane (150 kDa) and the filtrate collection. Analyses that were carried out during this study a) in the feed water, b) in/on the biofilm, and c) in the filtrate are listed on the left side of this overview.

and autofluorescence of flagellates. If necessary, and to avoid particle coincidence, samples were diluted with sheath fluid (2.5 g $L^{-1}$ NaCl; filtered by 0.2 µm-pore-size). Sample volume was calculated from the analysed sample weight. Data obtained by flow cytometry were analysed with the custom-made software ViiGate 1.0a. Bacterial cells were identified using side scattered light (SSC) versus DAPI fluorescence (431 nm), flagellates were determined on the basis of SSC versus green fluorescence (531 nm). Bacterial aggregates were operationally defined by their DAPI fluorescence and scatter properties equal to or higher than that of flagellates [22]. Cyanobacterial cells of the *Microcystis aeruginosa* PCC 7806 culture were identified using SSC versus their auto-fluorescence at 692 nm (without prior staining).

## Extraction of microcystins

The entire filtrate was collected for solid phase extraction of MCs. The C18 cartridges (1 g, 60 mL, Mega Bond Elute, Varian, Agilent Technologies, Basel, Switzerland) were first equilibrated with 10% MeOH before adding the filtrate water. Afterwards, the MCs were eluted with 100% methanol. The samples were dried in a vacuum rotary evaporator at 40°C and 35 mbar. The residues were re-suspended in 1 ml 60% MeOH, the microcystin quantification was performed by HPLC in duplet. The MC concentrations in the filtrate were always measured 24 h after injection and were expressed as percentage of the MCs that were removed from the system (removal efficiency) or as MC removal rates (ug $L^{-1}$ $d^{-1}$).

At the end of the experiment, GDM polyethersulfone membranes were first frozen at $-23°C$ for 3 h to isolate the MCs from the biofilm. After thawing, the biofilms were extracted twice with 5 mL 60% MeOH for 1 h, and both extracts were combined. Accordingly, the solvent was evaporated (40°C and 35 mbar), and the samples were prepared for HPLC analysis. The residues were re-suspended in 1 ml 60% MeOH, and centrifuged for 5 min at 10'000 rpm. Afterwards, the supernatants were taken for microcystin quantification by HPLC (as described above).

## 454 tag pyrosequencing analysis

Prior to the analysis, the filter parts of both replicates were pooled. The DNA extraction of the biofilm bacteria was

performed using the UltraClean Water DNA isolation kit (MO BIO Laboratories, Inc.). Subsamples of 300 µl of DNA suspension (final concentration 4–10 ng $µL^{-1}$) of all three DNA extractions (CON, LMA, and DMA) were sent to Research and Testing Laboratory, Inc. (Lubbock, TX, USA) for further processing. Partial 16S rRNA gene encoding sequences were obtained from 454 pyrosequencing (Roche FLX platform) following Assay b.9 by using the primer pair 799F and 1115R that exclusively amplify DNA of heterotrophic bacteria and exclude cyanobacteria from the process [23]. Raw data (68'981 Reads; mean raw read length 378.4 base pairs) were processed by a custom-made pipeline on a local computer cluster consisting of 16 units (each equipped with an 8 core AMD FX-8150 CPU, 16 GB RAM and a 128 GB SSD hard disk) and a separate control workstation. The program was developed in DELPHI and run under Windows 7. The processing of the raw data is extensively described elsewhere [24]. In brief, reads were denoised at the level of flowgrams according to Quince and co-workers [25]. Afterwards, quality filtering strategies were applied to finally end up with the number of 27'391 sequences (raw reads reduced by 60%) corresponding to 956 operational taxonomic units (OTUs, 3% similarity). A distance matrix was calculated and the OTUs were produced after the pairwise alignment (Needleman-Wunsch algorithm) by average linkage at similarity levels of 97%. OTUs were assigned to taxonomic entities on the level of similarity of the OTU to the most closely related sequence in the SILVA reference data base (release 109) [26]. OTUs were grouped into different levels of sequence identity: $\leq$ 3% divergence in 16S rRNA gene sequence corresponds to species level, $\leq$5% to genus level, and $\leq$10% to family level. Finally, the OTUs of the both treatments were compared with each other. Only OTUs were included that were >0.5% of sequences per sample (CON or LMA+DMA) and that occurred at least ten times more frequently in one treatment than the other. Shannon's diversity index was estimated using the formula $H' = -\Sigma(P_i * \ln P_i)$, where $P_i$ is the relative abundance of the sequences per sample. The index $H'$ is used to characterize the diversity of species or species-like units (OTUs) in a community.

## Results

### Physical parameters: Permeate Flux and structure of the biofilms

Flux stabilization was observed approximately after eight to ten days of the experiment in all three treatments albeit great differences between the control and both Microcystis treatments (Figure 2). A mean flux of 4.7 L $m^{-2}$ $h^{-1}$ was measured after 12 days in the CON treatment. In one of the two replicates, the flux stayed constant until the end of the experiment. In the second replicate, the flux increased slowly to 6.9 L $m^{-2}$ $h^{-1}$ on day 21. Accordingly, the mean thickness (as assessed by OTC measurements) of the biofilms of both control replicates at the end of the experiment were slightly different. The biofilm of the first replicate had a thickness of about 125 ($\pm$23) µm; the biofilm of the second replicate was 96 ($\pm$17) µm thick. The second replicate was less heterogeneous than the first replicate, but both exhibited low relative roughness values of 0.49 and 0.35, respectively. The permeate flux in both Microcystis treatments showed a similar trend. Stabilization could be observed at a mean flux of 1.6 L $m^{-2}$ $h^{-1}$ in the LMA and of 2.0 L $m^{-2}$ $h^{-1}$ in the DMA treatment. In both treatments, mean permeate flux decreased further to 1.0 L $m^{-2}$ $h^{-1}$ and 1.36 L $m^{-2}$ $h^{-1}$, as measured at the end of the experiment. Thus, the mean flux in the Microcystis replicates was about 80% lower than in the CON treatment. Biofilms in the DMA treatment were about six to seven times thicker as in the CON treatment with values of 625 ($\pm$33) µm and 796 ($\pm$29) µm for both replicates. Unfortunately, a quantification of the biofilm thickness in the LMA treatment could not be carried out.

### Microcystin removal

The microcystin removal efficiency of the biofilms was calculated from the amount of MCs that was injected via living or dead Microcystis cells into the systems and the amount that was measured 24 hours later in the filtrates (Figure 3, upper panel). In the LMA treatment, both replicates were working similarly, showing already high removal efficiency of almost 70% at the beginning. After 10 days of the experiment, the biofilms of both replicates showed nearly 100% removal efficiency; complete removal was achieved after 15 days and remained constant until the end of the experiment. Both replicates of the DMA treatment started with a low removal efficiency of about 10% during the first

three days of the experiment. One of the replicates developed a biofilm that was able to remove the MCs completely after 15 days. The MC removal efficiency of the biofilm of the second replicate decreased again to less than 80% (Figure 3, upper panel). The MC removal rates in both treatments increased during the course of the experiment up to 440, respectively 300 µg $L^{-1}$ $d^{-1}$ (Figure 3, lower panel).

Altogether, $4.6 \times 10^9$ living or dead cells of M. aeruginosa (containing 262.5 µg MCs) were added to each MC-replicate of the GDM systems throughout the duration of the experiment (Table 1). About 10% (27 µg) of the added MCs were found again in the filtrates of each replicate of the LMA treatments during the entire experiment, 96.5 µg (36.8%) were found on the LMA filters at the end of the experiment. Thus, 140 µg of MCs were removed in both replicates during the course of the experiment. Only 0.5 µg MC was found on the filter in the DMA treatments, 99 µg, respectively 121 µg have been collected in the DMA filtrates of both replicates; 141–163 µg of the MCs were degraded during the course of the experiment.

### Cell numbers

Bacterial single cell numbers in the CON filtrates constantly increased during the course of the experiment (Figure 4, upper panel) up to 0.1 and $0.3 \times 10^6$ cells $mL^{-1}$, respectively. Similarly, flagellate and bacterial aggregate numbers increased slowly, but stayed on comparable low mean levels of 280 flagellates $mL^{-1}$ and 78 aggregates $mL^{-1}$ on day 21 of the experiment (Figure 4). In contrast, both MC treatments showed higher cell numbers as

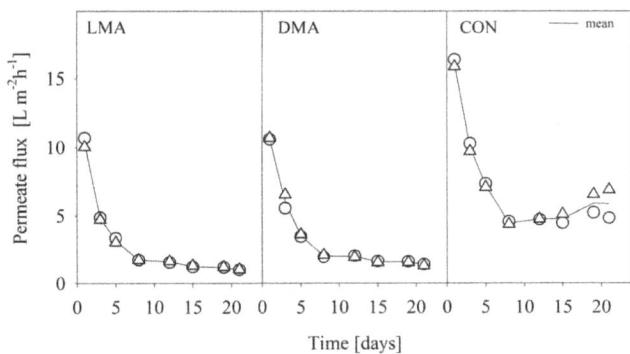

**Figure 3. Microcystin removal and removal rates.** Time course of the microcystins (MCs) removal efficiency of the GDM system (upper panel) and the MC removal rate (lower panel) in the two *Microcystis* treatments (LMA and DMA according to Figure 1). The two replicates per system are shown as circles and triangles, and are connected by the mean.

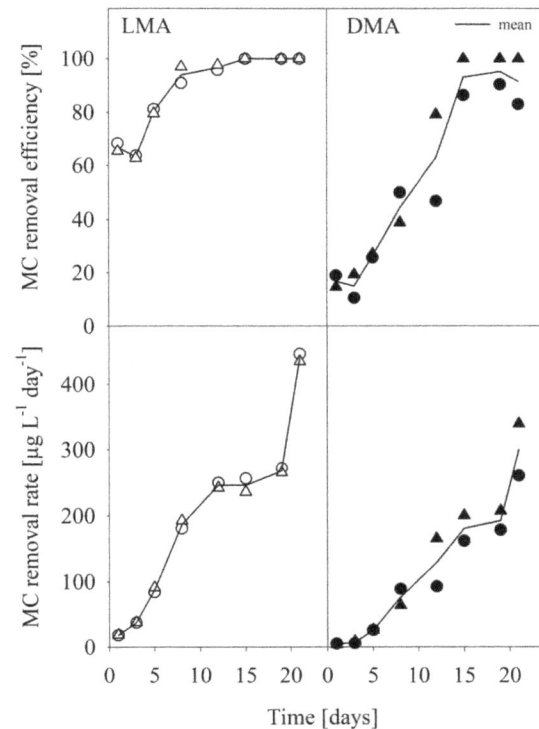

**Figure 2. Evolution of the permeate flux.** The flux is shown in L $m^{-2}$ $h^{-1}$ for the filtration of differently treated feed water sources (LMA, DMA, and CON according to Figure 1). The two replicates per system are shown as circles and triangles, and are connected by the mean.

**Table 1.** Microcystin concentrations.

| | | LMA | | DMA | |
| --- | --- | --- | --- | --- | --- |
| | | **Replicate 1** | **Replicate 2** | **Replicate 1** | **Replicate 2** |
| Total amount of MCs injected over 21 days | | 262.5 (100) | 262.5 (100) | 262.5 (100) | 262.5 (100) |
| Amount of MCs in the filtrate on day 21 | | 0 (0) | 0 (0) | 0 (0) | 2.9 (1.11) |
| MCs on the filter at the end of the experiment (on day 21) | | 96.5 (36.8) | 96.5 (36.8) | 0.5 (0.2) | 0.5 (0.2) |
| Total amount of MCs in the filtrates, collected for 21 days | | 28 (10.6) | 26 (9.9) | 99 (37.7) | 121 (46.1) |
| MCs loss | | 138 (52.6) | 140 (53.3) | 163 (62.1) | 141 (53.7) |
| Threshold concentration of less than 1 µg L$^{-1}$ was reached on day | | 15 | 15 | 15 | Not reached |

Overview about the amount of MCs [µg; (%)] totally injected, found on the biofilm or in the filtrate, and the estimated loss of MCs during the experiment. Last row shows the day, at which the threshold concentration of less than 1 µg L$^{-1}$ was reached (LMA and DMA according to Figure 1).

compared to the CON treatment. Between 0.8 and $0.5 \times 10^6$ bacterial single cells mL$^{-1}$ were found in both replicates of the LMA treatment at the end of the experiment. Comparable numbers of bacterial single cell numbers were determined also in the replicate 1 of the DMA treatment ($0.4 \times 10^6$ mL$^{-1}$). The bacterial single cell numbers in the corresponding second replicate of the DMA treatment were about ten times higher ($3.9 \times 10^6$ mL$^{-1}$). Flagellate numbers in the filtrates of the LMA treatment were comparable high at the end of the experiment (mean value $0.67 \times 10^4$ mL$^{-1}$), as well as bacterial aggregates (mean value $0.8 \times 10^3$ mL$^{-1}$). Comparable numbers of flagellates were found in the filtrate of the first replicate of the DMA treatment ($0.54 \times 10^4$ mL$^{-1}$) as well as the highest amount of aggregates ($4.0 \times 10^3$ mL$^{-1}$) at the end of the experiment. However, rather low numbers of flagellates and aggregates were found in the second replicate of the DMA treatment, $0.15 \times 10^4$ mL$^{-1}$ and $0.1 \times 10^3$ mL$^{-1}$, respectively (Figure 4).

## Phylogenetic analyses

A total of 27'391 sequences, assigned to 956 OTUs, were evaluated after removing low quality reads and chimeric sequences. For the bacterial communities of the MC-treated biofilms, about 452 OTUs (8'064 sequences; LMA) and 378 OTUs (8'048 sequences; DMA) were determined (Figure 5A); slightly more sequences were obtained for the CON bacterial community (11'279 sequences; 551 OTUs). The bacterial communities of the three biofilms shared only 11% of all OTUs (105 OTUs), but comprised 54% of all sequences (Figure 5A and 5B). The OTUs therein were large, and consisted of 140 sequences on average. This core community consisted predominantly of Sphingobacteriales (Bacteroidetes; 48% of the shared OTUs) and Comamonadaceae (Betaproteobacteria, 45% of the shared OTUs). The CON treatment consisted of the largest amount of OTUs (35% of all OTUs) that were unique to this special treatment, but comprised only 1764 sequences (5.2 sequences per OTU). LMA and DMA shared 108 OTUs, the average sizes of these OTUs were about 50 sequences per OTU (Figure 5A and 5B).

Most of the sequences in the combined LMA and DMA (MA) communities received taxonomic assignments at least to the genus level (93% in the LMA treatment, 96% in the DMA treatment), affiliation to families could be determined for 97%, respectively 99% of these sequences. However, only 44% of all sequences in the CON communities were assigned to the genus level, and 52% to the family level. With a high divergence in similarity of more than 12% to the closest known relative, two large OTUs were

assigned to either Myxococcales (1'990 sequences) or to Fibrobacteres (2'357 sequences).

## Bacterial taxa favoured by Microcystis addition

Addition of Microcystis cells led to compositional differentiation between the communities of both MC-treated and the CON biofilms. Over one third (38.3%) of all sequences in the CON assemblage was affiliated with Deltaproteobacteria, a class that was significantly less abundant in the combined LMA and DMA (MA) communities (5.0%) (Figure 6). The Fibrobacteres (36.6%) and the Alphaproteobacteria (19.4%) were the second and third most abundant taxa affiliated with the CON assemblage, both taxa were underrepresented in MA assemblage as well. These three classes comprised almost 95% of all sequences in the CON assemblage. However, more than one third (40.4%) of all sequences in the MA assemblage was affiliated with Betaproteobacteria that were found only marginally in the CON communities. Firmicutes (22.4%) and Gammaproteobacteria (12.8%) were the second and third most abundant taxa in the MA assemblage, but were not found in the communities of the CON treatment. These three classes comprised 75.6% of all sequences. Additionally, minor fractions of Candidate division TM7 and Bacteroidetes were present in both assemblages, Spirochaetes only to a minor extent in the MA assemblage (1.3%).

Most of the bacterial genera found in the CON assemblage were typically isolated from microbial biofilms in drinking or freshwater reservoirs or pipelines, such as Hirschia, Phenylobacterium, or Comamonas. Haliangium or the Myxococcales were present in anaerobic filter sediments or suboxic freshwater ponds (Table S1 in File S1). The Fibrobacteres were isolated before from upper sediment layers and biofilm samples. Bacterial sequences affiliated with Sphingomonas containing MC degradation proteins were only found in the CON assemblage, where they made up 12% of all sequences (Table S1 in File S1). Other genera of the Alphaproteobacteria that contain MC degradation proteins or MC dependent proteins have been found only in the MA assemblage, such as Azospirillum or Magnetospirillum, as well as the Rheinheimera (belonging to the Gammaproteobacteria) or Spirochaetales (Table S2 in File S1). Some other bacterial genera have been repeatedly found in the Planktothrix layer of Lake Zurich, such as Variovorax (5.4% of all sequences in the MA assemblage) or unknown genera belonging to the Sphingobacteriales (6.5%). Paucibacter known to degrade MCs has been only found in the MA assemblage (3.2%). The Firmicutes (22.4% of all sequences of the MA assemblage) comprised mainly of Acidaminobacter that are typical inhabitants of suboxic freshwater ponds.

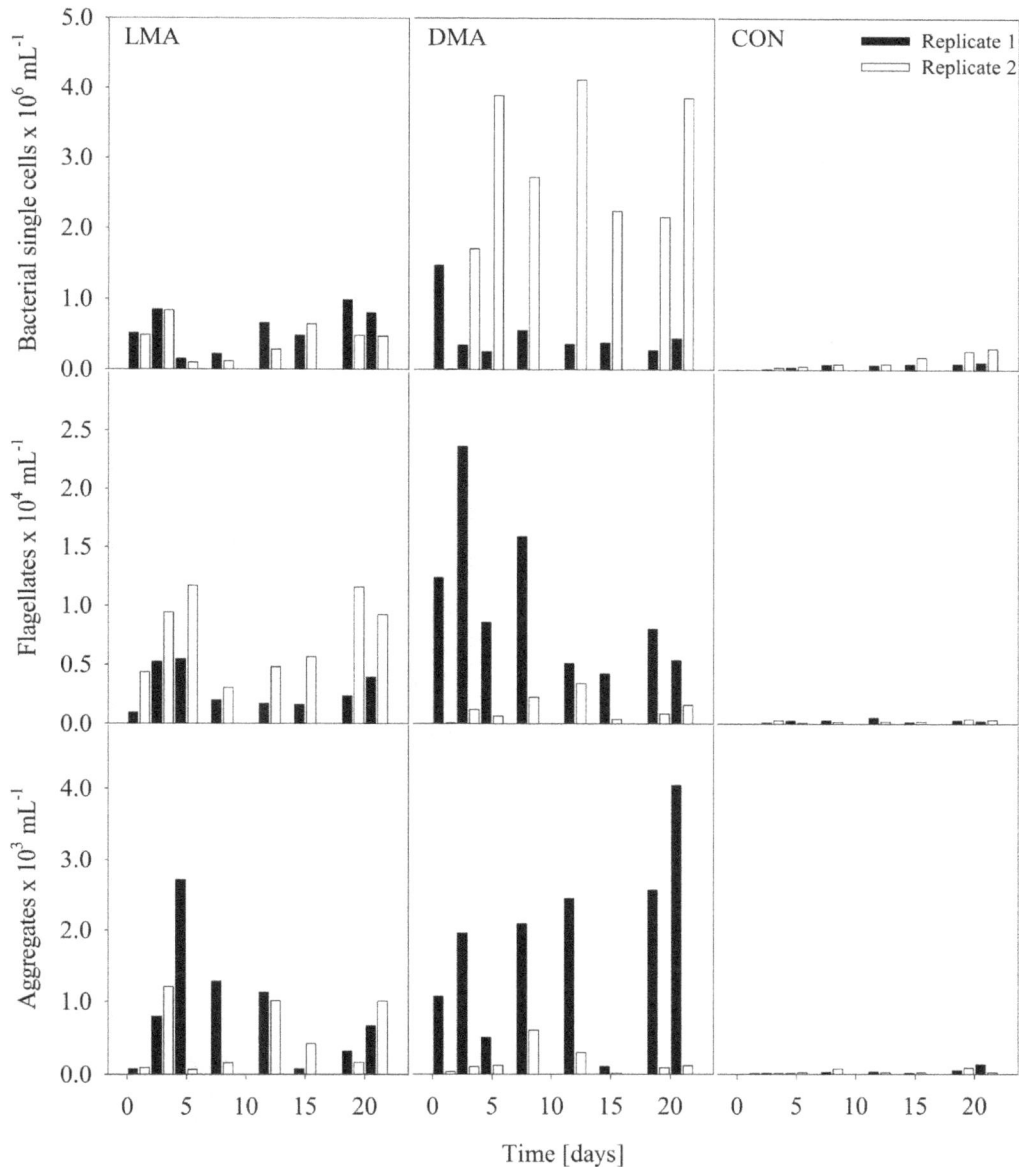

**Figure 4. Cell abundances in the filtrate.** Regrowth of bacterial single cells (upper panel), flagellates (middle panel) and flagellate-inedible bacterial aggregates (lower panel) in the filtrate of the three different treatments (LMA, DMA, and CON according to Figure 1).

## Discussion

### Performance of the GDM system and fate of microcystins

Two factors have a strong impact on the performance of GDM systems: the composition of the microbial community (bacteria and predators) and the organic carbon content of the feed water [12]. In the latter study, the almost complete absence of metazoan organisms resulted in smooth and homogeneous biofilm structures. In our study, this was indicated by the comparably low values of relative roughness of both CON replicates. In the absence of predation, the total organic carbon (TOC) content governs the permeability of the biofilms. The decrease of the permeate flux with increasing TOC content is due to, both, a higher accumulation of particulate matter (originated from the influent) and a higher bacterial growth because of the higher nutrient loads upon cell lysis [12]. Cyanobacterial blooms are such a source of

high TOC. The enormous load of cyanobacterial biomass in both Microcystis treatments resulted in the massive increase in biofilm thickness. Thus, both Microcystis treatments lost 80% of their performance, which was shown by a much lower flux and permeate, respectively.

At the same time, these biofilms were able to degrade MCs. A complete removal of the MCs by the biofilms in our study could be achieved after 15 days, which was one week after the stabilization of the flux, i.e., after the development of a stable biofilm. It is conceivable that after that time bacteria capable of the degradation of MCs had accumulated on the biofilm. The key role of the bacterial biofilm in MC degradation can be confirmed by comparing the performance of the GDM system during the initial phase of the experiment and after flux stabilization. A reduction of 10% during the first three days of the experiment (Figure 3) cannot be attributed to the not yet developed biofilm but rather to

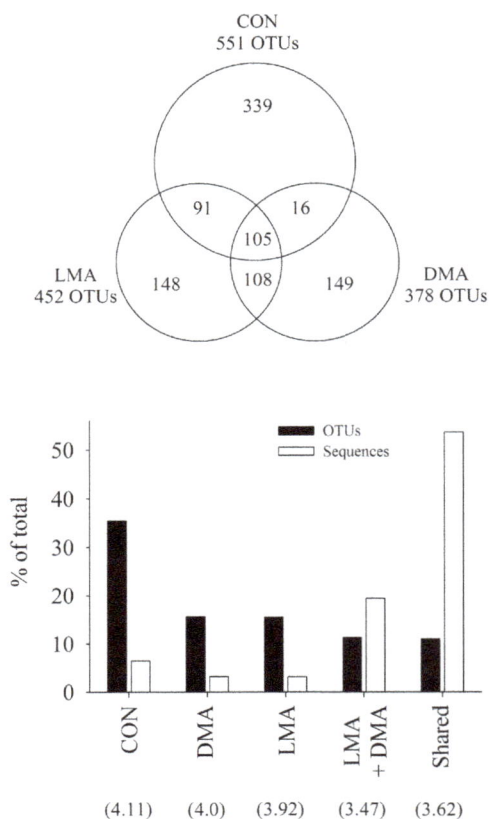

**Figure 5. Overview of microbial diversity analysis.** Upper panel: Venn diagram of shared OTUs among the biofilms of the three treatments. Lower Panel: percentage of operational taxonomic units (OTUs) and sequences found in the different treatments. Number in brackets marks the Shannon's diversity index (*H'*) for each treatment (LMA, DMA, and CON according to Figure 1).

adsorption processes, either to the filtration units or to cyanobacterial cells that retained on the filter membrane.

However, the potential of biofilms to remove MCs does not only depend on the presence and fast accumulation of possible MC degrading bacteria, but also on their efficiencies to biodegrade these MCs. Degradation rates of various bacterial strains might range between 1.5 up to several $10'000$ µg $L^{-1}$ $d^{-1}$ [16] Already after 15 days, the added MCs were completely removed. In our study, MC removal rates increased constantly after a short lag phase, but did not reach a plateau. Therefore, the degradation capability of the aged biofilm might have further increased and even higher concentrations of MCs might be successfully degraded. The previous natural exposure of the feed water to cyanobacteria and MCs (as is the case for water from Lake Zurich) seems to be a significant driver for a fast performance of a MC degrading biofilm and thus its accelerated maturation [27]. Lake Zurich contains large populations of Planktothrix rubescens, a filamentous microcystin-producing cyanobacterium, which is frequently found in natural pre-alpine lakes [2]. It should be noted that the biofilm in our study was selectively grown under Microcystis bloom conditions. It remains to be investigated, how a bacterial biofilm without prior presence of MCs would respond to Microcystis addition. Another interesting aspect, which was beyond the scope of our experiment, would be the comparison of the bacterial composition in the biofilms by comparing MC-producing with MC-non-producing strains. This would allow for a more specific assessment, which shifts in microbial community compositions were caused by other substrates provided by the addition of cyanobacterial biomass.

## Drinking water quality

Pro- and eukaryotic microbes can regrow during drinking water treatment and the distribution of non-chlorinated potable water [28]. The total bacterial cell counts in tap and mineral water may reach values between $1.5 \times 10^5$ to $5 \times 10^5$ cells $mL^{-1}$ [29,30]. In our study, bacterial abundances in the CON treatment were within this range. Bacteria most likely regrew in biofilms underneath the filter membranes or on materials downstream the filtration process and dropped into the filtrate [29]. Inadequate disinfection, hydraulic retention time, flow regime, pipe material, temperature, source of water or corrosion could lead to the development of a microbial biofilm. However, excessive regrowth

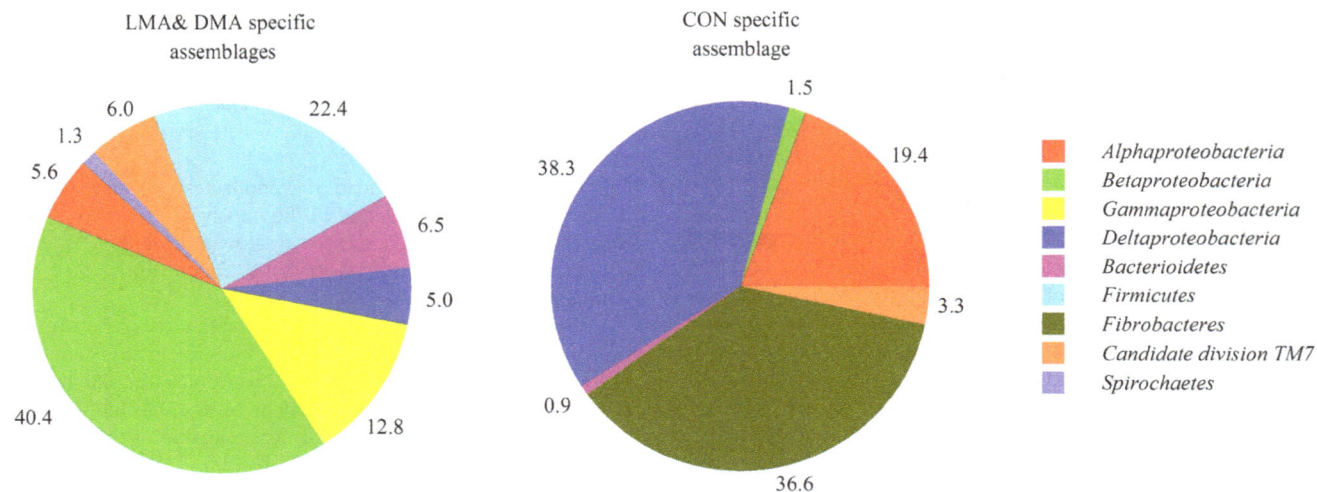

**Figure 6. Phylogenetic composition of bacterial biofilms.** Phylogenetic composition [%] of bacteria specific for the (combined) LMA and DMA assemblage (left) and for the CON (right) assemblage. Specificity for either assemblage was defined by the size of the OTU (>0.5% of sequences per sample) and by a ten times higher occurrence in one category than the other (LMA, DMA, and CON according to Figure 1).

in drinking water supply systems can be also triggered by nutrient introduction. It is conceivable that the massive load of cyanobacterial biomass in our study led to higher bacterial cell numbers, not only under the membrane filter but also in the filtrate. The increased availability of various substrates released from broken cells in the DMA treatment resulted in even higher cell numbers in the corresponding filtrates. As the drinking water quality may be deteriorated by the presence of pathogenic bacteria, an adequate storage of the drinking water is important to avoid re-contamination [31]. The focus in our study was on the removal of cyanobacterial toxins. However, the effect of other components of cyanobacterial cells that enhance regrowth remains to be investigated.

Drinking water storage or distribution systems represent functional ecosystems with well-established and structured microbial communities [32]. The increasing numbers of bacteria support the succession of protists: on average $10^5$ cell $L^{-1}$ were found in the water phase and $10^3$ cells $cm^{-2}$ in different biofilms [32]. As a result, some bacteria may aggregate to overcome the predation pressure by protists as it was shown in field and laboratory studies before [15]. The regrowth of bacteria underneath the filter membranes or on materials downstream the filtration process of the LMA treatments in our study stimulated the growth of bacterivorous flagellates, followed by an increase in bacterial aggregates (Figure 4). Interestingly, the DMA replicates showed two different responses: high numbers of free-living planktonic bacteria were established in the absence of flagellates in one replicate of the DMA treatment, whereas the numbers of free-living bacteria in the second replicate were low, as bacteria either were grazed by flagellates or formed aggregates in order to resist predation. This illustrates that there might be variability in the primary microbial assemblages of such filtrates that may have consequences for the development of drinking water quality.

## Differences in microbial biofilms

Comparative phylogenetic analyses of 16S rDNA has increased our understanding of microbial diversity in environmental samples, since only few of the identifiable major phyla within the domain Bacteria have cultivable representatives [33]. Many of these uncultivated bacteria are found in diverse habitats in extraordinarily high abundances and might be at best only remotely related to strains that have been characterized by phenotype or by genome sequencing [33]. This in turn implies that there is very limited understanding of their respective physiologies, e.g., their substrate degradation potential. Calculating distances to sequences in a well-curated database [26] allowed us to classify OTUs to the closest taxonomic level. The CON treatment was the most diverse of all treatments, as reflected by the high number of small OTUs and an H' index of 4.11. With the exception of river and stream habitats [34] biofilms that occur under more oligotrophic conditions (as in the CON treatment) seem to be rather understudied. As a consequence, more "exotic" species were found that had a high distance to the closest known relatives (CKR), such as uncultured Myxococcales (11.3–12.1% distance to CKR) and Fibrobacteres (12.7% distance to CKR) (Table S1 in File S1).

The massive load of cyanobacterial biomass selected for a few large OTUs (H' = 3.47 of the combined MA) and the bacterial taxa represented by these OTUs were from a more "known" bacterial diversity (Table S2 in File S1). An intriguing example for typical freshwater bacteria that were present in our MA samples is a set of OTUs with <3% distance to known culturable genera, such as Azospirillum, Pelomonas, or Undibacterium (Table S2 in File S1). Moreover, oxygen subsaturation in the thick biofilms of

the MA treatment was suggested by the presence of obligate anaerobic bacteria such as Desulfovibrio, Acidaminobacter, Fusibacter, or Spirochaetales. The ability to remove MCs appears to be not only common for aerobic bacteria, but can also be a feature of anaerobic microorganisms (not otherwise specified) that were found in lake sediments or sediments of water recharge facilities [35]. It seems that several Spirochaeta contain the MC-LR degradation protein MlrC (Table S2 in File S1). However, further studies are needed to determine possible MC degrading bacteria in anoxic environments. Interestingly, Bacteroidetes such as Sphingobacteriales that are common in the Planktothrix layer of Lake Zurich [36] were also present in high proportions in the MA treatments only. The natural co-occurrence of these bacteria with MC-producing cyanobacteria might indicate a possible and so far unknown role in the MC-degradation process.

Previous investigations have suggested the Sphingomonadales (Alphaproteobacteria) being the major MC degraders in aquatic environments: genetic studies on Sphingomonadaceae revealed the distinct gene cluster mlrABCD to be involved in MC removal [37], because it encodes for an enzymatic ring cleavage and thus a linearization of the MCs. However, during a massive cyanobacterial bloom in Lake Erie only ~1% of the total bacterial community could be attributed to Sphingomonadales [38] and also the metagenomic identification of bacterioplankton taxa involved in MC degradation revealed only a minor importance of these bacteria [20]. Our data also suggest that Sphingomonadales may not necessarily be relevant in MC degradation, as they only enriched in the CON treatment (Figure 6, Table S1 in File S1).

Our data indicate that Betaproteobacteria may be more important amongst the major MC degradation bacteria as they constituted the major fraction (>40%) in the MA biofilms but were hardly found in the CON assemblages. OTUs were found in high quantities that were closely related to Paucibacter, capable of degrading MCs [39] and Variovorax, containing MC-LR degradation proteins, (Table S2 in File S1). The importance of Betaproteobacteria (mainly Burkholderiales and Methylophilales) has already been suggested before based on laboratory microcosms experiments amended with MCs [38]. Interestingly, recent studies revealed that Betaproteobacteria such as Methylophilales were capable of degrading MCs but lacked the mlr cluster, thereby providing an alternative and so far unknown means of MC removal [20]. Studies focussing on the detection of the mlrA gene as the only marker for MC degradation bacteria might therefore underestimated the possible presence of other MC degrading bacterial taxa [16,20,21].

## Conclusion

i. We demonstrated that GDM ultrafiltration systems provide a fast and efficient way to remove MCs from drinking water. However, it should be noted that complete MC degradation only took place one week after establishment of a stable biofilm.

ii. Addition of live or dead Microcystis cells led to remarkable differences between the bacterial communities of both MC-treated and the CON biofilms.

iii. Betaproteobacteria were identified as potentially important taxa for MC degradation in the MA biofilms. Additionally, Spirochaeta and Bacteroidetes such as Sphingobacteriales were enriched in these biofilms, and might indicate their so far unknown role in the MC degradation process.

## Supporting Information

**File S1** File S1 contains two supplemental tables: **Table S1, Phylogenetic composition.** Affiliation of bacteria in the CON assemblage (OTUs specific for CON treatment), number of OTUs and sequences, and phylogenetic distances of OTUs and associated sequences in the CON assemblage to the most closely related genotype in the SILVA reference database. **Table S2, Phylogenetic composition.** Affiliation of bacteria in the LMA & DMA assemblage (OTUs specific for the *Microcystis* treatment), number of OTUs and sequences, and phylogenetic distances of OTUs and associated sequences in the LMA & DMA assemblage to the most closely related genotype in the SILVA reference database.

## References

1. Paerl HW, Huisman J (2008) Climate - Blooms like it hot. Science 320: 57–58.
2. Posch T, Köster O, Salcher MM, Pernthaler J (2012) Harmful filamentous cyanobacteria favoured by reduced water turnover with lake warming. Nat Clim Chang 2: 809–813.
3. Carmichael WW, Azevedo S, An JS, Molica RJR, Jochimsen EM, et al. (2001) Human fatalities from cyanobacteria: Chemical and biological evidence for cyanotoxins. Environ Health Perspect 109: 663–668.
4. WHO (2011) Guidelines for Drinking-water quality. Fourth edition. Geneva.
5. Ueno Y, Nagata S, Tsutsumi T, Hasegawa A, Watanabe MF, et al. (1996) Detection of microcystins, a blue-green algal hepatotoxin, in drinking water sampled in Haimen and Fusui, endemic areas of primary liver cancer in China, by highly sensitive immunoassay. Carcinogenesis 17: 1317–1321.
6. Sabart M, Pobel D, Briand E, Combourieu B, Salencon MJ, et al. (2010) Spatiotemporal variations in microcystin concentrations and in the proportions of microcystin-producing cells in several *Microcystis aeruginosa* populations. Appl Environ Microbiol 76: 4750–4759.
7. Dyble J, Fahnenstiel GL, Litaker RW, Millie DF, Tester PA (2008) Microcystin concentrations and genetic diversity of *Microcystis* in the lower Great Lakes. Environ Toxicol 23: 507–516.
8. Jones GJ, Orr PT (1994) Release and degradation of microcystin following algicide treatment of a *Microcystis aeruginosa* bloom in a recreational lake, as determined by HPLC and protein phosphatase inhibition assay. Water Res 28: 871–876.
9. Campinas M, Rosa MJ (2010) Removal of microcystins by PAC/UF. Sep Purif Technol 71: 114–120.
10. Grutzmacher G, Wessel G, Klitzke S, Chorus I (2010) Microcystin elimination during sediment contact. Environ Sci Technol 44: 657–662.
11. Hoeger SJ, Dietrich DR, Hitzfeld BC (2002) Effect of ozonation on the removal of cyanobacterial toxins during drinking water treatment. Environ Health Perspect 110: 1127–1132.
12. Derlon N, Koch N, Eugster B, Posch T, Pernthaler J, et al. (2013) Activity of metazoa governs biofilm structure formation and enhances permeate flux during Gravity-Driven Membrane (GDM) filtration. Water Res 47: 2085–2095.
13. Peter-Varbanets M, Hammes F, Vital M, Pronk W (2010) Stabilization of flux during dead-end ultra-low pressure ultrafiltration. Water Res 44: 3607–3616.
14. Peter-Varbanets M, Zurbrugg C, Swartz C, Pronk W (2009) Decentralized systems for potable water and the potential of membrane technology. Water Res 43: 245–265.
15. Christoffersen K, Lyck S, Winding A (2002) Microbial activity and bacterial community structure during degradation of microcystins. Aquat Microb Ecol 27: 125–136.
16. Dziga D, Wasylewski M, Wladyka B, Nybom S, Meriluoto J (2013) Microbial degradation of microcystins. Chem Res Toxicol 26: 841–852.
17. Park HD, Sasaki Y, Maruyama T, Yanagisawa E, Hiraishi A, et al. (2001) Degradation of the cyanobacterial hepatotoxin microcystin by a new bacterium isolated from a hypertrophic lake. Environ Toxicol 16: 337–343.
18. Ho LN, Gaudieux AL, Fanok S, Newcombe G, Humpage AR (2007) Bacterial degradation of microcystin toxins in drinking water eliminates their toxicity. Toxicon 50: 438–441.
19. Jiang YG, Shao JH, Wu XQ, Xu Y, Li RH (2011) Active and silent members in the *mlr* gene cluster of a microcystin-degrading bacterium isolated from Lake Taihu, China. FEMS Microbiol Lett 322: 108–114.
20. Mou XZ, Lu XX, Jacob J, Sun SL, Heath R (2013) Metagenomic identification of bacterioplankton taxa and pathways iInvolved in microcystin degradation in Lake Erie. Plos One 8.
21. Shimizu K, Maseda H, Okano K, Hiratsuka T, Jimbo Y, et al. (2013) Determination of microcystin-LR degrading gene *mlrA* in biofilms at a biological

(DOCX)

## Acknowledgments

We thank Eugen Loher for his valuable help with the experimental setup, and M. M. Salcher for searching the worldwide ARB for close relatives. C. Ewert is being acknowledged for her valuable help in the accomplishment of the experiment and in the microcystin measurements.

## Author Contributions

Conceived and designed the experiments: EK TP ND EM JP JFB. Performed the experiments: EK JV TP ND TS. Analyzed the data: EK JV TP ND EM JP JFB. Contributed reagents/materials/analysis tools: EM JP. Contributed to the writing of the manuscript: EK TP ND EM JP JFB.

    drinking water treatment facility. Maejo Int J Sci Technol 7 (Special Issue): 22–35.
22. Blom JF, Zimmermann YS, Ammann T, Pernthaler J (2010) Scent of Danger: Floc formation by a freshwater bacterium is induced by supernatants from a predator-prey coculture. Appl Environ Microbiol 76: 6156–6163.
23. Chelius MK, Triplett EW (2001) The diversity of archaea and bacteria in association with the roots of *Zea mays* L. Microb Ecol 41: 252–263.
24. Shabarova T, Villiger J, Morenkov O, Niggemann J, Dittmar T, et al. (2014) Bacterial community structure and dissolved organic matter in repeatedly flooded subsurface karst water pools. FEMS Microbiol Ecol 89: 111–126.
25. Quince C, Lanzen A, Davenport RJ, Turnbaugh PJ (2011) Removing noise from pyrosequenced amplicons. Bmc Bioinformatics 12: Article Number 38.
26. Pruesse E, Quast C, Knittel K, Fuchs BM, Ludwig WG, et al. (2007) SILVA: a comprehensive online resource for quality checked and aligned ribosomal RNA sequence data compatible with ARB. Nucleic Acids Res 35: 7188–7196.
27. Li JM, Shimizu K, Maseda H, Lu ZJ, Utsumi M, et al. (2012) Investigations into the biodegradation of microcystin-LR mediated by the biofilm in wintertime from a biological treatment facility in a drinking-water treatment plant. Bioresour Technol 106: 27–35.
28. Leclerc H, Moreau A (2002) Microbiological safety of natural mineral water. Fems Microbiol Rev 26: 207–222.
29. Hammes F, Berney M, Wang YY, Vital M, Köster O, et al. (2008) Flow-cytometric total bacterial cell counts as a descriptive microbiological parameter for drinking water treatment processes. Water Res 42: 269–277.
30. Yamaguchi N, Torii M, Uebayashi Y, Nasu M (2011) Rapid, semiautomated quantification of bacterial cells in freshwater by using a microfluidic device for on-chip staining and counting. Appl Environ Microbiol 77: 1536–1539.
31. Roberts L, Chartier Y, Chartier O, Malenga G, Toole M, et al. (2001) Keeping clean water clean in a Malawi refugee camp: a randomized intervention trial. Bull World Health Organ 79: 280–287.
32. Sibille I, Sime-Ngando T, Mathieu L, Block JC (1998) Protozoan bacterivory and *Escherichia coli* survival in drinking water distribution systems. Appl Environ Microbiol 64: 197–202.
33. Rappe MS, Giovannoni SJ (2003) The uncultured microbial majority. Annu Rev Microbiol 57: 369–394.
34. Battin TJ, Kaplan LA, Newbold JD, Hansen CME (2003) Contributions of microbial biofilms to ecosystem processes in stream mesocosms. Nature 426: 439–442.
35. Holst T, Jørgensen NOG, Jørgensen C, Johansen A (2003) Degradation of microcystin in sediments at oxic and anoxic, denitrifying conditions. Water Res 37: 4748–4760.
36. Van den Wyngaert S, Salcher MM, Pernthaler J, Zeder M, Posch T (2011) Quantitative dominance of seasonally persistent filamentous cyanobacteria (*Planktothrix rubescens*) in the microbial assemblages of a temperate lake. Limnol Oceanogr 56: 97–109.
37. Shimizu K, Maseda H, Okano K, Kurashima T, Kawauchi Y, et al. (2012) Enzymatic pathway for biodegrading microcystin LR in *Sphingopyxis* sp C-1. J Biosci Bioeng 114: 630–634.
38. Mou XZ, Jacob J, Lu XX, Robbins S, Sun SL, et al. (2013) Diversity and distribution of free-living and particle-associated bacterioplankton in Sandusky Bay and adjacent waters of Lake Erie Western Basin. J Gt Lakes Res 39: 352–357.
39. Rapala J, Berg KA, Lyra C, Niemi RM, Manz W, et al. (2005) *Paucibacter toxinivorans* gen. nov., sp nov., a bacterium that degrades cyclic cyanobacterial hepatotoxins microcystins and nodularin. Int J Syst Evol Microbiol 55: 1563–1568.

# 24

# Removal of Fast Flowing Nitrogen from Marshes Restored in Sandy Soils

**Eric L. Sparks**[1,2]*, **Just Cebrian**[1,2], **Sara M. Smith**[1]

**1** Dauphin Island Sea Lab, Dauphin Island, Alabama, United States of America, **2** Marine Sciences, University of South Alabama, Mobile, Alabama, United States of America

## Abstract

Groundwater flow rates and nitrate removal capacity from an introduced solution were examined for five marsh restoration designs and unvegetated plots shortly after planting and 1 year post-planting. The restoration site was a sandy beach with a wave-dampening fence 10 m offshore. Simulated groundwater flow into the marsh was introduced at a rate to mimic intense rainfall events. Restoration designs varied in initial planting density and corresponded to 25%, 50%, 75% and 100% of the plot area planted. In general, groundwater flow was slower with increasing planting density and decreased from year 0 to year 1 across all treatments. Nevertheless, removal of nitrate from the introduced solution was similar and low for all restoration designs (3–7%) and similar to the unvegetated plots. We suggest that the low $NO_3^-$ removal was due to sandy sediments allowing rapid flow of groundwater through the marsh rhizosphere, thereby decreasing the contact time of the $NO_3^-$ with the marsh biota. Our findings demonstrate that knowledge of the groundwater flow regime for restoration projects is essential when nutrient filtration is a target goal of the project.

**Editor:** Fei-Hai Yu, Beijing Forestry University, China

**Funding:** Funding for this project provided by the Alabama Department of Conservation and Natural Resources, State Lands Division, Coastal Section, in part, by a grant from the National Oceanic and Atmospheric Administration, Office of Ocean and Coastal Resource Management, Award #10NOS4190206. The funders had no role in study design, data collection and analysis, decision to publish, or preparation of the manuscript.

* Email: esparks4040@gmail.com

## Introduction

Marsh restoration is a ubiquitous practice for mitigation of global marshland loss [1]. However, marsh restoration is expensive and labor intensive [2,3]. Compounded with the costly nature of marsh restoration, there is often inconsistency and discrepant outcomes among different techniques and designs [3]. Some studies have been conducted to evaluate cost-effectiveness of vegetative growth for restored marshes [2]; however, evaluations of the ecosystem services provided by different marsh restoration designs is scant, but should be evaluated to inform managers interested in maximizing the effectiveness of restoration projects [4–6].

Marshes provide important ecosystem services [7–10], and it has been suggested that nutrient filtration is the most economically valuable ecosystem service [4]. Processes such as denitrification and plant uptake can remove a large portion of nutrient inputs into marshes as groundwater percolates through the marsh rhizosphere [11,12].

Most nutrient filtration studies for marshes are conducted in mature natural marshes that are subjected to low to moderate flows of groundwater [12–14]. These studies have demonstrated that the presence of marsh plants increases nutrient removal through direct plant uptake as well as facilitating bacterial processes responsible for outgassing nitrogen (e.g., denitrification and anammox; [15]). However, marshes are subjected to varying groundwater flow rates from upland sources [16–18] and are dependent on factors such as rainfall intensity and soil permeability. In general, when areas are subjected to intense flow events (e.g., heavy rain), it is likely a smaller portion of the nutrients carried in these events can be removed than when the site is subjected to lower flow rates [19]. Along the northern Gulf of Mexico (nGOM) coast, there are frequent and intense rain events [20], thereby subjecting these marshes to a mixture of fast flow events, during and immediately after these rain events, and lower flow between events [16]. Assessment of nutrient removal by restored marshes under different scenarios of groundwater flow is important to improve the effectiveness of marsh restoration efforts targeting nutrient filtration as a primary goal.

In this study, we use black needlerush (*Juncus roemerinaus*) as our restored marsh plant. Black needlerush marshes are dominant on the nGOM coast [21] and have suffered significant loss over past decades primarily attributed to coastal development [22]. Due to the losses of marshes along the nGOM coast and prevalence of black needlerush, this marsh plant is the target for many restoration projects [2,23–25].

In this study, we compare groundwater flow rate and nitrate ($NO_3^-$) removal from an introduced groundwater solution in five black needlerush marsh restoration designs, varying in initial plant density, with unvegetated controls immediately after planting and one year after planting. Utilization of these marsh planting designs allows for comparisons of $NO_3^-$ removal from fast flowing groundwater across designs that vary in the effort required to plant (i.e., time and cost). The groundwater plume introduced into the marsh mimics a pulse of groundwater derived from an intense rainfall event percolating through porous sediments. Expectations were groundwater flow rates, through the marsh rhizosphere, would decrease and $NO_3^-$ removal would increase with increasing planting density over time. Results from this study can inform managers interested in maximizing restoration efficiency with the goal of reducing nutrient pollution into water bodies.

## Materials and Methods

### 1. Site Construction

On June 11, 2010, we planted a black needlerush marsh on the outskirts of Camp Beckwith (30°23′16″ N, 87°50′31″ W) located on the eastern coast of Weeks Bay in Fairhope, Alabama, USA. The staff of this privately owned camp gave us permission to conduct this work on their property and they should be contacted for future permissions. The planting site was situated on a stretch of sandy beach with natural marsh nearby. This sandy beach was subjected to high wave energy from boat wakes. To reduce wave action at this site, a fence was constructed, prior to marsh restoration, ten meters offshore of the restoration site to reduce shoreline wave energy and erosion. The fence consisted of a wooden frame filled up with dead tree branches and trunks along with other natural debris. Black needlerush sods (approximately 20 cm long, 20 cm wide and 20 cm deep) were harvested from an adjacent marsh and planted at the restoration site. Individual sods had a black needlerush shoot density typical of nGOM salt marshes, with ranges from 1400–1800 shoots m$^{-2}$ [2]. For experimental setup, we used a randomized block design with 3 blocks consisting of 6 plots each, yielding a total of 18 experimental plots (Fig. 1). Blocks were separated by 2 m and each plot had dimensions of 40 cm wide and 170 cm long (Fig. 1). Plots represented different restoration designs in terms of initial plant density (25%, 50%, 50%A, 75% and 100%) plus non-vegetated controls (0%) rendering 6 designs with 3 replicated plots each. Each plot contained 16 sod-sized units and the number of sods planted in the plot corresponded to the planting density treatment (e.g., 4 sods for the 25% planting density; Fig. 1). Plots planted in the 50%A design were arranged in an alternating "checkerboard" pattern (Fig. 1). To contain the introduced groundwater solution, each plot was enclosed on the top (i.e., upland) and the two lateral sides with vertical placement of rigid plastic sheeting. Thirty cm of the sheet height was buried below the sediment surface with 10 cm of sheet height above the sediment surface. Porewater collection wells, screened from 5 cm to 30 cm below the sediment surface, were placed at the bottom of each plot (experimental porewater wells) and 3 (natural porewater wells) on each lateral side of every block (Fig. 1). A diffuser plate was buried in the sand 10 cm from the upland planting edge to help disperse the introduced solution.

### 2. Experimental methodology and sampling

On June 21, 2010 and June 30, 2011 fluorescein tracing tests were conducted by releasing 15 L of a 20 mg L$^{-1}$ fluorescein solution into each plot. Fluorescein was used because it is not actively removed through biological processes at high rates; therefore, the only factor that can change the fluorescein concentration is dilution [26]. These attributes of fluorescein allow it to be used to assess travel time and dilution rates of groundwater plumes through the plots [26]. The fluorescein solution was released during receding tides when the tide line was at the upland edge of the plots. It took approximately 15 min for the 15 L of solution to disperse through the diffuser plate. This quick pulse was intended to mimic groundwater inputs from an

intense rain event over porous sediments and was equivalent to a 1.5 cm rainfall event drained off of a 1 hectare area through 100 linear meters of fringing marsh over a 12 hour period. Porewater samples were taken from each well in 15 min intervals after tracer release for 90 minutes and stored on ice in a cooler for transport back to Dauphin Island Sea Lab for analysis. All fluorescein samples were analyzed on a Turner Designs-700 fluorometer.

To determine how effective the marsh designs were at removing nutrient pollution from a fast flowing groundwater solution, four rounds of $NO_3^-$ solution release and subsequent sampling were performed after the initial planting (June 27, July 8, 9, and 13, 2010) and 1 year post-planting (July 14, 21, 29 and August 2, 2011). The releases were done during receding tides when the tide line was at the upland edge of the plots (i.e., the same timing as the fluorescein tracing tests). For each round we released 15 L of a 200 μM $NO_3^-$ solution into each plot. Water samples were taken from each input container and porewater well at the bottom of the plots. The sampling timeline for porewater samples was determined by the fluorescein experiments (i.e., sample at time fluorescein peak for each plot). To determine background nutrient levels, we also took samples from the natural porewater wells located outside of each block on every sampling day. Porewater samples were filtered in the field and transported to the lab on ice for analysis. $NO_3^-$ concentrations were analyzed using cadmium reduction azo dye assays [27].

### 3. Calculations and statistical analyses

**3.1. Flow rates.** Flow rates were analyzed by recording the time when peak fluorescein concentration was observed in each plot. These peak times were analyzed using an ANOVA (treatment × year × block). If block and the interaction between treatment and year were found to be insignificant factors, data were pooled across block and reanalyzed with an ANOVA (treatment × year). For all statistical analyses, tests were conducted using Sigma Stat 3.5 and significance was considered at p<0.05 [28].

**3.2. $NO_3^-$ removal.** As the introduced solution travels through the plots, it will be subjected to dilution through mixing with natural porewater (i.e., not subjected to the introduced solution) with lower background concentrations of $NO_3^-$. To determine the $[NO_3^-]$ in the portion of the porewater derived from the introduced solution, a dilution correction must be applied that accounts for the $[NO_3^-]$ found in natural porewater. Applying a dilution correction allows for the removal of $NO_3^-$ from the introduced solution to be calculated as the introduced solution (*Input*) travels to the downland edge of the marsh. To calculate removal of $NO_3^-$ from the introduced solution, we subtracted the dilution corrected $[NO_3^-]$ at the downland porewater well from the $NO_3^-$ concentration in the input using the following equation: *Removal = Input - (Downland well – Natural × Dilution) ÷ (1-Dilution)*. The previous equation will be referred to as equation 1 throughout the manuscript. The term *Downland well* in equation 1 is the $[NO_3^-]$ measured at restored marsh porewater wells subjected to the simulated pollution plume. The *Natural* term is the $[NO_3^-]$ in the porewater wells outside of the restored marsh that was not subjected to the simulated pollution plume (Fig. 1). The *Dilution* term in equation 1 is the proportion of sample derived from natural porewater and was calculated as the proportional decrease in [fluorescein] from the input to the porewater collection well (i.e., peak % tracer contribution in Fig. 2). $NO_3^-$ removal was converted to percent removal by dividing it by $[NO_3^-]$ of the input. Similar to the flow rate study, percent $NO_3^-$ removal was first analyzed using an ANOVA (treatment × year × block). If block and the interaction between treatment and year were found to be insignificant factors,

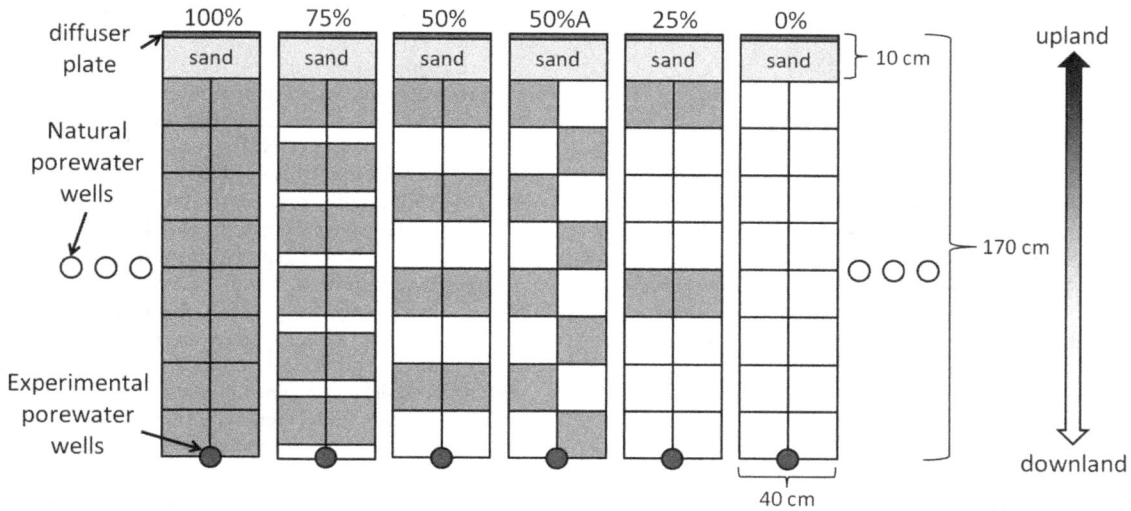

**Figure 1. Schematic of 1 block of 6 marsh restoration designs (0%, 25%, 50%, 50%A, 75% and 100%).** Shaded squares represent planted sods. There were a total of 3 blocks with each block consisted of a randomized arrangement of all 6 restoration designs. The groundwater solution was introduced at the diffuser plate and flowed down the plots toward the porewater collection well.

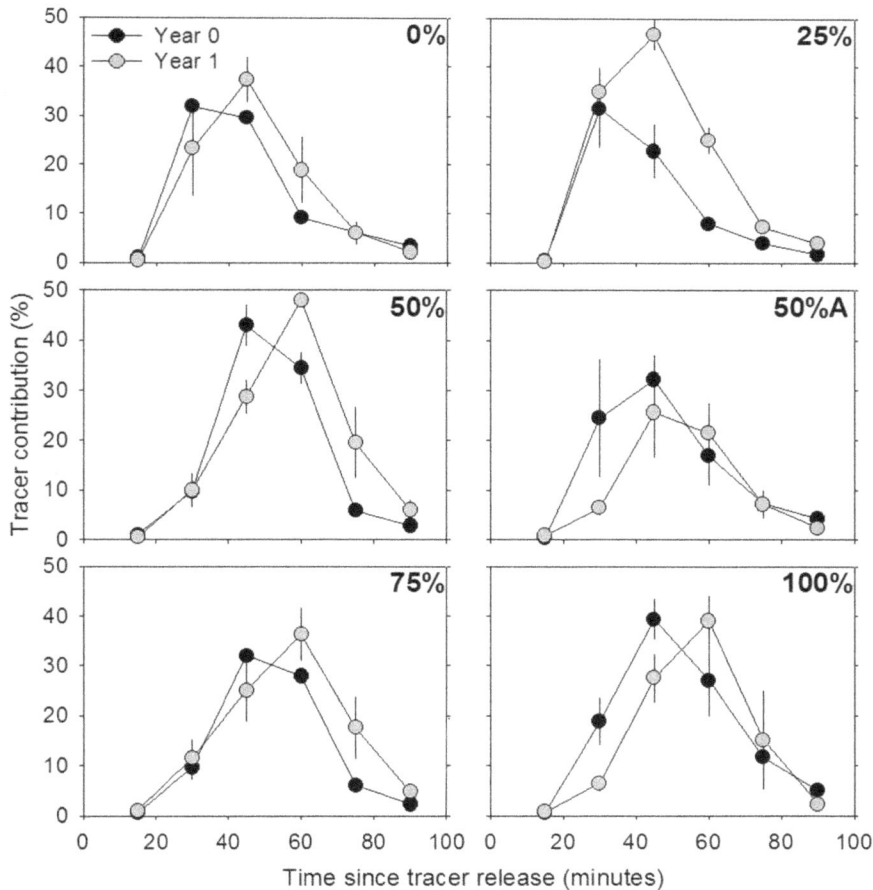

**Figure 2. Fluorescein tracer porewater contribution (%) at the downland well through time.** Black circles represent year 0 and grey circles represent year 1 samples. Percentages in the top right portion of each plot represents the planting density (0%, 25%, 50%, 50%A, 75% and 100%). Error bars indicate ±1 SE.

the data was pooled across block and reanalyzed with an ANOVA (treatment × year).

## Results

For flow rate and $NO_3^-$ removal, block was never a significant factor (flow rate - p = 0.40; $NO_3^-$ removal - p = 0.14) and there were no significant interactions between treatment and year (flow rate - p = 0.34; $NO_3^-$ removal - p = 0.99). Therefore, data were pooled across blocks and analyzed with an ANOVA for the effects of treatment and time (Table 1). Only the results of the ANOVA on the data pooled across blocks will be further discussed.

Most of the fluorescein solution traversed the plots in less than one hour across all treatments (Fig. 2). In general, the flow of the introduced solution decreased with increasing planting density (Table 1), as indicated by the later peaks in the dilution curves (Fig. 2). Furthermore, it took longer for the solution to cross the plots one year after planting than two weeks after planting (Table 1; Fig. 2). Longer retention time of the introduced solution within the plots at one year after planting than two weeks after planting implies that flow rates of the introduced solution, through the plots, decreased over time.

Concentrations of $NO_3^-$ in porewater collections wells outside the plots ranged from 0.5 μM to 1.5 μM (i.e., natural or ambient porewater) and were low when compared to concentrations within the plots (i.e., subjected to the introduced solution). As expected, the input had $[NO_3^-]$ of 200 μM ± 10 μM, whereas wells at the downland edge of the plots had lower $[NO_3^-]$ ranging from 44 μM to 126 μM. When combining the observed changes in $[NO_3^-]$ with a dilution factor (equation 1), our calculations showed only a small percentage of $NO_3^-$ was removed from the introduced solution (3–7%; Table 2). These small percentages of $NO_3^-$ removal were similar across all treatments (Table 1) and sampling years (Table 1; Fig. 3). While $NO_3^-$ processing increased slightly over time and with plant cover (Table 2) these differences were not statistically significant (Table 1).

## Discussion

In this study, we found that the introduced groundwater traveled slowest through the most vegetated planting designs and slower one year after planting than immediately after planting. These results offer evidence that the presence of marsh plants increases the time required for groundwater to flow through the restoration area, likely through the presence of the marsh rhizosphere and accumulation of finer grained sediments [29]. As the planted plots mature and density increases, they will likely continue to decrease groundwater flow rates through binding sediments, expansion of the marsh rhizosphere [30] and reduction

of wave energy that aids in the accumulation of finer grained sediments [31].

Increases in plant density and the accumulation of finer grained sediment are conducive to increased nutrient removal in marshes [15]. We did find some suggestive evidence that increases in vegetated area increased $NO_3^-$ removal (Table 2), albeit not statistically significant (Table 1). Our range in planting density was large (0%–100%) and we measured $NO_3^-$ removal in all of these planting densities twice over one year. If planting density was a primary driver of $NO_3^-$ removal from this fast flowing solution, we would have captured it with this sampling design. Given no effect of planting density and overall findings that only a small portion of the input $NO_3^-$ was removed across all of the planting designs through time (3–7%), it appears that the groundwater was traveling too quickly through the marsh rhizosphere for plants to have an impact on $NO_3^-$ removal. In studies where groundwater traveled slower through the marsh rhizosphere, marsh plants had time to uptake and facilitate bacterial processes that fueled large removals of nitrogen [17]. Comparing these studies to our study suggests that flow rate can influence how effective marshes are at removing $NO_3^-$ from groundwater. With these results, it is likely that nutrient processing in our restored marsh will remain similar to the unvegetated plots for several years to come [5]. An additional factor contributing to the negligible increases in $NO_3^-$ removal over time is the typical slow growth of black needlerush [21]. This slow growth pattern was evident by visual observations of marginal increases in vegetated area (<5%) for the vegetated designs and no colonization of the 0% planting design at one year after planting. A timeframe when these marshes will decrease groundwater flow enough to allow for large proportions of input nutrients to be removed is unknown; however, previous studies have indicated restoring other ecosystem functions in restored marshes to natural levels takes many years [2,32,33].

A probable explanation to the high groundwater flow rates and low nutrient filtration is that the sediment at the restoration site was sandy, which is also the case for many restored marshes (e.g., Jamaica Bay marsh islands in New York, USA, Grand Bay National Estuarine Research Reserve boat launch marsh in Mississippi, USA, and the Labranche wetlands in Louisiana, USA). Groundwater flows more quickly through sand than sediments with higher mud and silt content typical of mature marshes [34]. Most other studies of nutrient filtration in salt marshes have been conducted mature natural marshes with fine sediments (mainly mud and silt). Finer sediments have smaller interstitial spaces and slow down groundwater flow in relation to the flow in sandy sediments [12–14,35]. Slower flow rates increase contact time between nutrients and reactive areas of the sediment and rhizosphere, thereby increasing opportunities for nutrient processing [19]. In addition, sandy sediments are primarily composed of inorganic material (e.g. quartz) and typically contain

**Table 1.** Results of ANOVA for flow rates and $NO_3^-$ removal.

| Test | Effect | Degrees of Freedom | F value | P value |
|------|--------|--------------------|---------|---------|
| Flow rates | Treatment | 5 | 21.545 | <0.001 |
| | Year | 1 | 76.818 | <0.001 |
| | Treatment × Year | 5 | 1.200 | 0.339 |
| $NO_3^-$ removal | Treatment | 5 | 1.195 | 0.317 |
| | Year | 1 | 0.197 | 0.658 |
| | Treatment × Year | 5 | 0.020 | 0.999 |

**Table 2.** Mean percent $NO_3^-$ removal across 6 restoration designs directly after planting and 1 year after planting ($\pm 1$ SE).

| | Treatment | | | | | |
|---|---|---|---|---|---|---|
| Year | 0% | 25% | 50% | 50% A | 75% | 100% |
| 0 | 3.22 ($\pm$0.87) | 4.29 ($\pm$0.94) | 5.44 ($\pm$1.81) | 6.23 ($\pm$1.54) | 5.95 ($\pm$1.22) | 6.47 ($\pm$2.18) |
| 1 | 3.73 ($\pm$1.19) | 4.28 ($\pm$1.25) | 5.99 ($\pm$1.27) | 6.27 ($\pm$2.68) | 6.89 ($\pm$1.12) | 6.97 ($\pm$2.49) |

little organic matter, which limits the biological processing of incoming nutrients [15].

While we did not find evidence for a strong role of these marsh planting designs as filters of runoff nutrient pollution, they provide other important services such as habitat and shoreline stabilization. We did not quantify these services, but we would expect these services to be small in magnitude for these young marshes and increase as the marshes age [6]. Similarly, we expect that these restored marshes will increase vegetated area and become effective nutrient filters through time.

## Conclusions

In conclusion, our results suggest that groundwater flow rates and sediment type should be considered when planning marsh restoration efforts with the specific goal of runoff pollution

removal. Despite finding slower groundwater flow rates with increasing vegetation, our introduced groundwater did still flow quickly through the restored plots (30–60 minutes). This fast flow is most likely attributable to the coarse texture of the sediment. We calculated that on average only 3–7% of the $NO_3^-$ entering the restored marshes was processed by the marshes within one year since planting, and it appears that $NO_3^-$ processing by these restored marshes will remain similar to the unvegetated plots for many years to come. However, the slower groundwater flow in the more vegetated plots suggests they will likely become effective nutrient filters prior to the less vegetated plots. Understanding the nature, control, and extent of nutrient processing by restored marshes requires additional research, such as more restoration designs across many different marsh environments, in order to help managers decide on best restoration practices given budget constraints.

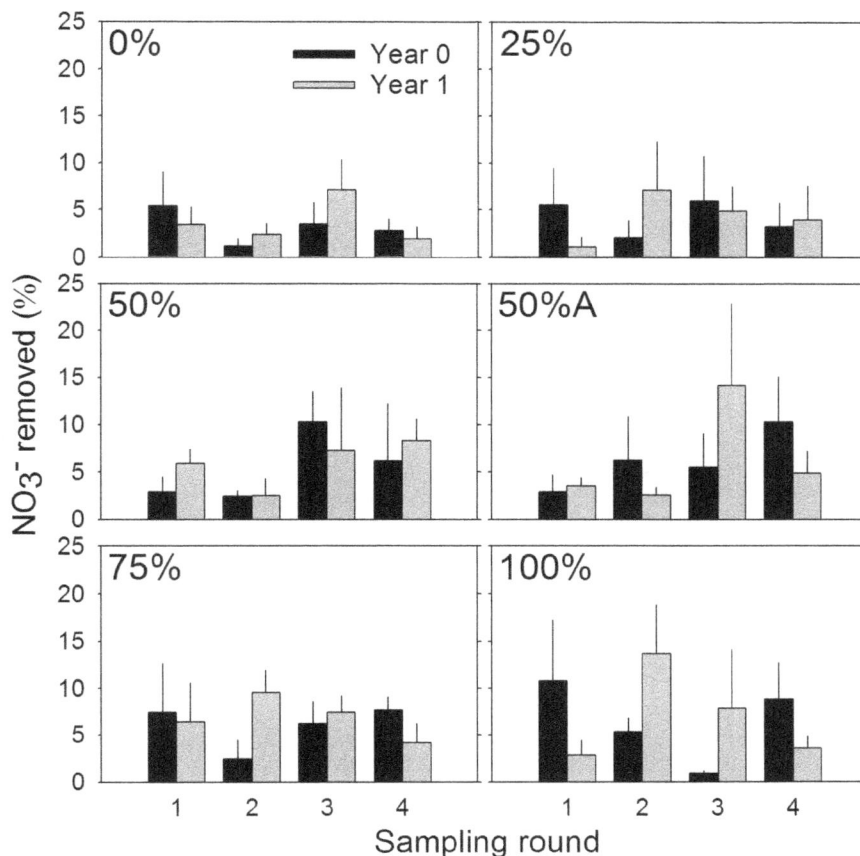

**Figure 3. Percentage of $NO_3^-$ removed from introduced solution directly after planting and 1 year post-planting for each sampling round.** Black bars represent year 0 and grey bars represent year 1 samples. Percentages in the top left portion of each plot represent initial planting densities. Error bars indicate $\pm 1$ SE.

## Supporting Information

**Data Set S1**   Data for flow rate and dilution calculations. (XLSX)

**Data Set S2**   Data for $NO_3^-$ removal calculations. (XLSX)

## Acknowledgments

We would like to thank Jason Howard, Jelani Reynolds, Jennifer Hemphill and Amanda Pratt of the Ecosystems Lab at the Dauphin Island Sea Lab for their field assistance.

## Author Contributions

Conceived and designed the experiments: ELS JC SMS. Performed the experiments: ELS SMS. Analyzed the data: ELS. Contributed reagents/materials/analysis tools: JC. Contributed to the writing of the manuscript: ELS JC.

## References

1.  Bromberg Gedan K, Silliman BR, Bertness MD (2009) Centuries of Human-Driven Change in Salt Marsh Ecosystems. Annu Rev of Mar Sci 1: 117–141.
2.  Sparks EL, Cebrian J, Biber PD, Sheehan KL, Tobias CR (2013) Cost-effectiveness of two small-scale salt marsh restoration designs. Ecol Eng 53(2013): 250–256.
3.  Chapman MG, Underwood AJ (2000) The need for a practical scientific protocol to measure successful restoration. Wetlands (Australia) 19(1): 28–49.
4.  Costanza R, d'Arge R, de Groot R, Farber S, Grasso M, et al. (1997) The value of the world's ecosystem services and natural capital. Nature 387: 253–260.
5.  Ehrenfeld JG (2000) Defining the Limits of Restorations: The Need for Realistic Goals. Restor Ecol 8: 2–9.
6.  Hilderbrand RH, Watts AC, Randle AM (2005) The myths of restoration ecology. Eco Soc 10: (online) URL: http://www.ecologyandsociety.org/vol10/iss1/art19/.
7.  Beck M, Heck K Jr, Able K, Childers D, Eggleston D, et al. (2001) The identification, conservation, and management of estuaries and marine nurseries for fish and invertebrates. Biosci 51(8): 633–641.
8.  Chmura GL, Anisfeld SC, Cahoon DR, Lynch JC (2003) Global carbon sequestration in tidal, saline wetland soils. Glob Biogeochem Cycles 17(44): 1–22.
9.  Moeller I, Spencer T, French JR (1996) Wind wave attenuation over saltmarsh surfaces: Preliminary results from Norfolk, England. J Coast Res 12: 1009–1016.
10. Valiela I, Cole ML (2002) Comparative Evidence that Salt Marshes and Mangroves May Protect Seagrass meadows from Land-derived Nitrogen Loads. Ecosystems 5(1): 92–102.
11. Hammersley MR, Howes BL (2005) Coupled nitrification-denitrification measured in situ in a *Spartina alterniflora* marsh with a 15NH4 tracer. Mar Ecol Prog Ser 299: 123–135.
12. Tobias C, Macko S, Anderson I, Canuel E, Harvey J (2001) Tracking the fate of a high concentration nitrate plume through a fringing marsh: A combined groundwater tracer and in-situ isotope enrichment study. Limnol Oceanogr 46(8): 1977–1989.
13. Drake DC, Peterson BJ, Galvan KA, Deegan LA, Hopkinson C, et al. (2009) Salt marsh ecosystem biogeochemical responses to nutrient enrichment: a paired $^{15}$N tracer study. Ecology 90(9): 2535–2546.
14. Tobias CR, Anderson IC, Canuel EA, Macko SA (2001) Nitrogen cycling through a fringing marsh-aquifer ecotone. Mar Ecol Prog Ser 210: 25–39.
15. Tobias CR, Neubauer SC (2009) Salt marsh biogeochemistry: An overview in Perillo G, Wolanski E, Cahoon D, Brinson M (eds). Coastal wetlands: An integrated ecosystem approach. Elsevier. p. 445–492.
16. Lehrter JC, Cebrian J (2009) Uncertainty propagation in an ecosystem nutrient budget. Ecol Appli 20: 508–524.
17. Tobias C, Harvey JW, Anderson IC (2001) Quantifying groundwater discharge through fringing wetlands to estuaries: Seasonal variability, methods comparison, and implication for wetland-estuary exchange. Limnol Oceanogr 46(3): 604–615.
18. Valiela I, Costa J, Foreman K, Teal JM, Howes B, et al. (1999) Transport of groundwater-borne nutrients from watersheds and their effects on coastal waters. Biodegradation 10(3): 177–197.
19. Barling RD, Moore ID (1994) Role of Buffer Strips in Management of Waterway Pollution: A Review Environ manag 18(4): 543–558.
20. Stout JP, Heck KL Jr, Valentine JF, Dunn SJ, Spitzer PM (1998) Preliminary Characterization of Habitat Loss: Mobile Bay National Estuary Program, MESC Contribution Number 301.
21. Eleuterius LN (1976) The distribution of *Juncus roemerianus* in the salt marshes of North America. Earth Environ Sci 17(4): 289–292.
22. Turner RE (1990) Landscape development and coastal wetland losses in the northern Gulf of Mexico Am Zool 30(1): 89–105.
23. LaSalle MW (1996) Assessing the functional level of a constructed intertidal marsh in Mississippi. Tech Rep WRP-RE-15, US Army Corps of Engineers, Waterways Experiment Station, Vicksburg, MS.
24. Lewis RR (1982) Creation & Restoration of Coastal Plant Communities. CRC Press, Boca Raton: 153–171.
25. Turner RE, Streever B (2002) Approaches to Coastal Wetland Restoration: Northern Gulf of Mexico. SPB Academic Publishing, Hague, The Netherlands.
26. Corbett DR, Dillon K, Burnett W (2000) Tracing groundwater flow on a barrier island in the north-east Gulf of Mexico. Estuar Coast Shelf Sci 51(2): 227–242.
27. Maynard DG, Kalra YP (1993) Nitrate and extractable ammonium nitrogen. In M.R. Carter, Ed. Soil Sampling and Methods of Analysis. Lewis Publishers, Boca Raton, FL, 25–38.
28. Quinn GP, Keough MJ (2002) Experimental design and data analysis for biologists. Experimental design and data analysis for biologists. Cambridge University Press, Cambridge, UK.
29. Vukovic M, Soro A (1992) Determination of Hydraulic Conductivity of Porous Media from Grain-Size Composition. Littleton, Colorado: Water Resources Publications.
30. Stumpf RP (1983) The process of sedimentation on the surface of a salt marsh. Estuar Coast Shelf Sci 17: 495–508.
31. Harrell J, Blatt H (1978) Polycrystallinity; effect on the durability of detrital quartz. J Sediment Res 48(1): 25–30.
32. Wilkins S, Keith DA, Adam P (2003) Measuring success: evaluating the restoration of a grassy eucalypt woodland on the Cumberland Plain, Sydney, Australia. Restor Ecol 11: 489–503.
33. Zedler JB, Callaway JC (1999) Tracking wetland restoration: do mitigation sites follow desired trajectories? Restor Ecol 7: 69–73.
34. Dingman SL (2008) Physical Hydrology. Waveland Press Inc, Long Grove, IL, USA.
35. Valiela I, Teal JM, Volkmann S, Shafer D, Carpenter EJ (1978) Nutrient and particulate fluxes in a salt marsh ecosystem: Tidal. Limnol Oceanogr 23(4): 798–812.

# Permissions

# List of Contributors

**Lynn Ranåker**
Department of Biology, Aquatic Ecology, Ecology Building, Lund University, Lund, Sweden

**Jens Persson**
Swedish Agency for Marin and Water Management, Gothenburg, Sweden

**Mikael Jönsson**
Department of Biology, Functional zoology, Biology Building, Lund University, Lund, Sweden

**P. Anders Nilsson**
Department of Biology, Aquatic Ecology, Ecology Building, Lund University, Lund, Sweden
Department of Environmental and Life Sciences, Biology, Karlstad University, Karlstad, Sweden

**Christer Brönmark**
Department of Biology, Aquatic Ecology, Ecology Building, Lund University, Lund, Sweden

**Diogo M. O. Ogawa**
Biotechnology and Natural Resources Program, University of the State of the Amazonas, Manaus, AM, Brazil
Laboratory of Biochemistry and Biotechnology, Institute for Marine Sciences, Federal University of Ceara, Fortaleza, CE, Brazil
Center for Environment and Biodiversity Studies, University of the State of the Amazonas, Manaus, AM, Brazil
RIKEN Center for Sustainable Resource Science, and Biomass Engineering Corporation Division, Yokohama, Japan

**Shigeharu Moriya**
RIKEN Center for Sustainable Resource Science, and Biomass Engineering Corporation Division, Yokohama, Japan
RIKEN Antibiotics Laboratory, Yokohama, Japan
Graduate School of Medical Life Science, Yokohama City University, Suehiro-cho, Tsurumi-ku, Yokohama, Japan

**Yuuri Tsuboi and Yasuhiro Date**
RIKEN Center for Sustainable Resource Science, and Biomass Engineering Corporation Division, Yokohama, Japan
Graduate School of Medical Life Science, Yokohama City University, Suehiro-cho, Tsurumi-ku, Yokohama, Japan

**Á lvaro R. B. Prieto-da-Silva**
Biotechnology and Natural Resources Program, University of the State of the Amazonas, Manaus, AM, Brazil
Center for Environment and Biodiversity Studies, University of the State of the Amazonas, Manaus, AM, Brazil
Laboratory of Genetics, Butantan Institute, Sao Paulo, SP, Brazil

**Gandhi Ra´ dis-Baptista**
Biotechnology and Natural Resources Program, University of the State of the Amazonas, Manaus, AM, Brazil
Laboratory of Biochemistry and Biotechnology, Institute for Marine Sciences, Federal University of Ceara, Fortaleza, CE, Brazil
Center for Environment and Biodiversity Studies, University of the State of the Amazonas, Manaus, AM, Brazil

**Tetsuo Yamane**
Biotechnology and Natural Resources Program, University of the State of the Amazonas, Manaus, AM, Brazil
Center for Environment and Biodiversity Studies, University of the State of the Amazonas, Manaus, AM, Brazil
Center of Biotechnology of Amazon, Manaus, AM, Brazil

**Jun Kikuchi**
RIKEN Center for Sustainable Resource Science, and Biomass Engineering Corporation Division, Yokohama, Japan
Graduate School of Medical Life Science, Yokohama City University, Suehiro-cho, Tsurumi-ku, Yokohama, Japan
Graduate School of Bioagricultural Sciences, Nagoya University, Nagoya, Japan

**Sebastian Emde and Thomas Kuhn**
Institute for Ecology, Evolution and Diversity, Goethe-University, Frankfurt am Main, Hesse, Germany

**Judith Kochmann**
Senckenberg Gesellschaft für Naturforschung, Biodiversity and Climate Research Centre, Frankfurt am Main, Hesse, Germany

**Martin Plath**
College of Animal Science and Technology, Northwest Agriculture & Forestry University, Yangling, Shaanxi Province, P. R. China

**Sven Klimpel**
Institute for Ecology, Evolution and Diversity, Goethe-University, Frankfurt am Main, Hesse, Germany Senckenberg Gesellschaft für Naturforschung, Biodiversity and Climate Research Centre, Frankfurt am Main, Hesse, Germany

**Alfredo Pauciullo, Angela Perucatti, Alessandra Iannuzzi, Domenico Incarnato, Viviana Genualdo and Leopoldo Iannuzzi**
Institute for Animal Production System in Mediterranean Environment, National Research Council, Naples, Italy

**Gianfranco Cosenza and Dino Di Berardino**
Department of Agriculture, University of Naples Federico II, Portici, Italy

**Tessa B. Francis**
University of Washington Tacoma, Puget Sound Institute, Tacoma, Washington, United States of America

**Elizabeth M. Wolkovic**
National Center for Ecological Analysis and Synthesis, University of California Santa Barbara, Santa Barbara, California, United States of America
The Biodiversity Research Centre, University of British Columbia, Vancouver, British Columbia, Canada

**Mark D. Scheuerell**
Fish Ecology Division, Northwest Fisheries Science Center, National Marine Fisheries Service, National Oceanic and Atmospheric Administration, Seattle, Washington, United States of America

**Stephen L. Katz**
Channel Islands National Marine Sanctuary, National Ocean Service, National Oceanic and Atmospheric Administration, Santa Barbara, California, United States of America

**Elizabeth E. Holmes**
Conservation Biology Division, Northwest Fisheries Science Center, National Marine Fisheries Service, National Oceanic and Atmospheric Administration, Seattle, Washington, United States of America

**Stephanie E. Hampton**
National Center for Ecological Analysis and Synthesis, University of California Santa Barbara, Santa Barbara, California, United States of America

**Maria Ilhéu and Paula Matono**
Departamento de Paisagem Ambiente e Ordenamento, Escola de Ciências e Tecnologia, Universidade de Évora, Évora, Portugal
Instituto de Ciências Agrárias e Ambientais Mediterraˆnicas, Universidade de Évora, Évora, Portugal

**João Manuel Bernardo**
Departamento de Paisagem Ambiente e Ordenamento, Escola de Ciências e Tecnologia, Universidade de Évora, Évora, Portugal

**Kurt W. Alt and Nicole Nicklisch**
Center for Natural and Cultural History of the Teeth, Danube Private University, Krems, Austria State Office for Heritage Management and Archaeology Saxony-Anhalt and State Museum of Prehistory, Halle, Germany
Institute for Prehistory and Archaeological Science, Basel University, Basel, Switzerland

**Corina Knipper**
Curt Engelhorn Centre Archaeometry gGmbH, Mannheim, Germany

**Daniel Peters**
Institut für Prähistorische Archäologie, Freie Universität Berlin, Berlin, Germany

**Wolfgang Müller**
Department of Earth Sciences, Royal Holloway University of London, London, United Kingdom

**Anne-France Maurer**
Laboratório Hercules, Universidade de Évora, Évora, Portugal

**Isabelle Kollig**
State Office for Heritage Management and Archaeology Saxony-Anhalt and State Museum of Prehistory, Halle, Germany

**Christiane Müller, Sarah Karimnia, Guido Brandt and Christina Roth**
Institute of Anthropology, University of Mainz, Mainz, Germany

**Martin Rosner**
IsoAnalysis UG, Berlin, Germany

**Balász Mende**
Archaeological Institute, Research Centre for Humanities, Hungarian Academy of Sciences, Budapest, Hungary

**Bernd R. Schöne**
Institute of Geosciences, University of Mainz, Mainz, Germany

**Tivadar Vida**
Department of Prehistory and Protohistory, Eötvös
Loránd University of Budapest, Budapest, Hungary

**Uta von Freeden**
German Archaeological Institute, Roman Germanic
Commission, Frankfurt a. M., Germany

**Florian Altermatt**
Department of Aquatic Ecology, Eawag: Swiss
Federal Institute of Aquatic Science and Technology,
Dübendorf, Switzerland
Department of Environmental Systems Science, ETH
Zentrum, Zürich, Switzerland
Institute of Evolutionary Biology and Environmental
Studies, University of Zurich, Zürich, Switzerland

**Roman Alther and Elvira Mächler**
Department of Aquatic Ecology, Eawag: Swiss
Federal Institute of Aquatic Science and Technology,
Dübendorf, Switzerland

**Cene Fišer and Marjeta Konec**
Department of Biology, Biotechnical Faculty, University
of Ljubljana, Ljubljana, Slovenia

**Jukka Jokela**
Department of Aquatic Ecology, Eawag: Swiss
Federal Institute of Aquatic Science and Technology,
Dübendorf, Switzerland
Department of Environmental Systems Science, ETH
Zentrum, Zürich, Switzerland

**Daniel Küry**
Life Science AG, Basel, Switzerland

**Pascal Stucki**
Aquabug, Neuchâtel, Switzerland

**Anja Marie Westram**
Department of Aquatic Ecology, Eawag: Swiss
Federal Institute of Aquatic Science and Technology,
Dübendorf, Switzerland
Animal and Plant Sciences, University of Sheffield,
Western Bank, Sheffield, United Kingdom

**Xuelian Zhang, Yanxia Li, Bei Liu, Jing Wang,
Chenghong Feng, Min Gao and Lina Wang**
State Key Laboratory of Water Environment Simulation,
School of Environment, Beijing Normal University,
Beijing, China

**Davi Pedroni Barreto**
Instituto de Microbiologia Prof. Paulo de Góes,
Universidade Federal do Rio de Janeiro, Rio de
Janeiro, Brazil

**Ralf Conrad, Melanie Klose and Peter Claus**
Max-Planck Institute for Terrestrial Microbiology,
Marburg, Hessen, Germany

**Alex Enrich-Prast**
Instituto de Biologia, Universidade Federal do Rio de
Janeiro, Rio de Janeiro, Brazil
Department of Water and Environmental Studies,
Linköping University, Linköping, Sweden

**Paloma M. Lopes, Vinicius F. Farjalla and Reinaldo
L. Bozelli**
Laboratório de Limnologia, Departamento de Ecologia,
Instituto de Biologia, Universidade Federal do Rio de
Janeiro, CCS, Cidade Universitária, Rio de Janeiro, RJ,
Brazil

**Luis M. Bini**
Departamento de Ecologia, Instituto de Ciências
Biológicas, Universidade Federal de Goiás, Goiânia,
GO, Brazil

**Steven A. J. Declerck**
Netherlands Institute of Ecology (NIOO-KNAW),
Department of Aquatic Ecology, Wageningen, The
Netherlands

**Ludgero C. G. Vieira**
Faculdade UnB Planaltina, Universidade de Brasília, Â
rea Universitária n. 1 - Vila Nossa Senhora de Fátima,
Planaltina, Distrito Federal, Brazil

**Claudia C. Bonecker and Fabio A. Lansac-Toha**
Núcleo de Pesquisas em Limnologia, Ictiologia e
Aqüicultura (NUPELIA), Universidade Estadual de
Maringá, Jd. Universitário, Maringá, Paraná, Brazil

**Francisco A. Esteves**
Laboratório de Limnologia, Departamento de Ecologia,
Instituto de Biologia, Universidade Federal do Rio de
Janeiro, CCS, Cidade Universitária, Rio de Janeiro, RJ,
Brazil
Núcleo de Pesquisas em Ecologia e Desenvolvimento
Sócio Ambiental de Macaé, Rodovia Amaral Peixoto,
Macaé, RJ, Brazil

**Josepha M. H. van Diggelen, Gijs van Dijk and
Alfons J. P. Smolders**
B-WARE Research Centre, Radboud University
Nijmegen, Mercator 3, Nijmegen, The Netherlands
Institute for Water and Wetland Research, Department
of Aquatic Ecology and Environmental Biology,
Radboud University Nijmegen, Nijmegen, The
Netherlands

**Leon P. M. Lamers and Jan G. M. Roelofs**
Institute for Water and Wetland Research, Department of Aquatic Ecology and Environmental Biology, Radboud University Nijmegen, Nijmegen, The Netherlands

**Maarten J. Schaafsma**
B-WARE Research Centre, Radboud University Nijmegen, Mercator 3, Nijmegen, The Netherlands

**Rui P. Rivaes, Patricia M. Rodríguez-González and Maria Teresa Ferreira**
Forest Research Center, Instituto Superior de Agronomia, Universidade de Lisboa, Lisbon, Portugal

**António N. Pinheiro**
CEHIDRO, Instituto Superior Técnico, Universidade de Lisboa, Lisbon, Portugal

**Emilio Politti and Gregory Egger**
Environmental Consulting Klagenfurt, Klagenfurt, Austria

**Alicia García-Arias and Felix Francés**
Research Institute of Water and Environmental Engineering, Universitat Politécnica de Valéncia, Valencia, Spain

**Xue-Dong Lou**
Chinese Research Academy of Environmental Sciences, Beijing, China
College of Life Sciences, Northwest Agriculture & Forestry University, Yangling, Shaanxi, China

**Sheng-Qiang Zhai and Li-Le Hu**
College of Life Sciences, Northwest Agriculture & Forestry University, Yangling, Shaanxi, China

**Bing Kang**
College of Life Sciences, Northwest Agriculture & Forestry University, Yangling, Shaanxi, China

**Ya-Lin Hu**
Institute of Applied Ecology, Chinese Academy of Sciences, Shenyang, China

**Julian D. Olden**
School of Aquatic and Fishery Sciences, University of Washington, Seattle, Washington, United States of America

**Mariana Tamayo**
Faculty of Life and Environmental Sciences, University of Iceland, Reykjavík, Iceland

**Małgorzata Dukowska and Maria Grzybkowska**
Department of Ecology and Vertebrate Zoology, Faculty of Biology and Environmental Protection, University of Łódź, Łódź, Poland

**André A. Padial**
Departamento de Botânica, Universidade Federal do Paraná, Curitiba, Paraná, Brazil
Programa de Pós-graduação em Ecologia e Conservação, Universidade Federal do Paraná, Curitiba, Brazil

**Fernanda Ceschin**
Programa de Pós-graduação em Ecologia e Conservação, Universidade Federal do Paraná, Curitiba, Brazil

**Steven A. J. Declerck**
Department of Aquatic Ecology, Netherlands Institute of Ecology (NIOO-KNAW), Wageningen, The Netherlands

**Luc De Meester**
KU Leuven, University of Leuven, Laboratory of Aquatic Ecology, Evolution and Conservation, Leuven, Belgium

**Cláudia C. Bonecker, Fabio A. Lansac-Tôha, Liliana Rodrigues, Luzia C. Rodrigues, Sueli Train and Luiz F. M. Velho**
Núcleo de Pesquisa em Limnologia, Ictiologia e Aqüicultura (Nupelia), Universidade Estadual de Maringá, Maringá, Brazil

**Luis M. Bini**
Departamento de Ecologia, Universidade Federal de Goiás, Goiânia, Brazil

**Curtis J. Hayden and J. Michael Beman**
Life and Environmental Sciences and Sierra Nevada Research Institute, University of California Merced, Merced, California, United States of America

**David B. McWethy**
Department of Earth Sciences, Montana State University, Bozeman, Montana, United States of America

**Janet M. Wilmshurst**
Landcare Research, Lincoln, New Zealand
School of Environment, University of Auckland, Auckland, New Zealand

**Cathy Whitlock**
Department of Earth Sciences, Montana State University, Bozeman, Montana, United States of America
Institute on Ecosystems, Montana State University, Bozeman, Montana, United States of America

**Jamie R. Wood and Matt S. McGlone**
Landcare Research, Lincoln, New Zealand

**Jacob T. Westhoff**
Missouri Cooperative Fish and Wildlife Research Unit, Department of Fisheries and Wildlife Sciences, University of Missouri, Columbia, Missouri, United States of America

**Craig P. Paukert**
U.S. Geological Survey, Missouri Cooperative Fish and Wildlife Research Unit, University of Missouri, Columbia, Missouri, United States of America

**Gilberto F. Barroso and Fábio da C. Garcia**
Department of Oceanography and Ecology, Federal University of Espírito Santo, Vitória, Espírito Santo, Brazil

**Monica A. Gonçalves**
Espírito Santo State Water Resources Agency, Vitória, Espírito Santo, Brazil

**François De Vleeschouwer, Heleen Vanneste and Gaël Le Roux**
Université de Toulouse, INP, UPS, EcoLab (Laboratoire Ecologie Fonctionnelle et Environnement), ENSAT, Castanet Tolosan, France
CNRS, EcoLab, Castanet Tolosan, France

**Dmitri Mauquoy**
School of Geosciences, University of Aberdeen, Aberdeen, United Kingdom

**Natalia Piotrowska**
Department of Radioisotopes, Institute of Physics, Silesian University of Technology, Gliwice, Poland

**Fernando Torrejón**
Environmental Sciences Center EULA-Chile, University of Concepció n, Concepció n, Chile

**Thomas Roland**
Geography, College of Life and Environmental Sciences, University of Exeter, Exeter, United Kingdom
Palaeoenvironmental Laboratory (PLUS), Geography and Environment, University of Southampton, Southampton, United Kingdom

**Ariel Stein**
NOAA/Air Resources Laboratory, R/ARL - NCWCP, College Park, Maryland, United States of America

**Esther Kohler, Jörg Villiger, Thomas Posch, Jakob Pernthaler, Judith F. Blom and Tanja Shabarova**
Limnological Station, Institute of Plant Biology, University of Zurich, Kilchberg, Switzerland

**Nicolas Derlon**
Eawag: Swiss Federal Institute of Aquatic Science and Technology, Dübendorf, Switzerland

**Eberhard Morgenroth**
Eawag: Swiss Federal Institute of Aquatic Science and Technology, Dübendorf, Switzerland
Institute of Environmental Engineering, ETH Zurich, Zurich, Switzerland

**Eric L. Sparks and Just Cebrian**
Dauphin Island Sea Lab, Dauphin Island, Alabama, United States of America
Marine Sciences, University of South Alabama, Mobile, Alabama, United States of America

**Sara M. Smith**
Dauphin Island Sea Lab, Dauphin Island, Alabama, United States of America

# Index

www.ingramcontent.com/pod-product-compliance
Lightning Source LLC
Chambersburg PA
CBHW080504200326
41458CB00012B/4075